INNOVATIONS IN STEM CELL TRANSPLANTATION

Edited by **Taner Demirer**

Innovations in Stem Cell Transplantation
http://dx.doi.org/10.5772/56082
Edited by Taner Demirer

Contributors

Laurent Lecanu, Oluwatoyin Olatundun Ilesanmi, Francisco Barriga, Hugo F. Fernandez, Ashraf Badros, Bhavana Bhatnagar, Yongzhi Xi, Yuying Sun, Helgi Van De Velde, Pier Paolo Piccaluga, Stefania Paolini, Felicetto Ferrara, Giuseppe Visani, Anna Gazzola, Alessandro Broccoli, Vittorio Stefoni, Jeane Eliete Laguila Visentainer, Amanda Marangon, Daniela Cardozo, Ana Maria Sell, Miroslaw Markiewicz, Wanming Da, Toshihisa Tsuruta, Xiang Gu, Gülsan Sucak, Zeynep Arzu Yegin, Şahika Zeynep Akı, Taner Demirer, Mustafa ÇETİN, Leylagül Kaynar, Ali ÜNAL

Published by InTech

Janeza Trdine 9, 51000 Rijeka, Croatia

Copyright © 2013 InTech

Notice

Publishing Process Manager Ana Pantar
Technical Editor InTech DTP team
Cover InTech Design team

Additional hard copies can be obtained from orders@intechopen.com

Innovations in Stem Cell Transplantation, Edited by Taner Demirer
p. cm.
ISBN 978-953-51-0980-8

For EU product safety concerns:
IN TECH d.o.o., Prolaz Marije Krucifikse Kozulić 3, 51000 Rijeka, Croatia,
info@intechopen.com or visit our website at intechopen.com.

Meet the editor

Dr. Demirer graduated from Ankara University Medical School in Turkey in 1984 and trained in the USA from 1987 to 1997. He performed research in basic hematologic techniques and applications of flow cytometry at the Chicago Medical School between 1987 and 1989 and did Internal Medicine residency at the Medical College of Wisconsin in 1989-1992. He did Hematology/Oncology and Bone Marrow Transplant Fellowship at the University of Washington, Fred Hutchinson Cancer Research Center (FHCRC) in Seattle,1992-1997, and trained under Professors Don Thomas (Nobel laureate), Dean Buckner, Frederick Appelbaum and Rainer Storb in the clinical division. He is a diplomate of the American Board of Internal Medicine and Board certified in Internal Medicine and Medical Oncology. He was given the title of 'Fellow of American College of Physicians (FACP)' by ACP in July 1996. He has conducted many clinical studies at the FHCRC as principal investigator or co-investigator mainly related to peripheral blood stem cell mobilization and high-dose chemotherapy (HDC) in patients with solid tumors and hematologic malignancies. During his career he has written many papers in medical journals and books with regard to stem cell mobilization kinetics, factors influencing the stem collection and engraftment as well as HDC in patients with multiple myeloma, breast and ovarian cancer. He was chairman of the EBMT Solid Tumors Working Party (STWP) between 2001 and 2007. He has conducted, chaired and published studies related to HDC in patients with solid tumors at the EBMT STWP. He was the president of the 29th EBMT Annual Congress which was held in Istanbul in 2003. He will also be the president of EBMT Annual Meeting in 2015 which will be held in Istanbul. He is the member of the editorial board of Bone Marrow Transplantation. Dr. Demirer is currently the professor of Medicine and Hematology/Oncology at the Ankara University Medical School in Ankara, Turkey, and a full member of Turkish Academy of Sciences.

Contents

Preface

This book documents the increased number of stem cell related research, basic and clinical applications as well as views for the future. The book covers a wide range of issues related to new developments and innovations in cell-based therapies containing basic and clinical chapters from the respected authors involved in stem cell studies and research around the world. It thereby complements and extends the basic coverage of stem cells such as immunogenetics, neuron replacement therapy, cover hematopoietic stem cells, issues related to clinical problems, advanced HLA typing, alternative donor sources as well as gene therapy that employs novel methods in this field. Clearly, the treatment of various malignancies and biomedical engineering will depend heavily on stem cells, and this book is well positioned to provide comprehensive coverage of these developments.

This book will be the the main source for clinical and preclinical publications for scientists working toward cell transplantation therapies with the primary goal of replacing diseased cells with donor cells of various organs and transplanting those cells close to the injured or diseased target. With the increased number of publications related to stem cells and *Cell Transplantation*, we felt it was important to take this opportunity to share these new developments and innovations describing stem cell research in the cell transplantation field with our world-wide readers.

Stem cells have a unique ability; they are able to self renew limitlessly allowing them to replenish themselves as well as other cells. Another ability of stem cells is that they are able to differentiate to any cell type. A stem cell does not differentiate directly to a specialized cell, however. There are often multiple intermediate stages. A stem cell will first differentiate to a progenitor cell – a progenitor cell is similar to a stem cell, although they are limited in the number of times they can replicate and they are also restricted in which cells they can further differentiate to. Serving as a sort of repair system for the body, they can theoretically divide without limit to replenish other cells as long as the person or animal is still alive. When a stem cell divides, each new cell has the potential to either remain a stem cell or become another type of cell with a more specialized function, such as a muscle cell, a red blood cell or a brain cell.

During this last decade, the number of published articles or books investigating the role of stem cells in cell transplantation or regenerative medicine increased remarkably across all sections of the stem cell related journals. The largest number of stem cell articles was published mainly in the field of clinical transplantation, neuroscience, followed by the bone, muscle, and cartilage and hepatocytes. Interestingly, in recent years, the number of stem cell articles describing the potential use of stem cell therapy and islet transplantation in the diabetes has also slowly been increasing, even though this field of endeavor could have one of the greatest clinical and societal impacts.

It will be exciting and interesting for our readers to follow the recent developments in the field of basic and clinical aspects of stem cells and cell transplantation. Although we are close to finding pathways for stem cell therapies in many medical conditions, scientists need to be careful how they use stem cells ethically and should not rush into clinical trials without carefully investigating the side effects. Focus must be on Good Manufacturing Procedures (GMP) and careful monitoring of the long-term effects of transplanted stem cells in the host.

In conclusion, *Cell Transplantation* is bridging cell transplantation research in a multitude of disease models as methods and technology continue to be refined. The use of stem cells in many therapeutic areas will bring hope to many patients awaiting replacement of malfunctioning organs or repair of damaged tissue. We hope that this book will be an important tool and reference guide for all scientists worldwide who work in the field of stem cells and cell transplantation, and that it will shed light upon many important debatable issues in this field.

I would like to thank all authors who contributed to this book with excellent and up-to-date chapters relaying the recent developments to our readers in the field of stem cell transplantation. I would like to give a special thanks to Ana Pantar, Publishing Process Manager, and all InTech staff for their valuable contribution in making this book available.

Taner DEMİRER, MD, FACP
Professor of Medicine, Hematology/Oncology
Dept. of Hematology
Ankara University Medical School
Ankara, TURKEY

Section 1

Basic Aspects of Stem Cell Transplantation

Chapter 1

Immunogenetics of Hematopoietic Stem Cell Transplantation

Amanda Vansan Marangon, Ana Maria Sell,
Daniela Maira Cardozo and Jeane E. L. Visentainer

Additional information is available at the end of the chapter

http://dx.doi.org/10.5772/54281

1. Introduction

There are few hematopoietic stem cells (HSCs) in the bone marrow of adult mammals; these are required throughout life to replenish the short-lived mature blood cells of specific hematopoietic lineages. HSCs have several biological functions including homeostasis control, regeneration, immune function and response to microorganisms and inflammation.

The regenerative potential of human HSCs is best illustrated by successful stem cell transplantation in patients with a variety of genetic disorders, acquired states of bone marrow failure and cancer [1].

The first bone marrow transplantation took place in 1949 with studies that demonstrated the protection provided to the spleen of mice given a dose of irradiation that would otherwise be lethal. In 1960, studies in dogs provided important information about bone marrow transplantation in exogamic species, results that are applicable to humans. It was demonstrated that dogs could bear 2-3 times the lethal dose of total body irradiation with an infusion of bone marrow cells collected and cryopreserved before irradiation [2,3].

At the same time that animal experiments were being carried out, a number of attempts were made to treat humans with chemotherapy or irradiation associated with bone marrow infusions [4].

The first successful allogeneic bone marrow graft was achieved in a patient with leukemia, although the patient died due to the complications of chronic graft-versus-host disease (GVHD) [5].

Currently, bone marrow transplantation is the treatment of choice for many hematologic diseases with the course of transplant being dependent on several factors, including the stage of the disease at transplant, the conditioning regimen, source of cells, genetic factors, and the development of GVHD. The goal to this chapter is to show some genetic factors that have a strong influence on hematopoietic stem cell transplantation (HSCT) outcomes, such as the genes of the human leukocyte antigen (HLA) system located in the major histocompatibility complex (MHC), and other genetic factors, including non-HLA genes that seem to influence transplant outcomes and that are being studied to optimize donor selection. Non-HLA genes mainly include killer cell immunoglobulin-like receptor (KIR) genes, cytokine genes and receptors, MHC class I-related chain (MIC) genes and human minor histocompatibility antigens (mHAgs).

2. HLA immunogenetics and its influence on hematopoietic stem cell transplantation

Histocompatibility

The immune system is the result of germline selection and thymic education (self vs. non-self) through contact with pathogenic life and is thus a characteristic that is unique to each individual and specific to a given point in time; like all other physiological systems, the immune system is affected by disease, stress, trauma and environmental events [6].

An important cell lineage within this system is represented by T lymphocytes. The main functions of T lymphocytes are defense against intracellular microorganisms and the activation of other cells including macrophages and B lymphocytes.

T lymphocytes are capable of interacting with other cells because the antigen receptors on T cells recognize antigens that are presented by other cells; presentation is achieved by specialized proteins that are encoded by genes in a MHC locus [7]. The MHC system has the greatest diversity of all functional genetic systems at the population level [6]. The MHC glycoprotein family, also referred to as HLAs, presents endogenous and exogenous antigens to T lymphocytes for recognition and response.

This system was discovered in mice by Peter Gorer and George Snell. These researchers discovered an antigen which was involved in tumour rejection and subsequently they showed that similar antigens in other strains of mice were probably alleles of the same "tumour-resistant" gene [8].

Experiments show that transplants of tissue between animals from the same population (endogamic) were successful, while the consequence of transplants between animals from different populations (exogamous) was the rejection of tissue. The result of these studies was the discovery of MHC genes which are capable of recognizing foreign antigens and presenting them to T lymphocytes.

Antibodies induced by transfusions or pregnancy and which react with leukocyte antigens were first recognized in 1954. Studies showed that kidney transplant patients who suffered rejection have circulating antibodies reactive to antigens present in leukocytes; as these antigens are expressed on leukocytes they were named HLAs [9,7].

Many studies were conducted over the next few years to understand and characterize the immunogenicity of these antigens.

Structure and function

The MHC, contained within 4.2 Mbp of DNA on the short arm of chromosome 6 at 6p21.3, has more than 200 genes, most of which have functions related to immunity. It is divided into three main regions [10].

The HLA-A, -B and -C classic genes and -E, -F and -G non-classic genes, as well as other genes and pseudogenes are located in the HLA Class I region near to the telomere. The HLA Class II region, near to the centromere, contains the HLA-DR, -DQ and -DP genes. The HLA-DR sub-region, includes the DRA gene that encodes the alpha chain is non-polymorphic and can bind with any beta chain to encode for DRB genes [11].

Located between class I and II regions, the class III region has C2, C4A, C4B and B factor, that encode complement proteins and the tumour necrosis factor (TNF) [10,11].

HLA molecules are polymorphic membrane glycoproteins found on the surface of nearly all cells. Multiple genetic loci within the MHC encode these proteins with each individual simultaneously expressing several polymorphic forms from a large pool of alleles in the population. The overall structure of HLA class I and class II molecules is similar, with most of the polymorphisms found in the peptide binding groove (PBG) where antigens are recognized [12].

Class I molecules are made up of one heavy chain (45kD) encoded within the MHC and a light chain called β2- microglobulin (12kD) whose gene is on chromosome 15. Class II molecules consist of one α (34kD) and one β chain (30kD) both within the MHC [10]

The class I heavy chain has three domains with the membrane-distal α1 and α2 domains being polymorphic. Within these domains, polymorphisms concentrate in three regions: positions 62 to 83, 92 to 121, and 135 to 157. These areas are called hypervariable regions. The two polymorphic domains are encoded by exons 2 and 3 of the class I gene. Diversity in these domains is very important because these two domains form the antigen binding cleft or PBG of MHC class I molecules [13,14].

The sides of the antigen binding cleft are formed by α_1 and α_2, while the floor of the cleft is comprised of eight anti-parallel β sheets. The antigenic peptides of eight to ten amino acids (typically nonamers) bind to the cleft with low specificity but high stability. The α_3 domain contains a conserved seven amino acid loop (positions 223 to 229) which serves as a binding site for CD8 [12,15-17].

Class II molecules consist of two transmembrane glycoproteins, the α and β chains which are restricted to cells of the immune system (e.g. B cells, dendritic cells - DCs), but can be induced by other cell types during immune response. The PBG of class II molecules has open ends which allow the peptide to extend beyond the groove at both ends and therefore to be longer (12-24 amino acids). The peptide is presented to CD4 T cells [10].

Generally both the α and β chains in class II molecules are polymorphic. In these chains, the $\alpha1$ and $\beta1$ domains are of the PBG and therefore the diversity is found mainly in these domains. These domains are encoded by exon 2 of their class II A or B genes and the hypervariable regions tend to be found in the walls of the groove [16].

T-cell activation occurs following recognition of peptide/MHC complexes on an antigen-presenting cell (APC). T-cell activation can be viewed as a series of intertwined steps, ultimately resulting in the ability to secrete cytokines, replicate and perform various effector functions. During antigen presentation, CD4 and CD8 are intimately associated with the T-cell receptor and bind to the MHC molecule. Besides this interaction between T cells and APCs, ligation between counter-receptors on the T cell and accessory molecules on the APC is required as additional signals for T-cell activation [18].

Haplotype, Linkage Disequilibrium and Expression of HLA genes

HLA genes are transmitted following Mendel's law of segregation, so the allelic variant is codominantly expressed. The set of alleles present in the HLA loci located in a single chromosome of a chromosome pair is called a haplotype. The probability that two siblings having the same HLA haplotype is 25%; in this situation, it is considered that they are matching [11].

Moreover, a fact called linkage disequilibrium occurs in HLA genes. This means that certain alleles occur together at a higher frequency than would normally be expected by chance (gametic association). Consequently, some combinations of alleles appear more or less commonly in a population than would normally be expected from a random formation of haplotypes from alleles based on their frequencies [10].

For example, if a determined population has genic frequencies of 14% and 9% for HLA-A*01 and HLA-B*08, respectively, the expected frequency of a haplotype with this combination would be 1.26% (0.14 x 0.09). However, the true frequency may be 8.8% in this population, that is, higher than expected, characterizing a positive linkage disequilibrium [11].

Examples can be seen in studies of linkage disequilibrium related to bone marrow donation. A strong linkage disequilibrium has been reported for HLA-B*39:13 with the DRB1*04:02, DRB1*08:07 and A*31:12 haplotypes in the Brazilian population [19].

Other reports for unrelated donors involved HLA-A*01 and HLA-B*08, HLA-A*03 and HLA-B*35 and HLA-A*02 and HLA-B*12. This type of results suggests that these data have clinical application, such as in the selection of unrelated donors for bone marrow transplantation [20].

HLA compatibility of donors

The genetic origin of patients for whom bone marrow transplantation has been proposed, is a key determinant in the possibility of identifying compatible unrelated and sibling donors and consequently in the possibility of performing the procedure.

The strict HLA compatibility that is required for bone marrow transplantation increases the difficulties in finding donors. A patient has one chance in four of having a compatible donor among his brothers and sisters. This chance becomes one in a million, on average, in unrelated donors [21].

Different methods are used to identify HLA antigens. In the past, HLA antigens for bone marrow transplantation were identified by serological methods based in mixed lymphocyte culture. However this technique is not as sensitive as molecular biology methods which can define HLA antigens at the allele level.

In molecular analysis, HLA genes can be identified by polymerase chain reaction (PCR) using the Specific Sequence Primers (SSP), Specific Sequence Oligonucleotides (SSO) or sequencing techniques. These methods are the most commonly used due to its specificity and sensibility that can define HLA genes only (low resolution) or genes and alleles (high resolution).

These results are very important in bone marrow transplantation in order to choose the best matched donor. The probability of finding a well-matched unrelated donor is improved if high resolution typing is available for the patient prior to the search. Therefore typing must ideally be done by DNA methods to avoid hidden mismatches, particularly in the case of antigenically silent alleles, and should include the HLA-A, -B, -C and -DRB1 genes at least [10].

Matched or mismatched donors

There are many studies which try to show that better outcomes in bone marrow transplantation are linked to full donor matches. In 2004 the National Marrow Donor Program (NMDP) published the results on the outcomes of 1874 unrelated donor transplants. This study showed a highly significant survival advantage for 8/8 matched pairs compared to those with one or two mismatches [22].

Moreover, the study of the Center for International Blood and Marrow Transplant Research (CIBMTR) examined clinical outcomes in recipients of both sibling and unrelated donors for chronic myeloid leukemia (CML) in the first chronic phase. There were 1052 recipients of unrelated transplants; 531 were matched for 8/8 alleles, 252 mismatched for 1 (7/8) allele and 269 mismatched for multiple alleles [22]. The overall survival (OS) at 5 years was 55% for 8/8 matched transplant recipients, 40% for those with a 7/8 matched graft and 21-34% for those with various multiple mismatched combinations. The recipients of stem cell matched related donors, predominantly siblings, have lower risk of infections, of the reactivation of cytomegalovirus and of mortality than the latter group. Additionally, T-cell immunity reconstitution is delayed in mismatched sibling donors and the unrelated group [23, 24].

Graft rejection, GVHD and delayed immune recovery, the major obstacles to successful allogeneic HSCT, are more severe with unrelated donors than in HLA-identical sibling transplants. Because identical donors are available to only about 30% of patients, the identification of a suitable unrelated donor by better, more precise HLA matching of donor and recipient is necessary [25].

Studies have shown strong negative effects of HLA mismatching of the HLA-A, -B, -C, -DRB1 or -DQB1 loci on OS. The presence of multiple mismatches was worse for survival and for severe acute GVHD (grade III-IV). Other trials analyzed the incidence of chronic GVHD in patients who survived more than 100 days after transplantation. It became evident that a HLA-A/B mismatch induces a significantly higher incidence of chronic GVHD and lower OS rate. The same was not confirmed for a HLA-DQ/DR mismatch that showed no association with the occurrence of chronic GVHD [25;26].

High resolution HLA typing can help in the characterization of donors; there are differences in the outcomes of bone marrow transplantation if the mismatching of donors was defined by low or high resolution. Studies show that high resolution matching of HLA-A, -B, -C and -DRB1 between volunteer HSC donors and recipients is associated with a better survival. Additionally, single HLA-B and -C mismatches appear to be better tolerated than single HLA-A or -DRB1 mismatches [27].

Other studies affirm that survival after unrelated HSCT for severe acquired aplastic anemia has improved significantly over the last 15 years mainly due to better HLA matching at the allelic level [28].

HLA and bone marrow transplantation

The outcome of transplantation using unrelated donors is highly influenced by HLA matching between the donor and recipient. Two particular individuals always differ in their genome structure in respect to minor histocompatibility antigens, killer cell immunoglobulin-like receptor (KIR) genes and several other groups of genes.

However, the most potent transplantation antigens are the HLAs encoded by genes located in the MHC. HLA-C is a class I gene locus, yet its importance in transplantation has been less validated than the HLA-A and B loci [10].

However, studies that analyzed structure and peptide-binding for HLA-C, show that divergence in peptide-binding specificity may be a contributor to the risk of mortality after transplantation perhaps due to the alloreactivity of donor T cells towards peptides presented by patient HLA molecules but not by donor antigens presenting cells during T-cell development in the thymus [29].

There are two main reasons for the HLA antibodies to result in graft failure and GVHD. The first is the rapid increase in the number of HLA-mismatched HSCT, including in cord blood transplantation, haploidentical HSCT and unrelated HSCT. The second is the technical advance in the methods of HLA Ab testing, which have attained a rapid, accurate and objective identification and qualification of specific HLA antibodies [30].

HLA, sibling and unrelated hematopoietic stem cell transplantation

Matched or mismatched

The use of stem cells from HLA-matched unrelated volunteer donors is an accepted option for patients without a matching sibling donor providing comparable outcomes to matched sibling donor HSCT. Many studies have been performed to compare the results between sibling and unrelated transplantation. Other studies show single results about sibling and unrelated transplantation, the importance of HLA compatibility and what effect HLA mismatches may have on GVHD, graft failure and relapse.

Research has shown that HLA class II DRB1*15 (*15:01 and *15:02) are important in the outcome to sibling matched transplants for patients who have aplastic anemia. In multivariate analysis, the secondary graft failure rate at two years was lower in patients who were HLA-DR15+[31].

Recent studies in a Chinese population show that the outcome of unrelated donor transplantation matched for HLA-A, HLA-B and HLA-DRB1 but unknown for HLA-C antigens was associated with a significant risk of mortality and that this risk was higher with HLA-A, B or DRB1 mismatches compared to an 8/8 match [32].

Other studies confirm that there is no association between HLA mismatching of unrelated donors with the cumulative incidence of grade II-IV or grade III-IV acute GVHD. Similarly, there was no association with chronic GVHD, but the incidence of graft failure was higher in HLA-mismatched unrelated transplants [33]. Trials highlight the importance of defining HLA by high resolution techniques to improve the outcomes in pediatric transplants using unrelated donors. The patients that suffered graft failure were mismatched for HLA-C by a high resolution technique [34].

However, studies show that in unrelated transplantations, the outcome is improved when the patients are HLA-C and HLA-DPB1 mismatched in some combinations, thus resulting in lower risk of relapse. Probably some combinations increase the graft-versus-leukemia (GvL) effect [35].

One study analyzed the impact of HLA class I and II high-resolution matching of 1874 unrelated donors and found that HLA-C mismatching was most strongly associated with graft failure, HLA-A mismatching was associated with significantly increased risk of grade III/IV acute and chronic GVHD and mismatches of HLA-A, B, C and DR were associated with death [36].

HLA and Haploidentical HSCT

When no matched sibling or unrelated donor exists, the potential curative option is haplo-HSCT, that is, transplant with a donor who shares only one haplotype with the recipient. Haploidentical stem cell transplants are increasingly used in the treatment of malignancies, and immune and hematologic diseases. As multiple mismatched related donors may be available for transplantation, it is important to select a donor that is most likely to produce a successful outcome [37].

There are studies that correlate the HLA-B mismatch effect in haplo-HSCT. Studies analyzed the impact of HLA-A, -B, -DRB1, -DRB3, -DRB4 and -DRB5 and demonstrated that a HLA-B mismatch not only has a significant effect on GVHD and transplant-related mortality but was also associated with reduced OS and leukemia-free-survival [38].

There is an important point in haploidentical transplants that should be considered: the conditioning regimen. Many protocols have been performed to improve the outcomes of transplantation and to minimize the effect of HLA incompatibility. For example, studies on nonmyeloablative HLA-haploidentical bone marrow transplantation with high-dose post-transplant cyclophosphamide. The results showed that HLA mismatch was not associated with relapse or GVHD [39].

In bone marrow transplantation with mismatch of the HLA-DRB1 antigen in the GVHD direction and two or more HLA Class I (HLA-A, -B and -Cw) mismatches in either direction were found to be associated with decreased incidences of relapse without an increased incidence in nonmyeloablative conditioning with post-transplant cyclophosphamide [40].

HLA and cord blood transplantation

The use of umbilical cord blood transplantation (UCBT) for patients with hematological malignances or hereditary diseases is becoming increasingly more common. In October 2006, the International NETCORD Foundation maintained an inventory of more than 124,000 umbilical cord blood (UCB) units and documented more than 4900 unrelated UCBT [41]. Several studies have shown that the number of cells is the most important factor for engraftment, while some degree of HLA mismatch is acceptable.

For example, studies show that unrelated UCBT is comparable to a 1-antigen mismatch at the HLA-A, HLA-B or HLA-DR loci in respect to GVHD, relapse and OS [42]. For other studies on UCBT, HLA-A and -B are defined by low-resolution and HLA-DRB1 by high-resolution, with minimum compatibility of 4/6. It is important to apply the rules of equivalence of serological groups for HLA-B*14, -B*15, -B*40, and -B*50 as determined by molecular methods [43].

Clinical comparison studies of UCBT and HLA-A, -B and -DRB1 6/6 allele-matched bone marrow transplantation or single mismatched for leukemia from unrelated donors in adult recipients showed similar results. There was no significant increase of relapse rates among UCB recipients when compared with DRB1 single-mismatched bone marrow recipients. The OS for UCB recipients was similar too when compared with DRB1 single-mismatched bone marrow recipients [44].

Korean pediatric studies also show that the results of UCBT are promising. One study compared the outcomes of acute leukemia children submitted to transplantation using UCB, bone marrow and peripheral blood stem cells from HLA-matched or unrelated donors. The results confirm that survival after UCBT was similar to survival after matched related donor and unrelated donor transplantations. In conclusion, for patients lacking an HLA matched donor, the use of UCB is a suitable alternative [45].

Additionally, studies show that recipients of UCB transplants from HLA-identical siblings have lower incidences of acute and chronic GVHD than recipients of bone marrow transplants from HLA-identical siblings [46]. Hence, studies on the distribution of HLA alleles and haplotypes in different ethnic populations are also important to find a suitable unrelated cord blood donor for a patient. One study investigated the frequencies of alleles and HLA-A, -B and -DRB1 haplotypes with high-resolution typing data in a total of 710 Taiwanese UCB units [47]. The most common alleles found for HLA-A were

A*11:01, A*24:02, A*33:03 and A*02:01; for HLA-B they were B*40:01, B*46:01, B*58:01 and B*13:01 and for DRB1 they were DRB1*09:01, DRB1*12:02, DRB1*15:01 and DRB1*03:01. Moreover, the five most frequently found haplotypes were A*11:01, B*35:05, DRB1*11:02; A*24:07, B*35:05, DRB1*12:02; A*01:01, B*5701, DRB1*09:01; A*11:01, B*40:01, DRB1*09:01 and A*11:01, B*46:01, DRB1*09:01. These haplotypes are common in Taiwanese and Asian American populations [47].

Ethnic studies carried out in London showed that the most common alleles in 1500 UCB units were HLA-A*34, A*36, A*80, HLA-B*75, B*61, B*53, B*78, B*81 and B*82. This kind of study should help to increase the chances of obtaining acceptably HLA-matched donors for patients from ethnic minorities [48].

3. Non-HLA immunogenetics and its influence on hematopoietic stem cell transplantation

Natural killer cells and Killer immunoglobulin-like receptors

Human natural killer (NK) cells are components of the innate immune response that comprise approximately 10-15% of all peripheral blood lymphocytes and play a major role in immunity against viral infections and tumors [49-51]. Years of intensive research in mice and humans have shown a special importance of NK cells in the hematological diseases and in mediating favorable HSCT outcomes [52-57]. NK cells were first identified by their *in vitro* capacity to kill tumor cells without the requirement of prior immune sensitization of the host [58-59].

The function of NK cells is regulated by a diverse array of cell-surface receptors including KIR, NKG2D and DNAM-1. The KIR receptors, in the setting of HSCT, seem to be the most important NK cell receptor family. These receptors can either inhibit or activate NK cells with the difference between inhibitory and activating KIRs lying mainly in their intracytoplasmic tail. Inhibitory KIRs have long cytoplasmic tails (KIR-L) and activating KIRs have short cytoplasmic tails (KIR-S) with KIRs having two or three Ig-domains (KIR2D or KIR3D) [60-61].

In humans, the gene family encoding the *KIR* is located in the leukocyte receptor complex (LRC) on chromosome 19q13.4. To date, 15 genes have been well characterized, of which 9 are NK cell inhibitors (*KIR2DL1, KIR2DL2, KIR2DL3, KIR2DL4, KIR2DL5A, KIR2DL5B, KIR3DL1, KIR3DL2* and *KIR3DL3*), 6 are activating (*KIR2DS1, KIR2DS2, KIR2DS3, KIR2DS4, KIR2DS5* and *KIR3DS1*), and 2 are pseudogenes (*KIR2DP1* and *KIR3DP1*) [60-61]. An exception is *KIR2DL4* that although it has long tail it has an amino acid in the transmembrane region that allows an association with an accessory protein, FceRI-g, which confers an activating signal [62].

Individuals differ in the number and type of inherited *KIR* genes and the *KIR* haplotypes are divided into two groups, A and B. The A or AA haplotype has a fixed number of genes, all of which are inhibitory except for one activating gene (*KIR2DS4*). Haplotypes with additional activating *KIR* genes (*KIR2DS1, KIR2DS2, KIR2DS3, KIR2DS5, KIR3DS1)* or with *KIR2DL5* are either AB or BB and are grouped together as KIR Bx haplotypes. Often, the *KIR2DS4* gene is present in a deleted form and is not believed to be expressed at the cell surface. The "framework genes", *KIR3DL2, KIR3DP1, KIR2DL4,* and *KIR*3DL3, are common to both groups of haplotypes [60-61, 63].

The KIRs interact with some HLA class I antigens on target cells. HLA-Bw4 and distinct allotypes of HLA-C (C1 and C2 groups) are the main ligands for most KIRs [64]. HLA-C alleles are classified as C1 or C2 KIR ligand groups, depending on two amino acid positions encoded in exon 2. HLA-C1 allotypes have serine at position 77 and asparagine at position 80 and are ligands for the *KIR2DL2* and *KIR2DL3* inhibitory receptors. HLA-C2 allotypes have asparagine and lysine at positions 77 and 80, respectively and are ligands for the *KIR2DL1* inhibitory receptor and thought to be the ligand for the *KIR2DS1* activating receptor [65-66]. HLA-Bw4 allotypes are characterized by at least 5 different patterns of amino acids at positions 77 and 80-83 and are ligands for *KIR3DL1.* Some HLA-A alleles, namely 23:01, 24:02 and 32:01, are also ligands for *KIR3DL1* [67-71]. In addition, HLA-A3 and HLA-A11 are ligands for *KIR3DL2;* and HLA-A11 and some C1 and C2 allotypes are ligands for *KIR2DS4* [64, 72-74].The *KIR* gene and respective ligands are listed in Table 1.

KIR	FUNCTION	LIGAND
KIR2DL1	Inhibitory	HLA-C group 2
KIR2DL2	Inhibitory	HLA-C group 1
KIR2DL3	Inhibitory	HLA-C group 1
KIR2DL4	Inhibitory, activating	HLA-G
KIR2DL5	Inhibitory	Unknown
KIR3DL1	Inhibitory	HLA-B Bw4 and some HLA-A Bw4*
KIR3DL2	Inhibitory	HLA-A3 and HLA-A11
KIR2DS1	Activating	HLA-C group 2
KIR2DS2	Activating	Unknown
KIR2DS3	Activating	Unknown
KIR2DS4	Activating	HLA-A11 and subsets of HLA-C group 1 and group 2
KIR2DS5	Activating	Unknown
KIR3DS1	Activating	Unknown

* HLA-A*23:01, HLA-A*24:02 and HLA-A*32:01

Table 1. KIR receptors and their HLA ligands

The mechanism of recognition of a target cell by NK cells differs from others lymphocytes. [59] The NK cells are able to recognize a reduction or absence of self HLA class I ligands, as a form of distinguishing normal cells from target cells: this is the "missing-self recognition". It is well established that cancer cells and some infected cells develop various mechanisms to escape lysis by T cells [75-76]. An effective mechanism is to decrease or remove completely the HLA expression. The downregulation of HLA class I expression leads to resistance to lysis by T lymphocytes but, as a consequence, can lead to a susceptibility to lysis by NK cells [77-80]. During development, NK cells become licensed or educated by interaction with self-HLA molecules to maintain tolerance to normal tissues. NK cells that do not express inhibitory receptors for self are retained in an anergic or hypofunctional state and those which express inhibitory KIRs for self-HLA ligands are functionally active and thus can sense the lack of expression of self HLA molecules on target cells which triggers lysis of these cells. This is thought to be the main mode of action of NK cells [81-86].

Natural killer cell alloreactivity in hematopoietic stem cell transplantation

The clinical significance of missing-self recognition is especially evident in allogeneic HSCT. In HSCT the NK cell alloreactivity is determined by an analysis of the donor's *KIR* gene profile and by differences in MHC class I genes between the donor and the recipient. This can be better explained by the presence in the donor of NK cells expressing inhibitory KIRs that are not engaged by any of the HLA class I alleles present on the receptor [87]. As a consequence, donor NK cells become uninhibited and may display alloreactivity against mismatched allogeneic targets [81-86].

Furthermore, NK cells are relevant in the setting of HSCT because they are the first lymphoid cell subset to reconstitute after transplantation at a time when the adaptive immune system is impaired. They have been detected *in vivo* in recipients within 1 to 3 months after transplantation and up to 3 years after [88-91].

KIR model studies

Considering: 1) a strong correlation between the presence of *KIR* genes and their HLA ligands and cytotoxicity and 2) the advent of methods of precise genetic characterization, it is possible to determine the contributions of the various inhibitory and activating *KIR* genes in HSCT [92]. There are several models to define NK alloreactivity by *KIR* incompatibility or *KIR* mismatching, most of which are based on the analysis of *KIR* and HLA class I alleles. In the ligand-ligand model, the *KIR* expression is assumed following HLA typing. In this model, *KIR* ligands in recipients and donors are analyzed and at least one group of donor *KIR* ligands must to be absent in the recipient's *KIR* ligand repertoire. In the receptor-ligand model, the *KIR* genes are typed for the donor and the HLA alleles are analyzed for recipients and at least one of the inhibitory *KIR*s of the donor is not engaged in the recipient's ligand repertoire. Moreover some studies perform phenotypic analysis of inhibitory KIRs and CD94/NKG2A in donor NK cells and also functional assays which can provide more information about the degree of alloreactivity of NK cells [87,93-94]. It is difficult to know which model is the most adequate to select the *KIR* mismatch donor. Some authors suggest that an increasing number of receptor-ligand mismatch pairs increase the potency of the anti-leukemia effect and also suggest that the receptor-ligand model could improve the accuracy of the prediction of relapse rather than the ligand-ligand model in patients with lymphoid malignancies [94]. However, it has not been well established and further studies are needed to confirm this hypothesis.

In addition, a novel observation emerged that NK cells of maternal donors of HSCT provided better protection from leukemia relapse than other related donors [95]. According to the authors, the better outcome of mother-to-child transplantation may be the result of the contact of maternal immune cells with the semi-allogeneic placenta during pregnancy. It was suggested that if further studies confirm the better outcomes of mother donors, it may be incorporated as a donor selection criterion.

Another interesting aspect was shown in a recent study with patients that received unrelated unmanipulated peripheral blood progenitor cells. The authors indicate that four-digit allele matching of HLA-C may have effects on the HSCT outcome dependent on the presence of C1 and C2 KIR epitopes in the patients [96] suggesting the importance of analysis of HLA-C at allele level for donor selection. While there are no common rules to select the best donor according *KIR* compatibilities, all the findings must be analyzed.

KIR genes and haploidentical hematopoietic stem cell transplantation

Full-haplotype mismatched (haploidentical) HSCT is an option to treat patients lacking a matched donor or a suitable UCB unit. In haploidentical HSCT (haplo-HSCT), the T cells present on allogeneic hematopoietic grafts are important to promote engraftment and mediate the GvL effect. However, they can also mediate GVHD [97-98]. These T-cell responses can be controlled by an appropriate intensity of immunosuppression by the conditioning regimen. The T-cell depletion of the graft help prevent GVHD but, as a consequence, T-cell depleted haplo-HSCT increases the risk of graft rejection and leukemic relapse. In this context, the presence of NK cell alloreactivity in the GVH direction seems to influence the prevention of leukemia relapse and has been investigated in several preclinical and clinical trials [89,99-100]. It has been observed that *KIR*-HLA mismatches can promote clinical benefits in haplo-HSCT especially in patients with acute myeloid leukemia (AML). In the first studies published by Ruggeri et al. [89,99] and the more recent updates in 2007 [100] appropriate KIR-Ligand incompatibilities were associated with a reduction in the risk of relapse of leukemia and graft rejection, and also protection against GVHD in patients with AML. These results were supported by animal models, in which the presence of NK alloreactivity was suggestive of a low incidence of acute GVHD due to the killing of host APCs, which are critical for inducing donor T-cell activation [101]. Similarly, experimental data suggest that the engraftment rate was improved as a result of the lysis of residual host T lymphocytes by alloreactive donor NK cells [89,99,101-102] and also that this contributed to the eradication of leukemia blasts that escaped from the conditioning regimen. These studies showed very good results and led to a novel concept of mismatch to search for a transplant donor. Since then, several investigations based on *KIR* mismatching have been carried out with different outcomes.

In a study of patients that received haplo-HSCT with T-cell depletion [94] *KIR* incompatibility (*KIR*-mismatch) was related to lower relapse rates in children with AML and also in children with acute lymphoid leukemia (ALL). Interesting, in the studies of Ruggeri et al., patients with ALL were not susceptible to *KIR*-ligand mismatched haplo-HSCT. The different result in these two groups may be related to the fact that Ruggeri et al. studied only adult patients. On the other hand, another recent study found no impact of *KIR* mismatch on children with chemoresistant ALL that received T-cell depleted Haplo-HSCT [103]. In general, NK cell alloreactivity seems to positively influence patients with myeloid malignancies and may benefit childhood ALL patients, but further studies are needed to confirm this.

A positive effect of NK cells on the outcome of haplo-HSCT in paedriatic patients was demonstrated in another study. The authors analyzed 21 children with different hematologic malignances and found anti-leukemia activity of alloreactive NK cells in most transplanted patients. They found that the NK cells derived from the donor were capable to selectively killing C1/C1 target cells, including the patient's leukemia blasts. Additionally, *KIR2DL2/3+* NK cells that co-expressed *KIR2DS1* killed C2/C2 leukemic blasts. These data suggest that the presence of *KIR2DS1* in alloreactive NK cells may mediate potent cytotoxicity [91]. In agreement, in another study, the *KIR2DS1* expression in alloreactive NK cells conferred an advantage in the ability of NK cells to kill C2/C2 or C1/C2 myelomonocytic DCs and T-cell blasts [104]. Another recent study examined 86 patients with advanced hematologic malignancies who received nonmyeloablative, HLA-haploidentical HSCT with high-dose, post-transplantation cyclophosphamide. The inhibitory *KIR* gene mismatches between donor and recipient, or *KIR* haplotype AA transplant recipients of *KIR* genotype Bx donors, were associated with lower relapse and non-relapse mortality (NRM) and improved OS and event-free survival [105].

Nevertheless, other studies failed in find any association with *KIR* incompatibilities in the GVHD direction [106], or found worse outcomes of transplantation for donors with the potential to NK alloreactivity [107]. Using the ligand-ligand model in a study of 62 patients with ALL, AML and CML, the *KIR* mismatch was associated to considerably lower OS and a higher incidence of GVHD [107].

KIR genes and unrelated hematopoietic stem cell transplantation

The impact of KIR-ligand mismatching in HSCT using unrelated donors has been associated with controversial results. Beneficial outcomes have been shown in some studies. Unrelated HSCT KIR-ligand incompatibility was associated with a reduced incidence of grade III-IV acute GVHD and a better OS and disease free survival (DFS) in an analysis of 130 patients with different hematological malignances. The conditioning regimen included anti-thymocyte globulin (ATG) for T-cell depletion and the association with DFS remained significant even when patients with myeloid diseases were analyzed separately [108]. KIR-HLA incompatibilities were also associated with low rates of leukemic relapse in a study of 374 patients with myeloid malignances submitted to T-cell replete unrelated HSCT [109]. In this study, in spite of this beneficial result, the rates of graft failure were higher and there were no significant differences in DFS or transplant-related mortality. A large study described an advantage of donor NK alloreactivity; the authors analyzed 1770 patients of several centers and found that the absence of HLA-C2 or HLA-Bw4 ligands but not mismatches were associated with a decreased risk of relapse in recipients receiving unmanipulated grafts from unrelated donors [110].

Although most studies focus on the effect of the lack of inhibitory *KIR*s by their HLA class I pair, some studies have shown interesting results about the role of activating *KIR*s and *KIR* genotypes in unrelated HSCT. Certain B haplotype *KIR* groups have also been found to favorably affect the outcome after T-cell depleted HLA-identical sibling transplants [111]. In 2009 a study showed that donor group B haplotypes significantly improve graft survival in AML patients submitted to T-replete unrelated HSCT [112]. The same group in a recent report reconfirmed the influence of the B haplotype on transplantation outcome. They published a large study of 1409 unrelated transplants for AML and ALL and analyzed centromeric and telomeric gene-content motifs in both group A and B *KIR* haplotypes. They suggest that centromeric and telomeric motifs present in B haplotypes could promote protection against leukemic relapse, as well as, improve survival. Moreover, they found a reduced relapse in those patients whose donors had 2 or more B gene-content motifs [113]. In addition, in a prospective study, the presence of the B *KIR* genotype in donors was also related to fewer bacterial infections at six months post transplant in recipients of unrelated HSCT [114]. In fact it has been observed that some activating genes present in Haplotype B may have an influence on unrelated HSCT. The presence of *KIR3DS1* in the donor has been associated with reduced grade II-IV acute GVHD and a lower transplantation-related mortality rate [115-116]. Donor *KIR2DS1* in isolation or in association to *KIR2DS2* appears to provide protection against relapse in unrelated HSCT [116-117]. On the other hand, in an analysis of patients and their respective HLA-identical sibling or unrelated donors, *KIR2DS1* in the donor and the absence of this gene in the receptor was associated with increased risk of acute GVHD, *KIR2DS3* was associated to chronic GVHD and *KIR2DS5* was associated to relapse [118]. Another study also demonstrated deleterious effects of activating *KIR*s; an increased number of donor activating *KIR* genes was suggested to be a significant factor in the probability of relapse. The *KIR*-ligand mismatch pairs were a risk factor for transplant-related mortality [119]. The effect of activating *KIR*s was mainly found in AML and myelodysplastic syndrome (MDS) patients. The conditioning regimen included using ATG for *in vivo* T-cell depletion.

As discussed above, there are several studies describing improved outcomes based on *KIR*-ligand mismatching, however, most studies have reported no advantage [120-123] or worse outcomes for *KIR*-ligand mismatch donors in unrelated HSCT. Deleterious results included lower OS in patients with myeloid malignances submitted to *KIR*-ligand mismatch HSCT [120] increased infection rates [124]; increased probability of leukemic relapse [125] increased rates of rejection and association with acute Grade III and IV GVHD [126].

KIR genes and sibling hematopoietic stem cell transplantation

On applying *KIR* genotyping, some studies investigated the effect of *KIR* in sibling HSCT. A study of 220 donor-recipient pairs in HLA-matched sibling HSCTs found that patients with myeloid disease who were homozygous for the C2 group had worse OS than patients who were either homozygous or heterozygous for a C1 group. This effect was seen only in patients who received a graft from a donor carrying the *KIR2DS2* gene and only for patients with myeloid disease (no effect was seen in patients with lymphoid disease) [127]. In another study the *KIR*-ligand mismatch was associated to better DFS and OS and lower incidence of relapse in patients with AML and MDS that received T-cell depleted HLA-identical sibling transplants.

AML and MDS patients who lacked two HLA ligands for donor-inhibitory *KIR* had the highest DFS and OS. Interesting, these results were found only for AML and MDS patients and not for CML or ALL patients [128]. Benefits were also described for AML and MDS patients in another study; the authors found a reduced risk for relapse in patients undergoing HLA-identical sibling HSCT who both received a high (above-median) NK cell number and lacked at least one HLA-B or HLA-C ligand of the donor's inhibitory *KIR*s. Transplants with more than two different activating donor *KIR*s were associated with an increased risk for non-relapse mortality [129]. In another study, *KIR*-genotyping of 246 T-cell depleted HSCTs with HLA-identical sibling donors was performed; the *2DL5A*, *2DS1*, and *3DS1 KIR* genes were associated with significantly less relapse in patients with AML but not in patients with other myelogenous or lymphoid malignancies. All these findings suggest that NK cells have implications in donor selection for myeloid diseases especially for AML patients [130].

Some studies have investigated *KIR* genes in respect to post-transplant infections in sibling HSCT. In one study, additional activating *KIR* genes in the donor compared to the recipient's genotype were associated with lower transplant related mortality, better survival, and a reduced incidence of cytomegalovirus (CMV) reactivation [131]. In another study of T-cell replete HSCT from matched sibling donors, the presence of donor *KIR* haplotype B was associated with a 65% reduction in CMV reactivation [132]. Moreover, in another the presence of specific activating *KIR* haplotypes in the donor was associated with protection from CMV reactivation in patients submitted to sibling and unrelated HSCT [133]. Other researchers analyzed patients according to the combination of group A and B *KIR* haplotypes in the transplant donor and recipient and found a higher OS when the donor lacked and the recipient had group B *KIR* haplotypes. Moreover, the poorest OS rate and increased relapse and acute GVHD were recorded when the donor had and the recipient lacked group B *KIR* haplotypes and both were homozygous for the C1 *KIR* ligand. The presence of the Bw4 ligand was also associated with increased acute GVHD. In contrast, the presence of both *KIR3DL1* and its cognate Bw4 ligand was associated with decreased non-relapse mortality. An analysis of *KIR* genes individually revealed *KIR2DS3* as a protective factor for chronic GVHD [134]. In another study, 60 AML patients submitted to T-cell replete HLA-matched related donor allogeneic bone marrow transplants were analyzed. Heterozygous C1/C2 patients had significantly worse survival than those homozygous for C1 or C2 and the C1/C2 group had a higher relapse rate. Multivariate analysis found C1/C2 status to be an independent predictor for mortality. Since C1/C2 heterozygotes have a greater opportunity to engage inhibitory *KIR*s than C1 or C2 homozygotes, they may more effectively inhibit *KIR*-positive NK cell and T cell populations involved in GvL responses [135].

KIR genes and autologous stem cell transplantation

Few research groups have demonstrated the influence of *KIR* genes in autologous stem cell transplantation (ASCT). The interest in the role of *KIR* genes in the setting of ASCT is mainly related to preventing relapse, the main cause of treatment failure. Some studies have shown that rapid and early NK cell recovery following ASCT is associated with a better progression-free survival (PFS) in some diseases. An analysis of 182 patients with myeloma multiple submitted to ASCT showed a worse outcome in patients who were *KIR3DS1+*. The

KIR3DS1 genotype was associated with a shorter PFS with the effect being more notable in patients who received a transplant while in complete or partial remission after induction chemotherapy and those who lacked HLA-Bw4 [136].

Similarly, in a study of 169 neuroblastoma patients treated by ASCT, a survival advantage was shown in patients lacking HLA class I ligands for autologous inhibitory KIRs. Those who lacked the HLA-C1 ligand for *KIR2DL2/ KIR2DL3* had the highest 3-year survival rate [137]. Another study analyzed the influence of KIR mismatch in ASCT by the receptor-ligand mismatch model. The study, involving 16 patients who were submitted to ASCT for non-Hodgkin's lymphoma and solid tumours, found a reduced relapse rate for patients with an inhibitory KIR-HLA mismatch [138]. On the other hand, another study of 67 patients with solid tumors or lymphomas who were treated with ASCT did not find any effect of KIR-ligand interactions on the outcomes of ASCT [139].

KIR genes and unrelated umbilical cord blood transplantation

Unrelated UCBT has proved to be a viable treatment option. An advantage of using UCB is the relatively low risk of acute GVHD due to a lower number of mature donor T cells and thus an increased possibility of using HLA-mismatched units. Moreover, UCBT, as in haplo-HSCT, is characterized by a rapid post-transplant recovery of NK cells. An analysis of 218 patients with AML or ALL showed that patients who received a single UCBT unit from a *KIR* ligand incompatible donor showed a lower incidence of relapse, and increased DFS and OS [140]. Additionally, as was seen in the Ruggery studies, the benefits were significantly more marked in patients with AML. However, another study failed to observe any benefit of KIR-ligand mismatch in 155 recipients of UCB after myeloablative conditioning. In fact, in 102 patients who received UCB after nonmyeloablative conditioning, *KIR*-Ligand mismatch was associated with an increased rate of acute GVHD and higher treatment-related mortality [141].

Altogether these data show that simple assessments of the *KIR* genotype might help in the selection of donors for HSCT. *KIR* mismatches seem to be effective in haplo-HSCT and mainly in patients with myeloid diseases. The contradictory results reported about the influence of *KIR* mismatches in the diverse types of HSCT can certainly be explained by differences in the transplant protocols employed. Differences like number of patients analyzed, type of disease studied, stage of the disease, patient age, conditioning regimen, stem cell source, GVHD prophylaxis and variability in the definition of *KIR* mismatch can influence transplant outcomes. Factors like T-cell depletion and no post-transplant immune suppression seem to be important in maximizing NK cell alloreactivity [142].

Cytokines genes and receptors in HSCT

There are many other genetic factors that influence to outcome of transplant, independent of whether the transplant is autologous, allogeneic, matched or mismatched, sibling or unrelated donor, or haploidentical and of whether the cell source is bone marrow, peripheral blood or UCB.

The goal of the majority to studies is to know what kind of influence these genetic factors and HLA compatibility have and what effect they have on the course of the transplant: acute and chronic GVHD, relapse, OS and mortality.

http://dx.doi.org/10.5772/54281

One important factor is the polymorphisms within the regulatory sequences of cytokine genes. Proinflammatory cytokines, receptors and related inhibitors have been implicated in a large number of immune diseases. The main role of cytokine genes is related to the immunopathogenesis of GVHD [143].

Studies on cytokine genes in the transplant setting involve receptors of the TNF, IL-10, the IL-1 gene family, IL-2, IL-6, interferon TNF-γ, TGF-β1 and TGF-β1 [28, 144-145].

Tissue injury, including of the mucosa and liver, occurs during the conditioning regimen. This process causes the secretion of the TNF-α, IL-6 and IL-1 pro-inflammatory cytokines that increase HLA antigens, thus increasing the antigens recognized by donor T-cell receptors in allogeneic transplantation. Moreover, during the activation to donor cells, T cells produce IL-2 and INF-γ (Th1) that trigger GVHD and are balanced for Th2 cytokines such as IL-4 and IL-10 [146-147].

Studies on allogeneic HSCT, investigated 16 patients with chronic GVHD by a systematic clinical examination of the oral cavity, and by biopsies of the buccal mucosa and labial salivary glands. The findings demonstrated that the mRNA expression of IL-2, INF-γ, IL-4 and IL-5 in the buccal mucosa of chronic GVHD patients was greater than in control individuals. A similar result was detected for the labial salivary glands with the addition of IL-10 [148].

Studies show that IL-2 and INF-γ were detected more frequently in patients with acute GVHD. Additionally, IL-12 and IL-18 were increased while IL-10 was decreased in the same group, and IL-4 did not present a significant difference between the control and patient groups [144]. Other studies show high IL-10 gene expression in the recipient that may be related to a reduced incidence of grades II to IV acute GVHD and a reduced graft-versus-tumor effect after HSCT with nonmyeloablative conditioning [145].

On the other hand, studies affirm that IL-4 producing cells inhibit the development of acute GVHD and the increased percentage of IL-4 secreting cells may be responsible for the unexpected low incidence of acute GVHD after peripheral blood HSCT, despite the presence of large numbers of mature T cells in the donor infusion [148].

Many studies show that polymorphisms of cytokine genes influence to outcome of transplants, such as with the development of GVHD. One example is that the *IL17+197*^{a} allele was associated with increased risk of grade III and IV acute and chronic GVHD. Other studies demonstrate clinically important relationships between genetic polymorphisms in TNF-α and the severity of acute GVHD [147,149]. There are many other associations of polymorphisms of cytokine genes that course to acute and chronic GVHD.

Major histocompatibility complex class I-related chain genes and HSCT

The MHC class I-related chain (*MIC*) genes have been the subject of interest in the setting of HSCT. This family of genes, located in the MHC classical class I region, was first described in 1994 [150-151]. These genes are very polymorphic, but not as much as the classical HLA class I genes. Humans have seven *MIC* genes, named *MICA* to *MICG* but only two *MIC* genes are functional, the MHC class I-related chain A (*MICA*) and B (*MICB*) genes. The MIC proteins are similar to the HLA class I gene products however they are not associated with β-2-microglobulin and also do not bind peptides to present to T cells [150,152]. MIC proteins appear to be induced by stress [153] and are expressed on the cell surface of fibroblasts and endothelium cells [154]. They are ligands for NKG2D [155], a receptor present on NK cells and some T cells, and because of this they can co-stimulate NK cells and T cells and can therefore determine the outcome of certain effector functions that are related to GVHD. In fact, *MIC* genes have been show to be attractive targets in diverse cancers, autoimmune diseases and in organ rejection after transplantation.

Several studies have demonstrated that the MICA may be a target molecule in allograft rejection because MICA can elicit antibody production after solid organ transplantation [156-163]. Some studies have reported diverse outcomes in HSCT related to *MIC* genes. It was suggested that *MIC* genes play a role in GVHD in HLA-matched HSCT because a higher rate of grade II-IV acute GVHD was found as was more gastrointestinal GVHD in MICA mismatched patients [164]. In addition, matches of MICA and MICB loci were shown to increase patient survival in a study of 44 patients who received unrelated HSCT [165].

Some polymorphisms in *MICA* genes have also been associated to outcomes in transplants. A change at position 129 of the α2-heavy chain domain of *MICA* can denote the strength of interaction with the NKG2D receptor. The presence of methionine at position 129 of the *MICA* gene characterizes a strong binder, and the presence of valine characterizes a weak binder [166]. Hence, the *MICA*-129 valine genotype and soluble *MICA* serum level were considered risk factors for chronic GVHD in a study of 211 HLA-identical sibling pairs of HSCT while before transplantation, the presence of anti-*MICA* antibodies that can neutralize soluble *MICA* confers protection [167]. Altogether, these data suggest that *MIC* genes, in particular the *MICA* genes, could be used as biomarkers for chronic GVHD and should be studied further.

Minor histocompatibility antigens and HSCT

The human minor histocompatibility antigens (mHAgs) are another group of immunogenic peptides, distinct from the MHC system, which seem to have a role in HSCT outcomes. They are derived from intracellular polymorphic proteins and are presented by HLA class I and II restricted T cells [168-170]. Accumulated evidence suggests that they can elicit allogeneic T-cell mediated immune response in HLA-matched allogeneic HSCT and because of this have been investigated in order to understand their possible role in the control of GVHD and GvL.

Diverse minor histocompatibility antigens of various genetic and cellular origins have been described. More than 40 different genes that encode mHAgs recognized by either CD8+ or CD4+ T cells have been identified [171-174]. Most of the mHAgs are result of non-synonymous single nucleotide polymorphisms in autosomal genes while others are encoded by the sex chromosomes. At least 6 genes in the Y chromosome encode male-specific MHAgs (so-called HY antigens). Additionally, mHAgs may also be caused by gene deletions and genetic variations in non-coding regions affecting gene transcription [175-178].

The best-characterized minor histocompatibility antigen is encoded by the Y chromosome (HA-1). The mHAgs related to gender seems to be involved in HSCT outcomes because their absence in women can lead to a response to male antigens; female-to-male transplants seem to be more susceptible to GVHD [168-170, 179-180,181-185]. Antibody responses to HY proteins are also associated with both chronic GVHD and the maintenance of remission [186], but whether these antibody responses contribute meaningfully to GVHD, or simply serve as markers for it, remains unclear. In spite of female-to-male immune responses being more common, the opposite can also happen [187-188].

Some mHAgs are expressed only in the hematopoietic system while others are also expressed in normal tissues. mHAgs whose expression is limited to hematopoietic tissue may be recognized by specific donor T cells and may selectively contribute to a GvL effect and those with broad tissue expression may mediate GVHD [189].

Several studies have associated the presence of mHAg-specific T cells post-transplantation with graft rejection [179, 190], GVHD [191-194], and the GVL effect [195-197]. Mismatches between patient and donor for HA-1, HA-2, HA-4 and HA-5 are associated with an increased incidence in GVHD [191].

The role and the mechanisms of alloreactions related to mHAgs are not fully understood, but these data suggest that they may be relevant in determining post-transplantation outcomes.

4. Conclusion

Genetic differences between donor and recipient are crucial factors capable of influencing transplantation outcomes. Much has been learned about the HLA and non-HLA genes, their expression, their polymorphisms and their role in mediating GvL and GVHD responses. A better understanding of these genes may permit more refined donor selection criteria and consequently a more accurate assessment of transplant-related complications.

Author details

Amanda Vansan Marangon, Ana Maria Sell, Daniela Maira Cardozo and
Jeane E. L. Visentainer

Immunogenetics Laboratory, Department of Basic Sciences, Maringá State University - UEM, Maringá, Paraná, Brazil

References

[1] Apperley J, Carreras E, Gluckman E, Masszi T. Haematopoietic Stem Cell Transplantation 2012. European School of Haematology, 74-89; 2012.

[2] Jacobson LO, Marks EK, Robson MJ. Effect of spleen protection on mortality following x-irradiation. J Lab Clin Med 1949;34:1538-43.

[3] Van Bekkun DW, DeVries MJ. Radiation Chimeras 1967. New York: Academic Press, 1967.

[4] Thomas ED, Lochte HL, Lu WC, Ferrebe JW. Intravenous infusion of bone marrow in patients receiving radiation and chemotherapy. N Engl J Med 1957;257:491-6. http://www.ncbi.nlm.nih.gov/pubmed/13464965 (accessed 02 August 2012).

[5] Mathé G, Amiel J.L, Schwarzenberg L, Catan A, Schneider M. Adoptive immunotherapy of acude leukemia : experimental and clinical results. Cancer Res 1965;25:1525-31. http://www.ncbi.nlm.nih.gov/pubmed/5323965 (accesses 02 August 2012).

[6] Charron D. HLA, immunogenetics, pharmacogenetics and personalized medicine. Vox Sanguinis, 2011; 100: 163-6. http://www.ncbi.nlm.nih.gov/pubmed?term=HLA%20immunogenetics%20pharmacogenetics%20and%20personalized%20medicine (accessed 08 August 2012).

[7] Abbas A. K, Celular and Molecular Immunology. 5th ed. 2005.

[8] Lee SJ, Klein J, Haagenson M et al. High-resolution donor-recipient HLA matching contributes to the success of unrelated donor marrow transplantation. Blood 2007; 110:4576-83. http://bloodjournal.hematologylibrary.org/content/110/13/4576.long (accessed 10 August 2012).

[9] Miescher PP, Fauconnet M. Mise en evidence de different groups leucocytaires chez l'homme. Schweiz Med Wochenschr 1954;84:597-9. http://www.ncbi.nlm.nih.gov/pubmed/13168327 (accessed 13 August 2012)

[10] Jawdat D, Al Saleh S, Sutton P, Al Anazi H, Shubaili A, Uyar FA, et al. HLA-C Polymorphisms in two cohorts of donors of bone marrow transplantation. Saudi J Kidney Dis Transpl.2012;23:467-70. http://www.sjkdt.org/article.asp?issn=1319-2442;year=2012;volume=23;issue=3;spage=467;epage=470;aulast=Jawdat (accessed 13 August 2012).

[11] Voltarelli J. C, Pasquini R, Ortega E. T. T, Transplante de Células-Tronco Hematopoiéticas. Ed. Atheneu 2010

[12] Bjorkman PJ, Saper MA, Samraoui B, Bennett WS, Strominger JL, Wiley DC. Structure of the human class I histocompatibility antigen, HLA-A2. J Immunol 2005;174:6-9. http://www.jimmunol.org/content/174/1/6.long (accessed 13 August 2012).

[13] Steinmetz M, Hood L. Genes of the major histocompatibility complex in mouse and man. Science 1983; 222: 727-33. http://www.sciencemag.org/content/222/4625/727.long (accessed 14 August 2012)

[14] Le Bouteiller P. HLA class I chromosomal region, genes, and products: facts and questions. Critical Reviews in Immunology 1994; 14: 89-129. http://

www.ncbi.nlm.nih.gov/pubmed?term=HLA%20class%20I%20chromosomal%20region%2C%20genes%2C%20and%20products%3A%2 (accessed 16 August 2012).

[15] Bjorkman PJ, Saper MA, Samraoui B, Bennett WS, Strominger JL, Wiley DC. The foreign antigen binding site and T cell recognition regions of class I histocompatibility antigens. Nature 1987;329:512-8. http://www.nature.com/nature/journal/v329/n6139/abs/329512a0.html (accessed 18 August 2012).

[16] Rammensee HG, Friede T, Stevanovic S. MHC ligands and peptide motifs: first listing. Immunogenetics 1995;41:178-28. http://www.ncbi.nlm.nih.gov/pubmed?term=MHC%20ligands%20and%20peptide%20motifs%3A%20first%20listing (accessed 29 August 2012).

[17] Salter RD, Benjamin RJ, Wesley PK, et al. A binding site for the T-cell co-receptor CD8 on the alpha 3 domain of HLA-A2. Nature 1990;345:41-6. http://www.nature.com/nature/journal/v345/n6270/abs/345041a0.html (accessed 30 August 2012).

[18] Croft M, Dubey C. Accessory molecule and costimulation requirements for CD4 T cell response. Crit Rev Immunol 1997; 17:89-118. http://www.ncbi.nlm.nih.gov/pubmed?term=Accessory%20molecule%20and%20costimulation%20requirements%20for%20CD4%20T%20cell%20response (accessed 02 September 2012).

[19] Moraes M. E, Romero M, Salomão I, Correa M, Gouvea N, Romero M et. al, Strong linkage disequilibrium between HLA-B*3913 and DRB1*0807 in Brazilians. Transplant. Proc. 2004; 36: 823-4. http://www.transplantation-proceedings.org/article/S0041-1345(04)00328-8/abstract (accessed 03 September 2012).

[20] Vojvodic S, Popovic S, and Macukanovic G. L. Implications of HLA linkage disequilibrium phenomenon in finding unrelated volunteer bone marrow donors. Med. Preql. 2007; 60: 178-82. http://www.ncbi.nlm.nih.gov/pubmed?term=Implications%20of%20HLA%20linkage%20disequilibrium%20phenomenon%20in%20finding%20unrelated%20volunteer%20bone% (accessed 04 September 2012).

[21] Marry E. Genetics diversity and bone marrow transplantation. Bull Soc Pathol Exot. 2012; May;105(2):137-42. http://www.springerlink.com/content/y48r78g80384716x/ (accessed 13 September 2012).

[22] Flomenberg N, Baxter-Lowe LA, Confer D, Fernandez-Vina M, Filipovich A, Horowitz M et al. Impact of HLA class I and class II high-resolution matching on outcomes of unrelated donor bone marrow transplantation: HLA-C mismatching is associated with strong adverse effect on transplantation outcome. Blood, 2004; 104: 1923-30. http://bloodjournal.hematologylibrary.org/content/104/7/1923.long (accessed 14 September 2012).

[23] Arora M, Weisdorf DJ, Spellman SR, Haagenson MD, Klein JP, Hurley CK et al. HLA-identical sibling compared with 8/8 matched and mismatched unrelated donor bone marrow transplant for chronic phase chronic myeloid leukemia. J Clin Oncol.

2009; 27:1644-1652. http://www.ncbi.nlm.nih.gov/pmc/articles/PMC2668970/ (Accessed 13 August 2012).

[24] Cheuk DKL, Wang P, Lee TL, Chiang AK, Ha SY, Lau YL, et al, Risk factors and mortality predictors of hepatic veno-occlusive disease, after pediatric hematopoietic stem cell transplantation. Bone Marrow Transplant, 2007; 40: 935-44. http://www.nature.com/bmt/journal/v40/n10/full/1705835a.html (accessed 29 September 2012).

[25] Loiseau P, Busson M, Balere M.L, Dormoy A, Bignon JD, Gagne K, et al. HLA association with Hematopoietic Stem Cell Transplantation Outcome: The number of mismatches at HLA-A, -B, -C, -DRB1, or –DQB1 Is strongly associated with overall survival. Biology of blood and marrow transplantation. 2007; 13:965-74. http://www.bbmt.org/article/S1083-8791(07)00252-2/abstract (accessed 01 September 2012).

[26] Morishima Y, Sasazuki T, Inoko H, Juji T, Akaza T, Yamamoto K et al. The clinical significance of human leukocyte antigen (HLA) allele compatibility in patients receiving a marrow transplant from serologically HLA-A, HLA-B, and HLA-DR matched unrelated donors. 2012; 99:4200-6. http://bloodjournal.hematologylibrary.org/content/99/11/4200.long (accessed 06 August 2012).

[27] Lee S. J, Klein J, Haagenson M, Confer DL, Eapen M, Fernandez-Vina M et. al. High-resolution donor recipient HLA matching contributes to the success of unrelated donor marrow transplantation. 2007; Blood. 110:4576-83. http://bloodjournal.hematologylibrary.org/content/110/13/4576.long (accessed 23 September 2012).

[28] Maury S, Appert B. L. M, Chir Z, Boiron J, , Galambrun C, Yakouben K, Bordigoni P et. al. Unrelated stem cell transplantation for severe acquire aplastic anemia: improved outcome in the era of high-resolution HLA matching between donor and recipient. 2007. Hematologica Journal 2007; 92: 589-96. http://www.haematologica.org/content/92/5/589.long (accessed 24 August 2012).

[29] Yanover C, Petersdorf EW, Malkki M, Gooley T, Spellman S, Velardi A et al. HLA mismatches and hematopoietic cell transplantation: structural simulations assess the impact of changes in peptide binding specificity on transplant outcome. Immunome Research. 2011; 7:2:4. http://immunome-research.net/journal/index.php/immunome/article/view//24 (accessed 14 September 2012).

[30] Yoshihara S, Tanigushi K, Ogawa H, Saji H, The role of HLA antibodies in allogeneic SCT: is the type and screen strategy necessary not only for blood type but also for HLA? Nature 2012; 1-8. http://www.nature.com/bmt/journal/vaop/ncurrent/full/bmt2011249a.html (accessed 03 September 2012).

[31] Battiwalla M, Wang T, Carreras J, Deeg HJ, Ayas M, Bajwa RP et al, HLA-matched sibling transplantation for severe aplastic anemia: Impact of HLA DR15 Antigens status on Engrafment, graft-versus-host disease, and overall survival. Biol Blood Mar-

row Transplant. 2012;18:1401-6. http://www.bbmt.org/article/S1083-8791(12)00095-X/abstract (accessed 13 September 2012).

[32] Wang T.F, Huang H, Tzeng CH, Wang PN, Wu T, Sun J, et al, Impact of donor characteristics and HLA matching on survival of Chinese patients with hematologic malignancies undergoing unrelated hematopoietic stem cell transplantation. Biol Blood Marrow Transplant. 2012; 1-6. http://www.bbmt.org/article/S1083-8791(12)00304-7/abstract (accessed 13 September 2012).

[33] Horan J, Wang T, Haagenson M, Spellman SR, Dehn J, Eapen M et al, Evaluation of HLA Maching in Unrelated Hematopoietic Stem Cell Transplantation for Nonmalignant Disorders. Blood 2012; http://bloodjournal.hematologylibrary.org/content/120/14/2918.long (accessed 14 September 2012).

[34] Giebel S, Giorgiani G, Martinetti M, Zecca M, Maccario R, Salvaneschi L et. al, Low incidence of severe acute graft-versus-host-diseasein children given haematopoietic stem cell transplantation from unrelated donors prospectively matched for HLA class I and II alleles with high-resolution molecular tiping. Bone Marrow Transplant. 2003; 31: 987-93.

[35] Kawase T, Matsuo K, Kashiwase K, Inoko H, Saji H, Ogawa S et al, HLA mismatch combinations associated with decreased risk of relapse: implications for the molecular mechanism. Blood 2009; 113: 2851-8. http://bloodjournal.hematologylibrary.org/content/113/12/2851.long (accessed 15 September 2012).

[36] Flomenberg N, Baxter-Lowe L, Confer D, Fernandez-Vina M, Filipovich A, Horowitz MA, et. al, Impact of HLA class I and class II high-resolution matching on outcomes of unrelated donor bone marrow transplantation: HLA-C mismatching is associated with a strong adverse effect on transplantation outcome. Blood. 2004;104:1923-30. http://bloodjournal.hematologylibrary.org/content/104/7/1923.long (accessed 20 September 2012).

[37] Ciurea S. O, Chaplim MDR. Donor selection in T-cell replete haploidentical stem cell transplantation: Knowns, unknowns and controversies. Biol Blood Marrow Transplant. 2012; 319-9. http://www.bbmt.org/article/S1083-8791(12)00319-9/abstract (accessed 20 September 2012).

[38] Huo MR, Xu LP, Li D, Liu DH, Liu KY, Chen H. et al, The effect of HLA disparity on clinical outcome after HLA-haploidentical blood and marrow transplantation. Clin Transplant 2012; 26:284-91. http://onlinelibrary.wiley.com/doi/10.1111/j.1399-0012.2011.01499.x/abstract (accessed 20 September 2012).

[39] Kasamon Y. L, Luznik L, Leffell MS, Kowalski J, Tsai HL, Bolaños-Meade J et al, Nonmyeloblative HLA-haploidentical BMT with high-dose posttransplantation cyclophosphamide: effect of HLA disparity on outcome. Biol. Bone Marrow Transplant. 2010; 16: 482-489. http://www.ncbi.nlm.nih.gov/pmc/articles/PMC2998606/ (accessed 20 September 2012).

[40] Munchel A, Kesserwan C, Symons HJ, Luznik L, Kasamon YL, Jones RJ, et al. Non-myeloblative HLA-haploidentical bone marrow transplantation with higth dose, post transplantation cyclophosphamide. 2011 Pediatric Reports 2011;3:2-15. http://www.ncbi.nlm.nih.gov/pmc/articles/PMC3206539/ (accessed 01 September 2012).

[41] Liao C, Wu J. Y, et al, Indiscernible benefit of high-resolution HLA typing in improving long-term clinical outcome of unrelated umbilical cord blood transplant. Bone Marrow Transplantation 2007; 40: 201-8. http://www.nature.com/bmt/journal/v40/n3/full/1705711a.html (accessed 02 September 2012).

[42] Kanda J, Ichinohe T, Kato S, Uchida N, Terakura S, Fukuda T et al, Unrelated cord blood transplantation vs related transplantation with HLA 1-antgen mismatch in the graft-versus-host direction. Leukemia 2012; 203. http://www.nature.com/leu/journal/vaop/ncurrent/full/leu2012203a.html (accessed 03 September 2012).

[43] Rodrigues C. A, Pereira N. F, et. al, Umbilical Cord blood transplantation. Ver. Bras. Hematol. Hemoter. 2010.

[44] Atsuta Y, Morishima Y, Suzuki R, Nagamura-Inoue T, Taniguchi S, Takahashi S et. al, Comparison of unrelated cord blood transplantation and HLA-mismatched unrelated bone marrow transplantation for adults with leukemia. Biol. Blood Marrow Transplant 2012;18:780-7. http://www.bbmt.org/article/S1083-8791(11)00411-3/abstract (accessed 20 August 2012).

[45] Sang Yi E, Lee SH. Hematopoietic stem cell in children with acute leukemia: similar outcomes in recipients of umbilical cord blood versus marrow or peripheral blood stem cells from related or unrelated donors. Korean Journal Pediatric 2012;55:93-9. http://www.ncbi.nlm.nih.gov/pmc/articles/PMC3315625/ (accessed 03 September 2012).

[46] Rocha V, John MD. Graft-Versus-Host Disease in Children who have received a cord blood or bone marrow transplant from an HLA-identical sibling. N. Engl. J. Med 2000; 342:1846-54. http://www.nejm.org/doi/full/10.1056/NEJM200006223422501 (accessed 20 August 2012).

[47] Wen S. H, Lai M. J, Yang K. L. Human Leukocyte antigen-A, -B, -DRB1 haplotypes of cord blood units in the Tzu Chi Taiwan Cord Blood Bank. Human immunology. 2008; 69: 430-6. http://www.sciencedirect.com/science/article/pii/ (accessed 20 September 2012).

[48] Brown J, Poles A, Brown CJ, Contreras M, Navarrete CV. HLA-A, -B, and DR antigen frequencies of the London Cord Blood Bank units differ from those found in established bone marrow donor registries. Bone Marrow Transplan. 2000; 25: 475-81. http://www.nature.com/bmt/journal/v25/n5/full/1702197a.html (accessed 20 September 2012).

[49] Trinchieri G. 1989. Biology of natural killer cells. Adv. Immunol. 1989. 47:187–376.

[50] Cerwenka A, Lanier LL .Natural killer cells, viruses and cancer. Nat Rev Immunol 2001;1:41–9. http://www.uiowa.edu/~immuno/classMaterial/Advanced%20Topics%202004/Session%2010,%20Sept%2027/Cerwenka,%20Lanier.pdf (accessed 03 may 2012)

[51] Caligiuri MA. Human natural killer cells. Blood 2008;112:461–9.http://bloodjournal.hematologylibrary.org/content/112/3/461.short (accessed 02 may 2012)

[52] Street SE, Hayakawa Y, Zhan Y, Lew AM, MacGregor D, Jamieson AM, et al. Innate immune surveillance of spontaneous B cell lymphomas by natural killer cells and gammadelta T cells. J Exp Med 2004;199:879– 84 http://jem.rupress.org/content/199/6/879.short (accessed 02 may 2012)

[53] Verheyden S, Demanet C. NK cell receptors and their ligands in leukemia. Leukemia 2008; 22: 249–57.http://www.nature.com/leu/journal/v22/n2/full/2405040a.html (accessed 02 may 2012)

[54] Kulkarni S, Martin MP, Carrington M. The yin and yang of HLA and KIR in human disease. Semin Immunol 2008;20:343–52. http://www.sciencedirect.com/science/article/pii/S1044532308000523 (accessed 01 April 2009)

[55] Dupont B, Hsu KC. Inhibitory killer Ig-like receptor genes and human leukocyte antigen class I ligands in haematopoietic stem cell transplantation. Curr Opin Immunol 2004;16:634–43.http://www.sciencedirect.com/science/article/pii/S0952791504001190 (accessed 01 April 2009)

[56] Barao I, Murphy W. The immunobiology of natural killer cells and bone marrow allograft rejection. Biol Blood Marrow Transplant 2003; 9:727-41.http://www.sciencedirect.com/science/article/pii/S1083879103003343 (accessed 23 january 2008)

[57] Thielens, A; Vivier E; Romagne F. NK cell MHC class I specific receptors (KIR): from biology to clinical intervention. Current Opinion in Immunology 2012; 24:239–45.http://www.sciencedirect.com/science/article/pii/S0952791512000039 (accessed 02 may 2012)

[58] Kiessling R, Klein E, Wigzell H. "Natural" killer cells in the mouse. I. Cytotoxic cells with specificity for mouse Moloney leukemia cells. Specificity and distribution according to genotype. Eur J Immunol 1975;(5):112–7. http://onlinelibrary.wiley.com/doi/10.1002/eji.1830050208/abstract;jsessionid=682FA874FA648EE08CFD6607AE91874E.d01t02 (accessed 15 january 2006)

[59] Cooper MA, Fehniger TA, Caligiuri MA. The biology of human natural killer-cell subsets. Trends Immunol 2001;22:633–40.http://www.sciencedirect.com/science/article/pii/S1471490601020609 (accessed 02 may 2012)

[60] Uhrberg M, Valiante NM, Shum BP, Shilling HG, Lienert-Weidenbach K, Corliss B, et al. Human diversity in killer cell inhibitory receptor. Genes Immun 1997;7:753–63.http://www.sciencedirect.com/science/article/pii/S1074761300803945 (accessed 09 February 2006)

[61] Martin AM, Freitas EM, Witt CS, Christiansen FT. The genomic organization and evolution of the natural killer immunoglobulin-like receptor (KIR) gene cluster. Immunogenetics 2000;51:268 – 80. http://www.springerlink.com/content/100517/ (accessed 09 February 2006)

[62] Kikuchi-Maki A, Catina TL, Campbell KS. Cutting edge: KIR2DL4 transduces signals into human NK cells through association with the Fc receptor gamma protein. J Immunol 2005; 174:3859–63 http://www.jimmunol.org/content/174/7/3859.long (accessed 09 February 2006)

[63] Hsu KC,Liu XR, Selvakumar A, Mickelson E, O'Reilly RJ, Dupont B. Killer Ig-like receptor haplotype analysis by gene content: evidence for genomic diversity with a minimum of six basic framework haplotypes, each with multiple subsets. J Immunol 2002;169:5118 –129. http://www.jimmunol.org/content/169/9/5118.long (accessed 02 may 2012)

[64] Winter CC, Gumperz JE, Parham P, Long EO, Wagtmann N. Direct binding and functional transfer of NK cell inhibitory receptors reveal novel patterns of HLA-C allotype recognition. J Immunol 1998;161:571–7.http://www.jimmunol.org/content/161/2/571.long (accessed 05 July 2012)

[65] Colonna M, Borsellino G, Falco M, Ferrara GB, Strominger JL. HLA-C is the inhibitory ligand that det ermines dominant resistance to lysis by NK1- and NK2-specific natural killer cells. Proc Natl Acad Sci U S A 1993;90:12000- 4.http://www.ncbi.nlm.nih.gov/pmc/articles/PMC48113/ (accessed 05 July 2012)

[66] Wagtmann N, Rajagopalan S, Winter CC, Peruzzi M, Long EO. Killer cell inhibitory receptors specific for HLA-C and HLA-B identified by direct binding and by functional transfer. Immunity 1995;3:8 01-9.http://www.sciencedirect.com/science/article/pii/107476139 5900691 (accessed 05 July 2012)

[67] Thananchai H, Gillespie G, Ma rtin MP, Bashirova A, Yawata N, Yawata M, et al. Cutting edge: allele-specific and peptide-dependent interactions between KIR3DL1 and HLA- A and HLA-B. J Immunol 2007; 178:33-7.http://www.jimmunol.org/content/178/1/ 33.long (accessed 02 may 2012)

[68] Foley BA, De Santis D, Van Beelen E, Lathbury LJ, Christiansen FT, Witt CS. The reactivity of Bw4-positive HLA-B and HLA-A alleles with KIR3DL1: implications for pati ent and donor suitability for haploidentical stem cell transplants. Blood 2008; 112:435-443.http://bloodjournal.hematologylibrary.org/content/112/2/435.full (accessed 25 Jan uary 2012)

[69] Stern M, Ruggeri L, Capanni M, Mancusi A, Velardi A. Human leukocyte antigens A23, A24 and A32 but not A25 are ligands for KIR3DL1. Blood 2008; 112:708-1 0.http://bloodjournal.hematologylibrary.org/content/112/3/708.long (accessed 25 January 2012)

[70] O'Connor GM, Guinan KJ, Cunningham RT, Middleton D, Parham P, Gardiner CM. Functional polymorphism of the KIR3DL1/S1 receptor on human NK cells. J Immu-

nol 2007;178:235-41. http://www.jimmunol.org/content/178/1/235.long (accessed 1 7 April 2010)

[71] Muller CA, Engler-Blum G, Gekeler V, Steiert I, Weiss E, Schmidt H. Genetic and serological heterogeneity of the supertypic HLA-B locus specificities HLA-Bw4 and-Bw6 Immunogenetics 1989; 30: 200-7. http://www.springerlink.com/content/p43708943411367 7/ (accessed 05 August 2012)

[72] Biassoni R, Pessino A, Malaspina A, Cantoni C, Bottino C, Sivori S, et al. Role of amino acid position 70 in the binding affinity of p50.1 and p58.1 receptors for HLA-Cw4 molecules. Eur J Immunol 1997;27, 3095–9. http://onlinelibrary.wiley.com/doi/10.1002/eji.1830271203/abstract (accessed 25 January 2012)

[73] Parham, P. MHC class I molecules and KIRs in human history, health and survival. Nat Rev Immunol, 2005; 5(3):201-214. (http://www.ncbi.nlm.nih.gov/pubmed/15719024 (accessed 17 August 2009)

[74] Katz G, Markel G, Mizrahi S, Arnon TI, Mandelboim O. Recognition of HLA-Cw4 but not HLA-Cw6 by the NK cell receptor killer cell Ig-like receptor two-domain short tail number 4. J Immunol 2001;166:7260-7267.(accessed 05 August 2012)

[75] Orange JS, Fassett MS, Koopman LA, Boyson JE, Strominger JL. Viral evasion of natural killer cells. Nat Immunol 2002;3:1006 –12.

[76] Verheyden S, Ferrone S, Mulder A, Claas FH, Schots R, De Moerloose B, et al. Role of the inhibitory KIR ligand HLA-Bw4 and HLA-C expression levels in the recognition of leukemic cells by Natural Killer cells. Cancer Immunol Immunother 2009;58(6): 855–65.http://www.springerlink.com/content/v5158n27t2216718/ (accessed 05 August 2012)

[77] Karre K, Ljunggren HG, Piontek G, Kiessling R. Selective rejection of H-2-deficient lymphoma variants suggests alternative immune defense strategy. Nature 1986;319:675–8. http://www.nature.com/nature/journal/v319/n6055/abs/319675a0.html (accessed 05 August 2012)

[78] Ljunggren HG, Karre K. In search of the 'missing self': MHC molecules and NK cell recognition. Immunol Today 1990;11:237–44. http://www.ncbi.nlm.nih.gov/pubmed/2201309 (accessed 05 August 2012)

[79] Moretta L, Ciccone E, Moretta A, Hoglund P, Ohlen C, Karre K. Allorecognition by NK cells: nonself or no self? Immunol Today 1992;13:300–6. http://www.ncbi.nlm.nih.gov/pubmed/1380815 (accessed 05 August 2012)

[80] Van Bergen J, Thompson A, Retiere C, Trowsdale J, Koning F. Cutting edge: killer Ig-like receptors mediate "missing self" recognition in vivo. J Immunol 2009;182(5): 2569–72.http://www.ncbi.nlm.nih.gov/pubmed/19234149 (accessed 05 August 2012)

[81] Velardi A, Ruggeri L, Mancusi A, Aversa F, Christiansen FT. Natural killer cell allorecognition of missing self in allogeneic hematopoietic transplantation: a tool for im-

munotherapy of leukemia. Curr Opin Immunol 2009; 21:525–30.http://www.ncbi.nlm.nih.gov/pubmed/19717293 (accessed 05 August 2012)

[82] Parham P: Taking license with natural killer cell maturation and repertoire development. Immunol Rev 2006; 214:155-160. http://www.ncbi.nlm.nih.gov/pubmed/17100883 (accessed 05 August 2012)

[83] Joncker NT, Raulet DH. Regulation of NK cell responsiveness to achieve self-tolerance and maximal responses to diseased target cells. Immunol Rev 2008; 224:85-97. http://www.ncbi.nlm.nih.gov/pubmed/18759922 (accessed 05 August 2012)

[84] Yawata M, Yawata N, Draghi M, Partheniou F, Little AM, Parham P.
MHC class I-specific inhibitory receptors and their ligands structure diverse human NK-cell repertoires toward a balance of missing self-response. Blood 2008; 112:2369-80. http://www.ncbi.nlm.nih.gov/pubmed/18583565 (accessed 05 August 2012)

[85] Jonsson AH, Yokoyama WM. Natural killer cell tolerance licensing and other mechanisms. Adv Immunol 2009; 101:27-79. http://www.ncbi.nlm.nih.gov/pubmed/19231592 (accessed 05 August 2012)

[86] Brodin P, Karre K, Hoglund P. NK cell education: not an on–off switch but a tunable rheostat. Trends Immunol 2009; 30:143-9. http://www.ncbi.nlm.nih.gov/pubmed/19282243 (accessed 05 August 2012)

[87] Pegram HJ, Ritchie DS, Smyth MJ, Wiernik A, Prince HM, Darcy PK et al. Alloreactive natural killer cells in hematopoietic stem cell transplantation leukemia Research 2011;35:14–21.http://www.ncbi.nlm.nih.gov/pubmed/20719383 (accessed 05 August 2012)

[88] Kook H, Goldman F, Padley D, Giller R, Rumelhart S, Holida M, et al. Reconstruction of the immune system after unrelated or partially matched T-cell-depleted bone marrow transplantation in children: immunophenotypic analysis and factors affecting the speed of recovery. Blood 1996;88:1089–97.http://www.ncbi.nlm.nih.gov/pubmed/8704219 (accessed 05 August 2012)

[89] Ruggeri L, Capanni M, Urbani E, Perruccio K, Shlomchik WD, Tosti A et al. Effectiveness of donor natural killer cell alloreactivity in mismatched hematopoietic transplants. Science 2002; 295:2097–100.http://www.ncbi.nlm.nih.gov/pubmed/11896281 (accessed 05 August 2012)

[90] Aversa F, Terenzi A, Tabilio A, et al. Full haplotype-mismatched hematopoietic stem-cell transplantation: a phase II study in patients with acute leukemia at high risk of relapse. J Clin Oncol 2005;23:3447–54. http://www.ncbi.nlm.nih.gov/pubmed/15753458 (accessed 05 August 2012)

[91] Pende D, Marcenaro S, Falco M Martini S, Bernardo ME, Montagna D, et al. Anti-leukemia activity of alloreactive NK cells in KIR ligand-mismatched haploidentical HSCT for pediatric patients: evaluation of the functional role of activating KIR and

re-definition of inhibitory KIR specificity. Blood 2009; 113:3119–29.http://www.ncbi.nlm.nih.gov/pubmed/18945967 (accessed 05 August 2012)

[92] Sun JY, Gaidulis L, Miller MM, Goto RM, Rodriguez R, Forman SJ et al. Development of a multiplex PCR-SSP method for Killer-cell immunoglobulin-like receptor genotyping. Tissue Antigens 2004;64:462–8.http://www.ncbi.nlm.nih.gov/pubmed/15361123 (accessed 05 August 2012)

[93] Ciccone E, Pende D, Vitale M, Nanni L, Di Donato C, Bottino C et al. Self class I molecules protect normal cells from lysis mediated by autologous natural killer Cells. Eur J Immunol. 1994;24(4):1003-6. http://www.ncbi.nlm.nih.gov/pubmed/8149950 (accessed 05 August 2012)

[94] Leung W, Iyengar R, Turner V, Lang P, Bader P, Conn P, et al. Determinants of antileukemia effects of allogeneic NK cells. J Immunol 2004;172(1):644–50. http://www.ncbi.nlm.nih.gov/pubmed/14688377 (accessed 05 August 2012)

[95] Stern M, Ruggeri L, Mancusi A, Bernardo ME, De Angelis C, Bucher C, et al. Survival after T cell-depleted haploidentical stem cell transplantation is improved using the mother as donor. Blood 2008; 112:2990-5 http://www.ncbi.nlm.nih.gov/pubmed/18492955 (accessed 05 August 2012)

[96] Fischer JC; Kobbe G; Enczmann J; Haas R; Uhrberg M. The impact of HLA-C matching depends on the C1/C2 KIR ligand status in unrelated hematopoietic stem cell transplantation Immunogenetics 2012 Aug 26. [Epub ahead of print] http://www.ncbi.nlm.nih.gov/pubmed/22923051 (accessed 05 September 2012)

[97] Kärre K. Immunology. A perfect mismatch. Science. 2002;295 (5562):2029-31. Comment on: Science. 2002;295(5562):2097- 100; Science. 2002;295(5562):2094-7.http://www.sciencemag.org/content/295/5562/2029 (accessed 10 April 2012)

[98] Beatty PG, Clift RA, Mickelson EM, Nisperos BB, Flournoy N, Martin PJ, et al. Marrow transplantation from related donors other than HLA-identical siblings. N Engl J Med. 1985;313(13):765-71.http://www.ncbi.nlm.nih.gov/pubmed/3897863 (accessed 10 April 2012)

[99] Ruggeri L, Capanni M, Casucci M, Volpi I, Tosti A, Perruccio K et al. Role of natural killer cell alloreactivity in HLA-mismatched hematopoietic stem cell transplantation. Blood 1999;94:333–9. http://bloodjournal.hematologylibrary.org/content/94/1/333.full (accessed 10 April 2012)

[100] Ruggeri L, Mancusi A, Capanni M, Urbani E, Carotti A, Aloisi T et al. Donor natural killer cell allorecognition of missing self in haploidentical hematopoietic transplantation for acute myeloid leukemia: challenging its predictive value. Blood 2007;110:433–440. http://www.ncbi.nlm.nih.gov/pubmed/17371948 (accessed 05 August 2012)

[101] Shlomchik WD, Couzens MS, Tang CB, McNiff J, Robert ME, Liu J, et al. Prevention of graft versus host disease by inactivation of host antigen- presenting cells. Science

1999;285:412–5. http://www.sciencemag.org/content/285/5426/412.abstract (accessed 05 August 2012)

[102] RuggeriL, Mancusi A, Burchielli E, Perruccio K, Aversa F, Mertelli MF. Natural killer cell recognition of missing self and haploidentical hematopoietic transplantation. Seminars in Cancer Biology 2006; (16):404–11. http://www.sciencedirect.com/science/article/pii/S1044579X06000599 (accessed 05 August 2012)

[103] Klingebiel T, Cornish J, Labopin M, Locatelli F, Darbyshire P, Handgretinger R, et al. Pediatric Diseases and Acute Leukemia Working Parties of the European Group for Blood and Marrow Transplantation (EBMT). Results and factors influencing outcome after fully haploidentical hematopoietic stem cell transplantation in children with very high-risk acute lymphoblastic leukemia: impact of center size: an analysis on behalf of the Acute Leukemia and Pediatric Disease Working Parties of the European Blood and Marrow Transplant group Blood. 2010; 29;115(17):3437-46. http://bloodjournal.hematologylibrary.org/content/115/17/3437.short (accessed 05 August 2012)

[104] Sivori S, Carlomagno S, Falco M, Romeo E, Moretta L, Moretta A. Natural killer cells expressing the KIR2DS1-activating receptor efficiently kill T-cell blasts and dendritic cells: implications in haploidentical HSCT. Blood 2011; 117(16):4284-92. http://bloodjournal.hematologylibrary.org/content/117/16/4284.short (accessed 23 July 2012)

[105] Symons HJ, Leffell MS, Rossiter ND, Zahurak M, Jones RJ, Fuchs EJ. Improved survival with inhibitory Killer Immunoglobulin Receptor (KIR) gene mismatches and KIR haplotype B donors after nonmyeloablative, HLA-haploidentical bone marrow transplantation Biol Blood Marrow Transplant. 2010; 16(4): 533–42.http://www.sciencedirect.com/science/article/pii/S1083879109005825 (accessed 23 July 2012)

[106] Ciceri F, Labopin M, Aversa F, Rowe JM, Bunjes D, Lewalle P, et al. A survey of fully haploidentical hematopoietic stem cell transplantation in adults with high-risk acute leukemia: a risk factor analysis of outcomes for patients in remission at transplantation Blood, 2008;112(9): 3574-81.http://bloodjournal.hematologylibrary.org/content/112/9/3574.short (accessed 23 July 2012)

[107] Bishara A, De Santis D, Witt CC, Brautbar C, Christiansen FT, Or R, et al. The beneficial role of inhibitory KIR genes of HLA class I NK epitopes in haploidentically mismatched stem cell allografts may be masked by residual donor-alloreactive T cells causing GVHD Tissue Antigens. 2004;63(3):204-11.http://onlinelibrary.wiley.com/doi/10.1111/j.0001-2815.2004.00182.x/full (accessed 23 July 2012)

[108] Giebel S, Locatelli F, Lamparelli T, Velardi A, Davies S, Frumento G, et al. Survival advantage with KIR ligand incompatibility in hematopoietic stem cell transplantation from unrelated donors Blood, 2003; 102(3):814-9.http://bloodjournal.hematologylibrary.org/content/102/3/814.short (accessed 23 July 2012)

[109] Beelen DH, Ottinger HD, Ferencik S, Elmaagacli AH, Peceny R, Trenschel R, et al. Genotypic inhibitory killer immunoglobulin-like receptor ligand incompatibility enhances the long-term antileukemic effect of unmodified allogeneic hematopoietic

stem cell transplantation in patients with myeloid leukemias. Blood, 2005;105(6): 2594-600.http://bloodjournal.hematologylibrary.org/content/105/6/2594.short (accessed 23 July 2012)

[110] Hsu KC, Gooley T, Malkki M, Pinto-Agnello C, Dupont, B, Bignon JD et al. KIR Ligands and Prediction of Relapse after Unrelated Donor Hematopoietic Cell Transplantation for Hematologic Malignancy. Biology of Blood and Marrow Transplantation 2006;12:828-836.http://www.sciencedirect.com/science/article/pii/S108387910600317X (accessed 23 July 2012)

[111] Savani BN, Mielke S, Adams S, Uribe M, Rezvani K, Yong AS et al. Rapid natural killer cell recovery determines outcome after T-cell-depleted HLA-identical stem cell transplantation in patients with myeloid leukemias but not with acute lymphoblastic leukemia. Leukemia 2007;21:2145–52.http://www.nature.com/leu/journal/v21/n10/abs/2404892a.html (accessed 23 July 2012)

[112] Cooley S, Trachtenberg E, Bergemann TL, Saeteurn K, Klein J, Le CT et al. Donors with group B KIR haplotypes improve relapse-free survival after unrelated hematopoietic cell transplantation for acute myelogenous leukemia.Blood. 2009;113(3): 726-32. http://www.ncbi.nlm.nih.gov/pubmed/18945962 (accessed 23 July 2012)

[113] Cooley S, Trachtenberg E, Bergemann TL, Saeteurn K, Klein J, Le CT et al. Donor selection for natural killer cell receptor genes leads to superior survival after unrelated transplantation for acute myelogenous leukemia Blood. 2009;113(3):726-32.http://bloodjournal.hematologylibrary.org/content/116/14/2411.short (accessed 23 July 2012)

[114] Tomblyn M, Young JH, Haagenson MD, Klein JP, Trachtenberg EA, Storek J, et al. Decreased Infections in Recipients of Unrelated Donor Hematopoietic Cell Transplantation from Donors with an Activating KIR Genotype Biol Blood Marrow Transplant. 2010;16(8):1155-61.http://www.sciencedirect.com/science/article/pii/S1083879110000972 (accessed 23 July 2012)

[115] Venstrom JM, Gooley TA, Spellman S, Pring J, Malkki M, Dupont B, et al. Donor activating KIR3DS1 is associated with decreased acute GVHD in unrelated allogeneic hematopoietic stem cell transplantation. Blood 2010;115(15):3162-5.http://bloodjournal.hematologylibrary.org/content/115/15/3162.short (accessed 23 July 2012)

[116] Venstrom JM, Pittari G, Gooley TA, Chewning JH, Spellman S, Haagenson M et al. HLA-C–Dependent Prevention of Leukemia Relapse by Donor Activating KIR2DS1. N Engl J Med. 2012;367(9):805-16.http://www.nejm.org/doi/full/10.1056/NEJMoa1200503 (accessed 23 July 2012)

[117] Verheyden S, Schots R, Duquet W, Demanet C. A defined donor activating natural killer cell receptor genotype protects against leukemic relapse after related HLA-identical hematopoietic stem cell transplantation. Leukemia. 2005;19(8): 1446-51.http://www.nature.com/leu/journal/v19/n8/abs/2403839a.html (accessed 23 July 2012)

[118] Giebel S, Nowak I, Dziaczkowska J, Czerw T, Wojnar J, Krawczyk-Kulis M, et al. Activating killer immunoglobulin-like receptor incompatibilities enhance graft-versus-host disease and affect survival after allogeneic hematopoietic stem cell transplantation Eur J Haematol. 2009;83(4):343-56.http://onlinelibrary.wiley.com/doi/10.1111/j.1600-0609.2009.01280.x/full (accessed 23 July 2012)

[119] Kroger N, Binder T, Zabelina T, Wolschke C, Schieder H, Renges H, et al. Low number of donor activating killer immunoglobulin-like receptors (KIR) genes but not KIR-ligand mismatch prevents relapse and improves disease-free sur- vival in leukemia patients after in vivo T-cell depleted unrelated stem cell transplantation. Transplantation 2006;82(8):1024–30.http://journals.lww.com/transplantjournal/Abstract/2006/10270/Low_Number_of_Donor_Activating_Killer.4.aspx (accessed 23 July 2012)

[120] Davies SM, Ruggieri L, For TD, Wagner JE, Weisdorf DJ, Miller JS, et al. Evaluation of KIR ligand incompatibility in mismatched unrelated donor hematopoietic transplants Blood, 2002;100(10):3825-7. http://bloodjournal.hematologylibrary.org/content/100/10/3825.short (accessed 23 July 2012)

[121] Gagne K, Brizard G, Gueglio B, Milpied N, Herry P, Bonneville F, et al. Relevance of KIR Gene Polymorphisms in Bone Marrow Transplantation Outcome Hum Immunology 2002;63, 271–280. http://www.sciencedirect.com/science/article/pii/S0198885902003737 (accessed 23 July 2012)

[122] Farag SS, Bacigalupo A, Eapen M, Hurley C, Dupont B, Caligiuri MA et al.The effect of KIR ligand incompatibility on the outcome of unrelated donor transplantation: a report from the center for international blood and marrow transplant research, the European blood and marrow transplant registry, and the Dutch registry. Biol Blood Marrow Transplant. 2006;12:876–84. http://www.ncbi.nlm.nih.gov/pubmed/16864058 (accessed 23 July 2012)

[123] Loiseau P, Busson M, Balere M-L, Dormoy A, Bignon J-D, Gagne K et al. HLA Association with Hematopoietic Stem Cell Transplantation Outcome: The Number of Mismatches at HLA-A, -B, -C, -DRB1, or -DQB1 Is Strongly Associated with Overall Survival. Biol Blood Marrow Transplant 2007;13:965-974 http://www.ncbi.nlm.nih.gov/pubmed/17640601 (accessed 23 July 2012)

[124] Schaffer M, Malmberg KJ, Ringden O, Ljunggren HG, Remberger M. Increased infection-related mortality in KIR-ligand-mismatched unrelated allogeneic hematopoietic stem-cell transplantation. Transplantation 2004;78(7):1081–5.http://www.ncbi.nlm.nih.gov/pubmed/15480179 (accessed 23 July 2012)

[125] Bornhauser M, Schwerdtfeger R, Martin H, Frank KH, Theuser C, Ehninger G. Role of KIR ligand incompatibility in hematopoietic stem cell transplantation using unrelated donors. Blood 2004;103(7):2860–1, author reply 2862. http://bloodjournal.hematologylibrary.org/content/103/7/2860.short (accessed 23 July 2012)

[126] Santis DD, Bishara A, Witt
CS, Nagler A, Brautbar C, Slavin

S, et al. Natural killer cell HLA-C epitopes and killer cell immunoglobulin-like receptors both influence outcome of mismatched unrelated donor bone marrow transplants.Tissue Antigens 2005; 65:519–28.http://onlinelibrary.wiley.com/doi/10.1111/j.1399-0039.2005.00396.x/abstract;jsessionid=E8C4CEAE2AAE53DE255675692DAD0D7B.d04t03?deniedAccessCustomisedMessage=&userIsAuthenticated=false (accessed 23 July 2012)

[127] Cook MA, Milligan DW, Fegan CD, Darbyshire PJ, Mahendra P, Craddock CF. The impact of donor KIR and patient HLA-C genotypes on outcome following HLA-identical sibling hematopoietic stem cell transplantation for myeloid leukemia Blood 2004; 103(4):1521-6. http://bloodjournal.hematologylibrary.org/content/103/4/1521.short (accessed 23 July 2012)

[128] Hsu KC, Keever-Taylor CA, Wilton A, Pinto C, Heller G, Arkun K, et al Improved outcome in HLA-identical sibling hematopoietic stem-cell transplantation for acute myelogenous leukemia predicted by KIR and HLA genotypes Blood 2005;105(12):4878-84. http://bloodjournal.hematologylibrary.org/content/105/12/4878.short (accessed 23 July 2012)

[129] Clausen J, Wolf D, Petzer AL, Gunsilius E, Schumacher P, Kircher B, et al. Impact of natural killer cell dose and donor killer-cell immunoglobulin-like receptor (KIR) genotype on outcome following human leucocyte antigen-identical haematopoietic stem cell transplantation. Clinical and Experimental Immunology 2007;148: 520–8.http://onlinelibrary.wiley.com/doi/10.1111/j.1365-2249.2007.03360.x/full (accessed 23 July 2012)

[130] Stringaris K, Adams S, Uribe M, Eniafe R, Wu CO, Savani BN, et al. Donor KIR Genes 2DL5A, 2DS1 and 3DS1 Are Associated with a Reduced Rate of Leukemia Relapse After HLA-Identical Sibling Stem Cell Transplantation for Acute Myeloid Leukemia but Not Other Hematologic Malignancies Biol Blood Marrow Transplant 2010;16(9):1257-64. http://www.bbmt.org/article/S1083-8791(10)00109-6/abstract (accessed 23 July 2012)

[131] Chen C, Busson M, Rocha V, Appert M-L, Lepage V, Dulphy N et al. Activating KIR genes are associated with CMV reactivation and survival after non-T-cell depleted HLA-identical sibling bone marrow transplantation for malignant disorders. Bone Marrow Transplant 2006;38:437–444. http://www.nature.com/bmt/journal/v38/n6/abs/1705468a.html (accessed 23 July 2012)

[132] Cook M, Briggs D, Craddock C, et al. Donor KIR genotype has a major influence on the rate of cytomegalovirus reactivation following T-cell replete stem cell transplantation. Blood 2006;107:1230–2. http://bloodjournal.hematologylibrary.org/content/107/3/1230.abstract (accessed 23 July 2012)

[133] Zaia JA, Sun JY, Gallez-Hawkins GM, Thao L, Oki A, Lacey SF, et al. The effect of single and combined activating KIR genotypes on CMV infection and immunity after hematopoietic cell transplantation. Biol Blood Marrow Transplant. 2009;15(3):315–25.

http://www.sciencedirect.com/science/article/pii/S1083879108005806 (accessed 23 July 2012)

[134] McQueen KL, Dorighi KM, Guethlein LA, Wong R, Sanjanwala B, Parham P. Donor-Recipient Combinations of Group A and B KIR Haplotypes and HLA Class I Ligand Affect the Outcome of HLA-Matched, Sibling Donor Hematopoietic Cell Transplantation Hum Immunol. 2007; 68(5): 309–23. http://www.sciencedirect.com/science/article/pii/S0198885907000286 (accessed 23 July 2012)

[135] Sobecks RM, Ball EJ, Maciejewski JP, Rybicki LA, Brown S, Kalaycio M, et al. Survival of AML patients receiving HLA-matched sibling donor allogeneic bone marrow transplantation correlates with HLA-Cw ligand groups for killer immunoglobulin-like receptors. Bone Marrow Transplantation 2007;39:417–24. http://www.nature.com/bmt/journal/v39/n7/abs/1705609a.html (accessed 23 July 2012)

[136] Gabriel IH, Sergeant R, Szydlo R, Apperley JF, Lavallade H, Alsuliman A, et al. Interaction between KIR3DS1 and HLA-Bw4 predicts for progression-free survival following autologous stem cell transplantation in patients with multiple myeloma. Blood 2010;116(12):2033-9.http://bloodjournal.hematologylibrary.org/content/116/12/2033.short (accessed 23 July 2012)

[137] Venstrom JM, Zheng J, Noor N, Danis KE, Yeh AW, Cheung IY et al. KIR and HLA genotypes are associated with disease progression and survival following autologous hematopoietic stem cell transplantation for high-risk neuroblastoma. Clin Cancer Res. 2009;15(23): 7330-4. http://clincancerres.aacrjournals.org/content/15/23/7330.short (accessed 23 July 2012)

[138] Leung W, Handgretinger R, Iyengar R, Turner V, Holladay MS, Hale GA. Inhibitory KIR-HLA receptor-ligand mismatch in autologous haematopoietic stem cell transplantation for solid tumour and lymphoma. Br J Cancer. 2007;97(4):539-42. http://www.nature.com/bjc/journal/v97/n4/abs/6603913a.html (accessed 23 July 2012)

[139] Stern M, Paulussen M, Rischewski J, Tichelli A, Gratwohl A. Missing ligand model in autologous stem cell transplantation. Br J Cancer. 2008;98(4):852-3. http://www.nature.com/bjc/journal/v98/n4/full/6604153a.html (accessed 23 July 2012)

[140] Willemze R, Arrais Rodrigues C, Labopin M, Sanz G, Michel G, Socié G et al KIR-ligand incompatibility in the graft-versus-host direction improves outcomes after umbilical cord blood transplantation for acute leukemiaKIR-ligand incompatibility and umbilical cord blood stem cell transplantation Leukemia 2009; 23, 492-500 http://www.nature.com/leu/journal/v23/n3/full/leu2008365a.html (accessed 23 July 2012)

[141] Brunstein CG, Wagner JE, Weisdorf DJ, Cooley S, Noreen H, Barker JN et al. Negative effect of KIR alloreactivity in recipients of umbilical cord blood transplant depends on transplantation conditioning intensity. Blood 2009;113:5628–34. http://bloodjournal.hematologylibrary.org/content/113/22/5628.short (accessed 23 July 2012)

[142] Velardi A. Natural killer cell alloreactivity 10 years later. Curr Opin Hematol 2012, 19:000 – 000 DOI:10.1097/MOH.0b013e3283590395 http://www.ncbi.nlm.nih.gov/pubmed/22954728 (accessed 25 September 2012)

[143] Visentainer JE, Lieber SR, Persoli LB, Vigorito AC, Aranha FJ, Eid KA, et al. Serum cytokine levels and acute graft-versus-host disease after HLA-identical hematopoietic stem cell transplantation. Experimental Hematology 2003;31:703-10. http://www.exphem.org/article/S0301-472X(03)00264-9/abstract (accessed 20 September 2012).

[144] Ju XP, Xu B, Xiao ZP, Li JY, Chen L, Lu S et al. Cytokine expression during acute graft-versus-host disease after allogeneic peripheral stem cell transplantation. Bone Marrow Transplantation 2005;35:1179-1186. http://www.nature.com/bmt/journal/v35/n12/full/1704972a.html (accessed 20 August 2012).

[145] Petersen SL, Madsen HO, Ryder LP, Svejgaard A, Djckmeiss E, Vindeloy LL. Cytokine gene expression in peripheral blood mononuclear cells and alloreactivity in hematopoietic cell transplantation with nonmyeloblative conditioning.. Biol. Blood Marrow Transplantation 2006;12:48-60. http://www.bbmt.org/article/S1083-8791(05)00628-2/abstract (accessed 21 August 2012).

[146] Vizoni S, Lieber S, Souza CA, Sell AM, Visentainer JL. Role of cytokines in the immunophatogenesis of Graft-versus-host disease. Revista Brasileira de Hematologia e Hemoterapia 2008;30:142-152. http://www.scielo.br/scielo.php?script=sci_arttext&pid=S1516-84842008000200013 (accessed 30 September 2012).

[147] Espinoza JL, Takami A, Onizuka M, Kawase T, Sao H, Akiyama H, et al. A single nucleotide polymorphism of IL-17 gene in the recipient is associated with acute GVHD after HLA-matched unrelated BMT. Bone Marrow Transplantation 2011;46:1455-1463. http://www.nature.com/bmt/journal/v46/n11/full/bmt2010325a.html (accessed 30 September 2012).

[148] Takabayashi M, Kanamori H, Takasaki H, Yamaji S, Koharazawa H, Taguchi J et al. A possible association between the presence of interleukin-4 secreting cells and a reduction in the risk of acute graft-versus-host disease. Experimental Hematology 2005;33: 251-257. http://www.exphem.org/article/S0301-472X(04)00358-3/abstract (accessed 12 August 2012).

[149] Goyal RK, Lin Y, Schultz KR, Ferrell RE, Kim Y, Fairfull L, et al. Tumor necrosis factor-alpha gene polymorphisms are associated with severity of acute graft-versus-host disease following matched unrelated donor bone marrow transplantation in children: a Pediatric Blood and Marrow Transplant Consortium study. Biology of Blood and Marrow Transplantation 2010;16:927-936. http://www.bbmt.org/article/S1083-8791(10)00025-X/abstract (accessed 13 September 2012).

[150] Bahram S, Bresnahan M, Geraghty DE, Spies T. A second lineage of mammalian major histocompatibility complex class I genes. Proc Natl Acad Sci U S A 1994;

91:6259-63. http://www.ncbi.nlm.nih.gov/pmc/articles/PMC44180/ (accessed 24 August 2012).

[151] Leelayuwat C, Townend DC, Degli-Esposti MA, Abraham LJ, Dawkins RL. A new polymorphic and multicopy MHC gene family related to nonmammalian class I. Immunogenetics 1994; 40: 339–51. http://www.ncbi.nlm.nih.gov/pubmed?term=A%20new%20polymorphic%20and%20multicopy%20MHC%20gene%20family%20related%20to%20nonmammalian%20class%20I (accessed 29 September 2012).

[152] Zwirner NW, Dole K, Stastny P. Differential surface expression of MICA by endothelial cells, fibroblasts, keratinocytes, and monocytes. Human Immunol 1999; 60:323-330. http://www.sciencedirect.com/science/article/pii/S0198885998001281 (accessed 29 September 2012).

[153] Groh V, Bahram S, Bauer S, Herman A, Beauchamp M, Spies T. Cell stress-regulated human major histocompatibility complex class I gene expressed in gastrointestinal epithelium. Proc Natl Acad Sci U S A 1996; 93: 12445–50. http://www.ncbi.nlm.nih.gov/pmc/articles/PMC38011/ (accessed 29 September 2012).

[154] Zwirner NW, Fernandez-Vina MA, Stastny P. MICA, a new polymorphic HLA-related antigen, is expressed mainly by keratinocytes, endothelial cells, and monocytes. Immunogenetics 1998; 47: 139–48. http://www.springerlink.com/content/100517/ (accessed 30 August 2012).

[155] Bauer S, Groh V, Wu J et al. Activation of NK cells and T cells by NKG2D, a receptor for stress-inducible MICA. Science 1999; 285: 727–9. http://www.sciencemag.org/content/285/5428/727.long (accessed 23 August 2012).

[156] Zwirner NW, Marcos CY, Mirbaha F, Zou Y, Stastny P. Identification of MICA as a new polymorphic alloantigen recognized by antibodies in sera of organ transplant recipients. Hum Immunol 2000;61:917–24. http://www.sciencedirect.com/science/article/pii/S0198885900001622 (accessed 20 September 2012).

[157] Mizutani K, Terasaki PI, Shih RN, Pei R, Ozawa M, Lee J. Frequency of MIC antibody in rejected renal transplant patients without HLA antibody. Hum Immunol. 2006;67(3):223-229. http://www.sciencedirect.com/science/article/pii/S0198885906000280 (accessed 13 August 2012).

[158] Hankey KG, Drachenberg CB, Papadimitriou JC, Klassen DK, Philosophe B, Bartlett ST. MIC expression in renal and pancreatic allografts. Transplantation 2002;73:304-306. http://journals.lww.com/transplantjournal/pages/articleviewer.aspx?year=2002&issue=01270&article=00029&type=abstract (accessed 23 September 2012).

[159] Narayan S, Tsai EW, Zhang Q, Wallace WD, Reed EF, Ettenger RB. Acute rejection associated with donor specific anti-MICA antibody in a highly sensitized pediatric renal transplant recipient. Pediatric Transplantation. 2011;15(1): E1-7. http://onlinelibrary.wiley.com/doi/10.1111/j.1399-3046.2010.01407.x/abstract (accessed 12 September 2012).

[160] Panigrahi A, Gupta N, Siddiqui JA, Margoob A, Bhowmik D, Guleria S. Post transplant development of MICA and anti-HLA antibodies is associated with acute rejection episodes and renal allograft loss. Hum Immunol 2007; 68:362-7. http://www.sciencedirect.com/science/article/pii/S0198885907000195 (accessed 12 September 2012).

[161] Sumitran-Holgersson S. Relevance of MICA and other non-HLA antibodies in clinical transplantation
Curr Opi Immunol 2008; 20:607–13. http://www.sciencedirect.com/science/article/pii/S0952791508001234 (accessed 30 Aigust 2012).

[162] Terasaki PI, Ozawa M, Castro R. Four-year follow-up of a prospective trial of HLA and MICA antibodies on kidney graft survival. Am J Transplant 2007; 7:408-15. http://onlinelibrary.wiley.com/doi/10.1111/j.1600-6143.2006.01644.x/abstract (accesses 13 September 2012).

[163] Stastny P. Introduction: MICA/MICB in innate immunity, adaptive immunity, autoimmunity, cancer, and in the immune response to transplants. Hum Immunol 2006; 67:141-144. http://www.sciencedirect.com/science/article/pii/S0198885906000401 (accessed 24 September 2012).

[164] Parmar S, Del lima M, Zou Y, Patah PA, Liu P, Cano P et al. Donor-recipient mismatches in MHC class I chain-related gene A in unrelated donor transplantation lead to increased incidence of acute graft-versus-host disease. Blood 2009;114:2884–7. http://bloodjournal.hematologylibrary.org/content/114/14/2884.long (accessed 20 September 2012).

[165] Kitcharoen, K, Witt CS, Romphruk AV, Christiansen FT, Leelayuwat. MICA, MICB, and MHC beta block matching in bone marrow transplantation: relevance to transplantation outcome. Hum Immunol 2006;67:238–46. http://www.sciencedirect.com/science/article/pii/S0198885906000334 (accessed 23 August 2012).

[166] Steinle A, Li P, Morris DL, Groh V, Lanier LL, Strong RK et al. Interactions of human NKG2D with its ligands MICA, MICB, and homologs of the mouse RAE-1 protein family. Immunogenetics 2001;53(4):279-87. http://www.springerlink.com/content/100517/ (accessed 12 August 2012).

[167] Boukouaci W, Busson M, Peffault de Latour R, Rocha V, Suberbielle C, Bengoufa D, et al. MICA-129 genotype, soluble MICA, and anti-MICA antibodies as biomarkers of chronic graft-versus-host disease. Blood 2009;114(25):5216-24. http://bloodjournal.hematologylibrary.org/content/114/25/5216.long (accessed 12 August 2012).

[168] Goulmy E, Termijtelen A, Bradley BA, van Rood JJ. Alloimmunity to human H-Y. Lancet 1206, 1976. http://www.sciencedirect.com/science/article/pii/S014067367691727X (accessed 24 August 2012).

[169] Goulmy E. Human minor histocompatibility antigens. Curr Opin Immunol. 1996;8(1):75-81.

[170] Goulmy E Minor Histocompatibility Antigens: from Transplantation Problems to Therapy
of Cancer. Human Immunology 2006; 67, 433–438 (2006). http://www.sciencedirect.com/science/article/pii/S0198885906000528 (accessed 24 August 2012).

[171] Warren EH, Zhang XC, Li S, Fan W, Storer BE, Chien JW, et al. Effect of MHC and non-MHC donor/recipient genetic disparity on the outcome of allogeneic HCT. Blood. 2012; Aug 2. [Epub ahead of print]. http://bloodjournal.hematologylibrary.org/content/120/14/2796.long (accessed 24 September 2012).

[172] Spierings E, Drabbels J, Hendriks M, Pool J, Spruyt-Gerritse M, et al. A uniform genomic minor histocompatibility antigen typing methodology and database designed to facilitate clinical applications. PloS One. 2006;1(1):e42. doi:10.1371/journal.pone.0000042. http://www.ncbi.nlm.nih.gov/pmc/articles/PMC1762400/ (accessed 24 September 2012).

[173] Feldhahn M, Dönnes P, Schubert B, Schilbach K, Rammensee HG, Kohlbacher O. miHA-Match: Computational detection of tissue-specific minor histocompatibility antigens. J Immunol Methods. 2012; Sep 14. pii: S0022-1759(12)00275-X. doi: 10.1016/j.jim.2012.09.004. http://www.sciencedirect.com/science/article/pii/S002217591200275X (accessed 24 September 2012).

[174] Miller JS, Warren EH, Van den Brink MR, Ritz J, Shlomchik WD, Murphy WJ, et al. NCI First International Workshop on The Biology, Prevention, and Treatment of Relapse After Allogeneic Hematopoietic Stem Cell Transplantation: Report from the Committee on the Biology Underlying Recurrence of Malignant Disease following Allogeneic HSCT: Graft-versus-Tumor/Leukemia Reaction. Biol Blood Marrow Transplant. 2010;16(5):565-86. http://www.ncbi.nlm.nih.gov/pmc/articles/PMC2858433/ (accessed 13 August 2012).

[175] Spaapen R, Mutis T. Targeting haematopoietic-specific minor histocompatibility antigens to distinguish graft-versus-tumour effects from graft-versus-host disease. Best Pract Res. 2008;21: 543-57. http://www.bprch.com/article/S1521-6926(08)00052-2/abstract (accessed 14 September 2012).

[176] Kawase T, Akatsuka Y, Torikai H, Morishima S, Oka A, Tsujimura A et al. Alternative splicing due to an intronic SNP in HMSD generates a novel minor histocompatibility antigen. Blood 2007;110:1055-63. http://bloodjournal.hematologylibrary.org/content/110/3/1055.long (accessed 14 August 2012).

[177] Mullally A, Ritz J, Beyond HLA. the significance of genomic variation for allogeneic hematopoietic stem cell transplantation. Blood. 2007;109:1355-62. http://bloodjournal.hematologylibrary.org/content/109/4/1355.long (accessed 23 August 2012).

[178] Murata M, Warren EH, Riddell SR. A human minor histocompatibility antigen resulting from differential expression due to a gene deletion. J Exp Med 2003;197:1279-89. http://www.ncbi.nlm.nih.gov/pmc/articles/PMC2193779/ (accessed 25 August 2012).

[179] Goulmy E, Termijtelen A, Bradley BA, et al. Y-antigen killing by T cells of women is restricted by HLA. Nature 1977;266:544-5.http://www.ncbi.nlm.nih.gov/pubmed?term=Y-antigen%20killing%20by%20T%20cells%20of%20women%20is%20restricted%20by%20HLA (accessed 24 August 2012).

[180] Verdijk RM, Kloosterman A, Pool J, Van de Keur M, Naipal AMIH, Brand A et al. Pregnancy induces minor Histocompatibility antigen specific cytotoxic T cells: implications for stem cell transplantation and immunotherapy. Blood 2004;103:1961–4. http://bloodjournal.hematologylibrary.org/content/103/5/1961.long (accessed 24 August 2012).

[181] Gratwohl A, Hermans J, Niederwieser D, van Biezen A, van Houwelingen HC, Apperley J. Female donors influence transplant-related mortality and relapse incidence in male recipients of sibling blood and marrow transplants. Hematol J. 2001;2(6): 363-70. http://www.ncbi.nlm.nih.gov/pubmed?term=Female%20donors%20influence%20transplant-related%20mortality%20and%20relapse%20incidence%20in%20male%20recipients%20of%20sib (accessed 01 August 2012).

[182] Kollman C, Howe CWS, Anasetti C, Antin JH, Davies SM, Filipovich AH et al. Donor characteristics as risk factors in recipients after transplantation of bone marrow from unrelated donors: the effect of donor age. Blood. 2001;98(7):2043-51. http://bloodjournal.hematologylibrary.org/content/98/7/2043.long (accessed 24 September 2012).

[183] Randolph SS, Gooley TA, Warren EH, Appelbaum FR, Riddell SR. Female donors contribute to a selective graft-versus-leukemia effect in male recipients of HLA-matched, related hematopoietic stem cell transplants. Blood. 2004;103(1):347-52. http://bloodjournal.hematologylibrary.org/content/103/1/347.long (accessed 24 September 2012)

[184] Loren AW, Bunin GR, Boudreau C, Champlin RE, Cnaan A, Horowitz MM et al. Impact of donor and recipient sex and parity on outcomes of HLA-identical sibling allogeneic hematopoietic stem cell transplantation. Biol Blood Marrow Transplant. 2006;12(7):758-69. http://www.bbmt.org/article/S1083-8791(06)00290-4/abstract (accessed 23 August 2012).

[185] Stern M, Brand R, de Witte T, et al. Female-versus-male alloreactivity as a model for minor histocompatibility antigens in hematopoietic stem cell transplantation. Am J Transplant. 2008;8(10):2149-57. http://onlinelibrary.wiley.com/doi/10.1111/j.1600-6143.2008.02374.x/abstract (accessed 24 August 2012).

[186] Miklos DB, Kim HT, Miller KH, Guo L, Zorn E, Lee SJ et al. Antibody responses to H-Y minor histocompatibility antigens correlate with chronic graft-versus-host disease and disease remission. Blood. 2005;105(7):2973-8. http://www.ncbi.nlm.nih.gov/pmc/articles/PMC1350982/ (Access 20 August 2012).

[187] Storb R, Prentice RL, Thomas ED. Treatment of aplastic anemia by marrow transplantation from HLA identical siblings. Prognostic factors associated with graft ver-

sus host disease and survival. J Clin Invest. 1977;59(4):625-32. http://www.ncbi.nlm.nih.gov/pmc/articles/PMC372266/ (accessed 05 September 2012).

[188] Stern M, Passweg JR, Locasciulli A, Socié G, Schrezenmeier H, Bekássy AN et al. Influence of donor/recipient sex matching on outcome of allogeneic hematopoietic stem cell transplantation for aplastic anemia. Transplantation. 2006;82(2):218-26. http://www.ncbi.nlm.nih.gov/pubmed?term=Influence%20of%20donor%2Frecipient%20sex%20matching%20on%20outcome%20of%20allogeneic%20hematopoietic%20stem (accesse 02 September 2012).

[189] Hambach L, Spierings E, Goulmy E. Risk assessment in haematopoietic stem cell transplantation: minor histocompatibility antigens. Best Pract Res. 2007;20:171-87. http://www.bprch.com/article/S1521-6926(06)00062-4/abstract (accessed 02 September 2012).

[190] Voogt PJ, Fibbe WE, Marijt WA, Veenhol, WFJ, Hamilton M, Zwaan FE. Rejection of bone-marrow graft by recipient-derived cytotoxic T lymphocytes against minor histocompatibility antigens. Lancet 1990;335:131-4. http://www.ncbi.nlm.nih.gov/pubmed?term=Rejection%20of%20bone-marrow%20graft%20by%20recipient-derived%20cytotoxic%20T%20lymphocytes%20against%20minor%20histocompatibility%20antigens (accessed 02 September 2012).

[191] Goulmy E, Schipper R, Pool J, Blokland E, Falkenburg F, Vossen J et al. Mismatches of minor histocompatibility antigens between HLA-identical donors and recipients and the deveopment of graft-versus-host disease after bone marrow transplantation. N Engl J Med. 1996;334: 281-5. http://www.nejm.org/doi/full/10.1056/NEJM199602013340501 (accessed 03 September 2012).

[192] Perez-Garcia A, De la Camara R, Torres A, Gonzalez M, Jimenez A, Gallardo. Minor histocompatibility antigen HA-8 mismatch and clinical outcome after HLA-identical sibling donor allogeneic stem cell transplantation. Haematologica. 2005;90:1723-4. http://www.haematologica.org/content/90/12/1723.long (accessed 04 September 2012).

[193] Akatsuka Y, Warren EH, Gooley TA, Brickner AG, Lin MT, Hansen JA et al. Disparity for a newly identified minor histocompatibility antigen, HA-8, correlates with acute graft-versus-host disease after haematopoietic stem cell transplantation from an HLA-identical sibling. Br J Haematol. 2003;123:671-5. http://www.ncbi.nlm.nih.gov/pubmed?term=Disparity%20for%20a%20newly%20identified%20minor%20histocompatibility%20antigen%2C%20HA-8%2C%20correlates%20with%20acute%20graft-versus-host%20disease%20after%20haematopoietic%20stem%20cell%20transplantation%20from%20an%20HLA-identical%20sibling. (accessed 04 September 2012).

[194] McCarroll SA, Bradner JE, Turpeinen H, Volin L, Martin PJ, Chilewski SD et al. Donor-recipient mismatch for common gene deletion polymorphisms in graft-versus-host disease. Nat Genet. 2009;41(12):1341-4. http://www.ncbi.nlm.nih.gov/pmc/articles/PMC2804745/ (accessed 04 September 2012).

[195] de Rijke B, van Horssen-Zoetbrood A, Beekman JM, Otterud B, Maas F, Woestenenk R et al. A frameshift polymorphism in P2X5 elicits an allogeneic cytotoxic T lymphocyte response associated with remission of chronic myeloid leukemia. J Clin Invest. 2005;115:3506-16. http://www.ncbi.nlm.nih.gov/pmc/articles/PMC1297240/ (accessed 23 August 2012).

[196] Van Bergen CA, Kester MG, Jedema I, Heemskerk MHM, Van Luxemburg-Heijset ASP, Kloosterboer FM et al. Multiple myeloma– reactive T cells recognize an activation-induced minor histocompatibility antigen encoded by the ATP-dependent interferon-responsive (ADIR) gene. Blood 2007;109:4089-96. http://bloodjournal.hematologylibrary.org/content/109/9/4089.long (accessed 23 August 2012).

[197] Stumpf AN, van der Meijden ED, van Bergen CA, Willemze R, Frederik F, Griffioen M et al. Identification of 4 new HLA-DR-restricted minor histocompatibility antigens as hematopoietic targets in antitumor immunity. Blood 2009;114:3684-92. http://bloodjournal.hematologylibrary.org/content/114/17/3684.long (accessed 23 August 2012).

Chapter 2

The Advanced HLA Typing Strategies for Hematopoietic Stem Cell Transplantation

Sun Yuying and Xi Yongzhi

Additional information is available at the end of the chapter

http://dx.doi.org/10.5772/53084

1. Introduction

The occurrence of graft rejection and/or graft-versus-host disease (GVHD) after allogeneic hematopoietic stem cell transplantation (allo-HSCT) is largely depended on whether the recipient and the donor have matched HLA types. Under normal circumstances, the individual with completely matched HLA antigens can be the donor. However, due to the high level of HLA polymorphism, the major obstacle in the allogeneic hematopoietic stem cell transplantation is to find a donor with HLA antigens that are a perfect match. This can prove to be quite problematic.

In 1954, an organ transplantation team led by Dr. Merri at Harvard University successfully completed a kidney transplantation between identical twins for the first time. From then on, the importance of histocompatibility in organ transplantation has been well recognized. The first human bone marrow transplantation between identical twins in 1957 provided a new approach for the treatment of leukemia and other hematologic malignancies. As a result, the basic research on HLA as well as the HLA typing techniques gained much attention over the next 20 years. The short-term survival rate of organ transplantation has been greatly improved since the 1980s due to the clinical application of immunosuppressive agents such as CsA. These successes,as well as the defects and limitations in serotyping and cellular typing of HLA, the clinical value of HLA typing has been largely ignored in the medical community.

With the advance of research in immunology and transplantation immunology, particularly in the structure and function of HLA in the 1990s, new technology for HLA typing has emerged and continues to improve. Terasaki and Opelz analyzed a large amount of organ transplantation cases performed in major transplantation centers around the world. The role, status and importance of HLA typing in hematopoietic stem cell transplantation have been recognized once again. Overall, HLA typing is required in hematopoietic stem cell transplantation. HLA compatibility not only significantly reduces the incidence of acute rejection, but also significantly reduces the incidence of chronic rejection. HLA compatibility is one of the most critical factors that affect the long-term survival of the graft.

HLA loci are the most genetically variable gene loci in human. Two hundred and twenty four loci of HLA complex have been identified so far. Among these, 128 are functional loci that encode proteins, and 39.8% of HLA genes are related to the immune system, particularly those belong to class II loci. Almost all these genes display immune-related functions. Approximately 100 HLA genes loci have been cloned and named, and at least 18 of them have alleles. Since these loci have various amounts of alleles and each allele encodes a corresponding HLA antigen, the HLA complex has the most abundant genetic polymorphism in the human immune system.

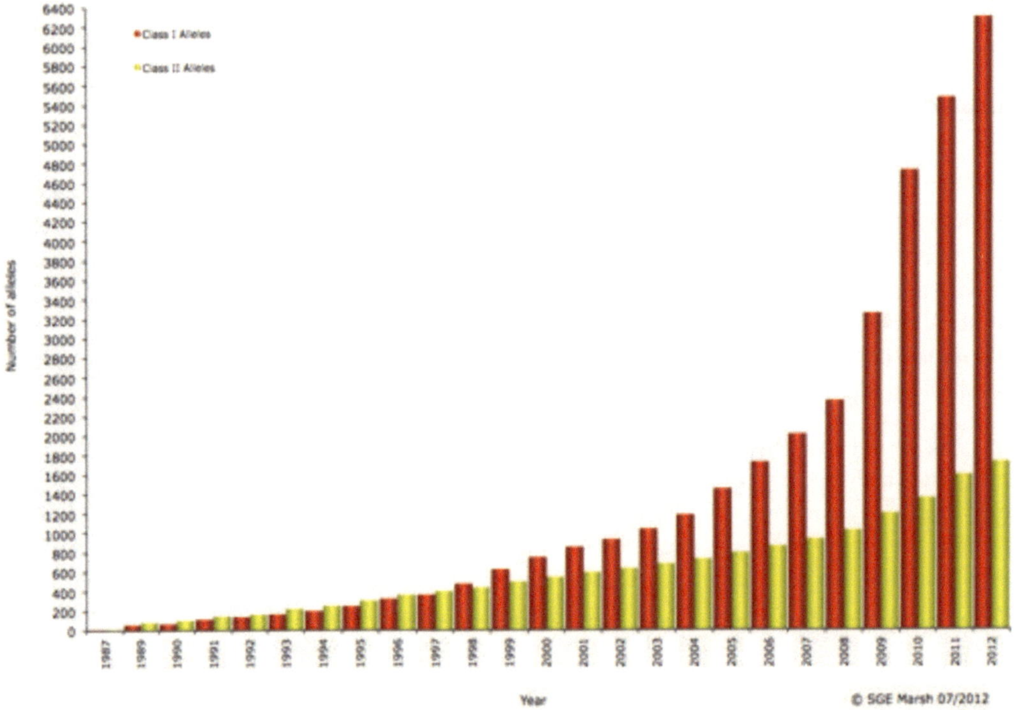

Figure 1. Increasing number of HLA alleles from 1987 to July 2012

Systemic investigations of the alleles in HLA loci began in 1987. There were just over 10 identified alleles at that time. The allele numbers in HLA-I and HLA-II loci were increased to 100 and 50 respectively in 1989. The allele number of HLA-I and HLA-II reached 1028 in 2000. As of July 2012, the total allele number of HLA loci has reached 8016. HLA-A, HLA-B and HLA-C loci have 2013, 2605 and 1551 alleles respectively. DRA site has 34 and DRB site has 1260 alleles. DQA1 and DQB1 sites have 47 and 176 alleles respectively, and DPA1 and

DPB1 sites have 34 and 155 alleles, respectively (Fig 1). Theoretically, it is very difficult to find an unrelated donor with a perfectly matched HLA genotype (at the allele level) in the general population.

The polymorphism of HLA makes it difficult to find a match between unrelated donor and recipient in the allotransplantation. Currently, the most commonly used HLA typing in organ transplantations around the world is based on HLA-A, B, C and DR genes. There are up to 7400 alleles in these genes corresponding to more than 100 specific antigens. With the increasing number of patients who need hematopoietic stem cell transplantation, the lack of appropriate donors has become a significant challenge. Therefore, there is an urgent need to develop novel scientific, practical, and feasible HLA typing methods in the field of hematopoietic stem cell transplantation.

2. Principles for HLA typing strategy in allogeneic hematopoietic stem cell transplantation

The first successful human bone marrow transplantation between identical twins in 1957 has provided a new approach for the treatment of leukemia and other hematologic malignancies. After the successful hematopoietic stem cell transplantation between unrelated donor and recipient with matched HLA, a bone marrow donor registry was established in 1988 (National Marrow Donor Program, NMDP) in the USA. Later on, a public cord blood bank was established. According to the World Marrow Donor Association (WMDA), as of July 2012, the association has 68 bone marrow banks in 49 countries and regions. It also has 46 cord blood banks in 30 countries and regions. The registered bone marrow and umbilical cord blood donors have exceeded 20 million. Meanwhile, the technology of HLA typing has been transformed from simple serotyping to more accurate genotyping. Although there are hundreds of reports regarding the effect of HLA matching degree on the efficacy of hematopoietic stem cell transplantation, these results are not consistent due to the differences in sample size, disease type and stage, and HLA typing. In addition, the interpretation of HLA genotyping results and their biological significance is becoming increasingly complicated. It is challenging for the clinicians outside of the HLA field to select an unrelated donor with the best-matched HLA. To meet this challenge, WMDA, NMDP of the USA and European Federation of Immunogenetics (EFI) have provided guidelines for HLA typing.

2.1. Correlation between HLA allele and HLA antigen specificity

There is a fundamental difference in the result and biological significance between HLA serotyping and genotyping. In the HLA serotyping, HLA antibodies are used to identify the HLA antigens on the surface of lymphocytes. HLA antigens are proteins that can be recognized by the host immune system during blood transfusion, organ transplantation, as well as pregnancy. Specific antibodies against HLA antigens are the basis of the identification of the HLA antigens. The HLA antisera used in serotyping, regardless of whether they are from the same species or different species, are all produced by immune stimulation with HLA antigens or peptides. In the HLA genotyping analysis, a specific HLA gene fragment is amplified *in vitro* from an individual's genomic DNA using synthetic oligonucleotide probes or primers. The genetic difference caused by variant HLA gene alleles is reflected by the variation in the DNA sequence. Therefore, HLA genotyping can identify all HLA alleles at the DNA level while HLA serotyping can only detect part of variants. The efficacy of bone marrow transplantation is closely related to the matching level of HLA between the donor and recipient. However, the HLA genotyping result does not directly reflect the antigen that causes immune rejection after the transplantation. Therefore, for the purpose of clinical relevance, the result of HLA genotyping should be converted to the HLA specificity. To this end, the NMPD and the University of California in Los Angels (UCLA) established the International Cell Exchange program, through which correlations between the HLA alleles and HLA antigen specificities are established by comparing a large amount of testing results worldwide. The dictionary of HLA alleles and their corresponding antigen specificities is under constant updating. As of 2008, 70% of HLA alleles have been correlated to HLA antigen specificities. The rest 30% alleles are rare alleles with a frequency less than 1 in 10,000. Therefore, their clinical values are relatively low. The HLA genotyping result can be easily converted to the HLA antigen specificity by using this HLA dictionary.

2.2. The number of donor with matched HLA gene types is much lower than that with matched HLA antigens

The criteria of matched HLA between the donor and recipient are different for the HLA genotyping and HLA serotyping in the bone marrow transplantation. From the HLA dictionary, one can tell that the HLA antigen specificity is unique, while a unique antigen may have one or more corresponding HLA alleles. For example, HLA-DR10 antigen only corresponds to HLA-DRB1*1001 allele, while HLA-DR11 antigen corresponds to 21 alleles such as HLA-DRB1*1101, 1102 and 1103. Therefore, the choice of donor for bone marrow transplantation may differ, depending on the method of HLA typing. For example, a donor and recipient listed in Table 1 may have matched HLA according to antigen specificity. However, their HLA genotypes may not match. Which method is more accurate for bone marrow transplantation is currently under investigation. Statistical analysis indicates that the chance of finding matched HLA genotypes in a random population is much lower than finding matched HLA antigens. For instance, as of February 2002, HLA-A, B and DR have 93 specific antigens. HLA-A, B and DR have 25, 50 and 18 loci respectively, which can generate 2.2×10^{4} haplotypes. The genotype number of these haplotypes can be up to 2.5×10^{13}. Currently, there are 2100 alleles have been identified in HLA-A, B and DRB1 genes. Their combination will yield 3.4×10^{7} haplotypes. As a result, the number of HLA-A, B and DRB1 genotypes in a population can be up to 5.78×10^{15}, making it almost impossible to find the matched HLA genotype in a random population. In other words, the HLA genotypes of the donor and the recipient are always more or less mismatched in bone marrow transplantation. Because of this, the concept of permissible HLA mismatches has been introduced.

http://dx.doi.org/10.5772/53084

matching status		Recipient's HLA type				Donor's HLA type			
antigen	gene	Antigen		gene		Antigen		gene	
matched	matched	A2	A11	A* 0202	A*1101	A2	A11	A*0202	A*1101
		B60	B62	B*4001	B*1501	B60	B62	B*4001	B*1501
		DR4	DR8	DRB1*0402	DRB1*0801	DR4	DR8	DRB1*0402	DRB1*0801
matched	unmatched	A2	A11	A* 0202	A*1101	A2	A11	A*0205	A*1102
		B60	B62	B*4001	B*1501	B60	B62	B*4007	B*1504
		DR4	DR8	DRB1*0402	DRB1*0801	DR4	DR8	DRB1*0404	DRB1*0803
unmatched	unmatched	A2	A11	A* 0202	A*1101	A2	A30	A*0201	A*3001
		B60	B62	B*4001	B*1501	B61	B62	B*4002	B*1501
		DR4	DR8	DRB1*0402	DRB1*0801	DR4	DR11	DRB1*0401	DRB1*1102

Table 1. Examples of HLA antigen matching and allele matching between the recipient and the donor in bone marrow transplantation

2.3. Permissible HLA mismatches

In the case of permissible HLA mismatches, the donor and the recipient have mismatched HLA in a bone marrow transplant. However, the mismatch does not cause a significantly increased rate of GVHD or graft failure, and is acceptable for bone marrow transplantation. Results from retrospective analyses suggest that mismatched alleles in HLA class I antigens as well as alleles in HLA-DQ and DP loci have minimal impact on the efficacy of bone marrow transplantation.

2.3.1. HLA class I antigen or allele mismatch

Petersdorf *et al* had investigated the effect of matching level of HLA class I antigens and alleles on the success rate of bone marrow transplantation in 471 patients. The transplant failure rate is 0.7% in 280 cases with matched HLA-A, B and C genes, and is 0% in 47 cases with one of mismatched heterozygous HLA-A, B or C gene. However, the failure rate in 51 cases with one of mismatched HLA-A, B or C antigens is 14%, which is significantly higher than that in the control group. In 76 cases with 2 or more mismatched antigens or genes, the transplant failure rate is 17%. These results indicate that a single mismatched allele in the HLA class I gene does not increase the transplant failure rate, while a single mismatched antigen, or 2 or more mismatched antigens or genes can significantly increase the transplant failure rate. These results support Petersdorf's hypothesis that the immune response caused by mismatched HLA class I alleles is lower than that caused by mismatched antigens. Therefore, mismatched HLA class I genes are permissible in the bone marrow transplantation, as long as HLA antigens match. Rubinstein et al also believes that transplantation can be considered if there is only one mismatched allele. For example, the recipient's genotype is HLA-A*0202 while the donor's genotype is HLA-A*0203. This kind of mismatch does not increase the rate of immune rejection. Further analysis indicates that whether a single allele mismatch is allowed in the transplantation also depends on the type of corresponding mismatched amino acid and the position of that amino acid in the HLA class I antigen. HLA class I molecules consist of a covalently bound heavy chain molecule and a • 2 microglobulin. The extracellular fragment of the heavy chain has 3 activity domains

(▪ 1, ▪ 2 and ▪ 3), and the ▪ 1 and ▪ 2 domains form the peptide-binding region. The complex of HLA and its bound peptide on the cell surface constitutes the ligand for the T-cell receptor (TCR), thereby inducing an immune response. If there is only one mismatched allele between the donor and recipient, the number of mismatched amino acids will be much lower, and may rarely involve the amino acids for TCR binding. On the other hand, if the donor and the recipient have a mismatched antigen, it may have many mismatched amino acids, and some of these amino acids may be involved in peptide binding and TCR binding. This may explain why the matching of HLA class I antigen is more important than the matching of genotype (Fig 2).

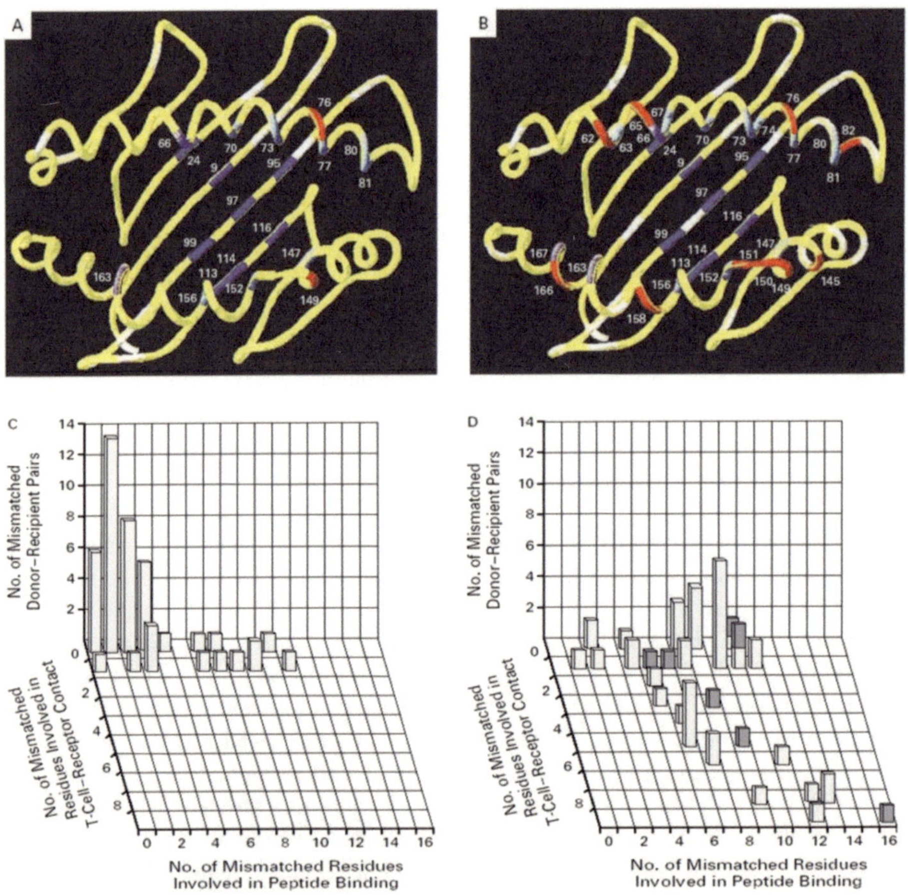

Figure 2. Spatial structure as well as the position and number of mismatched amino acid in the class I HLA with mismatched donor-recipient genotypes (A) or mismatched donor-recipient antigens (B). The position of amino acid residue is labeled according to its position in the whole protein. Amino acid residues with dark blue color are located in the ▪ -sheet and involved in peptide binding. Amino acid residues located on the ▪ helix are colored with light blue. Amino acid residues involved in TCR binding are in red. White amino acids are involved in neither TCR binding nor peptide binding, while gray amino acids are involved in both TCR and peptide binding. Number of mismatched amino acid involved in peptide binding and TCR binding in the class I HLA with mismatched donor-recipient genotypes (C) and mismatched antigens (D). In panel D, patients with a transplant failure are in dark gray block.

http://dx.doi.org/10.5772/53084

Further analysis by Petersdorf *et al* shows that one mismatched HLA-A, B or C antigen causes 71% transplant failure in 7 HLA homozygous patients, but causes 7% failure rate in 98 heterozygous patients, suggesting that for HLA homozygous patients, when a matched donor is not available, the homozygous donor with other matched heterozygous alleles should be chosen. For example, the recipient's HLA type is HLA-A2, B44 and DR8/9. Donor 1's HLA type is HLA-A2, B51, B46 and DR8/9. Donor 2's HLA type is HLA-A2/11, B44, B46 and DR8/9. In this case, donor 1 is preferred.

2.3.2. *HLA class II antigen or allele mismatch*

HLA class II genes encode antigens such as HLA-DR, DQ and DP. In order to understand the importance of HLA class II genes in unrelated bone marrow transplantation, McGlave *et al* have investigated the effect of mismatched DR, DQ and DP alleles on the result of transplantation. Data from NMDP that contain 831 chronic myeloid leukemia (CML) patients received bone marrow transplantation between 1988 and 1997 were analyzed. 696 patients have matched HLA-A and B base on serotyping results. Among them, 565 (81%) have matched DRB1 genotypes. Data analysis shows that matched HLA-DRB1 alleles can significantly improve graft survival and patient survival. While, mismatched HLA-DQA1, DQB1, DPA1 and DPB1 genes do not significantly affect the GVHD incidence or transplantation result. This result suggests that matching HLA-DRB1 alleles is an important factor in bone marrow transplantation (Fig 3).

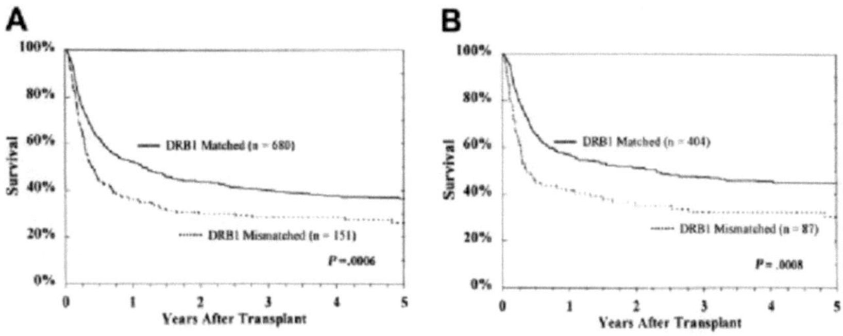

Figure 3. Effect of DRB1 matching level on patient's survival rate (A) CML patient, (B) Patients with matched HLA-A, B serotypes

2.4. HLA typing standard for hematopoietic stem cell transplantation

According to the guideline of World Bone Marrow Donor Association (WBDA) and European Federation for Immunogenetics (EFI), HLA typing of the donor in a large-scale bone marrow center is generally limited to 2 digits after the asterisk in the WHO HLA nomenclature, corresponding to the subtype of a specific HLA antigen. However, high-resolution HLA typing should be performed for recipients and donors with matched HLA. In addition, the typing of HLA class I genes should also include the locus C. Due to the increasingly recognized role of locus C in the immune rejection, the typing of HLA-C should be performed.

When choosing a donor, the HLA-DRB1 gene of the donor and the recipient should have 4 identical digits after the asterisk in the WHO HLA nomenclature.

Although most commonly used methods for HLA genotyping cannot cover all genes, it does not limit their applications in HLA typing for bone marrow transplantation. Among thousands of identified HLA alleles, most of them are rare alleles. Therefore, it is not necessary to type all HLA alleles. For instance, 244 expressing genes have been identified in DRB1 loci. Among them, 148 (60%) alleles have corresponding specific DR antigens, while 96 alleles (40%) do not. According to the NMDP, result of HLA-DRB1 typing in 65,752 donors shows that 86 alleles have 0 frequency and the frequency of another 105 alleles is lower than 0.0002. In addition, the total frequency of 10 alleles without corresponding antigens is 0.000084. Therefore, identification of the rest 43 DRB1 alleles will cover 99.6% of HLA-DR antigen specificities, which is sufficient for the screen of donor in hematopoietic stem cell transplantation.

3. PCR based HLA genotyping methods

The technology for HLA typing has evolved from the serological level to the cellular level, to the molecular level. Serotyping was the mainstream method for HLA type and has played a critical role in organ transplantations before 1990s. However, most HLA antisera are polyclonal and often have cross-reactions, making it difficult to distinguish antigens with subtle structural differences, and leading to misidentifications. Further more, many factors, such as a prolonged transportation time of the blood sample and excessive amount of immature cells, may affect the result of serotyping and cellular typing. These are the limitations of traditional HLA typing methods. The development of polymerase chain reaction (PCR) and its application in biomedical sciences has made the HLA typing at the DNA level possible. Therefore, using molecular methods to type HLA at the DNA level has gradually replaced serotyping and cellular typing. Commonly used DNA based HLA typing methods include PCR based sequence specific primers (PCR-SSP), and PCR based restriction fragment length polymorphism (PCR-RFLP), single-strand conformation polymorphism (PCR-SSCP), sequence-specific oligonucleotide (PCR-SSO) and single nucleotide polymorphism (PCR-SNP).

3.1. PCR-SSP (sequence specific primers)

To identify point mutations in a DNA molecule, Newton invented the amplification refractory mutation system (ARMS) for *in vitro* DNA amplification. The technique requires an allele sequence specific 3′ primer for the PCR amplification. Otherwise the PCR reaction will not be effective. This is because the Taq DNA polymerase used in the PCR reaction has 5′ to 3′ polymerase activity and 5′ to 3′ exonuclease activity but lacking 3′ to 5′ exonuclease activity. Therefore, the enzyme cannot repair the single mismatched nucleotide in the 3′ primer. In order to amplify the allele with a specific sequence, the primer with the corresponding sequence is designed. The conditions for PCR reaction are strictly controlled so that the amplification of the fragment with its sequence perfectly matching the primer is much more effective than the sequence with one or more mismatched nucleotide. One mismatched nucleotide between the 3′ primer and the template is sufficient to prevent the amplification.

The PCR product is further analyzed by electrophoresis to determine whether the amplicon corresponds to the anticipated primer-specific product. Since the DNA sequence of HLA class I and class II genes are known, PCR primers can be designed based on the specific sequence of each allele for PCR-SSP genotyping.

The encoding allele sequences of various HLA antigens can be amplified with sequence specific primers. By controlling the conditions of PCR, a specific primer can only amplify its corresponding allele, and not other alleles. Therefore, the presence of a PCR product can be used to determine the presence or absence of a specific allele. The specificity of PCR product can be further determined by agarose gel electrophoresis. Fig 4 shows the principle of PCR-SSP.

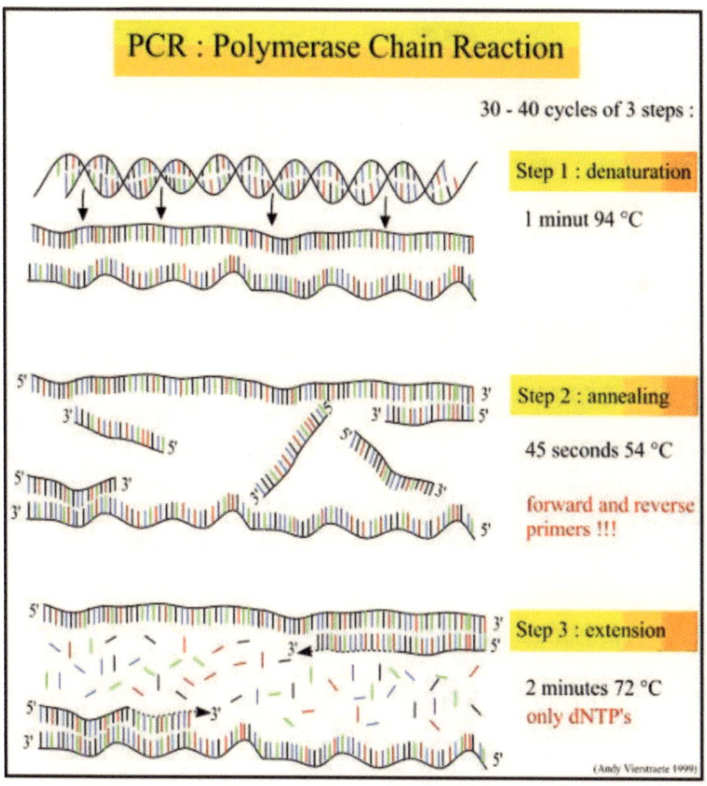

Figure 4. The diagram of PCR reaction

In the first step of PCR reaction, double-stranded DNA is denatured into single-stranded DNA. In the second step, specific primers anneal to the template DNA. In the third step, double stranded DNA is generated by TaqDNA polymerase by incorporating 4 types of dNTP into the newly synthesized DNA strand. After 30-40 cycles of amplification, the target gene is increased to 10^8 fold.

The main advantage of this method is that it is simple and fast, and the result is easy to interpret. The heterozygosity can be easily detected as well. Therefore, PCR-SSP is the currently most used method for HLA typing. There are several FDA approved high-resolution and low-resolution detection kits available for HLA class I and class II typing. Many clinical laboratories in China have been using this method for accurate pre-transplantation HLA typing. The procedure of PCR-SSP is shown in Fig 5. One disadvantage of this method is that it requires multiple primers in order to amply all relevant alleles.

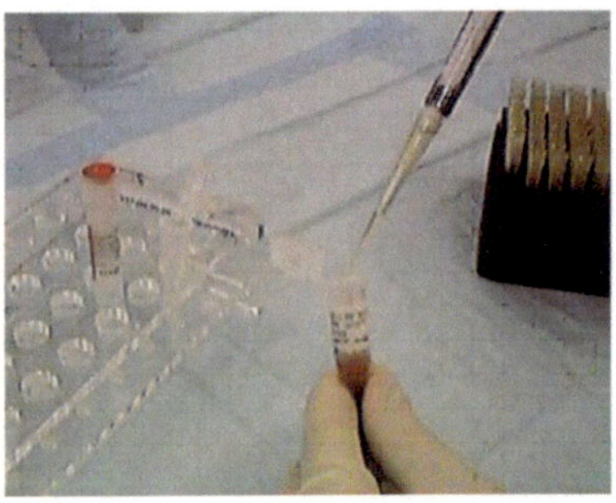

A. DNA polymerase and DNA sample are added to the tube containing PCR reaction buffer and dNTP.

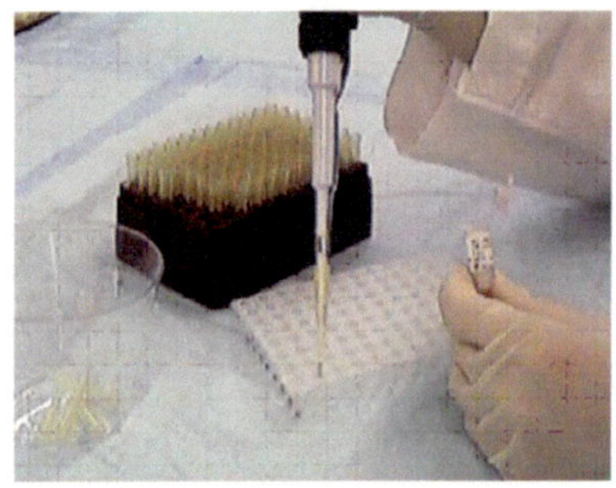

B. 10 ul of mixture of DNA and D-mix is added to the SSP kit. The negative control does not have this mixture.

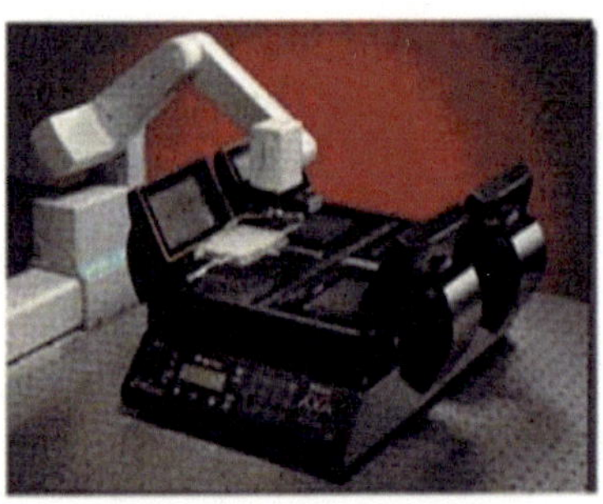

C. PCR amplification

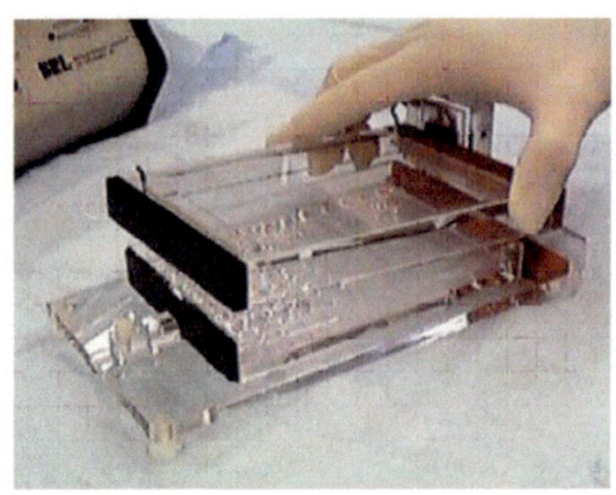

D. electrophoresis (2-4 min)

E. Automatic gel imaging system

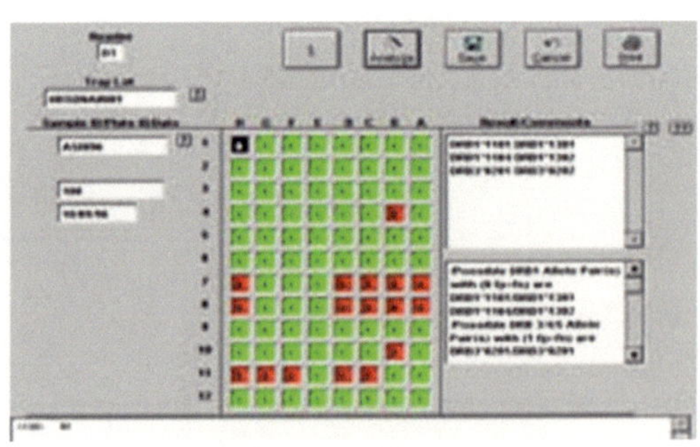

F. Analysis of the result by software

Figure 5. The experimental procedure of PCR-SSP

3.2. PCR-RFLP (restriction fragment length polymorphism)

Restriction endonucleases have unique recognition sites. Using computer analysis, restriction endonucleases that can recognize HLA sequence polymorphism are chosen to digest the PCR product. Because of sequence difference among the alleles, enzyme digestion will yield DNA fragments with unique patterns of length, which can be distinguished by electrophoresis.

Compared to serotyping, PCR-RFLP method is specific, simple and rapid and does not require probes. It can accurately detect single nucleotide difference and two linked polymorphic sites. The disadvantage of this method is that if the enzyme cannot completely digest the PCR product, the DNA fragments with similar lengths may be difficult to distinguish after electrophoresis. In addition, alleles need to have endonuclease recognition sites. Furthermore, PCR-RFLP cannot distinguish certain HLA heterozygosities. It requires multiple endonucleases for those alleles with high polymorphism such as HLA-DRB1, and may produce complicated restriction maps. For these reasons, this method is rarely used for HLA typing currently.

3.3. PCR-SSCP (single-strand conformation polymorphism)

Suzuki *et al* in Japan have found that a single-stranded DNA fragment has complex spatial conformation. The three-dimensional structure is generated by the intramolecular interactions among the base pairs. The changing of one nucleotide may affect the spatial conformation of the DNA strand. Single stranded DNA molecules have their unique size exclusion characters in polyacrylamide gels due to their molecular weights and three-dimensional structures. Therefore, they can be separated by non-denature polyacrylamide gel electrophoresis (PAGE). This method is sensitive enough to distinguish molecules with subtle structural differences, and it is called single-stranded conformation polymorphism (SSCP). The authors later applied SSCP in the detection of mutations in PCR products and developed PCR-SSCP technique, which has further improved the sensitivity and simplicity for mutation detection.

This method is simple, rapid, sensitive, requiring no special equipment, and is suitable for clinical applications. However, this method can only detect mutations. The location and the type of the mutation need to be determined by sequencing. In addition, the conditions of electrophoresis need to be tightly controlled. Furthermore, point mutations in certain locations may have no to little effect on the DNA conformation. Therefore, different DNA molecules may not be able to separate by PAGE due to these reasons and other factors. Nevertheless, this method has a relatively high detection sensitivity compared with other methods. It can detect mutations in unknown locations in the DNA molecule. Takao has demonstrated that SSCP can detect 90% of single nucleotide mutations in a DNA fragment smaller than 300bp. He believes that most known single nucleotide mutations can be detected by this method. Mutant DNA molecules can be separated and purified by PAGE due to the different migration rates, and the mutation can be eventually identified by DNA sequencing.

In SSCP analysis, the separation of single stranded DNA by non denature PAGE is not just based on their molecular weights and electric charges, but also on the retention force caused by their spatial conformations. Therefore, the migration rate of a DNA fragment does not reflect its molecular size. Since the wild type and mutant DNA molecules may migrate very closely and are difficult to distinguish, it is generally required for DNA molecules to migrate for more than 16-18 cm in the gel. Mobility is calibrated using reference DNA as an internal control. Because of these reasons, this method cannot clearly determine the HLA genotype.

3.4. PCR-SSO (sequence specific oligonucleotide)

In PCR-SSO, specific probes are synthesized according to the sequence in the HLA polymorphic region. The target DNA fragment is amplified *in vitro* first. Then a specific probe will be hybridized to the PCR product under certain conditions based on base pair complements. The hybridized product can be detected by radioactive or non-radioactive signals. There are two types of SSOP method, direct hybridization and reverse hybridization. In the direct hybridization, the PCR product is fixed on the membrane while in the reverse hybridization, the probe is fixed on membrane. Figure 6 is the diagram of PCR-SSO.

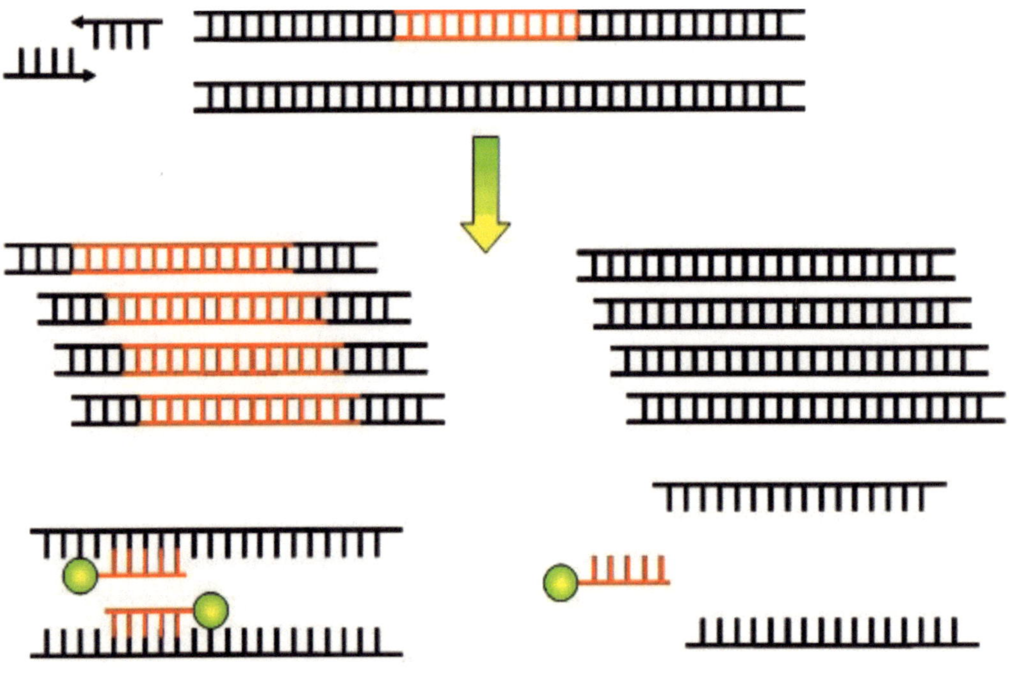

Figure 6. The diagram of PCR-SSO process

In 1986, Saiki *et al* were the first to report the analysis of DQA1 polymorphism using PCR and 4 allelic specific oligonucleotide (ASO) probes. MicKelson has typed the DR loci by serotyping and PCR-SSOP in 268 specimens. The success rate of serotyping is 91.0% while the success rate of PCR-SSOP is 97.0%. Overall, PCR-SSOP has a high success rate, a wide source of reagents, a high specificity and resolution. It can detect the difference of one nucleotide. In addition, PCR-SSOP can be used for a large number of samples with accurate and reliable results. However, this method is time consuming. It often takes a few days and needs a large amount of probes. In addition, it is difficult to detect heterozygous alleles, particularly those of the complicated HLA-DRB1 genes.

Overall, PCR-SSO is an accurate HLA genotyping method, and can identify all known HLA alleles for accurate analysis of HLA polymorphism. HLA is a super gene family and new alleles are continuously been identified. SSO probes can only be designed based on the sequences of known alleles. Although PCR-SSO may discover new HLA polymorphism through its hybridization pattern, dot-hybridization often leads to false positives. In addition, when an allele is identified in the sample, it is difficult to determine whether the allele is homozygous or heterozygous. Therefore, the HLA allele frequency and haplotype frequency cannot be precisely determined by this method.

3.5. PCR-SNP (single nucleotide polymorphism)

Single nucleotide polymorphism (SNP) is the inheritable and stable biallelic single nucleotide difference. In the human genome, every 1000 base pairs have one to 10 SNPs. SNP may have some regulatory functions in gene expression and protein activity. High SNP density has been found in HLA class I genes with one SNP in every 400bp, setting the basis for high-throughput MHC-SNP analysis. Compared with other methods, SNP is less time consuming and with a low cost. Gou *et al* have developed a simple and effective oligonucleotide microarray to detect SNPs in the coding sequence of HLA-B locus. Based on the known polymorphism in the exon 2 and 3 of HLA-B genes, 137 specific probes were designed. In a double-blind experiment, these probes were used in the PCR-SNP analysis of 100 specimens from unrelated individuals. The result showed that this method could explicitly identify all SNPs in the HLA-B locus. Bu Ying *et al* have established a rapid, efficient, and cost effective SNP detection method using a single tube. In this method, 4 primers are used for the PCR amplification. Two primers are used to amplify the DNA fragment containing the SNP region, and the other two primers are SNP specific. The primer extension error is significantly reduced when 4 primers simultaneously carry out the PCR reaction, thereby the accuracy of SNP analysis is greatly improved. With the development of third-generation genetic markers, it is expected to find a series of single nucleotide polymorphisms in the HLA complex, and generate high-density SNP maps. In order to develop SNP technology into a simple and effective HLA typing method, production of high-density SNP maps in the HLA regions and development of HLA-SNP genotyping kits have been proposed in the 13th IHWC conference.

4. Reference-strand-mediated conformation analysis (RSCA)

Arguello *et al* devised the double-stranded conformation analysis (DSCA) technique in 1998 for the detection and analysis of gene mutations and complex polymorphic loci. Based on this technique, reference strand mediated conformation analysis (RSCA) has been developed. This is a major technical breakthrough in HLA typing. This technique combines sequencing and conformational analysis to overcome the limitations of the methods that just employ DNA sequencing or conformational analysis. The concept behind RSCA is that a fluorescent labeled reference strand is hybridized with the amplified product of a specific gene to form stable double stranded DNA with unique conformation. After non-denature polyacrylamide gel electrophoresis or capillary electrophoresis, HLA alleles can be detected by laser scanning and computer software based analysis. Figure 7 is the basic procedure of RSCA.

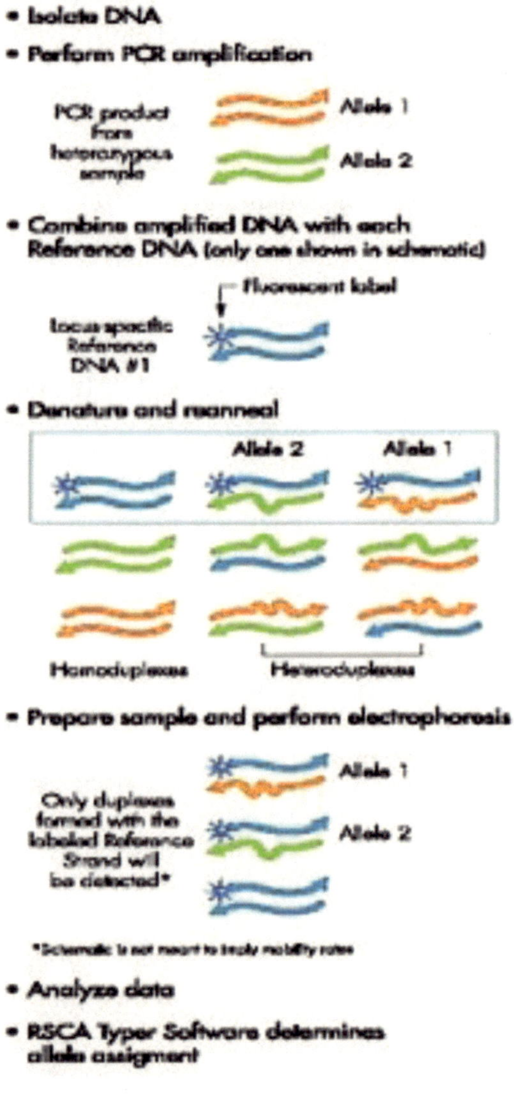

Figure 7. The basic procedure of RSCA

Compared with PCR-SSP, the most commonly used method of HLA genotyping, RSCA has the following advantages: (1) high resolution and sensitivity. RSCA is based on the differential migration rate of fluorescent-labeled double stranded DNA during the electrophoresis.

Alleles with different sequences will produce DNA duplexes with different spatial structures after hybridization with their fluorescent labeled probes. Two alleles with one nucleotide difference will cause a change in the spatial structure of a hybridized duplex, resulting in an altered migration rate in electrophoresis. Therefore, RSCA can distinguish the alleles with a single nucleotide difference. For example, HLA-A*0207 and A*0209 alleles only differ from one nucleotide at the site 368 of exon 3. In this site, A*0207 has a G while A*0207 has an A. Likewise, HLA-A*0224 and A*0226 only differ from one nucleotide. These alleles all can be distinguished by RSCA. (2) high reproducibility. In RSCA, each lane in the non-denature polyacrylamide gel has markers and each gel has a DNA ladder. Therefore, the alteration caused by different gels or lanes can be eliminated. (3) new allele or mutation identification. RSCA is based on the electrophoretic mobility difference caused by different spatial structure of the duplexes after allele-FLR hybridization. New alleles or mutations will have electrophoretic mobility different from that of known alleles. (4) RSCA can be applied at a large scale with a low cost.

The disadvantages of RSCA are (1) time-consuming for a single sample; (2) requiring high quality samples; PCR-SSP requires 10-100ng/ml of DNA, which can be obtained with a regular DNA purification kit from patients even with a low amount of white blood cells. However, RSCA requires 50-100ng/ml of DNA. It may require an increased amount of blood sample for patients with low levels of white blood cell in order to obtain sufficient DNA; and (3) insufficient database.

5. Pyrosequencing: A high-resolution method for HLA typing

Pyrosequencing is a new HLA genotyping technology based on real time sequencing during DNA amplification. The reaction system contains 4 enzymes (DNA polymerase, ATP sulfurylase, luciferase and apyrase), a substrate (APS: adenosine 5′ phosphosulfate), fluorescein (luciferin), primers and the single stranded DNA template. After one type of dNTP (dATP, dTTP, dCTP and dGTP) is added to the reaction system, it will be incorporated into the newly synthesized chain if it is complementary to the nucleotide on the template. Incorporation of dNTP will generate the same molar amount of pyrophosphate (PPi). ATP sulfurylase converts APS and PPi into ATP, which provides energy for luciferase to oxidate luciferin and emit light. The amount of light signal is proportional to the amount of ATP. The optical signal is detected by a CCD (charge couple device) camera and generates a peak in the pyrogram. The principle of Pyrosequencing is shown in Fig 8.

The height of each signal's peak is proportional to the number of nucleotides incorperated. Unincorporated dNTPs and excessive ATP are converted to dNDPs, which are further converted to dNMPs by apyrase. The optical signal is quenched and the system is regenerated for the next reaction. The next dNTP can be added to the system to start the next reaction after the unincorporated dNTPs and excessive ATP are removed. The reaction cycle continues until the complementary DNA strand is synthesized. Under the room temperature, it takes 3-4 seconds from polymerization to light detection. In this system, 1 pmol of DNA will generate 6x10^{11} pmol of ATP, which in turn yields 6x10^{9} pmol of photon with a wavelength of 560nm. The signal can be easily detected by a CCD camera. For the analysis of DNA with an unknown sequence by Pyrosequencing, a cyclic nucleotide dispensation order (NDO) is used. dATP, dGTP, dTTP and dCTP are sequentially added to the reaction. After one nucleotide is incorporated, the other three will be degraded by the apyrase. For the DNA with a known sequence, non-cyclic NDO can be used and will yield a predicted pyrogram. The sequence of the complementary DNA strand can be determined based on the NOD and peak value in the pyrogram.

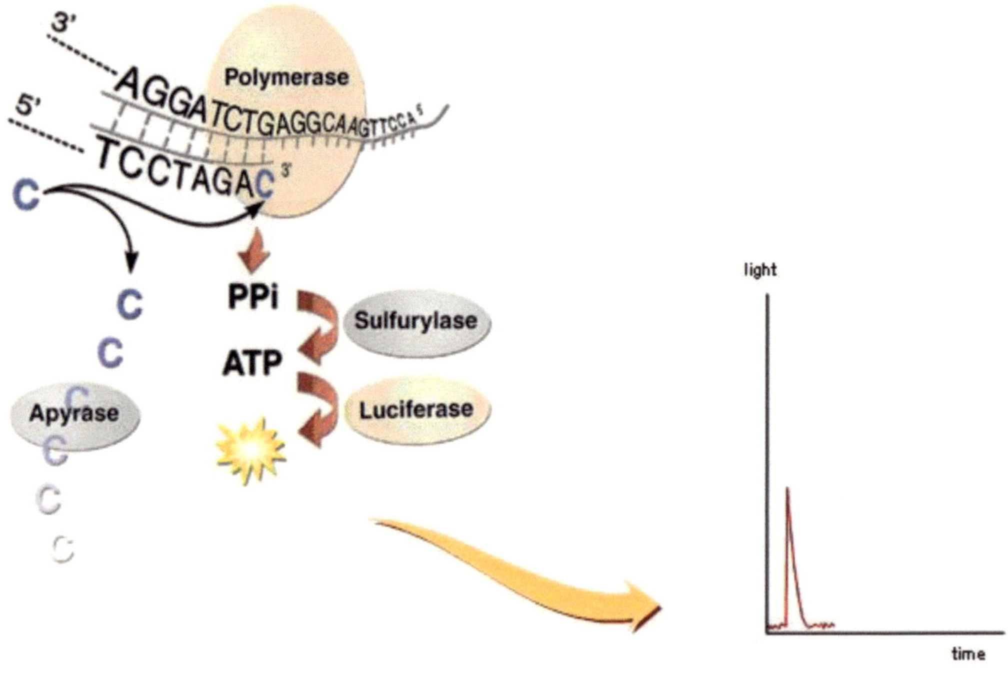

Figure 8. The principle of Pyrosequencing

Since nucleotides are differentially incorporated, Pyrosequencing can produce high-resolution results. Typing HLA-DRB1*04, 07 and DRB4* in the donor's DRB genes by Pyrosequencing not only yields the same result as using the SSOP typing kit, but also produces the result with a higher resolution. Compared with SSP, SSOP, direct or reverse hybridization, Pyrosequencing can be used to solve ambiguous allele combinations of HLA-DQ and HLA-A/B in a short time. The types of HLA-DQB1 and HLA-DRB alleles have been accurately determined by Pyrosequencing.

An inherent problem with this technology is the de novo sequencing of polymorphic region in heterozygous DNA, although polymorphism can be detected in most cases. When the nucleotide in the polymorphic region is altered, synchronized extension can be achieved by the addition of the substituted nucleotides. If there is a deletion or insertion in the polymorphic region,

http://dx.doi.org/10.5772/53084

and the deleted or inserted nucleotide is the same as the adjacent nucleotide on the template, the sequence after the polymorphic region will be synchronized. However, if the deleted or inserted nucleotide is different from the adjacent nucleotide on the template, the sequence reaction can be out of phase, making the subsequent sequence analysis difficult. Another issue with this technology is the difficulty in determining the number of incorporated nucleotides at the homopolymeric region. The light signal will become nonlinear after the incorporation of more than 5-6 nucleotides. Studies on the polymerization efficiency of the homopolymeric region have shown that it is possible to incorporate less than 10 identical nucleotides in the presence of apyrase. However, it needs a specific software algorithm of signal integration to determine the precise number of incorporated nucleotides. For re-sequencing, the nucleotide is added twice to ensure complete polymerization in the homopolymeric region. Another limitation of this technology is the length of the sequencing.

6. Application of flow cytometry in HLA typing

Flow cytometry has failed to become a main method for HLA typing since it was applied to the field of immunology for the first time in 1977. This is mainly due to the large number of specific probes required for HLA typing. The flow analyzer LABScan100 that combines the flow cytometry and reverse SSO technology is trending to replace three conventional methods, SSO, SSP and SBT (sequence-based typing, direct sequencing), in HLA typing.

On a suspension platform, multiple types of color-coded beads conjugated with SSO probes specifically bind to the single stranded DNA. Each type of bead has its unique spectral characteristics due to the different amount of fluorescent dye conjugated to the beads. When beads pass through a flow cytometer, the difference in the light scattering pattern from various angles can distinguish HLA genotypes.

Currently, LabType ™ SSO is a relatively more mature technique compared with others in HLA typing. Its unique advantage is that thousands of molecules can be simultaneously analyzed in a matter of seconds. Therefore, this technique can be used for a large-scale analysis. Overall, this technique has the following advantages: (1) It has increased accuracy due to the automated detection system. (2) The workload and reagent consumption are reduced. One reaction tube can have 100 different SSO probes, thus greatly reducing the workload and reagent consumption. (3) It produces rapid and objective results. The ambiguous results can be avoided with Specialty Probe Technology ™ (SP Technology). (4) Unlike regular flow cytometry that requires fresh samples, this technique can examine the sample at any time upon request or retrospectively. DNA samples can be analyzed right after extraction or stored at –20°C for more than 1 year without affecting the results. (5) The technique can analyze multiple HLA loci with low, medium and high resolutions. (6) It can be used in laboratories with large or small samples. More than 100 probes can be put in one test tube for one sample or in a 96-well plate for 96 samples. The analysis of 96 samples takes less than 90 min after amplification. (7) The pollution to the environment and potential harm to the staff are reduced because electrophoresis is not required in this method.

7. Gene chip or DNA microarray

In gene chip or DNA microarray, large amount of probe molecules (usually 6x10^4 molecules/cm^2) are attached to a solid surface. Labeled DNA samples are hybridized to the probes. The amount and sequence information of the target can be determined by the intensity of the hybridization signal. Gene chip or DNA microarray technology was first developed by Affymetrix in the USA, and has been improved significantly within a few years. The technology is based on the principle of reverse dot hybridization. Thousands of oligonucleotide probes representing different genes are spotted on a solid surface by a robot. These probes will bind to radioactive isotope or fluorescent dye labeled DNA or cDNA through complementary sequences. After autoradiography or fluorescence detection, signals are processed and analyzed by computer software. The intensity and distribution of hybridization signal reflect the expression level of the gene in the sample. The operation process of microarray is shown in Fig 9. Balazs *et al* spotted amplified DNA samples on silicon chips and compared the microarray results with PCR-SSO results in 768 specimens. It has been found that microarray has a high sensitivity and specificity. The consistent rate of genotyping results from microarray and PCR-SSO is 99.9%.

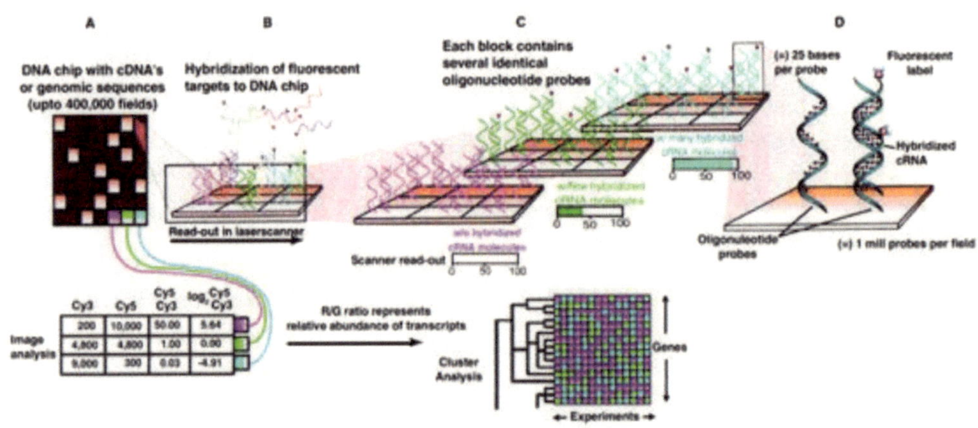

Figure 9. The procedure of gene chip/microarray analysis

Compared with existing genotyping methods, gene chip or microarray has the following advantages: (1) High intensity. The dot intensity on a chip can be higher than 6x10^4/cm^2. Therefore, probes to thousands of HLA-A, B, C, DR, DQ and DP sequences can be spotted on a tiny chip of several square centimeters to obtain the information of individual HLA genes simultaneously. (2) High resolution. It can obtain information at the allele level. (3) Simple operation. The results are generated by fluorescence scanning instead of gel electrophoresis, which greatly simplifies the procedure and shortens operation time. (4) High sensitivity. Signals are amplified twice, first, PCR amplification of the template DNA and second, amplification of fluorescence signal. Therefore, the sensitivity is greatly improved. (5)High accuracy. The intensity of the fluorescent signal generated by the perfect pairing of the probe and the sample is 5 to 35 times higher than the signal generated by the probe and the sample with one or two mismatched nucleotides. Accurate detection of fluorescent signal intensity is the basis of the detection specificity. Studies have shown that the consistency between microarray and Sanger sequencing in the detection of mutations and polymorphism is 99.9%. (6) High efficiency. The whole process is highly automatic, which saves manpower and time for data analysis. Genotyping of genes such as HLA-A, B, DR and DQ in multiple samples can be done with one PCR reaction and hybridization on one chip. (7) High level of standardization. Using a variety of multi-point synchronized hybridization and automated analysis, the human error is minimized to ensure the specificity and objectivity. (8) Low cost. Since the chip fabrication and signal detection are all automatic, only a small amount of probes and samples are required. One chip can be used for the analysis of samples from multiple individuals, which further reduces the cost. The biggest drawback of microarray analysis is its expensive equipment, which prevents it from becoming widely used. Only institutions with a large program can afford the equipment.

8. DNA sequencing technology

For the analysis of gene structure, sequencing is the most direct and accurate method. In this case, the DNA fragment is amplified by PCR and followed by sequencing. The basic process of this method is shown in Figure 10. Since the entire nucleotide sequence of the amplified fragment is obtained, this is the most reliable and thorough genotyping method. It can not only identify the sequence and genotype, but also lead to the discovery of new genotypes. Currently, the newly identified HLA alleles can only be verified by sequencing. It has been reported that if the HLA type cannot be determined by serotyping or the results from PCR-SSP and PCR-SSOP are inconsistent, sequence-based typing (SBT) often can yield accurate and reliable results with a high resolution. Hurley *et al* have typed HLA alleles by PCR-STB in 1775 bone marrow transplant patients and unrelated donors in NMDP, USA. The study has found that the degree of HLA allele mismatching between the recipient and donor of bone marrow transplantation is much higher than previously thought after examining the antigen matching results of HLA-A, HLA-B and HLA-DR.

The advantage of SBT over PCR-SSP and PCR-SSOP is its ability to analyze the entire gene sequence including the non-polymorphic region. SBT can be used not only for DNA sequencing but also for cDNA sequencing to determine gene expression. With increasing popularity of DNA sequencing technology, the PCR-SBT method has gained much attention for genotyping. PCR-SBT has advantages over other typing methods in terms of accuracy, efficiency and the degree of automation. Specialized software and solid phase sequencing kits with automatic loading are available for HLA typing. In addition, the cost of DNA sequencing has been greatly reduced. Therefore, PCR-SBT is an ideal method for HLA typing in researches. With further decrease in the cost of automatic sequencing, this genotyping method will be widely used.

Currently, PCR-SBT is the gold standard of HLA typing. This method has several advantages: (1) It can accurately determine gene type in the exon 8 by a high-resolution sequencing, sufficient to meet the need in researches and clinics. (2) It can analyze more than 15,000 samples every month with high throughput detection. (3) Automated SOP and advanced data management system can reduce human error. (4) It has high quality assurance. Ten percent blind samples are used repeatedly as internal quality control and 100% accuracy is achieved for 10 consecutive times using UCLA external quality assurance samples. The results are confirmed by SSP. (5). It may lead to the discovery of new alleles. (6) HLA genotype can be updated by re-analyzing the sequence after the HLA database is updated.

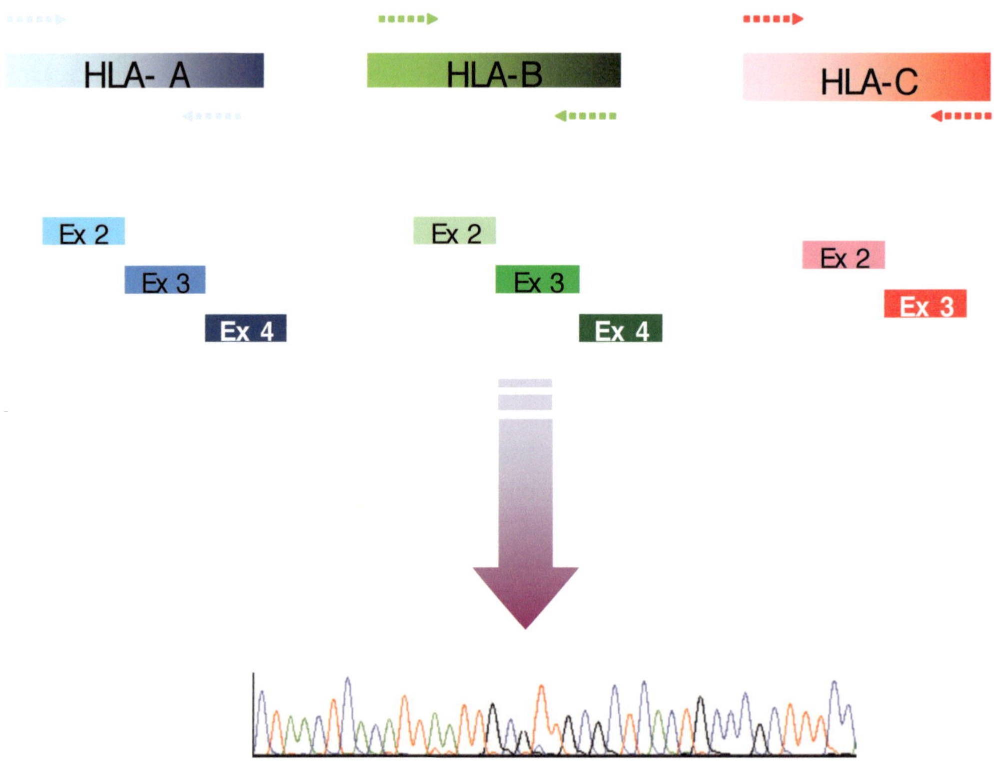

Figure 10. The diagram of DNA sequencing

9. Conclusion

Hematopoietic stem cell transplantation has become one of the most effective treatments for a variety of hematologic malignancies. However, graft-versus-host disease (GVHD) is still inevitable in some cases. This is mainly due to the difference in the major histocompatibility complex (human leukocyte antigen, HLA) between the recipient and the donor. Other known and unknown factors that may cause GVHD include minor histocompatibility anti-

http://dx.doi.org/10.5772/53084

gen (mHA) and tissue specific antigens. GVHD is the main cause of transplant failure in the allogeneic transplantation. Therefore, GVHD is the most significant challenge in allogeneic hematopoietic stem cell transplantation in clinics. It has been proven that whether the graft can survive largely depends on the degree of HLA matching between the recipient and the donor. Therefore, HLA typing of the recipient and the donor before the transplantation is particularly important.

Currently, PCR-SSP genotyping is a commonly used method for HLA typing in clinical laboratories worldwide. Like SSP method, PCR-SSP method depends on specific primers for genotyping. Although the process is simple and rapid, high-resolution genotyping requires a large number of sequence specific primers, which leads to a high cost and prolonged operation time. Similarly, SSO technique is based on the sequence-specific oligonucleotide probes. High-resolution genotyping by SSO significantly increases the cost and complexity. Therefore, it is rarely used for HLA typing today. PCR-SNP is a simple and fast method with a high resolution, and PCR-SNP is expected to become more popular in HLA typing as the technology continues to improve. Although RSCA and Pyrosequencing can achieve high-resolution results, their applications in HLA typing will be gradually eliminated as the technology of gene chip and sequencing continues to improve and the cost continues to decrease. HLA-chip genotyping is still largely dependent on the known sequence. It cannot identify new alleles with unknown sequence. At this moment, PCR-SBT technology has significant advantages over other HLA typing methods in terms of accuracy, efficiency and automation. There are specialized software and automatically loaded sequencing reagents for HLA typing by PCR-SBT. In addition, the operation cost has been greatly reduced. In conclusion, PCR-SBT technology with HLA-chip is the best method for HLA typing in research. With the reduction in the cost of automated nucleic acid sequencing, this genotyping method will be widely used in the field of basic research as well as in clinical transplantation.

Author details

Sun Yuying and Xi Yongzhi*

*Address all correspondence to: xiyz@yahoo.com

Department of Immunology and National Center for Biomedicine Analysis, Beijing 307 Hospital Affiliated to Academy of Medical Sciences, Beijing, P. R. China

References

[1] Opelz G, Terasaki PI. International study of histocompatibility in renal transplantation. Transplantation. 1982, 33:87-95.

[2] Marsh SGE, Albert ED, Bodmer WF, Bontrop RE, Dupont B, Erlich HA, Fernández-Viña M, Geraghty DE, Holdsworth R, Hurley CK, Lau M, Lee KW, Mach B, Maiers

M, Mayr WR, Müller CR, Parham P, Petersdorf EW, Sasazuki T, Strominger JL, Svejgaard A, Terasaki PI, Tiercy JM, Trowsdale J. Nomenclature for factors of the HLA system, 2010. Tissue Antigens. 2010, 75:291-455.

[3] http://hla.alleles.org/

[4] Hansen JA, Petersdorf EW, Lin MT, Wang S, Chien JW, Storer B, Martin PJ. Genetics of allogeneic hematopoietic cell transplantation. Role of HLA matching, functional variation in immune response genes. Immunol Res. 2008, 41:56-78.

[5] Rubinstein P. HLA matching for bone marrow transplantation: How much is enough?. N Engl J Med. 2001, 345: 1842-1844.

[6] Petersdorf EW, Hansen JA, Martin PJ, Woolfrey A, Malkki M, Gooley T, Storer B, Mickelson E, Smith A, Anasetti C. Major histocompatibility complex class I alleles and antigens in hematopoietic cell transplantation. N Engl J Med. 2001, 345:1794-1800.

[7] Petersdorf EW, Kollman C, Hurley CK, Dupont B, Nademanee A, Begovich AB, Weisdorf D, McGlave P. Effect of HLA class II gene disparity on clinical outcome in unrelated donor hematopoietic cell transplantation for chronic myeloid leukemia: the US national marrow donor program experience. Blood. 2001, 98:2922-2929.

[8] McGlave PB, Shu XO, Wen W, Anasetti C, Nademanee A, Champlin R, Antin JH, Kernan NA, King R, Weisdorf DJ. Unrelated donor marrow transplantation for chronic myelogenous leukemia: 9 years' experience of the national marrow donor program. Blood. 2000, 95:2219-2225.

[9] Hadhoud A, Abdulaziz AM, Menawi LA, Shaheen FA, Abdulghaffar A, Abas FA, Al Mobrak MF. The relationship between HLA typing and HCV infection and outcome of renal transplantation in HCV positive patients. Exp Clin Transplant. 2003, 1:19-25.

[10] Erlich HA, Opelz G, Hansen J. HLA DNA typing and translation. Immunity, 2001, 14: 347- 356.

[11] Brown M, Wittwer C. Flow cytometry: principles and clinical application in hematology. Clin Chem. 2000, 46:1221-1229.

[12] Terasaki PI, Cho Y, Takemoto S, Cecka M, Gjertson D. Twenty-year follow-up on the effect of HLA matching on kidney transplant survival and prediction of future twenty year survival. Transplant Proc. 1996, 28:1144-1145.

[13] Takemoto S, Terasaki PI. HLA compatibility can be predicted by matching only three residues with outward oriented sidechains.Transplant Proc. 1996, 28:1264-1266.

[14] Takemoto SK. HLA amino acid residue matching. Clin Transpl. 1996, 28:397-425.

[15] Takemoto SK, Terasaki PI, Gjertson DW, Cecka JM. Twelve years' experience with national sharing of HLA-matched cadaveric kidneys for transplantation. N Engl J Med. 2000,343: 1078-1084.

[16] Hata Y, Cecka JM, Takemoto S, Ozawa M, Cho YW, Terasaki PI. Effects of changes in the criteria for nationally shared kidney transplants for HLA-matched patients. Transplantation. 1998, 65: 208-212.

[17] Takemoto S, Cecka JM, Terasaki PI. Benefits of HLA-CREG matching for sensitized recipients as illustrated in kidney regraftes. Transplant Proc. 1997, 29:1417-1418.

[18] Orofino MG, Argiolu F, Sanna MA, Rosatelli MC, Tuveri T, Scalas MT, Badiali M, Cossu P, Puddu R, Lai ME, Cao A. Fetal HLA typing in beta thalassaemia: implications for haemopoietic stem-cell transplantation. Lancet. 2003, 362: 41-42.

[19] Karabon L, Polak M, Pacuszko T, Karabon L, Polak M, Pacuszko T. HLA typing for donor-recipient matching in unrelated donor hematopotetic stem cell transplantation.Transplant Proc. 2002, 34: 668-670.

[20] Torío A, Moya-Quiles MR, Muro M, Montes-Ares O, Ontañon J, Minguela A, Marin L, Alemany JM, Sánchez-Bueno F, Alvarez-López MR. Discrepancies in HLA-C typing in transplantation: comparison of PCR-SSP and serology results. Transplant Proc. 2002, 34: 419-420.

[21] Arrieta A, Maruri N, Rinon M, Riñón M, Fernández JR, Ortiz de Urbina J, García Masdevall MD. Confirmation of graft-versus-host disease by HLA typing after liver transplantation. Transplant Proc. 2002, 34: 278-279.

[22] Olerup O, Zetter H. HLA-DR typing by PCR amplification with sequence-specific primers (PCR-SSP) in 2 hours: an alternation to serological DR typing in clinical practice including donor-recipient matching in cadaveric transplantation. Tissue Antigens. 1992, 39: 225-235.

[23] Takemoto S, Terasaki PI. Refinement of permissible HLA mismatches. In: Terasaki PI, Cecka JM, eds. Clinical Transplants. Los Angeles: UCLA Tissue Typing Laboratory. 1994, 451.

[24] Stern LJ, Brown JH, Jardetzky TS, Gorga JC, Urban RG, Strominger JL, Wiley DC. Crystal structure of the human class II MHC protein HLA-DR1 complexed with an influenza virus peptide. Nature. 1994, 368:215-221.

[25] Smith PA, Brunmark A, Jackson MR, Potter TA. Peptide-independent recognition by alloreactive cytotoxic T lymphocytes (CTL). J Exp Med, 1997, 185: 1023-1033.

[26] Viret C, Wong FS, Janeway CA Jr. Designing and maintaining the mature TCR repertoire: the continuum of self-peptide: self-MHC complex recognition. Immunity. 1999, 10: 559.

[27] The MHC sequencing consortium: complete sequence and the gene map of a human major histocompatibility complex. Nature. 1999, 401:921-923.

[28] Jordan F, McWhinnie AJ, Turner S, Gavira N, Calvert AA, Cleaver SA, Holman RH, Goldman JM, Madrigal JA. Comparison of HLA-DRB1 typing by DNA-RFLP, PCR-

SSO and PCR-SSP methods and their application in providing matched unrelated donors for bone marrow transplantation. Tissue Antigens. 1995, 45, 103-110.

[29] Zetterquist H, Bengtsson M, Backstrom G, Bäckström G, Egle-Jansson I, Ekdahl AM, Grunnet N, Gustafsson I, Knutsen I, Kuhle A, Rydberg L, Spurkland A, Steffensen R, Storgärds M, Szojmer E, Söderholm G, Thuresson B, Turesson H, Olerup O. Report from the HLA class II typing by PCR-SSP Multicentre Study. Eur J Immunogenet. 1997, 24:191-199.

[30] Bryan CF, Harrell KM, Nelson PW, Pierce GE, Ross G, Shield CF 3rd, Warady BA, Aeder MI, Helling TS, Landreneau MD, Luger AM. HLA-DR and DQ typing by polymerase chain reaction using sequence-specific primer mixes reduces the incidence of phenotypic homozygosity (blanks) over serology. Transplantation. 1996, 62: 1819-1824.

[31] Suzuki Y, Orita M, Shiraishi M, Hayashi K, Sekiya T. Detection of ras gene mutations in human lung cancers by single-strand conformation polymorphism analysis of polymerase chain reaction products. Oncogene. 1990, 5:1037-1043.

[32] Hoshino S, Kimura A,Fukuda Y, Dohi K, Sasazuki T. Polymerase chain reaction-single strand conformation polymorphism analysis of polymorphism in DPA1 and DPB1 genes :A simple,economical and rapid method for histocompatibility testing. Hum Immunol. 1992, 33, 98-107.

[33] Pursall MC, Clay TM, Bidwell JL. Combined PCR-heteroduplex and PCR-SSCP analysis for matching of HLA-A, -B and -C allotypes in marrow transplantation. Eur J Immunogenet. 1996, 23: 41-53.

[34] Bradley BA, Bidwell JL. HLA-DQA1 and DQB1 genotyping by PCR-RFLP, heteroduplex and homoduplex analysis. Eur J Immunogenet, 1996. 23, 107-120.

[35] Saiki RK, Bugawan TL, Horn GT, ET AL. Analysis of enzymatically amplified beta-globin and HLA-DQ alpha DNA with allele-specific oligonucleotide probes. Nature. 1986, 324:163-166.

[36] Mickelson E, Smith A, McKinney S, Anderson G, Hansen JA. A comparative study of HLA-DRB1 typing by standard serology and hybridization of non-radioactive sequence-specific oligonucleotide probes to PCR-amplified DNA. Tissue Antigens. 1993, 41:86-93.

[37] Guo Z, Gatterman MS, Hood L, Hansen JA, Petersdorf EW. Oligonucleotide arrays for high-throughput SNPs detection in the MHC class I genes: HLA-B as a model system. Genome Res. 2002, 12:447-457.

[38] Consolandi C, Frosini A, Pera C, Ferrara GB, Bordoni R, Castiglioni B, Rizzi E, Mezzelani A, Bernardi LR, De Bellis G, Battaglia C. Polymorphism analysis within the HLA-A locus by universal oligonucleotide array.Hum Mutat. 2004, 24:428-434.

[39] Li AX, Seul M, Cicciarelli J, Yang JC, Iwaki Y. Multiplexed analysis of polymorphisms in the HLA gene complex using bead array chips.Tissue Antigens. 2004,63:518-528.

[40] Arguello JR, Little AM, Pay AL, Gallardo D, Rojas I, Marsh SG, Goldman JM, Madrigal JA. Mutation detection and typing of polymorphic loci through double-strand conformation analysis. Nat Genet. 1998, 18:192-194.

[41] Arguello JR, Little AM, Bohan E, Gallardo D, O'Shea J, Dodi IA, Goldman JM, Madrigal JA. A high resolution HLA class I and class II matching method for bone marrow donor selection. Bone Marrow Transplantation. 1998, 22:527-534.

[42] Lenz TL, Eizaguirre C, Becker S, Reusch TB. RSCA genotyping of MHC for high-throughput evolutionary studies in the model organism three-spined stickleback Gasterosteus aculeatus. BMC Evol Biol. 2009, 9:57.

[43] Sun Y, Zhao D, Jin L, Liu N, Liang F, Kong F, Xi Y. Human leukocyte antigens A and B Loci genotyping by reference strand-mediated conformation analysis in hematopoietic stem cell transplantation donor selection. Int J Hematol. 2007, 86:77-83.

[44] Shibata H, Yasunami M, Obuchi N, Takahashi M, Kobayashi Y, Numano F, Kimura A. Direct determination of single nucleotide polymorphism haplotype of NFKBIL1 promoter polymorphism by DNA conformation analysis and its application to association study of chronic inflammatory diseases. Hum Immunol. 2006, 67:363-373

[45] Kennedy LJ, Quarmby S, Fretwell N, Martin AJ, Jones PG, Jones CA, Ollier WE. High-resolution characterization of the canine DLA-DRB1 locus using reference strand-mediated conformational analysis. J Hered. 2005, 96:836-842.

[46] Ronaghi M, Elahi E. Pyrosequencing for microbial typing. J Chromatogr B. 2002, 782:67-72.

[47] Díaz S, Echeverría MG, It V, Posik DM, Rogberg-Muñoz A, Pena NL, Peral-García P, Vega-Pla JL, Giovambattista G. Development of an ELA-DRA gene typing method based on pyrosequencing technology. Tissue Antigens. 2008,72:464-468.

[48] Lank SM, Golbach BA, Creager HM, Wiseman RW, Keskin DB, Reinherz EL, Brusic V, O'Connor DH. Ultra-high resolution HLA genotyping and allele discovery by highly multiplexed cDNA amplicon pyrosequencing. BMC Genomics. 2012, 13:378.

[49] Lank SM, Wiseman RW, Dudley DM, O'Connor DH. A novel single cDNA amplicon pyrosequencing method for high-throughput, cost-effective sequence-based HLA class I genotyping. Hum Immunol. 2010, 71:1011-1017.

[50] Gabriel C, Danzer M, Hackl C, Kopal G, Hufnagl P, Hofer K, Polin H, Stabentheiner S, Pröll J. Rapid high-throughput human leukocyte antigen typing by massively parallel pyrosequencing for high-resolution allele identification. Hum Immunol. 2009, 70: 960-964.

[51] Lu Y, Boehm J, Nichol L, Trucco M, Ringquist S. Multiplex HLA-typing by pyrosequencing. Methods Mol Biol. 2009, 496:89-114.

[52] Ringquist S, Styche A, Rudert WA, Trucco M. Pyrosequencing-based strategies for improved allele typing of human leukocyte antigen loci. Methods Mol Biol. 2007, 373:115-134.

[53] Loginova MA, Paramonov IV, Trofimova NP. Experience in using the LABType SSO reagent kits in the practice of a HLA typing laboratory. Klin Lab Diagn. 2011, 4:45-47.

[54] Testi M, Iannelli S, Testa G, Troiano M, Capelli S, Fruet F, Federici G, Bontadini A, Andreani M. Evaluation of DRB1 high resolution typing by a new SSO-based Luminex method. Mol Biol Rep. 2012, 39:13-16.

[55] Feng C, Putonti C, Zhang M, Eggers R, Mitra R, Hogan M, Jayaraman K, Fofanov Y. Ultraspecific probes for high throughput HLA typing. BMC Genomics. 2009, 10:85.

[56] Lee KR, Park E, Moon SH, Kim JM, Kwon OJ, Kim MH, Sohn YH, Ko SY, Oh HB. Development and clinical evaluation of a microarray for HLA-A and -DRB1 genotyping. Tissue Antigens. 2008, 72:568-577.

[57] Consolandi C. High-throughput multiplex HLA-typing by ligase detection reaction (LDR) and universal array (UA) approach. Methods Mol Biol. 2009, 496:115-127.

[58] Consolandi C, Frosini A, Pera C, Ferrara GB, Bordoni R, Castiglioni B, Rizzi E, Mezzelani A, Bernardi LR, De Bellis G, Battaglia C. Polymorphism analysis within the HLA-A locus by universal oligonucleotide array. Hum Mutat. 2004, 24:428-434.

[59] Pröll J, Danzer M, Stabentheiner S, Niklas N, Hackl C, Hofer K, Atzmüller S, Hufnagl P, Gülly C, Hauser H, Krieger O, Gabriel C. Sequence capture and next generation resequencing of the MHC region highlights potential transplantation determinants in HLA identical haematopoietic stem cell transplantation. DNA Res. 2011, 18:201-210.

[60] Noreen HJ, Trachtenberg E, Williams TM, Baxter-Lowe LA, Begovich AB, Petersdorf E, Selvakumar A, Stastny P, Hegland J, Hartzman RJ, Carston M, Gandham S, Kollman C, Nelson G, Spellman S, Setterholm M. A high degree of HLA disparity arises from limited allelie diversity: analysis of 1775 unrelated bone marrow transplant donor-recipient pairs. Hum mmunol. 2007, 68: 30-40.

Chapter 3

Neuron Replacement and Brain Repair; Sex Does Matter

Laurent Lecanu

Additional information is available at the end of the chapter

http://dx.doi.org/10.5772/25649

1. Introduction

The brain is one of the main targets of gonadal steroid hormones. In addition, it contains many of the steroid metabolizing enzymes. The effect of gonadal steroids on brain development and maturation has been well documented [1,2]. The vast literature on the subject has introduced the common belief that gonadal steroids may be the only effectors of the brain sexual differentiation, overshadowing other key elements. Although there is no doubt about its importance, the dogma of the gonadal origin of somatic differentiation, including neuronal cells, usually implies that XX and XY cells, accordingly stem cells, are functionally equivalent unless gonadal secretions act on them in a sex-specific manner. The human Y chromosome encodes 27 different proteins [3] eight of which are expressed in the male brain and could have a male-specific effect on the brain, independent of any gonadal hormone influence [4,5]. Furthermore, XX cells contain an X chromosome that received a paternal genomic imprint, whereas XY cells do not, a fact that is likely to contribute to autonomous differences between male and female cells. De Vries and colleagues [6] generated mice in which the testis-determining gene *sry* was deleted from the Y chromosome and subsequently inserted onto an autosome. This experiment resulted in the generation of mice where the development of the testis occurred independently of the complement of X or Y chromosome. Although most of the sexual dimorphism correlated with the presence of testis or ovary (and therefore associated with gonadal hormones), XY mice (with testes or ovaries) were found to be more masculine than the XX mice (with testes or ovaries) in the density of vasopressin-immunoreactive fibers in the lateral septum, suggesting that sex chromosome genes contribute to the development of a sex difference in the brain. These results also suggest that one should not consider that female and male neural stem cells (NSC) are equal and react in the same manner to a specific environment or pharmacological agent. Furthermore, there are no data to support the *a priori* consideration that transplanting young female NSC in an old female brain would result in the same neural differentiation and functional recovery as transplanting young male NSC in an old male brain. Likewise, there is no evidence to *a priori* consider that male and female NSC neurogenic properties would evolve in a sex-independent manner throughout development and aging. To support this hypothesis, sexual dimorphism has already been described in various biological aspects of several types of stem cells [7-13]. In particular, we and others recently reported a sexual dimorphism in the neurogenic capacity of rat [14,15] and primate [11] NSC. Considering the dramatic and sex-specific hormonal changes occurring throughout development and aging one might expect a sex- and sex-through-aging-specific environment to be a prerequisite for successful neuro-genesis.

The chapter will first discuss the potential of stem cells for brain repair, tissue regeneration and function recovery, second the effect of sex on stem cell fate whether neural stem cells or peripheral stem cells, and third the potential translation in clinics. The goal of the present proposal is to discuss the therapeutic relevance of the largely under-explored sex- and age-based differences in the capacity of NSC to engage in neurogenesis programming. Indeed, beyond understanding the physiology and biochemistry of aging NSC, for both sexes, the overall objective is to discuss the potential foundation for future studies aimed at tailoring NSC transplantation strategies for brain repair as a function of sex/age, as well as considering sex- and age-specific pharmacological approaches towards the development of neurogenesis-inducing treatments.

2. Stem cells — Hope and hurdles

If one had to explain the excitement and exhilaration stem cell therapy triggers in the field of tissue repair and the hope it represents, a comparison that would make sense is the new horizons opened after the first man in flight or the first man in space. The hopes and hypes reflect the potential offered by stem cell therapy and in that sense, they are completely justifiable. Probing PubMed with 'stem cell' and 'therapy' as keywords in certainly the best way to understand how the interest of research community significantly evolved over 40 years. Indeed, in 1970, only 13 publications related to stem cell therapy were referenced in PubMed, compared to 7942 in 2010 (Figure 1). A slightly restricted search for 'stem cell', 'therapy' and 'brain' reveals that the number of publications related to brain repair, although increasing, is "plateauing" at 10% of the total number of publications focused on stem cell therapy for the past five years (Figure 1). This status reflects a greater difficulty to generate clear data on experimental and clinical stem cell therapy for brain repair than with any other organs. This situation is likely due to the unique architectonic of the brain, the way it interfaces with the peripheral compartment and the seemingly endless cell phenotypes that compose the brain.

http://dx.doi.org/10.5772/25649

clinical stem cell therapy for brain repair than with any other organs. This situation is likely due to the unique architectonic of the brain, the way it interfaces with the peripheral compartment and the seemingly endless cell phenotypes that compose the brain.

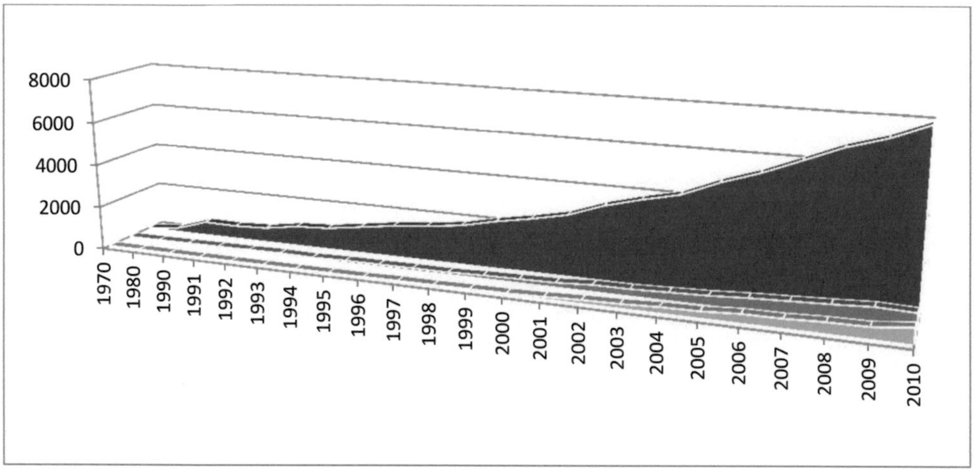

Figure 1. **Number of publications per keyword and per year.** Light grey "neural stem cells"; medium grey "stem cells AND therapy AND brain"; dark grey "stem cells AND therapy".

Figure 1. Number of publications per keyword and per year. Light grey "neural stem cells"; medium grey "stem cells AND therapy AND brain"; dark grey "stem cells AND therapy".

It is remarkable that the first report using the words 'neural stem cells' was published not that long ago in 1992, reaching the hundred only in 2001, catching up in 2010 with 714 references with the number of publications that include 'stem cell', 'therapy' and 'brain' (Figure 1). Although one cannot deny that the amount of data degenerated nowadays on NSC and brain repair is getting more important, the rather slow pace at which progresses are still reported does not prefigure any substantial impact in clinics to be made in the near future. Many hurdles lie on the road of brain stem cell therapy. The first one, and not necessarily by order of importance, is directly related to the various types of stem cells considered as a potential source for brain transplantation. Human embryonic stem cells (hESC) [16,17], fetal NSC [18,19], post-natal or adult NSC [20-25], induced pluripotent stem cells (iPS) [26-29], bone marrow stem cells (BMSC) [30-36], mesenchymal stem cells (MSC) [37,38], umbilical cord blood-derived stem cells [39-41] and adipose tissue-derived stem cells [42,43]. This luxury of choice turned out to become an increasingly difficult situation to handle, generating at best a non-consensual agreement as to which stem cell type should be used for what type of neurological disorder. Overall, it seems that potential stem cell sources have been insufficiently characterized, impeding clinical translation. Although it sounds like a very easy statement to make, we have to admit that our knowledge in long-term outcome of stem cell transplantation in the brain is extremely limited. In example, grafting stem cells not from neurodermal origin in an attempt to replace degenerated neurons supposes a transdifferentiation mechanism of the transplanted stem cells prior to undergo successful differentiation in the appropriate neuronal phenotype [44-49]. Very little literature is currently available in transdifferentiation mechanisms and even more so, recent data reported that transdifferentiation is reversible and may be restricted to certain type of stem cells [48,49], raising the question of what would happen in vivo. A major concern is the development of gliomas or other brain cancer types secondary to stem cell transplantation. Indeed, the property of stem cells to self-renew is exactly what confer them the capacity of being highly tumorigenic [50-53]. It is particularly true for the hESC which tumorigenic properties have been extensively studied [52]. In addition, iPS have been developed to exactly mimic embryonic stem cells neurogenic capacity without having the capacity to induce brain tumor formation. However, the iPS safety is still being debated [52]. A clinical case has been reported recently regarding a young boy who developed glioneuronal neoplasm four years after having received allogenic hESC transplantation in an attempt to improve his ataxia telangiectasia symptomatology [50]. It is remarkable that the tumor cells were identified as deriving from at least two of the donors composing the hESC pool from which the young boy received multiple transplantations. This specific data points out the possibility that using pool of donors may increase the risk of developing tumor or even adverse graft reaction [54-57]. Beyond the in vivo concern of stem cells tumorigenesis remains the question of inducing the proper neural phenotype needed to replace a specific type of neuron. Parkinson's disease is without any doubt the neurodegenerative disease for which brain transplant has been studied the most extensively. A consensual strategy seems to pre-differentiate the stem cells toward the dopaminergic fate prior to conduct the transplantation [58,59] and as a consequence, many studies have been conducted to develop suitable protocols [58-62]. The same question arises to ensure that stem would undergo GABAergic differentiation prior to be used in patients suffering from Huntington's disease [63]. However, despite many efforts to establish validated differentiation protocols, the phenotypic fate followed by transplanted stem cells still remain largely uncontrolled [64-66] and likely involves stem cells intrinsic properties, i.e. region or context dependency, sex or age [14,67-70]. In a medical context of growing interest for global personalized medicine in general [71] and focused on regenerative medicine in particular [27,72-74], answering the question "What stem cell for whom?" never appeared as critical.

It is remarkable that the first report using the words 'neural stem cells' was published not that long ago in 1992, reaching the hundred only in 2001, catching up in 2010 with 714 references with the number of publications that include 'stem cell', 'therapy' and 'brain' (Figure 1). Although one cannot deny that the amount of data degenerated nowadays on NSC and brain repair is getting more important, the rather slow pace at which progresses are still reported does not prefigure any substantial impact in clinics to be made in the near future. Many hurdles lie on the road of brain stem cell therapy. The first one, and not necessarily by order of importance, is directly related to the various types of stem cells considered as a potential source for brain transplantation. Human embryonic stem cells (hESC) [16,17], fetal NSC [18,19], post-natal or adult NSC [20-25], induced pluripotent stem cells (iPS) [26-29], bone marrow stem cells (BMSC) [30-36], mesenchymal stem cells (MSC) [37,38], umbilical cord blood-derived stem cells [39-41] and adipose tissue-derived stem cells [42,43]. This luxury of choice turned out to become an increasingly difficult situation to handle, generating at best a non-consensual agreement as to which stem cell type should be used for what type of neurological disorder. Overall, it seems that potential stem cell sources have been insufficiently characterized, impeding clinical translation. Although it sounds like a very easy statement to make, we have to admit that our knowledge in long-term outcome of stem cell transplantation in the brain is extremely limited. In example, grafting stem cells not from neurodermal origin in an attempt to replace degenerated neurons supposes a transdifferentiation mechanism of the transplanted stem cells prior to undergo successful differentiation in the appropriate neuronal phenotype [44-49]. Very little literature is currently available in transdifferentiation mechanisms and even more so, recent data reported that transdifferentiation is reversible and may be restricted to certain type of stem cells [48,49], raising the question of what would happen in vivo. A major concern is the development of gliomas or other brain cancer types secondary to stem cell transplantation. Indeed, the property of stem cells to self-renew is exactly what confer them the capacity of being highly tumorigenic [50-53]. It is particularly true for the hESC which

3. Stem cells transplantation in clinics — From hype to disillusion, keeping the faith alive

tumorigenic properties have been extensively studied [52]. In addition, iPS have been developed to exactly mimic embryonic stem cells neurogenic capacity without having the capacity to induce brain tumor formation. However, the iPS safety is still being debated [52]. A clinical case has been reported very recently regarding a young boy who developed glioneuronal neoplasm four years after having received allogenic hESC transplantation in an attempt to improve his ataxia telangiectasia symptomatology [50]. It is remarkable that the tumor cells were identified as deriving from at least two of the donors composing the hESC pool from which the young boy received multiple transplantations. This specific data points out the possibility that using pool of donors may increase the risk of developing tumor or even host-versus-graft reaction [54-57]. Beyond the trivial concern of stem cells tumorigenesis remains the question of inducing the proper neural phenotype needed to replace a specific type of neuron. Parkinson's disease is without any doubt the neurodegenerative disease for which brain transplant has been studied the most extensively. A consensual strategy seems to pre-differentiate the stem cells toward the dopaminergic fate prior to conduct the transplantation [58,59] and as a consequence, many studies have been conducted to develop suitable protocols [58-62]. The same question arises to ensure that stem would undergo GABAergic differentiation prior to be used in patients suffering from Huntington's disease [63]. However, despite many efforts to establish validated differentiation protocols, the phenotypic fate followed by transplanted stem cells still remain largely uncontrolled [64-66] and likely involves stem cells intrinsic properties, i.e. region or context dependency, sex or age [14,67-70]. In a medical context of growing interest for global personalized medicine in general [71] and focused on regenerative medicine in particular [27,72-74], answering the question "What stem cell for whom?" never appeared as critical.

3. Stem cells transplantation in clinics — From hype to disillusion, keeping the faith alive

The main reason of the arisen interest for stem cell therapy in central and peripheral nervous system lies in the fact that it addresses neuropathologies and conditions for which neurological damages are extensive, socially debilitating and irreversible and for which there is no 'magic pill'. Stem cell therapy for tissue repair is generating very high expectation to treat neurodegenerative diseases like Parkinson's disease [66,75], amyotrophic lateral sclerosis [76-81], Huntington's disease [18,19,82-84], Alzheimer's disease [85-87], multiple sclerosis [88-90], spinal cord injury [91,92] and retina degeneration [26,31,93]. Clinical translation has been, up to now and by far, most exclusively conducted in patients suffering from Parkinson's, Huntington's disease or amyotrophic lateral sclerosis and, to our knowledge, there is no data reported clinical evaluation of stem cell transplantation in Alzheimer's disease or retina degeneration.

3.1. Parkinson's disease

Two groups actively involved in conducting clinical studies assessing the therapeutic effectiveness of stem cell transplantation to treat Parkinson's disease reported conflicting results

http://dx.doi.org/10.5772/25649

[94-97]. In a first study, six Parkinson patients were to received bilateral transplantation of fetal nigral tissue the post-commissural putamen [95]. The fetal material was obtained from the mesencephalon of legally aborted fetus and used as solid grafts. Each patient received tissue pooled from 3 to 4 fetuses. Two years outcome measurement showed an improvement of the Unified Parkinson's Disease Rating Scale (UPDRS), and an increase of [^{18}F]-fluorodopa uptake in the putamen. On the downside, all patients started experience dyskinesia few weeks after the transplantation. When the same group performed the same type of study on a much larger cohort, the outcome was much more deceiving [97]. Out of the 34 patients followed up to 2 years, none of them displayed clinically relevant improvement although post-mortem analysis showed a robust survival of grafted dopaminergic neurons. These results led the authors to conclude that fetal nigral transplantation could not constitute a therapy for Parkinson's disease. One can argue that using solid grafts rather than well characterized stem cell/ neuroblast primary culture might have impaired the capacity of newly formed neurons to establish connections with the pre-existing network. This point is extensively discussed in a very recent review [66]. A more striking issue to us is the lack of information related to the sex of the donor suggesting that the tissue samples were pooled and grafted regardless of a potential sex-based difference in stem cell biology which could have an impact on the clinical outcome. In an other study, 40 patients were bilaterally grafted in the putamen with a mixture of stem cells/dopaminergic neuroblasts obtained from cultured mesencephalic tissue from 4 embryos [94]. Follow-up at one year post-transplantation showed some benefit but only for the patients 60 years of age or younger. In addition, the benefit was clinically significant only in some areas of the UPDRS and 15 percent of the patients became dyskinetic. A post-analysis of the one-year follow-up data revealed a sex-based difference in the graft outcome [96]. Indeed, the male patients displayed more clinical benefit than the female patients after the first year but the progress rate increased in the female to catch up with the male at the end of the second year of follow-up [96]. In addition, a follow-up performed over 4 years on 33 patients out of the 40 initially included showed that the age-based difference observed earlier [94] disappeared as the oldest patients' overall UPDRS was catching-up over the 4-years period of time on the youngest' one [96]. The authors concluded that some clinical benefit was still clearly present 4 years after transplantation with however no correlation with the [^{18}F]-fluorodopa uptake. Although the measured outcome seemed to be more encouraging than in the first series of experiments, the cell preparation may account for it, the results remain rather inconsistent and deceiving. On the same line, earlier clinical studies reported a somewhat beneficial effect of fetal mesencephalic neurons in patients affected by Parkinson's disease [75,98,99]. Several reasons have been proposed such inconsistency, among which the number of cells, the grafting site, the preparation of the cells, the use of immunosuppressant or the lack of functional rehabilitation associated with the transplantation [66,100] but no consensus has been reached so far. In addition, a recurrent issue that clinicians and patients are facing is the graft-induced dyskinesia. This locomotor alteration is a direct consequence of the transplantation and its causes and possible solutions remain rather elusive [75] and is seen as a permanent drawback to any progress made in stem cell transplantation to treat Parkinson's disease. As mentioned earlier, it is quite interesting that the cells of several donors were pooled with no reference to their sex as a factor of graft outcome. It is even more surprising since a sexual dimorphism

with functional implications has been established for human muscle stem cells [9,12,13] and monkeys mesenchymal stem cells [11].

3.2. Amyotrophic lateral sclerosis

Amyotrophic lateral sclerosis (ALS) is a late-onset neurodegenerative disease that selectively affects motor neurons in the cortex, brain stem and spinal cord [101] for which there is no treatment. Several clinical studies have been conducted which results raised some optimism but, overall, data related to the transplantation outcome remain, as with Parkinson's disease, contradictory and inconsistent. A very recent report based on thirteen clinical cases demonstrated the feasibility and the therapeutic relevance of performing autologous bone marrow-derived hematopoietic cell transplantation in patients displaying severe sporadic ALS [76]. One year after receiving CD34+ cells transplantation in the brain stem and the beginning of the spinal cord, most of the patients displayed an improvement of their status. The majority of the patients regained neuronal stimulatory capacity of their muscle, as measured by post-operative electroneuromyogram, and a better bulbar score. Some of them even recovered walking capacity compatible with a daily life. However, the majority of the patients who experienced a post-operative gain of function started to see their clinical status decline at one-year follow-up. This piece of data, rather than suggesting that bone marrow-derived stem cells cannot be used as a therapeutic tool suggests instead that the transplantation procedure should be repeated in order to obtain maximum functional recovery. An other study demonstrated the beneficial effect of stem transplantation as a cytotherapeutic tool for ALS. Thirty-three patients diagnosed with severe sporadic ALS were to receive autologous transplantation of peripheral blood mononuclear cells (PBMC) in the motor cortex and followed-up over one year [78]. All the patients that received CD133+ PBMC transplantation experienced a dramatic improvement of their clinical status as measured by the ALS Functional Rating Scale-Revised. The most striking beneficial effect resulting from the transplantation was the increase in the median survival from 19 months for the patients in the sham group to 66 months in the treated group. More remarkably, the patients' quality of life was also dramatically improved. Indeed, in the sham group, half of the patients had to undergo tracheotomy and gastrostomy at 12-months follow-up compared to the no tracheotomy and only one gastrostomy performed in the group of patients who received PBMC transplantation. Taken together, these results show promises for ALS cytotherapy. However, some more recent clinical studies reported that transplanting stem cells in ALS patients did not result in any beneficial effect [79,81]. In the first case, patients received autologous transplantation of mesenchymal stem cells isolated from bone marrow in the spinal cord at T4-T5 and T5-T6 level and were followed-up for 24 months [79]. Although MRI imaging ascertained graft survival, clinicians did not observe any clinical improvement. However, it is noteworthy that the patients recruited for this study were diagnosed with mild to moderate sporadic ALS unlike the two previous studies described above for which the patients were diagnosed with severe sporadic ALS. This apparent discrepancy may indicate that the stage of the disease at which the stem cell transplantation is to be performed has to be cautiously defined as it may critically affect the clinical outcome. In the second case, patients from 3 different centers received stem cells transplantation and were then evaluated at 12-months post-transplantation [81]. This study is difficult to interpret as it includes a total of 12 patients who received 3 different types of stem cells, depending on the clinical center of origin, and using intratechal route, intravenous or both. The stem cells used were embryonic olfactory ensheating stem cells, mesenchymal stem cells or CD34+ stem cells without further explanation as to how many embryos were used per patients or if the graft of adult stem cells was autologous or not. However, a common point shared with the previous reported study, beside its lack of success at restoring the patients' motor function, is the fact that most of the patients recruited were diagnosed with a mild to moderate sporadic ALS supporting the idea 1) that mild to moderate stage of ALS might not be the best moment to start stem cell therapy, or that 2) the types of stem cells that were proven to be therapeutically relevant in the severe cases might not be as relevant to treat mild or moderate cases of ALS. These results certainly advocate for a better personalized therapy to the patient.

3.3. Huntington's disease

Huntington's disease is a neurodegenerative condition which is clinically characterized by cognitive, motor and psychiatric deterioration and leads in 20 years maximum to the death of the patient. There is currently no treatment for this disease and stem cell therapy is being extensively investigated as a therapeutic tool. Many attempts of restoring some or all the lost functions by stem cells transplantation have been made but the long term results have been so far deceiving. Motor and cognitive functions improvement has been reported in some patients 1 year after the transplantation [82] but these improvements plateaued at 2 years and to finally completely reverse at 4-6 years [18]. Other studies reported successful engraftment and neuronal differentiation [19] associated with an improved cortical metabolism [102] unfortunately not accompanied by any clinical improvement. A major finding that is commonly reported on long-term post-mortem histological study is the poor survival of the transplanted stem cells [83,84] which is believed to be due to a reaction of the host immune system against the grafted cells [54,84]. Interestingly, to our knowledge, all the clinical studies conducted on Huntington patients reported having used human embryonic stem cells [18,19,54,82-84,102] which is a quite surprising situation if we compare to the variety of stem cell types used in Parkinson's disease and ALS. In addition, one is force to at least acknowledge the fact that the stem cells used are from embryonic origin may be the factor that triggers the immune response. Indeed, human embryonic stem cells are not devoid of immunogenicity and are capable of triggering a significant immune response leading to graft rejection [54,103]. In regard of what has been accomplished in Parkinson's disease and ALS, autologous transplantation of PBMC or bone marrow-derived stem cells seems to be more appropriate and one may wonder why it has not been tried yet. An other point that raises question is the fact that all the studies reported, except one [19], transplanted human embryonic stem cells pooled from several embryos, neither mentioning how many embryos were used per patient nor the sex of the embryos.

3.4. Other indications for stem cell therapy

Stem cell therapy, beyond the controversy of using embryonic stem cells, is a very seducing and highly promising therapeutic strategy for repairing the central nervous system. Using stem cell transplantation has become a very appealing topic for many researchers and clinicians around the world. However, developing and implementing such protocols to other neurological conditions than the ones discussed above reveals to be an even more challenging endeavor. Among the neurodegenerative disease that would benefit from stem cell therapy, Alzheimer's disease evidently represents a major interest. Probably because of the complexity of the disease and our very limited knowledge about it, stem cell therapy for Alzheimer's disease is still in a preclinical stage [85-87]. Stem cell therapy for spinal cord injury is certainly at the forefront of the preclinical research in order to establish valid therapeutic protocols to be translated in clinics, but once again translating the knowledge acquired in animal models facing several hurdles, among which cytotherapy specificity and safety [29,87,91,92,104,105]. Brain stroke is a neurological condition that could require extensive tissue reconstruction and results obtained in animal models are very encouraging [20,24,106]. So far, only one clinical study has been conducted in five patients by autologous transplantation of bone marrow-derived stem cells [36]. One year follow-up showed good safety profile and a trend to clinical improvement. Retina degeneration, whether idiopathic or post-traumatic, is a major problem as it is irreversible, without treatment and leads to blindness in most of the case. However, the retina architectonic and the differentiation process of retina stem cell represent two main problems to clinical translation. Indeed, many in vivo and in vitro studies have been conducted in various laboratories [26,30-32,35,37,45,107-117] showing promising results. However, although the post-transplantation cell survival was excellent in all cases, the retina structure revealed itself to be an obstacle to stem cell migration. In addition, the differentiation process that leads to the formation of mature retina cell is by far more complex than in the brain.

4. Looking at the forest instead of the tree — Turning the tide on clinical setbacks

We have recently published in a recent report that the estrogen receptor ERα is differentially expressed in male and female NSC. NSC isolated from 3 month-old rats display sexual dimorphism in the expression of estrogen receptor alpha and beta [14]. Male NSC contain one third of the ERα levels, whereas ERβ levels were 3 times greater than those expressed NSC isolated from same age females. Moreover, our data demonstrated that the ERα/ERβ ratio was close to 1 in the male but 10 fold higher in the female NSC. Interestingly, others previously reported that the expression of steroid receptors in the fish inner ear varies between sexes [118]. Such sexual dimorphism has been previously described in mature neurons of various brain structures [119,120] and has been shown to have a role in the differentiation of sexual behavior and gender identity [121]. Indeed, ERα has been shown to be primarily involved in masculinization, whereas ERβ is primarily involved in defeminization [122,123]. From birth and throughout life, sex hormones physiology and homeostasis are different between men and women which suggest that transplanting the "inadequate" type of NSC to the patient may not lead to the expected beneficial effect. Indeed, consequences in clinics may run from a lack of recovery to partial or inadequate recovery. Dissimilarity between NSC and the recipient tissue may cause these undesirable effects. Interestingly, in the same manner, NSCs isolated from 20 month old male and female rats displayed a dramatic increase in ERα and ERβ expression that was equivalent in both sexes, suggesting that male and female NSCs are not equal before aging. The effect of estrogens on neurogenesis has been extensively studied and it is commonly agreed that estrogens simultaneously promote NSC proliferation and differentiation [124-129]. There is increasing evidence on the estrogenic aspect of neurogenesis; however, the differential roles of ERα and ERβ in this process still remain to be fully characterized. Considering the currently known role of estrogens in NSC physiology and the regulation of the neurogenesis [124,126,128-132], the sexual dimorphism we observed in ERα and ERβ expression between male and female NSC [14] supports a sex-based intrinsic difference in the regulation of neurogenesis. In addition, ERα genotype has been recently reported to be responsible for the inter-individual variability of responses to estrogen and testosterone in mesenchymal stem cell-derived osteoblasts [133]. Moreover, estradiol has been described to alter neurogenesis in female, but not male rats [134]. We also provided evidence that male NSC expressed a dramatically higher level of CYP19 than female NSC [135] which supports a capacity for male NSC, unlike female NSC, to metabolize testosterone and in turn, to produce estradiol. Such biochemical sexual dimorphism may underlie a steroid-related pharmacological counterpart as male NSC may therefore have the ability to alter their local environment and modulate endogenous neurogenesis in a different manner than female NSC may do. Remarkably, an autocrine control loop in two different systems, the NSC kinin/kallicrein pathway [136] and the androgenic apparatus of human bone marrow stromal cells [137]. It is noteworthy that the sex-based differences unveiled in vitro translated in vivo as we recently reported [69]. Indeed, the outcome of NSC transplantation in brain was shown to tightly depend on the sex of the donor and the sex of the recipient. Interestingly, in some cases cross-sex grafting provided better cell survival results than same-sex transplantation. As others also demonstrated the occurrence of a sexual dimorphism in the neurogenic capacity of rhesus monkeys mesenchymal stem cells [11], we are the first to have shown a direct impact on the outcome of stem cell transplantation [69]. Nevertheless, stem cell sexual dimorphism has been previously demonstrated by others in various organs [9,12,13,67,134] and organisms [138], and it is our opinion that sex should be considered as a critical factor and integrated in the development of future clinical protocol.

There is a consensual agreement that age-associated alterations in the brain play an important role in the decline of neurogenesis reported in aging [11,139-141]. However, very little work has been done in defining the age-related alteration of NSC neurogenic properties. Old mice have been shown to have less NSC in the subventricular zone (SVZ) compared to young ones, and a similar reduction has been reported in the number of NSC maintained as neurospheres and recovered in culture *in vitro* [142]. Aging effect on neurogenesis may also be structure-specific as revealed by others showing a decrease of NSC proliferation has been described in the hippocampus of old rats but not in the SVZ of the same animal [143]. On an other hand, spatial redistribution and a delay in the migratory process seem to affect NSC in the SVZ rather than a decrease of the proliferation rate [144]. However, although some studies showed that NSC from old animals retain some or all of their neurogenic properties, to date, studies that have systematically explored age-based differences of NSC neurogenic properties in terms of neuronal phenotype and protein marker expression are extremely scarce

[11,14,69,135,138,145,146]. Surprisingly, NSC isolated from young animals did not perform better in term of survival rate after brain transplantation when compared to the ones isolated from old rats [69]. Thus, such a result provocatively raises the question of the rejuvenation process NSC isolated from old individuals may undergo after being transplanted in a younger environment as suggested by two recent works [147,148]. Understanding the role of the age of NSC as a critical factor modulating the neuronal fate specificity and the maturation level reached by the engrafted NSC is absolutely essential from the perspective of stem cell grafting into the brain. In addition, in agreement with our previous results showing that NSC age differently depending on their sex, it appears that sex cannot be dissociated from age as a determining factor of NSC capacity to lead to functional recovery following transplantation.

5. Conclusion

Stem cell therapy holds a lot of promises for brain repair and the recovery of impaired neurological functions. However, despite the generation of a large amount of encouraging results produced on various animal models, the translation at the patient's bed seems to be delayed. Facing mitigated results from the few clinical studies that have been conducted so far, we are constrained to a cautious optimism. Indeed, the low success rate encountered in clinics questions our knowledge on the topic and suggests that we go back to a more fundamental bench work. Indeed, the high number of stem cell types, to which have to be added the various engineering counterparts, turn a luxury of choice into a situation where it is extremely, if not impossible, to determine what cell type or cell line would be the best candidate for tissue repair. Compelling evidence suggests that sex may be a critical determining factor of stem cell transplantation outcome in the brain. Surprisingly, none of the clinical studies reported to date took this parameter into consideration. Furthermore, we foresee that NSC differentiation induced by neurogenic agents *in vitro* and *in vivo* will be an important part of brain repair procedures as brain repair therapy and optimal neuronal function restoration will likely require both exogenous NSC grafting and pharmacological stimulation of the endogenous neurogenesis. However, nothing is currently known about the effect of cell sex on NSC sensitivity to pharmacologically-induced differentiation, and we therefore strongly believe that acquiring such knowledge is critical for the development of neurogenic treatments.

Author details

Laurent Lecanu[1,2]

Address all correspondence to: laurent.lecanu@mcgill.ca

1 The Research Institute of the McGill University Health Centre, Montreal, Quebec, Canada

2 McGill University Health Centre, Department of Medicine, Quebec, Canada

References

[1] Phoenix CH, Goy RW, GERALL AA, Young WC. Organizing action of prenatally administered testosterone propionate on the tissues mediating mating behavior in the female guinea pig Endocrinology 1959; 65: 369-382.

[2] Smith ER, Davidson JM. Differential responses to hypothalamic testosterone in relation to male puberty Am J Physiol 1967; 212: 1385-1390.

[3] Skaletsky H, Kuroda-Kawaguchi T, Minx PJ, Cordum HS, Hillier L, Brown LG, Repping S, Pyntikova T, Ali J, Bieri T, Chinwalla A, Delehaunty A, Delehaunty K, Du H, Fewell G, Fulton L, Fulton R, Graves T, Hou SF, Latrielle P, Leonard S, Mardis E, Maupin R, McPherson J, Miner T, Nash W, Nguyen C, Ozersky P, Pepin K, Rock S, Rohlfing T, Scott K, Schultz B, Strong C, Tin-Wollam A, Yang SP, Waterston RH, Wilson RK, Rozen S, Page DC. The male-specific region of the human Y chromosome is a mosaic of discrete sequence classes Nature 2003; 423: 825-837.

[4] Maxson SC. Searching for candidate genes with effects on an agonistic behavior, offense, in mice Behav Genet 1996; 26: 471-476.

[5] Xu J, Burgoyne PS, Arnold AP. Sex differences in sex chromosome gene expression in mouse brain Hum Mol Genet 2002; 11: 1409-1419.

[6] De Vries GJ, Rissman EF, Simerly RB, Yang LY, Scordalakes EM, Auger CJ, Swain A, Lovell-Badge R, Burgoyne PS, Arnold AP. A model system for study of sex chromosome effects on sexually dimorphic neural and behavioral traits J Neurosci 2002; 22: 9005-9014.

[7] Herrmann JL, Abarbanell AM, Weil BR, Manukyan MC, Poynter JA, Wang Y, Coffey AC, Meldrum DR. Gender dimorphisms in progenitor and stem cell function in cardiovascular disease J Cardiovasc Transl Res 2010; 3: 103-113.

[8] Li WW, Penderis J, Zhao C, Schumacher M, Franklin RJ. Females remyelinate more efficiently than males following demyelination in the aged but not young adult CNS Exp Neurol 2006; 202: 250-254.

[9] Matsumoto T, Kubo S, Meszaros LB, Corsi KA, Cooper GM, Li G, Usas A, Osawa A, Fu FH, Huard J. The influence of sex on the chondrogenic potential of muscle-derived stem cells: implications for cartilage regeneration and repair Arthritis Rheum 2008; 58: 3809-3819.

[10] Payne KA, Didiano DM, Chu CR. Donor sex and age influence the chondrogenic potential of human femoral bone marrow stem cells Osteoarthritis Cartilage 2010; 18: 705-713.

[11] Yuan J, Yu JX, Ge J. Sexual dimorphism on the neurogenic potential of rhesus monkeys mesenchymal stem cells Biochem Biophys Res Commun 2010; 396: 394-400.

[12] Deasy BM, Lu A, Tebbets JC, Feduska JM, Schugar RC, Pollett JB, Sun B, Urish KL, Gharaibeh BM, Cao B, Rubin RT, Huard J. A role for cell sex in stem cell-mediated skeletal muscle regeneration: female cells have higher muscle regeneration efficiency J Cell Biol 2007; 177: 73-86.

[13] Deasy BM, Schugar RC, Huard J. Sex differences in muscle-derived stem cells and skeletal muscle Crit Rev Eukaryot Gene Expr 2008; 18: 173-188.

[14] Waldron J, McCourty AM, Lecanu L. Aging differentially affects male and female neural stem cell neurogenic properties Stem Cells Cloning Adv Appl 2010; 3: 119-127.

[15] Waldron J, McCourty AM, Lecanu L. Neural stem cell sex dimorphism in aromatase (CYP19) expression: a basis for differential neural fate. Stem Cells Cloning Adv Appl 2011; 3: 175-182.

[16] Cohen MA, Itsykson P, Reubinoff BE. Neural differentiation of human ES cells Curr Protoc Cell Biol 2007; Chapter 23: Unit.

[17] Joannides AJ, Chandran S. Human embryonic stem cells: an experimental and therapeutic resource for neurological disease J Neurol Sci 2008; 265: 84-88.

[18] Bachoud-Levi AC, Gaura V, Brugieres P, Lefaucheur JP, Boisse MF, Maison P, Baudic S, Ribeiro MJ, Bourdet C, Remy P, Cesaro P, Hantraye P, Peschanski M. Effect of fetal neural transplants in patients with Huntington's disease 6 years after surgery: a long-term follow-up study Lancet Neurol 2006; 5: 303-309.

[19] Capetian P, Knoth R, Maciaczyk J, Pantazis G, Ditter M, Bokla L, Landwehrmeyer GB, Volk B, Nikkhah G. Histological findings on fetal striatal grafts in a Huntington's disease patient early after transplantation Neuroscience 2009; 160: 661-675.

[20] Andres RH, Horie N, Slikker W, Keren-Gill H, Zhan K, Sun G, Manley NC, Pereira MP, Sheikh LA, McMillan EL, Schaar BT, Svendsen CN, Bliss TM, Steinberg GK. Human neural stem cells enhance structural plasticity and axonal transport in the ischaemic brain Brain 2011; 134: 1777-1789.

[21] Einstein O, Friedman-Levi Y, Grigoriadis N, Ben-Hur T. Transplanted neural precursors enhance host brain-derived myelin regeneration J Neurosci 2009; 29: 15694-15702.

[22] Ho AL, Keshavarzi S, Levy ML. Exploitation of genetically modified neural stem cells for neurological disease Adv Exp Med Biol 2010; 671: 74-92.

[23] Koch P, Kokaia Z, Lindvall O, Brustle O. Emerging concepts in neural stem cell research: autologous repair and cell-based disease modelling Lancet Neurol 2009; 8: 819-829.

[24] Lee IS, Jung K, Kim M, Park KI. Neural stem cells: properties and therapeutic potentials for hypoxic-ischemic brain injury in newborn infants Pediatr Int 2010; 52: 855-865.

[25] Mack GS. ReNeuron and StemCells get green light for neural stem cell trials Nat Biotechnol 2011; 29: 95-97.

[26] Carr AJ, Vugler AA, Hikita ST, Lawrence JM, Gias C, Chen LL, Buchholz DE, Ahmado A, Semo M, Smart MJ, Hasan S, da CL, Johnson LV, Clegg DO, Coffey PJ. Protective effects of human iPS-derived retinal pigment epithelium cell transplantation in the retinal dystrophic rat PLoS One 2009; 4: e8152.

[27] Chen LW, Kuang F, Wei LC, Ding YX, Yung KK, Chan YS. Potential application of induced pluripotent stem cells in cell replacement therapy for Parkinson's disease CNS Neurol Disord Drug Targets 2011; 10: 449-458.

[28] Okano H. Strategies toward CNS-regeneration using induced pluripotent stem cells Genome Inform 2009; 23: 217-220.

[29] Tsuji O, Miura K, Okada Y, Fujiyoshi K, Mukaino M, Nagoshi N, Kitamura K, Kumagai G, Nishino M, Tomisato S, Higashi H, Nagai T, Katoh H, Kohda K, Matsuzaki Y, Yuzaki M, Ikeda E, Toyama Y, Nakamura M, Yamanaka S, Okano H. Therapeutic potential of appropriately evaluated safe-induced pluripotent stem cells for spinal cord injury Proc Natl Acad Sci U S A 2010; 107: 12704-12709.

[30] Arnhold S, Absenger Y, Klein H, Addicks K, Schraermeyer U. Transplantation of bone marrow-derived mesenchymal stem cells rescue photoreceptor cells in the dystrophic retina of the rhodopsin knockout mouse Graefes Arch Clin Exp Ophthalmol 2007; 245: 414-422.

[31] Inoue Y, Iriyama A, Ueno S, Takahashi H, Kondo M, Tamaki Y, Araie M, Yanagi Y. Subretinal transplantation of bone marrow mesenchymal stem cells delays retinal degeneration in the RCS rat model of retinal degeneration Exp Eye Res 2007; 85: 234-241.

[32] Kociok N. Can the injection of the patient's own bone marrow-derived stem cells preserve cone vision in retinitis pigmentosa and other diseases of the eye? Graefes Arch Clin Exp Ophthalmol 2005; 243: 187-188.

[33] Lu B, Wang S, Girman S, McGill T, Ragaglia V, Lund R. Human adult bone marrow-derived somatic cells rescue vision in a rodent model of retinal degeneration Exp Eye Res 2010; 91: 449-455.

[34] Matsuse D, Kitada M, Ogura F, Wakao S, Kohama M, Kira JI, Tabata Y, Dezawa M. Combined Transplantation of Bone Marrow Stromal Cell-Derived Neural Progenitor Cells with a Collagen Sponge and Basic Fibroblast Growth Factor Releasing Microspheres Enhances Recovery After Cerebral Ischemia in Rats Tissue Eng Part A 2011.

[35] Smith LE. Bone marrow-derived stem cells preserve cone vision in retinitis pigmentosa J Clin Invest 2004; 114: 755-757.

[36] Suarez-Monteagudo C, Hernandez-Ramirez P, Alvarez-Gonzalez L, Garcia-Maeso I, Cuetara-Bernal K, Castillo-Diaz L, Bringas-Vega ML, Martinez-Aching G, Morales-

Chacon LM, Baez-Martin MM, Sanchez-Catasus C, Carballo-Barreda M, Rodriguez-Rojas R, Gomez-Fernandez L, Alberti-Amador E, Macias-Abraham C, Balea ED, Rosales LC, Del Valle PL, Ferrer BB, Gonzalez RM, Bergado JA. Autologous bone marrow stem cell neurotransplantation in stroke patients. An open study Restor Neurol Neurosci 2009; 27: 151-161.

[37] Joe AW, Gregory-Evans K. Mesenchymal stem cells and potential applications in treating ocular disease Curr Eye Res 2010; 35: 941-952.

[38] Joyce N, Annett G, Wirthlin L, Olson S, Bauer G, Nolta JA. Mesenchymal stem cells for the treatment of neurodegenerative disease Regen Med 2010; 5: 933-946.

[39] Ali H, Bahbahani H. Umbilical cord blood stem cells - potential therapeutic tool for neural injuries and disorders Acta Neurobiol Exp (Wars) 2010; 70: 316-324.

[40] Domanska-Janik K, Habich A, Sarnowska A, Janowski M. Neural commitment of cord blood stem cells (HUCB-NSC/NP): therapeutic perspectives Acta Neurobiol Exp (Wars) 2006; 66: 279-291.

[41] El-Badri NS. Cord blood stem cells: a fountain of youth for the brain Stem Cells Dev 2006; 15: 293-294.

[42] Erba P, Mantovani C, Kalbermatten DF, Pierer G, Terenghi G, Kingham PJ. Regeneration potential and survival of transplanted undifferentiated adipose tissue-derived stem cells in peripheral nerve conduits J Plast Reconstr Aesthet Surg 2010; 63: e811-e817.

[43] Erba P, Terenghi G, Kingham PJ. Neural differentiation and therapeutic potential of adipose tissue derived stem cells Curr Stem Cell Res Ther 2010; 5: 153-160.

[44] Krabbe C, Zimmer J, Meyer M. Neural transdifferentiation of mesenchymal stem cells--a critical review APMIS 2005; 113: 831-844.

[45] Liu Q, Guan L, Huang B, Li WH, Su Q, Yu M, Xu XP, Luo T, Lin SC, Sun X, Chen M, Chen XG. Adult peripheral blood mononuclear cells transdifferentiate in vitro and integrate into the retina in vivo Cell Biol Int 2011.

[46] Mohammad-Gharibani P, Tiraihi T, Arabkheradmand J. In vitro transdifferentiation of bone marrow stromal cells into GABAergic-like neurons Iran Biomed J 2009; 13: 137-143.

[47] Naghdi M, Tiraihi T, Namin SA, Arabkheradmand J. Transdifferentiation of bone marrow stromal cells into cholinergic neuronal phenotype: a potential source for cell therapy in spinal cord injury Cytotherapy 2009; 11: 137-152.

[48] Roybon L, Ma Z, Asztely F, Fosum A, Jacobsen SE, Brundin P, Li JY. Failure of transdifferentiation of adult hematopoietic stem cells into neurons Stem Cells 2006; 24: 1594-1604.

[49] Zurita M, Bonilla C, Otero L, Aguayo C, Vaquero J. Neural transdifferentiation of bone marrow stromal cells obtained by chemical agents is a short-time reversible phenomenon Neurosci Res 2008; 60: 275-280.

[50] Amariglio N, Hirshberg A, Scheithauer BW, Cohen Y, Loewenthal R, Trakhtenbrot L, Paz N, Koren-Michowitz M, Waldman D, Leider-Trejo L, Toren A, Constantini S, Rechavi G. Donor-derived brain tumor following neural stem cell transplantation in an ataxia telangiectasia patient PLoS Med 2009; 6: e1000029.

[51] Amariglio N, Rechavi G. On the origin of glioneural neoplasms after neural cell transplantation Nat Med 2010; 16: 157-158.

[52] Ben-David U, Benvenisty N. The tumorigenicity of human embryonic and induced pluripotent stem cells Nat Rev Cancer 2011; 11: 268-277.

[53] Saiz A, Graus F. Neurologic complications of hematopoietic cell transplantation Semin Neurol 2010; 30: 287-295.

[54] Capetian P, Dobrossy M, Winkler C, Prinz M, Nikkhah G. To be or not to be accepted: the role of immunogenicity of neural stem cells following transplantation into the brain in animal and human studies Semin Immunopathol 2011.

[55] Ubiali F, Nava S, Nessi V, Frigerio S, Parati E, Bernasconi P, Mantegazza R, Baggi F. Allorecognition of human neural stem cells by peripheral blood lymphocytes despite low expression of MHC molecules: role of TGF-beta in modulating proliferation Int Immunol 2007; 19: 1063-1074.

[56] Zheng XS, Yang XF, Liu WG, Pan DS, Hu WW, Li G. Transplantation of neural stem cells into the traumatized brain induces lymphocyte infiltration Brain Inj 2007; 21: 275-278.

[57] Al NF, Wennersten A, Holmin S, Meijer X, Wahlberg L, Mathiesen T. MHC expression after human neural stem cell transplantation to brain contused rats Neuroreport 2004; 15: 1871-1875.

[58] Barzilay R, Kan I, Ben-Zur T, Bulvik S, Melamed E, Offen D. Induction of human mesenchymal stem cells into dopamine-producing cells with different differentiation protocols Stem Cells Dev 2008; 17: 547-554.

[59] Fricker-Gates RA, Gates MA. Stem cell-derived dopamine neurons for brain repair in Parkinson's disease Regen Med 2010; 5: 267-278.

[60] Shim JW, Park CH, Bae YC, Bae JY, Chung S, Chang MY, Koh HC, Lee HS, Hwang SJ, Lee KH, Lee YS, Choi CY, Lee SH. Generation of functional dopamine neurons from neural precursor cells isolated from the subventricular zone and white matter of the adult rat brain using Nurr1 overexpression Stem Cells 2007; 25: 1252-1262.

[61] Iacovitti L, Donaldson AE, Marshall CE, Suon S, Yang M. A protocol for the differentiation of human embryonic stem cells into dopaminergic neurons using only chemi-

cally defined human additives: Studies in vitro and in vivo Brain Res 2007; 1127: 19-25.

[62] Liste I, Garcia-Garcia E, Martinez-Serrano A. The generation of dopaminergic neurons by human neural stem cells is enhanced by Bcl-XL, both in vitro and in vivo J Neurosci 2004; 24: 10786-10795.

[63] Bosch M, Pineda JR, Sunol C, Petriz J, Cattaneo E, Alberch J, Canals JM. Induction of GABAergic phenotype in a neural stem cell line for transplantation in an excitotoxic model of Huntington's disease Exp Neurol 2004; 190: 42-58.

[64] Givogri MI, Bottai D, Zhu HL, Fasano S, Lamorte G, Brambilla R, Vescovi A, Wrabetz L, Bongarzone E. Multipotential neural precursors transplanted into the metachromatic leukodystrophy brain fail to generate oligodendrocytes but contribute to limit brain dysfunction Dev Neurosci 2008; 30: 340-357.

[65] Thomas MG, Stone L, Evill L, Ong S, Ziman M, Hool L. Bone marrow stromal cells as replacement cells for Parkinson's disease: generation of an anatomical but not functional neuronal phenotype Transl Res 2011; 157: 56-63.

[66] Brundin P, Barker RA, Parmar M. Neural grafting in Parkinson's disease Problems and possibilities Prog Brain Res 2010; 184: 265-294.

[67] Lecanu L. Sex, the underestimated potential determining factor in brain tissue repair strategy Stem Cells Dev 2011.

[68] Meyer AK, Maisel M, Hermann A, Stirl K, Storch A. Restorative approaches in Parkinson's Disease: which cell type wins the race? J Neurol Sci 2010; 289: 93-103.

[69] Waldron J, Lecanu L. Age and sex-difference in neural stem cells transplantation outcome: a descriptive study in rats. Stem Cells Cloning Adv Appl 2011; 4: 25-37.

[70] Arias-Carrion O, Hernandez-Lopez S, Ibanez-Sandoval O, Bargas J, Hernandez-Cruz A, Drucker-Colin R. Neuronal precursors within the adult rat subventricular zone differentiate into dopaminergic neurons after substantia nigra lesion and chromaffin cell transplant J Neurosci Res 2006; 84: 1425-1437.

[71] Tonellato PJ, Crawford JM, Boguski MS, Saffitz JE. A national agenda for the future of pathology in personalized medicine: report of the proceedings of a meeting at the Banbury Conference Center on genome-era pathology, precision diagnostics, and preemptive care: a stakeholder summit Am J Clin Pathol 2011; 135: 668-672.

[72] Callaway E. Stem-cell progress Nature 2010; 468: 746.

[73] Helmy KY, Patel SA, Silverio K, Pliner L, Rameshwar P. Stem cells and regenerative medicine: accomplishments to date and future promise Ther Deliv 2010; 1: 693-705.

[74] Patel SA, King CC, Lim PK, Habiba U, Dave M, Porecha R, Rameshwar P. Personalizing Stem Cell Research and Therapy: The Arduous Road Ahead or Missed Opportunity? Curr Pharmacogenomics Person Med 2010; 8: 25-36.

[75] Lane EL, Bjorklund A, Dunnett SB, Winkler C. Neural grafting in Parkinson's disease unraveling the mechanisms underlying graft-induced dyskinesia Prog Brain Res 2010; 184: 295-309.

[76] Deda H, Inci MC, Kurekci AE, Sav A, Kayihan K, Ozgun E, Ustunsoy GE, Kocabay S. Treatment of amyotrophic lateral sclerosis patients by autologous bone marrow-derived hematopoietic stem cell transplantation: a 1-year follow-up Cytotherapy 2009; 11: 18-25.

[77] Fornai F, Meininger V, Silani V. Future therapeutical strategies dictated by pre-clinical evidence in ALS Arch Ital Biol 2011; 149: 169-174.

[78] Martinez HR, Gonzalez-Garza MT, Moreno-Cuevas JE, Caro E, Gutierrez-Jimenez E, Segura JJ. Stem-cell transplantation into the frontal motor cortex in amyotrophic lateral sclerosis patients Cytotherapy 2009; 11: 26-34.

[79] Mazzini L, Ferrero I, Luparello V, Rustichelli D, Gunetti M, Mareschi K, Testa L, Stecco A, Tarletti R, Miglioretti M, Fava E, Nasuelli N, Cisari C, Massara M, Vercelli R, Oggioni GD, Carriero A, Cantello R, Monaco F, Fagioli F. Mesenchymal stem cell transplantation in amyotrophic lateral sclerosis: A Phase I clinical trial Exp Neurol 2010; 223: 229-237.

[80] Silani V, Calzarossa C, Cova L, Ticozzi N. Stem cells in amyotrophic lateral sclerosis: motor neuron protection or replacement? CNS Neurol Disord Drug Targets 2010; 9: 314-324.

[81] Gamez J, Carmona F, Raguer N, Ferrer-Sancho J, Martin-Henao GA, Marti-Beltran S, Badia M, Gratacos M, Rodriguez-Gonzalez E, Seoane JL, Pallero-Castillo M, Burgos R, Puiggros C, Pasarin A, Bori-Fortuny I. Cellular transplants in amyotrophic lateral sclerosis patients: an observational study Cytotherapy 2010; 12: 669-677.

[82] Bachoud-Levi AC, Remy P, Nguyen JP, Brugieres P, Lefaucheur JP, Bourdet C, Baudic S, Gaura V, Maison P, Haddad B, Boisse MF, Grandmougin T, Jeny R, Bartolomeo P, Dalla BG, Degos JD, Lisovoski F, Ergis AM, Pailhous E, Cesaro P, Hantraye P, Peschanski M. Motor and cognitive improvements in patients with Huntington's disease after neural transplantation Lancet 2000; 356: 1975-1979.

[83] Cicchetti F, Soulet D, Freeman TB. Neuronal degeneration in striatal transplants and Huntington's disease: potential mechanisms and clinical implications Brain 2011; 134: 641-652.

[84] Keene CD, Sonnen JA, Swanson PD, Kopyov O, Leverenz JB, Bird TD, Montine TJ. Neural transplantation in Huntington disease: long-term grafts in two patients Neurology 2007; 68: 2093-2098.

[85] Lee HJ, Lee JK, Lee H, Shin JW, Carter JE, Sakamoto T, Jin HK, Bae JS. The therapeutic potential of human umbilical cord blood-derived mesenchymal stem cells in Alzheimer's disease Neurosci Lett 2010; 481: 30-35.

[86] Taupin P. Neurogenesis, NSCs, pathogenesis and therapies for Alzheimer's disease Front Biosci (Schol Ed) 2011; 3: 178-190.

[87] Zhongling F, Gang Z, Lei Y. Neural stem cells and Alzheimer's disease: challenges and hope Am J Alzheimers Dis Other Demen 2009; 24: 52-57.

[88] Scolding N. Adult stem cells and multiple sclerosis Cell Prolif 2011; 44 Suppl 1: 35-38.

[89] Uccelli A, Laroni A, Freedman MS. Mesenchymal stem cells for the treatment of multiple sclerosis and other neurological diseases Lancet Neurol 2011; 10: 649-656.

[90] Zeher M, Papp G, Szodoray P. Autologous haemopoietic stem cell transplantation for autoimmune diseases Expert Opin Biol Ther 2011.

[91] Bretzner F, Gilbert F, Baylis F, Brownstone RM. Target populations for first-in-human embryonic stem cell research in spinal cord injury Cell Stem Cell 2011; 8: 468-475.

[92] Dasari VR, Spomar DG, Gondi CS, Sloffer CA, Saving KL, Gujrati M, Rao JS, Dinh DH. Axonal remyelination by cord blood stem cells after spinal cord injury J Neurotrauma 2007; 24: 391-410.

[93] Adler R. Curing blindness with stem cells: hope, reality, and challenges Adv Exp Med Biol 2008; 613: 3-20.

[94] Freed CR, Greene PE, Breeze RE, Tsai WY, DuMouchel W, Kao R, Dillon S, Winfield H, Culver S, Trojanowski JQ, Eidelberg D, Fahn S. Transplantation of embryonic dopamine neurons for severe Parkinson's disease N Engl J Med 2001; 344: 710-719.

[95] Hauser RA, Freeman TB, Snow BJ, Nauert M, Gauger L, Kordower JH, Olanow CW. Long-term evaluation of bilateral fetal nigral transplantation in Parkinson disease Arch Neurol 1999; 56: 179-187.

[96] Ma Y, Tang C, Chaly T, Greene P, Breeze R, Fahn S, Freed C, Dhawan V, Eidelberg D. Dopamine cell implantation in Parkinson's disease: long-term clinical and (18)F-FDOPA PET outcomes J Nucl Med 2010; 51: 7-15.

[97] Olanow CW, Goetz CG, Kordower JH, Stoessl AJ, Sossi V, Brin MF, Shannon KM, Nauert GM, Perl DP, Godbold J, Freeman TB. A double-blind controlled trial of bilateral fetal nigral transplantation in Parkinson's disease Ann Neurol 2003; 54: 403-414.

[98] Lindvall O, Rehncrona S, Brundin P, Gustavii B, Astedt B, Widner H, Lindholm T, Bjorklund A, Leenders KL, Rothwell JC, Frackowiak R, Marsden D, Johnels B, Steg G, Freedman R, Hoffer BJ, Seiger A, Bygdeman M, Stromberg I, Olson L. Human fetal dopamine neurons grafted into the striatum in two patients with severe Parkinson's disease. A detailed account of methodology and a 6-month follow-up Arch Neurol 1989; 46: 615-631.

[99] Lindvall O, Brundin P, Widner H, Rehncrona S, Gustavii B, Frackowiak R, Leenders KL, Sawle G, Rothwell JC, Marsden CD, . Grafts of fetal dopamine neurons survive and improve motor function in Parkinson's disease Science 1990; 247: 574-577.

[100] Dobrossy M, Busse M, Piroth T, Rosser A, Dunnett S, Nikkhah G. Neurorehabilitation with neural transplantation Neurorehabil Neural Repair 2010; 24: 692-701.

[101] Mills KR. The natural history of central motor abnormalities in amyotrophic lateral sclerosis Brain 2003; 126: 2558-2566.

[102] Gaura V, Bachoud-Levi AC, Ribeiro MJ, Nguyen JP, Frouin V, Baudic S, Brugieres P, Mangin JF, Boisse MF, Palfi S, Cesaro P, Samson Y, Hantraye P, Peschanski M, Remy P. Striatal neural grafting improves cortical metabolism in Huntington's disease patients Brain 2004; 127: 65-72.

[103] Grinnemo KH, Sylven C, Hovatta O, Dellgren G, Corbascio M. Immunogenicity of human embryonic stem cells Cell Tissue Res 2008; 331: 67-78.

[104] Sobani ZA, Quadri SA, Enam SA. Stem cells for spinal cord regeneration: Current status Surg Neurol Int 2010; 1: 93.

[105] Zurita M, Otero L, Aguayo C, Bonilla C, Ferreira E, Parajon A, Vaquero J. Cell therapy for spinal cord repair: optimization of biologic scaffolds for survival and neural differentiation of human bone marrow stromal cells Cytotherapy 2010; 12: 522-537.

[106] Shen CC, Lin CH, Yang YC, Chiao MT, Cheng WY, Ko JL. Intravenous implanted neural stem cells migrate to injury site, reduce infarct volume, and improve behavior after cerebral ischemia Curr Neurovasc Res 2010; 7: 167-179.

[107] Aoki H, Hara A, Niwa M, Yamada Y, Kunisada T. In vitro and in vivo differentiation of human embryonic stem cells into retina-like organs and comparison with that from mouse pluripotent epiblast stem cells Dev Dyn 2009; 238: 2266-2279.

[108] Arnhold S, Klein H, Semkova I, Addicks K, Schraermeyer U. Neurally selected embryonic stem cells induce tumor formation after long-term survival following engraftment into the subretinal space Invest Ophthalmol Vis Sci 2004; 45: 4251-4255.

[109] Chen M, Chen Q, Sun X, Shen W, Liu B, Zhong X, Leng Y, Li C, Zhang W, Chai F, Huang B, Gao Q, Xiang AP, Zhuo Y, Ge J. Generation of retinal ganglion-like cells from reprogrammed mouse fibroblasts Invest Ophthalmol Vis Sci 2010; 51: 5970-5978.

[110] Francis PJ, Wang S, Zhang Y, Brown A, Hwang T, McFarland TJ, Jeffrey BG, Lu B, Wright L, Appukuttan B, Wilson DJ, Stout JT, Neuringer M, Gamm DM, Lund RD. Subretinal transplantation of forebrain progenitor cells in nonhuman primates: survival and intact retinal function Invest Ophthalmol Vis Sci 2009; 50: 3425-3431.

[111] Klassen H, Schwartz PH, Ziaeian B, Nethercott H, Young MJ, Bragadottir R, Tullis GE, Warfvinge K, Narfstrom K. Neural precursors isolated from the developing cat

brain show retinal integration following transplantation to the retina of the dystrophic cat Vet Ophthalmol 2007; 10: 245-253.

[112] Levkovitch-Verbin H, Sadan O, Vander S, Rosner M, Barhum Y, Melamed E, Offen D, Melamed S. Intravitreal injections of neurotrophic factors secreting mesenchymal stem cells are neuroprotective in rat eyes following optic nerve transection Invest Ophthalmol Vis Sci 2010; 51: 6394-6400.

[113] Lund RD, Ono SJ, Keegan DJ, Lawrence JM. Retinal transplantation: progress and problems in clinical application J Leukoc Biol 2003; 74: 151-160.

[114] Seiler MJ, Aramant RB, Thomas BB, Peng Q, Sadda SR, Keirstead HS. Visual restoration and transplant connectivity in degenerate rats implanted with retinal progenitor sheets Eur J Neurosci 2010; 31: 508-520.

[115] Wang DY, Wu SC, Lin SP, Hsiao SH, Chung TW, Huang YY. Evaluation of transdifferentiation from mesenchymal stem cells to neuron-like cells using microfluidic patterned co-culture system Biomed Microdevices 2011; 13: 517-526.

[116] Wang S, Girman S, Lu B, Bischoff N, Holmes T, Shearer R, Wright LS, Svendsen CN, Gamm DM, Lund RD. Long-term vision rescue by human neural progenitors in a rat model of photoreceptor degeneration Invest Ophthalmol Vis Sci 2008; 49: 3201-3206.

[117] Yang Z, Li K, Yan X, Dong F, Zhao C. Amelioration of diabetic retinopathy by engrafted human adipose-derived mesenchymal stem cells in streptozotocin diabetic rats Graefes Arch Clin Exp Ophthalmol 2010; 248: 1415-1422.

[118] Maruska KP, Fernald RD. Steroid receptor expression in the fish inner ear varies with sex, social status, and reproductive state BMC Neurosci 2010; 11: 58.

[119] Yamada S, Noguchi D, Ito H, Yamanouchi K. Sex and regional differences in decrease of estrogen receptor alpha-immunoreactive cells by estrogen in rat hypothalamus and midbrain Neurosci Lett 2009; 463: 135-139.

[120] Zhang JQ, Cai WQ, Zhou DS, Su BY. Distribution and differences of estrogen receptor beta immunoreactivity in the brain of adult male and female rats Brain Res 2002; 935: 73-80.

[121] Hisasue S, Seney ML, Immerman E, Forger NG. Control of cell number in the bed nucleus of the stria terminalis of mice: role of testosterone metabolites and estrogen receptor subtypes J Sex Med 2010; 7: 1401-1409.

[122] Kudwa AE, Bodo C, Gustafsson JA, Rissman EF. A previously uncharacterized role for estrogen receptor beta: defeminization of male brain and behavior Proc Natl Acad Sci U S A 2005; 102: 4608-4612.

[123] Kudwa AE, Michopoulos V, Gatewood JD, Rissman EF. Roles of estrogen receptors alpha and beta in differentiation of mouse sexual behavior Neuroscience 2006; 138: 921-928.

[124] Brannvall K, Korhonen L, Lindholm D. Estrogen-receptor-dependent regulation of neural stem cell proliferation and differentiation Mol Cell Neurosci 2002; 21: 512-520.

[125] Fowler CD, Johnson F, Wang Z. Estrogen regulation of cell proliferation and distribution of estrogen receptor-alpha in the brains of adult female prairie and meadow voles J Comp Neurol 2005; 489: 166-179.

[126] Kishi Y, Takahashi J, Koyanagi M, Morizane A, Okamoto Y, Horiguchi S, Tashiro K, Honjo T, Fujii S, Hashimoto N. Estrogen promotes differentiation and survival of dopaminergic neurons derived from human neural stem cells J Neurosci Res 2005; 79: 279-286.

[127] Martinez-Cerdeno V, Noctor SC, Kriegstein AR. Estradiol stimulates progenitor cell division in the ventricular and subventricular zones of the embryonic neocortex Eur J Neurosci 2006; 24: 3475-3488.

[128] Suzuki S, Gerhold LM, Bottner M, Rau SW, Dela CC, Yang E, Zhu H, Yu J, Cashion AB, Kindy MS, Merchenthaler I, Gage FH, Wise PM. Estradiol enhances neurogenesis following ischemic stroke through estrogen receptors alpha and beta J Comp Neurol 2007; 500: 1064-1075.

[129] Wang JM, Liu L, Brinton RD. Estradiol-17beta-induced human neural progenitor cell proliferation is mediated by an estrogen receptor beta-phosphorylated extracellularly regulated kinase pathway Endocrinology 2008; 149: 208-218.

[130] Okada M, Murase K, Makino A, Nakajima M, Kaku T, Furukawa S, Furukawa Y. Effects of estrogens on proliferation and differentiation of neural stem/progenitor cells Biomed Res 2008; 29: 163-170.

[131] Okada M, Makino A, Nakajima M, Okuyama S, Furukawa S, Furukawa Y. Estrogen Stimulates Proliferation and Differentiation of Neural Stem/Progenitor Cells through Different Signal Transduction Pathways Int J Mol Sci 2010; 11: 4114-4123.

[132] Sun H, Wang H, Hu S. Effects of estrogen on diverse stem cells and relevant intracellular mechanisms Sci China Life Sci 2010; 53: 542-547.

[133] Leskela HV, Olkku A, Lehtonen S, Mahonen A, Koivunen J, Turpeinen M, Uusitalo J, Pelkonen O, Kangas L, Selander K, Lehenkari P. Estrogen receptor alpha genotype confers interindividual variability of response to estrogen and testosterone in mesenchymal-stem-cell-derived osteoblasts Bone 2006; 39: 1026-1034.

[134] Barker JM, Galea LA. Repeated estradiol administration alters different aspects of neurogenesis and cell death in the hippocampus of female, but not male, rats Neuroscience 2008; 152: 888-902.

[135] Waldron J, McCourty AM, Lecanu L. Neural stem cell sex dimorphism in aromatase (CYP19) expression: a basis for differential neural fate. Stem Cells Cloning Adv Appl 2010; 3: 175-182.

[136] Martins AH, Alves JM, Trujillo CA, Schwindt TT, Barnabe GF, Motta FL, Guimaraes AO, Casarini DE, Mello LE, Pesquero JB, Ulrich H. Kinin-B2 receptor expression and activity during differentiation of embryonic rat neurospheres Cytometry A 2008; 73: 361-368.

[137] Sillat T, Pollanen R, Lopes JR, Porola P, Ma G, Korhonen M, Konttinen YT. Intracrine androgenic apparatus in human bone marrow stromal cells J Cell Mol Med 2009; 13: 3296-3302.

[138] Morgan DE, Crittenden SL, Kimble J. The C. elegans adult male germline: stem cells and sexual dimorphism Dev Biol 2010; 346: 204-214.

[139] Limke TL, Rao MS. Neural stem cells in aging and disease J Cell Mol Med 2002; 6: 475-496.

[140] Limke TL, Rao MS. Neural stem cell therapy in the aging brain: pitfalls and possibilities J Hematother Stem Cell Res 2003; 12: 615-623.

[141] Limke TL, Cai J, Miura T, Rao MS, Mattson MP. Distinguishing features of progenitor cells in the late embryonic and adult hippocampus Dev Neurosci 2003; 25: 257-272.

[142] Maslov AY, Barone TA, Plunkett RJ, Pruitt SC. Neural stem cell detection, characterization, and age-related changes in the subventricular zone of mice J Neurosci 2004; 24: 1726-1733.

[143] Kuhn HG, Dickinson-Anson H, Gage FH. Neurogenesis in the dentate gyrus of the adult rat: age-related decrease of neuronal progenitor proliferation J Neurosci 1996; 16: 2027-2033.

[144] Sato K, Hayashi T, Sasaki C, Iwai M, Li F, Manabe Y, Seki T, Abe K. Temporal and spatial differences of PSA-NCAM expression between young-adult and aged rats in normal and ischemic brains Brain Res 2001; 922: 135-139.

[145] Kudwa AE, Gustafsson JA, Rissman EF. Estrogen receptor beta modulates estradiol induction of progestin receptor immunoreactivity in male, but not in female, mouse medial preoptic area Endocrinology 2004; 145: 4500-4506.

[146] Kudwa AE, Harada N, Honda SI, Rissman EF. Regulation of progestin receptors in medial amygdala: estradiol, phytoestrogens and sex Physiol Behav 2009; 97: 146-150.

[147] Mayack SR, Shadrach JL, Kim FS, Wagers AJ. Systemic signals regulate ageing and rejuvenation of blood stem cell niches Nature 2010; 463: 495-500.

[148] Niikura Y, Niikura T, Wang N, Satirapod C, Tilly JL. Systemic signals in aged males exert potent rejuvenating effects on the ovarian follicle reserve in mammalian females Aging (Albany NY) 2010; 2: 999-1003.

Chapter 4

Potentiality of Very Small Embryonic-Like Stem Cells to Repair Myocardial Infarction

X. Gu, J. Gu, J.B. Sun, Y.X. Gu, L. Sun, Y. Zhang, Y. Cheng, Z.Y. Bao, F. Hang, X.M. Lu, R.S. Zhang and B.C. Li

Additional information is available at the end of the chapter

http://dx.doi.org/10.5772/54942

1. Introduction

Heart failure (HF) after myocardial infarction (MI) or the ischemic myocardiopathy (ICM) continues to be the most prevalent cause of morbidity and mortality worldwide. MI results in myocardial necrosis, scar formation, ventricular remodeling and eventually HF or death. Faced with the end stage of ICM, the most of present therapy protocols could only slow worsening of HF. Cardiac resynchronization therapy may be definite therapeutic effect to those cases with HF and complete left bundle-branch block. Heart transplantation can more efficiently improve the cardiac status, but, limited donor supply and organ rejection confine its widespread use. As a result, a significant proportion of survivors with ICM will still develop HF and have briefer life-span. Pathologically, HF and myocardial remodeling aggratate each other and a core pathogenic factor of ICM is loss of massive cardiomyocytes [1]. In fact, the myocardium itself posseses little capacity for self-regeneration. Although there are still considerable dispute in the clinical therapeutic effect based on stem cells (SCs), the positive results obtaned in the repair of damaged myocardium indicated it has become a promising strategy [2-4]. In this regard, an array of SCs types has been identified and applied, including bon marrow-derived mononuclear cell (BM-MNCs) [5] and umbilical cord blood-derived stem cells(UCB-SCs). At the current state of SCs clinical application there is no convincing data showing the superiority of any tissue committed monopotent stem cells (TCSCs), so heterogenous population of BM-MNCs is most often used [6]. Some recent studies have showed pluripotent stem cells (PSCs) are precursors of TCSCs during organ/tissue rejuvenation and a source of these cells in emergency situations when organs are damaged (e.g., MI or stroke). The application of PSCs has showed very encouraging results. PSCs includes induced pluripotent stem cells (iPSCs) using gene transfer [7-8] and very small embryonic-like embryonic/epiblast-like stem cells (VSELs) isolated from the adult tissues or UCB [9]. A rare Sca1+Lin−CD45− SCs population were initially identified, isolated and named as VSELs in adult mice using fluorescence activated cell sorting (FACS) [10]. Although VSELs are currently studied in a lot of laboratories worldwide, the research series of VSELs was mainly contributed by Kucia & Ratajczak and their colleagues. VSELs possess very primitive morphology and express PSCs markers (e.g., Oct4, Nanog, and SSEA-4) as well as the surface phenotype Sca1+/ CD133+Lin−CD45− in mice / humans. As VSELs can be mobilizated into PB following acute MI [11], improve heart function and alleviate cardiac remodeling[12,13], these cells seem to possibly become an optimal seed cells for cardiovascular repair. Recently, employing anti-CD133-conjugated paramagnetic beads followed by staining with Aldefluor has also been proposed for a faster large-scale VSELs isolation [14]. More recent evidences demonstrate that VSELs deposited in adults tissue share several markers with epiblast/germ line cells and play a role in rejuvenation of the TCSCs responsible for tissue regeneration/repair after organ injuries. Even, VSELs with maximum regenerative potential are recommended as the true PSCs in adult tissues, whereas the hematopoietic stem cells (HSCs) and mesenchymal stem cells (MSCs) are actually progenitor SCs arised from asymmetric cell division of VSELs [15]. As a promising candidate, their unique characteristics and potentiality may have very important pathophysiological and therapeutic implications for regenerative medicine including myocardial and endothelial repair.

2. Discovered history of VSELs

Small cells able to differentiate into cells from all three germ layers and called "spore-like stem cells" were isolated from adult mammalian tissues, however, it was not provided for these small SCs how to be purified and for their surface markers how to be expressed in the original paper [16]. Afterwords, small SCs expressing CXC chemokine receptor 4 (CXCR4+) and markers characteristic for embryonic stem cells (ESCs), epiblast stem cells (EPSCs), and primordial stem cells (PGCs) were purified from the murine BM and several adult organs. Based on their small size, presence of PSCs markers, distinct morphology (open-type chromatin, large nucleus, narrow rim of cytoplasm with multiple mitochondria) and ability to differentiate into all three germ layers, including mesoderm-derived cardiomyocytes, these cells were named as VSELs [9].

3. Found source of VSELs

The rare Sca-1+Lin−CD45−SCs population was initially discovered in mice BM 9. Phenotypically similar cells were subsequently identified and purified in murine peripheral blood (PB), fetal liver, brain, retina, kidneys, pancreas, skeletal muscles spleen, and thymus[17]. In humans, VSELs were identified in UCB, PB, BM, and cardiac tissue[18]. VSELs deposited in adult tissues seem to be a reserve pool for TCSCs [19].

4. Morphology of VSELs

The most common shape feature of VSELs is that they possess very primitive morphology and relatively small size. The distinctive morphology of VSELs was confirmed using confocal and transmission electron microscopy [9,10]. Comparison with other populations of cells, murine VSELs (4-6 μm) are smaller than HSCs, MNCs and granulocytes and erythrocytes, but, larger than platelets. Human VSELs (6–8 μm) are larger than murine. At the ultramicrostructural level, they show a very immature morphology, for example, possess a relatively large nucleus surrounded by a narrow rim of cytoplasm, a few mitochondria, scattered ribosomes, small profiles of endoplasmatic reticulum and a few vesicles [20].Recently, the high resolution of ImageStream system (ISS) analysis enables the identification of objects as small as 1 μm in diameter. Employing ISS analysis, murine VSELs are more precisely confirmed as ~3.6 μm in diameter [21].

5. Molecular biology and functional features of VSELs

VSELs not only possess the primitive morphology of early developmental cells but also express typical markers for PSCs

Characteristic markers of VSELs were confirmed using several complementary research tools including flow cytometry (FCM), ISS, direct immunofluorescence staining, confocal microscopy, reverse-transcription polymerase chain reaction (RT-PCR) and etc. Early embryonic markers (Oct-4, Nanog, SSEA-1, Rex1, Dppa3, Rif-1) were demonstrated at the protein /mRNA levels using immunofluorescent staining, ISS and FACS [21,22]. VSELs express SSEA-1 antigen on their surface and Oct-4 in their nuclei. Recent study indicates that the promoters of Oct4 and Nanog contain transcriptionally active chromatin in VSELs excluding the possibility of amplification of pseudogenes [23]. CD133+Lin−CD45− VSELs identified in Human UCB like their murine counterparts, i) highly express telomerase, ii) are diploid, and iii) are viable, as shown by their ability to exclude dye (7-aminoactinomycin D). Moreover, some of the CD133+Lin−CD45− VSELs, which represent only a very small subfraction among UCB Lin−CD45−non-hematopoietic cells, may co-express other stem cell markers, including CD34, CXCR4, and SSEA-4, may contain other stem cell types, including endothelial progenitor cells (EPCs) and mesenchymal stem cells (MSCs), and may be identified based on very small size (FSClow/SSClow) and co-expression of CD133, CD34, and, CXCR4 [23].

Also, there are some differences of the VSELs phenotype between mice and humans. Human VSELs surface markers consist of lin−CD45−CXCR4+, CD133+ and CD34− as confirmed on the mRNA level by RQ-PCR and protein level by IF and ISS, whereas Murine VSELs express Sca-1 antigen [24].

VSELs express chemokine receptor CXCR4 and are absent of pan-hematopoietic marker (CD45−) and hematopoietic lineage markers (Lin−) [9,10,25]. Further study demonstrated that VSELs can not primarily reveal hematopoietic ability immediately from isolation and expansion, but may eventually acquire hematopoietic potential following co-culture in hematopoiesis permissive environment over OP9 stroma feeder layer and reconstitute hematopoiesis in lethally irradiated mice 4–6 weeks after transplantation [26].

Freshly sorted VSELs can be expanded in coculture with C2C12 murine myoblast feeder layer. After 7 days of co-culture, approximately 5–10% of all VSELs form sphere-like clusters consisting of a few hundred cells resembling embryoid bodies (VSEL-derived spheres, VSEL-DSs). VSEL-DSs express placenta-like alkaline phosphatase. Expanded population of VSELs isolated from VSEL-DS retain the pluripotent capacity and have ability to differentiate into all three germ layers, including mesodermal cardiomyocytes, ectodermal neural cells and endodermal pancreatic cells [27].

The differentiation potency was also documented in circulating murine VSELs after injection of G-CSF. Rapidly mobilized VSELs showed up-regulation of PSC markers. These findings support not only the pluripotency of VSELs, but also their tissue repair function[10].Study also showed that VSELs possess diploid DNA. They do not express MHC-1 and human leukocyte antigen-D related (HLA-DR) antigens and are CD90–CD105–CD29–.Moreover, if plated over a C2C12 murine sarcoma cell feeder layer, ~5–10% of purified VSELs are able to form spheres that resemble embryoid bodies [9]. Similar spheres were also formed by VSELs isolated from murine fetal liver, spleen, and thymus [17]. Interestingly, VSELs are somewhat heterogenous developmentally. Although all of VSELs express the Oct-4, some of them express genes that are more closely related to genes expressed by EP-SCs and others to genes expressed by migrating PGCs [28].

In parallel, VSELs exhibits their potential biologic function. Oct-4+ SSEA-4+ SCs harvested from BM via elutriation, has been recently shown to give rise into functional insulin-producing cells in vivo in induced diabetic mice [29].In another report, VSELs purified from rat BM successfully repaired damaged myocardium in a model of MI [30].

Similarly, in vivo exposure to hypoxia in mice elicits chemoattractant a gradients that promote the mobilization of pluripotent very small embryonic-like stem cells from the bone marrow to peripheral blood. VSELs in the BM are the primary source of lung epithelial cells [31-33]. VSELs were also identified in neonatal retina and involved in optic nerve retinal regeneration in a rodent model [34]. VSELs may also play a major role as populations of cells that preferentially give rise to induced pluripotent stem cells (iPSCs) when BM-derived stromal cells are induced to pluripotency by genetic manipulation [35].In particular, VSELs derived by parthenogenesis have also been identified successfully [36].

6. Isolation strategies of VSELs

Isolation of VSLEs using FACS is dependent on gating strategy based on their small size, expression of PSC (Oct4, Nanog, and SSEA-4), surface markers (CXCR4, CD133 /Sca-1,

CD34) and absence of hematopoietic lineage markers (lin, CD45). Briefly, the first step is the lysis of red blood cells to obtain the fraction of nucleated cells. Erythrocyte lysis buffer is used instead of Ficoll centrifugation because the latter might deplete the population of very small cells [14]. Subsequently, cells are stained and sorted with antibodies against Sca-1 (murine VSELs) or CD133 (human VSELs), pan-hematopoietic antigen (CD45), hematopoietic lineages markers (lin), and CXCR4 [9]. Extended lymphocyte gate was used to include events with diameter 2–10 μm, approximately consisting of VSELs. The width of the gate was validated by using synthetic beads of predefined size (1–15 μm) [14]. Several other approaches to define the population of small cells were used, including ISS. Above standard procedure employing FACS, however, is time consuming, which usually requires up to 4 days to process and isolate VSELs from UCB MNCs in one entire cord blood unit (~50–100 ml). It is not very difficult for the future clinical application to take into consideration cell viability, the time of sorter usage, and the time commitment of a sorter operator.

In order to speed up isolation of VSELs, a faster large-scale isolation protocol based on anti-CD133-conjugated paramagnetic beads followed by staining with Aldefluor were recently employed. In this novel approach (i) A UCB research unit is lysed in a hypotonic ammonium chloride solution for 15 min at room temperature to deplete erythrocytes and washed twice in phosphate-buffered saline (1st step); (ii) A single-cell suspension of total nucleated cells was treated with antibodies against CD133 antigen-coated immunomagnetic beads and separated by a MACS Separator to obtan CD133+including VSELs (2nd step); and subsequently (iii) The CD133(+)cell fraction was reacted with the Aldefluor™ Kit reagent for detecting aldehyde dehydrogenase (ALDH). Cells were incubated with phycoerythrin (PE)-conjugated murine anti-human CD235a, PE-CY7-CD45, and allophycocyanin (APC)-conjugated CD133/2. Cells were washed and resuspended in cold Aldefluor buffer and sorted by FACS to obtain populations enriched in CD45–GlyA –CD133+ALDH^{low} VSELs.The whole isolation process takes approximately 2-3 h per UCB unit and these small Lin-CD45-CD133+cells isolated from human UCB highly express Oct-4, Nanog, and SSEA-4 at both the mRNA and protein levels [13,14].

This new isolation protocol was based on the following rationale.(i) Using erythrocytes lysis buffer was higher yield of VSELs than a Ficoll-Paque gradient centrifugation to remove erythrocytes [14]. (ii) On the other hand, CD133+ VSELs are highly enriched for PSC transcription factor expression (e.g., Oct-4 and SSEA-4) [14]. (iii)Small erythroblast GlyA+ that are present in UCB do not express CD45 antigen. Thus, selection for CD45–cells was used to enrich for these cells.

The isolation from one entire UCB unit can obtan ~10^3/100 ml of UCB for CD45–GlyA –CD133+ALDH^{low} cells and ~4×10^3/100 ml of UCB for CD45–GlyA–CD133+ALDH^{high} cells. Freshly isolated CD45-GlyA-CD133+ALDH^{high} VSELs express more hematopoietic transcripts (e.g., c-myb), CD45-GlyA-CD133+ALDH ^{low} VSELs exhibit higher levels of PSCs markers (e.g., Oct-4) [34,35].

7. Special potency and hypothesized role of VSELs

PSCs must correspond to certain in vitro and in vivo conditions. According to these criteria, PSCs should be provided with (i) giving rise to cells from all three germ layers, (ii) completing blastocyst development, and (iii) forming teratomas after inoculation into experimental animals. ESCs are generally known as PSCs. However, both VSELs and iPSCs are not different from ESCs in next two conditions. In special, There are some own unique superiority that VSELs deposited in various adult organs as a backup for primitive stem cells share several markers with epiblast/germ line cells, plays a role in rejuvenation of the pool of TCSCs involved in tissue regeneration, but, not complete blastocyst development and not form teratomas. During steady-state conditions, VSELs may be responsible for tissue rejuvenation and for processes of regeneration/repair after organ injuries. VSELs similarly as epiblast-derived PGCs change the epigenetic signature of some of the imprinted genes and therefore remain quiescent in adult tissues. This quiescence of VSELs is epigenetically regulated by DNA methylation of genomic imprinting [36].VSELs highly express growth-repressive genes (H19, p57KIP2, Igf2R) and downregulate growth-promoting ones (Igf2, Rasgrf1). The unique genomic imprinting pattern may explain the quiescent status of VSELs. Thus, VSELs may be progeny of epiblast cells to develope tissues and a reserve pool of PSCs to repair tissue. Furthermore, the quiescent state of VSELs may also be a physiological protective mechanism of preventing uncontrolled proliferation, tumor formation [28]. Furthermore, The bone-forming activity of VSELs, exceeded the activity of other populations of BM-purified cells tested in the same assay if embedded in gelatin sponges and implanted into living mice. Even, as few as 500 UCB VSELs was capable of forming bone-like structures in vivo [37]. Based on these finding, VSELs have been described as at the top of the hierarchy for the mesenchymal and endothelial lineages in BM [38,39]. Interestingly, the content of VSELs from mice BM at different ages (2 months-3 years) was evaluated employing FCM. the number of these cells gradually decreases over time from 0.052±0.018% to 0.003±0.002% between age of 2-months and 3-years, respectively. In another report, the concentration of VSELs is much higher in BM of long-lived (e.g., C57Bl6) as compared to short-lived (DBA/2J) mice. Especially, not only a number of these cells in adult organs decreases with the age but also their ability to form spheres containing VSEL-DS declaims with time. Whereas, a number of monopotent hemato/lymphopoiesis committed HSC increase in older animals [40,41].This age-dependent content and ability of VSELs in adult organs may explain that these cells could play a pivotal role in the normal cell turn over and the life span control of mammals.Moreover, a significantly higher number of VSELs in long-living murine strains (e.g., Laron dwarfs and Ames dwarfs), whose longevity is explained by low levels of circulating IGF1 and a decrease in IIS. By contrast, the number of VSELs is reduced in mice with high levels of circulating IGF1 and enhanced IIS (e.g., growth hormone-overexpressing transgenic mice) compared to normally aging littermates [42,43]. There was a envision that in future, VSELs could be isolated from the patient at young age and than inject back into same recipients several years latter to regenerate damaged organs and to expand life-span, in case of major health complications (e.g., heart infarct, stroke) due to aging.

8. Mobilization and cardiovascular repair of VSELs

VSELs express early cardiac and endothelial lineages markers (GATA-4, Nkx2.5/Csx, VE-cadherin, and von Willebrand factor), SDF-1 chemokine receptor CXCR4. Under steady-state conditions, VSELs circulate in PB is very rare, however, undergo rapid mobilization during acute MI [44].The processes are regulated by SDF-1, and its receptor CXCR4 as well as other important cytokine-receptor systems of regulating the stem cell mobilization and homing include leukemia inhibitory factor (LIF) – LIF receptor, hepatocyte growth factor (HGF) – c-met axis, stem cell factor-CD117 axes [45-48]. Interestingly, the number of these cells in PB is significantly higher in younger acute MI patients than in older ones. Number of VSELs was also correlated with left ventricular ejection fraction, troponin I and creatinine kinase-MB levels [44]. Further studies provided evidence that VSELs can be mobilized into PB in adult patients injected with granulocyte-colony stimulating factor (G-CSF) and their number could be of prognostic value [49].Consistently, a protocol which VSELs differentiate into cardiomyocytes in vitro has been developed. In the first step, VSELs are co-cultured with myoblast line (C2C12) where the cells expand and form VSEL-DS.Subsequently, VSELs isolated from VSEL-DS by FACS sorting are plated on cardiac media to differentiate them into cardiomyocytes. The period over 21 days in expression of early cardiac markers and cardiac structural proteins also resembles the maturation of cardimyocytes from ESCs [50].

Murin experiments in vivo also showed that expanded and subsequent cardiopoiesis-guided VSELs were markedly more effective than expanded and non-pre-differentiated cells. Interestingly, beneficial effects were observed despite use of only small number (10^4 cells) of VSELs. At the same time a much higher number of hematopoietic cells (10^5 cells) was not effective [37].

so far, a cochrane controlled trial called Myocardial Regeneration by Intracoronary Infusion of Selected Population of Stem Cells in Acute Myocardial Infarction (REGENT) study are available.The objective study was to assess the efficacy of intracoronary infusion of autologous BM-derived CD34+CXCR4+ progenitor cells in comparison to non-selected BMMNC on LVEF in patients with acute ST-segment elevation MI and reduced below 40% LVEF. CD34+CXCR4+ cells which is enriched for VSELs were isolated by two step immunomagnetic selection using the magnetic beads. The reslut demonstrated that the use of selected CD34+CXCR4+ cells or non-selected BMMNCs in patients with significantly reduced LV function is safe, feasible and not leads to a significant improvement of LVEF, there was however a trend in favour of cell therapy. In another, athough CD34+CXCR4+ cells were not pure population of VSELs, The use of a relatively small number of selected CD34+CXCR4+ cells is associated with similar trend as the use of 100 times higher number of non-selected BMMNCs (1.90×10^6 vs. 1.78×10^8) for improvement of LVEF. It also further showed which the activity of VSELs is more superior than that of BMMNC [52].Based on above finding, administration of VSELs after an acute MI increases LVEF and improves left ventricular structure, and these benefits remain stable during long-term follow-up. Although the mechanisms remain under investigation, paracrine effects, regeneration of cellular constituents, and stimulation of endogenous stem/progenitors may play combinatorial roles [49], because the rare VSEL-derived cardiac myocytes expressing cardiac markers were present in the recipients myocardium [50,52].

Thus, VSELs may serve as an ideal SCs source for cardiac repair by their ability to secrete various cardioprotective growth factors/cytokines, as well as their ability to differentiate into cardiomyocytes and endothelial cells

9. Remaining challenges of VSELs application

However, there are also some challenges and conflict coming out from recent VSELs' studies.At first, a recent study showed that VSELs from human UCB lack SCs characteristics and fail to expand in vitro under a wide range of culture conditions [53].We also found that it is very difficulte for human VSELs to be cultured or expanded in vitro using general culture conditions. Thus, it has to be further determined whether these cells are merely developmental remnants found in the adult tissue that cannot be harnessed effectively for regeneration or whether they are real SCs population for regeneration medicine.

Subsquently, the biological characteristics and role of VSELs were studied mostly in mice and human. No information of VSELs from large animal close to human has been reported. Future clinical studies using autologous VSELs are needed to validate those promising large animal experimental data.

Furthermore, it has seldomly obtained for parallel experiments to compare several populations of putative SCs to determine the similarities and differences between these cell populations.

10. Summary

Overall, as mentioned above, the importance of SSEA-1+Oct-4+Sca-1+/ CD133+CXCR4+Lin −CD45− pluripotent VSELs in adult tissue or UCB is now being stressed. New data from Kucia & Ratajczak group and other groups has provided mounting evidence on the existence and potential biological role of VSELs mostly in mice. VSELs would very possibly be a promising PSCs population for cardiac repair in future clinical application of patients with ICM. Their cardiogenic potential should be confirmed in large animal similar to humans and technical issues regarding their isolation, expansion and differentiation need to further be addressed.We also look forward to share how to higher efficiently isolate and expand these rare cells as well as to know about further information on their biology and in vitro and in vivo differentiation potential.

Acknowledgements

This study was supported by the National Natural Science Fund Committeethe (NSFC 81170104), Natural Science Foundation of Jiangsu Province (BK2012688) and and Program for Jiangsu Province Outstanding Medical Talented Leader (JS2006038).

Author details

X. Gu[1], J. Gu[1], J.B. Sun[1], Y.X. Gu[1], L. Sun[1], Y. Zhang[1], Y. Cheng[1], Z.Y. Bao[1], F. Hang[1], X.M. Lu[1], R.S. Zhang[1] and B.C. Li[2]

*Address all correspondence to: sbyygx@medmail.com.cn

1 Department of Cardiology, Northern People's Hospital, Medical College of Yangzhou University, Yangzhou,, China

2 Animal Science and Technical College, Yangzhou University, Yangzhou,, China

References

[1] Andre Alcon, Esra Cagavi Bozkulak,Yibing Qyang. Regenerating functional heart tissue for myocardial repair.Cell Mol Life Sci. 2012; 69 : 2635–2656.

[2] Orlic D, Kajstura J, Chimenti S, Jakoniuk I, Anderson SM, Li B, Pickel J, McKay R, Nadal-Ginard B, Bodine DM, Leri A, Anversa P. Bone marrow cells regenerate infarcted myocardium.Nature 2001;410:701–705.

[3] Rota M, Kajstura J, Hosoda T, Bearzi C, Vitale S, Esposito G, Iaffaldano G, Padin-Iruegas ME, Gonzalez A, Rizzi R, Small N, Muraski J, Alvarez R, Chen X, Urbanek K, Bolli R, Houser SR, Leri A, Sussman MA, Anversa P. Bone marrow cells adopt the cardiomyogenic fate in vivo. Proc Natl Acad Sci USA 2007;104:17783–17788.

[4] Amado LC, Saliaris AP, Schuleri KH, St John M, Xie JS, Cattaneo S, Durand DJ, Fitton T, Kuang JQ, Stewart G, Lehrke S, Baumgartner WW, Martin BJ, Heldman AW, Hare JM. Cardiac repair with intramyocardial injection of allogeneic mesenchymal stem cells after myocardial infarction. Proc Natl Acad Sci USA 2005;102:11474–11479.

[5] Cho HJ, Lee N, Lee JY, Choi YJ, Ii M, Wecker A, Jeong JO, Curry C, Qin G, Yoon YS. Role of host tissues for sustained humoral effects after endothelial progenitor cell transplantation into the ischemic heart. J Exp Med 2007; 204:3257–3269.

[6] Xiang Gu, Yong Xie, Jian Gu, Sun Lei, He Shenghu, Xu Rixin, Duan Junfei, Zhao Jianye, Hang Fei, Xu Houtian, Li Minghui, Cao Kejiang, Geng Yongjian. Repeated Intracoronary Infusion of Peripheral Blood Stem Cells With G-CSF in Patients With Refractory Ischemic Heart Failure: A Pilot Study. Circ J 2011; 75: 955-963.

[7] Templin C, Zweigerdt R, Schwanke K, Olmer R, Ghadri JR, Emmert MY, Müller E, Küest SM, Cohrs S, Schibli R, Kronen P, Hilbe M, Reinisch A, Strunk D,Haverich A, Hoerstrup S, Lüscher TF, Kaufmann PA, Landmesser U, Martin U.Transplantation and tracking of human-induced pluripotent stem cells in a pig model of myocardial infarction: assessment of cell survival, engraftment, and distribution by hybrid single

photon emission computed tomography/computed tomography of sodium iodide symporter transgene expression. Circulation. 2012;126:430-9.

[8] Carpenter L, Carr C, Yang CT, Stuckey DJ, Clarke K, Watt SM. Efficient differentiation of human induced pluripotent stem cells generates cardiac cells that provide protection following myocardial infarction in the rat. Stem Cells Dev. 2012;21:977-86.

[9] Kucia M, Reca R, Campbell FR, Zuba-Surma E, Majka M, Ratajczak J, Ratajczak MZ. A population of very small embryonic-like (VSEL) CXCR4+SSEA-1+Oct-4+ stem cells identified in adult bone marrow. Leukemia. 2006; 20:857–869.

[10] Zuba-Surma EK, Kucia M, Abdel-Latif A, Dawn B, Hall B, Singh R, Lillard JW, Jr., Ratajczak MZ. Morphological characterization of very small embryonic-like stem cells (VSELs) by ImageStream system analysis. J Cell Mol Med. 2008;12:292–303.

[11] Kucia M, Wysoczynski M, Wu W, Zuba-Surma EK, Ratajczak J, Ratajczak MZ.Evidence that very small embryonic-like stem cells are mobilized into peripheral blood.Stem Cells. 2008; 26: 2083–2092.

[12] Dawn B, Tiwari S, Kucia M, Zuba-Surma EK, Guo Y, Sanganalmath SK, Abdel-Latif A, Hunt G, Vincent RJ, Taher H, Reed NJ, Ratajczak MZ, BolliR Transplantationofbone marrow-derived very small embryonic-like stemcells attenuates leftventricular dysfunctionand remodeling after myocardial infarction.Stem Cells. 2008;26 :1646-55.

[13] Zuba-Surma EK, Guo Y, Taher H, Sanganalmath SK, Hunt G, Vincent RJ, Kucia M, Abdel-Latif A, Tang XL, Ratajczak MZ, Dawn B, Bolli R. Transplantation of expanded bone marrow-derived very small embryonic-like stem cells (VSEL-SCs) improves left ventricular function and remodelling after myocardial infarction. J Cell Mol Med. 2011; 15: 1319-28.

[14] Zuba-Surma EK, Klich I, Greco N, Laughlin MJ, Ratajczak J, Ratajczak MZ. Optimization of isolation and further characterization of umbilical-cord-blood-derived very small embryonic/epiblast-like stem cells (VSELs). Eur J Haematol. 2010; 84: 34–46.

[15] Bhartiya D, Shaikh A, Nagvenkar P, Kasiviswanathan S, Pethe P, Pawani H, Mohanty S, Rao SG, Zaveri K, Hinduja I. Very small embryonic-like stem cells with maximum regenerative potential get discarded during cord blood banking and bone marrow processing for autologous stem cell therapy.Stem Cells Dev. 2012; 21:1-6.

[16] Vacanti MP, Roy A, Cortiella J, Bonassar L, Vacanti CA. Identification and initial characterization of spore-like cells in adult mammals. J Cell Biochem. 2001; 80: 455–460.

[17] Zuba-Surma EK, Kucia M, Rui L, Shin DM, Wojakowski W, Ratajczak J, Ratajczak MZ. Fetal liver very small embryonic/epiblast like stem cells follow developmental migratory pathway of hematopoietic stem cells.Annals of the New York Academy of Sciences. 2009; 1176: 205–218.

[18] Ratajczak MZ, Kucia M, Ratajczak J, Zuba-Surma EK. A multi-instrumental approach to identify and purify very small embryonic like stem cells (VSELs) from adult tissues. Micron. 2009;40: 386–393.

[19] Ratajczak MZ, Liu R, Ratajczak J, Kucia M, Shin DM. The role of pluripotent embryonic-like stem cells residing in adult tissues in regeneration and longevity. Differentiation. 2011;81:153–161.

[20] Zuba-Surma EK, Ratajczak, MZ. Overview of very small embryonic-like stem cells (VSELs) and methodology of their identification and isolation by flow cytometric methods. Current Protocols in Cytometry, 2010, Chapter 9, Unit 9.29.

[21] Ortyn WE, Hall BE, George TC, Frost K, Basiji DA, Perry DJ, Zimmerman CA, Coder DC, Morrissey PJ. Sensitivity Measurement and Compensation in Spectral Imaging. Cytometry A.2006; 69A : 852–862.

[22] Zuba-Surma EK, Kucia M, Wu W, Klich I, Lillard JW, Jr, Ratajczak J, Ratajczak MZ. Very small embryonic-like stem cells are present in adult murine organs: ImageStream-based morphological analysis and distribution studies. Cytometry A. 2008; 73: 1116–1127.

[23] Shin DM, Zuba-Surma EK, Wu W, Ratajczak J, Wysoczynski M, Ratajczak MZ, Kucia M. Novel epigenetic mechanisms that control pluripotency and quiescence of adult bone marrow-derived Oct4(+) very small embryonic-like stem cells. Leukemia. 2009; 23: 2042–2051.

[24] Zuba-Surma EK, Wojakowski W, Ratajczak MZ, Dawn B. Very small embryonic-like stem cells: biology and therapeutic potential for heart repair. 2011;15:1821-34.

[25] Zuba-Surma EK, Ratajczak MZ. Overview of very small embryonic-like stem cells (VSELs) and methodology of their identification and isolation by flow cytometric methods. Curr Protoc Cytom Chapter 9: 29, 2010.

[26] Ratajczak J, Zuba-Surma EK, Klich I, Liu R,Wysoczynski M, Greco N, Kucia M, Laughlin MJ, and Ratajczak MZ. Hematopoietic differentiation of umbilical cord blood-derived very small embryonical/epiblast –like stem cells. Leukemia. 2011; 25: 1278–1285.

[27] Wojakowski W, Kucia M, Liu R, Zuba-Surma E, Jadczyk T, Bachowski R, Nabiałek E, Kaźmierski M, Ratajczak MZ, Tendera M. Circulating very small embryonic-like stem cells in cardiovascular disease. J Cardiovasc Transl Res. 2011;4 :138-44.

[28] Ratajczak MZ, Zuba-Surma EK, Shin DM, Ratajczak J, Kucia M. Very small embryonic-like (VSEL) stem cells in adult organs and their potential role in rejuvenation of tissues and longevity. Exp Gerontol 43: 1009-17, 2008.

[29] Iskovich S, Goldenberg-Cohen N, Stein J, Yaniv I, Fabian I, Askenasy N. Elutriated stem cells derived from the adult bone marrow differentiate into insulin-producing cells in vivo and reverse chemical diabetes. Stem Cells Dev. 2012;21:86–96.

[30] Tang XL, Rokosh DG, Guo Y, Bolli R.Cardiac progenitor cells and bone marrow-derived very small embryonic-like stem cells for cardiac repair after myocardial infarction[J]. Circ J. 2010 ;74(3):390-404.

[31] Gharib SA, Khalyfa A, Kucia MJ, Dayyat EA, Kim J, Clair HB, Gozal D. Transcriptional landscape of bone marrow-derived very small embryonic-like stem cells during hypoxia. Respir Res. 2011 May 10;12:63.

[32] Kassmer SH, Bruscia EM, Zhang PX, Krause DS. Non-Hematopoietic Cells are the Primary Source of Bone Marrow Derived Lung Epithelial Cells. Stem Cells. 2012;30:491-9.

[33] Sovalat H, Scrofani M, Eidenschenk A, Pasquet S, Rimelen V, Hénon P. Identification and isolation from either adult human bone marrow or G-CSF-mobilized peripheral blood of CD34(+)/CD133(+)/CXCR4(+)/Lin(-)CD45(-) cells, featuring morphological, molecular, and phenotypic characteristics of very small embryonic-like (VSEL) stem cells. Exp Hematol. 2011;39 :495-505.

[34] Liu Y, Gao L, Zuba-Surma EK, Peng X, Kucia M, Ratajczak MZ, Wang W, Enzmann V, Kaplan HJ, Dean DC. Identification of small Sca-1(+), Lin(-), CD45(-) multipotential cells in the neonatal murine retina. Exp Hematol. 2009;37:1096–107.

[35] Ratajczak J, Zuba-Surma E, Klich I, Liu R, Wysoczynski M, Greco N, Kucia M, Laughlin MJ, Ratajczak MZ. Hematopoietic differentiation of umbilical cord blood-derived very small embryonic/epiblast-like stem cells.Leukemia. 2011; 25: 1278-85.

[36] Mariusz Z. Ratajczak, Dong-Myung Shin, Rui Liu, Kasia Mierzejewska, Janina Ratajczak, Magda Kucia, Ewa K. Zuba-Surma. Very small embryonic/epiblast-like stem cells (VSELs) and their potential role in aging and organ rejuvenation – an update and comparison to other primitive small stem cells isolated from adult tissues. Aging. 2012; 4: 235–246.

[37] Russell S. Taichman, Zhuo Wang, Yusuke Shiozawa, Younghun Jung, Junhui Song, Alex Balduino, Jincheng Wang, Lalit R. Patel, Aaron M. Havens, Magdalena Kucia, Mariusz Z. Ratajczak, Paul H. Krebsbach. Prospective Identification and Skeletal Localization of Cells Capable of Multilineage Differentiation In Vivo. Stem Cells Dev. 2010; 19: 1557–1570.

[38] Anjos-Afonso F, Bonnet D. Nonhematopoietic/endothelial SSEA-1+ cells define the most primitive progenitors in the adult murine bone marrow mesenchymal compartment. 2007;109 :1298-306.

[39] Ratajczak MZ, Shin DM, Liu R, Mierzejewska K, Ratajczak J, Kucia M, Zuba-Surma EK. Very small embryonic/epiblast-like stem cells (VSELs) and their potential role in aging and organ rejuvenation--an update and comparison to other primitive small stem cells isolated from adult tissues. Aging (Albany NY). 2012 ;4:235-46.

[40] Ratajczak MZ, Zuba-Surma EK, Shin DM, Ratajczak J, and Kucia M. Very small embryonic-like (VSEL) stem cells in adult organs and their potential role in rejuvenation of tissues and longevity.Exp Gerontol. 2008; 43: 1009–1017.

[41] Chambers SM, Goodell MA. Hematopoietic stem cell aging: wrinkles in stem cell potential. Stem Cell Rev. 2007;3:201–211.

[42] Svensson J, Sjogren K, Faldt J, Andersson N, Isaksson O, Jansson JO, Ohlsson C. Liver-derived IGF-I regulates mean life span in mice. PLoS One. 2011; 6:e22640.

[43] Ratajczak MZ, Shin DM, Ratajczak J, Kucia M, Bartke A. A novel insight into aging: are there pluripotent very small embryonic-like stem cells (VSELs) in adult tissues overtime depleted in an Igf-1-dependent manner? Aging (Albany NY) 2010;2:875–883.

[44] Wojakowski W, Tendera M, Kucia M, Zuba-Surma E, Paczkowska E, Ciosek J, Halasa M, Krol M, Kazmierski M, Buszman P, Ochała A, Ratajczak J, Machaliński B, Ratajczak MZ. Mobilization of bone marrow-derived Oct-4+ SSEA-4+ very small embryonic-like stem cells in patients with acute myocardial infarction. J Am Coll Cardiol. 2009; 53: 1–9.

[45] Zuba-Surma EK, Wojakowski W, Ratajczak MZ, Dawn B. Very small embryonic-like stem cells: biology and therapeutic potential for heart repair. Antioxid Redox Signal. 2011 1;15:1821-34.

[46] Wojakowski W, Kucia M, Liu R, Zuba-Surma E, Jadczyk T, Bachowski R, Nabiałek E, Kaźmierski M, Ratajczak MZ,Tendera M.Circulating very small embryonic-like stem cells in cardiovascular disease. J Cardiovasc Transl Res. 2011;4 :138-44.

[47] Kucia M, Wojakowski W, Reca R, Machalinski B, Gozdzik J, Majka M, et al. The migration of bone marrow-derived non-hematopoietic tissue-committed stem cells is regulated in an SDF-1-, HGF-, and LIF-dependent manner. Archivum Immunologiae et Therapiae Experimentalis. 2006;54:121–135.

[48] Wojakowski W, Tendera M, Zebzda A, Michalowska A, Majka M, Kucia M, et al. Mobilization of CD34(+), CD117(+), CXCR4(+), c-met(+) stem cells is correlated with left ventricular ejection fraction and plasma NT-proBNP levels in patients with acute myocardial infarction. European Heart Journal.2006; 27: 283–289.

[49] Wojakowski W, Tendera M, Kucia M, Zuba-Surma E, Milewski K, Wallace-Bradley D, Kazmierski M, Buszman P, Hrycek E, Cybulski W, Kaluza G, Wieczorek P,Ratajczak J, Ratajczak MZ. Cardiomyocyte differentiation of bone marrow-derived Oct-4 + CXCR4 + SSEA-1+ very small embryonic-like stem cells. International Journal of Oncology. 2010; 37 :237–247.

[50] Zuba-Surma EK, Guo Y, Taher H, Sanganalmath SK, Hunt G, Vincent RJ, Kucia M, Abdel-Latif A, Tang XL, Ratajczak MZ, Dawn B, Bolli R.Transplantation of expanded bone marrow-derived very small embryonic-like stem cells (VSEL-SCs) improves left

ventricular function and remodeling after myocardial infarction. J Cell Mol Med. 2011 15:1319-28.

[51] Dawn B, Tiwari S, Kucia MJ, Zuba-Surma EK, Guo Y, Sanganalmath SK, Abdel-Latif A, Hunt G, Vincent RJ, Taher H, Reed NJ, Ratajczak MZ, Bolli R. Transplantation of bone marrow-derived very small embryonic-like stem cells attenuates left ventricular dysfunction and remodeling after myocardial infarction. Stem Cells. 2008;26 :1646–1655.

[52] Tendera M, Wojakowski W, Ruzyłło W, Chojnowska L, Kepka C, Tracz W, Musiałek P, Piwowarska W, Nessler J, Buszman P, Grajek S, Breborowicz P, Majka M,Ratajczak. Intracoronary infusion of bone marrow-derived selected CD34+CXCR4+ cells and non-selected mononuclear cellsin patients with acute STEMI and reduced left ventricular ejection fraction: results of randomized, multicentreMyocardial Regeneration by Intracoronary Infusion of Selected Population of Stem Cells in Acute MyocardialInfarction (REGENT) Trial. Eur Heart J. 2009;30 :1313-21.

[53] Danova-Alt R, Heider A, Egger D, Cross M, Alt R. Very small embryonic-like stem cells purified from umbilical cord blood lack stem cell characteristics. PLoS One. 2012;7:e34899.

Chapter 5

Recent Advances in Hematopoietic Stem Cell Gene Therapy

Toshihisa Tsuruta

Additional information is available at the end of the chapter

http://dx.doi.org/10.5772/53587

1. Introduction

Hematopoietic stem cell transplantation (HSCT) has a half-century history. It is currently an indispensable treatment for not only incurable blood diseases such as aplastic anemia and severe hemolytic anemia, but also malignant hematological diseases such as leukemia and lymphoma. Although allergenic HSCT is also used to treat hereditary diseases, its indications are restricted because of critical complications including regimen-related toxicities involving conditioning, infection, and graft-versus-host disease.

Studies in recent decades have shown that HSCT can have a long-term effect in the treatment of hereditary diseases involving a responsible gene in hematogenous cells. Although the first successful gene therapy using lymphocytes or bone marrow cells for a patient with adenosine deaminase (ADA) deficiency inspired great hope in the future of gene therapy [1-3], subsequent gene therapy using HSCs for patients with X-linked severe combined immunodeficiency (SCID-X1) resulted in tumorigenesis [4]. In addition to the self-renewal and multilineage differentiation capacities of tissue stem cells, HSCs exhibit cell-cycle dormancy, which complicates their use in gene therapy.

However, as technological advances have increased the safety and efficiency of introducing genes into HSCs, gene therapy with HSCs is attracting attention again. In this chapter, advances in the technology of HSC gene therapy, e.g., vector design to avoid genotoxicity and increase transgenic efficiency by taking advantage of the special characteristics of HSCs, are reviewed. In addition, recent studies on HSC gene therapy for various hereditary diseases, such as thalassemia, Fanconi anemia, hemophilia, primary immunodeficiency, mucopolysaccharidosis, Gaucher disease, and X-linked adrenoleukodystrophy (X-ALD) are discussed.

2. Characteristics of HSCs and gene therapy

The concept of the HSC was introduced by Till and McCulloch in 1961 [5]. Although a healthy adult produces approximately 1 trillion blood cells each day, they are considered to originate from a single HSC which can potentially be transplanted into a mouse [6, 7]. Generally stem cells are defined as cells capable of self-renewal and multilineage differentiation. In addition to these two characteristics, HSCs have the capability of cell-cycle dormancy, i.e. to enter a state of dormancy (G0 phase) in the cell cycle and can continue blood cell production over a lifetime while protecting themselves from various kinds of stress [8].

Fig. 1 shows HSC surface markers and the typical cytokines regulating HSCs. Stem cell factor (SCF) and thrombopoietin (TPO) are important direct cytokine regulators of HSCs. Although SCF promotes the proliferation and differentiation of hematopoietic progenitor cells, it is thought to not be essential for the initiation of hematopoiesis and HSC self-renewal [9]. TPO and its receptor, c-Mpl, are thought to play important roles in early hematopoiesis from HSCs. In contrast to the CD34^{+}CD38^{-}c-Mpl^{-} population, CD34^{+}CD38^{-}c-Mpl^{+} cells show significantly better HSC engraftment [10]. Mice lacking either TPO or c-Mpl have deficiencies in progenitor cells of multiple hematopoietic lineages [11]. TPO-mediated signal transduction for the self-renewal of HSCs is negatively regulated by the intracellular scaffold protein Lnk [12, 13]. A signal from angiopoietin-1 via Tie2 regulates HSC dormancy by promoting the adhesion of HSCs to osteoblasts in the bone marrow niche and maintains long-term repopulating activity [14]. Although cytokine-induced lipid raft clustering of the HSC membrane is essential for HSC reentry into the cell cycle, transforming growth factor-β (TGF-β) inhibits lipid raft clustering and induces p57Kip2 expression, leading to HSC dormancy [15, 16]. Recently, the hypoxic niche of HSCs has been demonstrated. It, along with the osteoblastic and vascular niches, are important for HSC dormancy [17-19]. They are targets in HSC gene therapy [20].

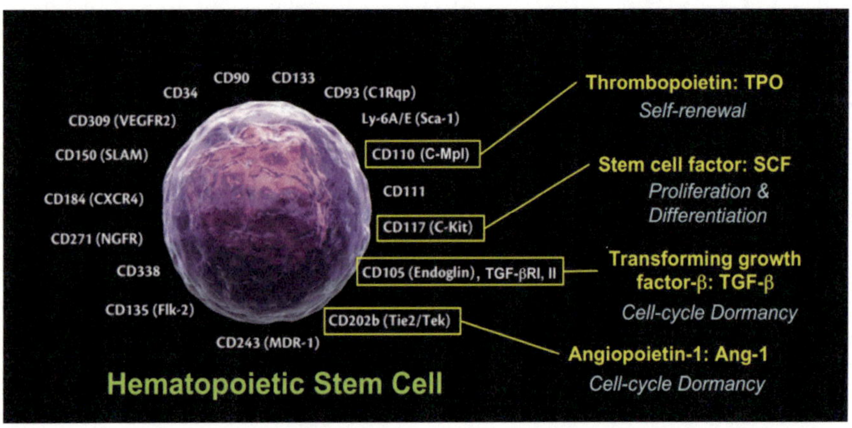

Figure 1. Hematopoietic stem cell (HSC) surface markers and typical cytokines that regulate HSCs. Stem cell factor (SCF) promotes the proliferation and differentiation of HSCs. Thrombopoietin (TPO) and its receptor, c-Mpl, play important roles in early hematopoiesis, especially self-renewal. Signals from angiotensin-1 via Tie2 and transforming growth factor -β via its receptors regulate HSC dormancy. (This figure is based on the illustration by BioLegend, Inc. San Diego, CA, U.S.A. http://www.biolegend.com/cell_markers)

While making a HSC with few opportunities for cell division into a transgenic target, it is important to design a safe and efficient vector for inserting a gene into the host chromosome. Furthermore, since a hematogenous cell also has many cells which exhibit its function in the specialization process to a mature effector cell, it is also important to select differentiation-specific or non-specific promoters or enhancers during the vector design process.

3. Vectors for HSC gene therapy

Vectors derived from the Retroviridae family, RNA viruses with reverse transcriptase activity, are widely used for inserting genes in host chromosomes. Although adeno-associated virus (AAV) vectors can also insert genes into host chromosomes, this process is inefficient and partial. Gammaretroviruses and lentiviruses are members of the Retroviridae family that are commonly used as vectors in HSC gene therapy. Generally, the former is called simply a retroviral vector and the latter is called a lentiviral vector. When a gene is inserted in the chromosome of an HSC with a Retroviridae vector, genotoxicity can occur.

3.1. Gammaretroviral (Retroviral) vectors

Retroviral vectors are commonly constructed from the Moloney murine leukemia virus (MoMLV) genome. Retroviral genomes have a *gag/pol* gene that codes for viral structure proteins, protease and reverse transcriptase, an *env* gene that codes for the envelope glycoprotein and the packaging signal. These genes are flanked by long terminal repeats (LTR) which contain enhancers and promoters. A retroviral vector consists of a packaging plasmid that does not have the packaging signal but does include the *gag/pol* gene, a transfer vector with the packaging signal, and the target gene cDNA. After transfection of these plasmids into producer cells (e.g., 297T cells, NIH3T3 cell, etc.), a target vector is obtained by collecting the culture solution.

Expression of a target gene can be inhibited by mechanisms such as methylation of CpG islands in the promoter region, insertion of a negative control region (NCR) into the LTR, and the presence of a repressor binding site (RBS) downstream of the 5′ LTR. Other vectors, such as the murine stem cell virus (MSCV) vector [21], the myeloproliferative sarcoma virus vector, the negative control region deleted (MND) vector [22], and the MFG-S vector [23] were developed to improve the efficiency of transgene expression; they are widely used in clinical applications of gene therapy involving HSCs.

Since the retroviral viral genome cannot cross the nuclear membrane, it can be incorporated into a chromosome only during the phase of mitosis when the nuclear membrane has disassembled. Since many HSCs are thought to exist in a dormant phase, insertions into the HSC genome with a retroviral vector require a proliferation stimulus by cytokines. Although various combinations of cytokines to suppress the decrease in HSC self-renewal have been studied, stem cell factor (SCF), fms-related tyrosine kinase-3 (Flt-3) ligand, interleukin-3 (IL-3), TPO, among others, are commonly used [24, 25].

3.2. Lentiviral vectors

Human immunodeficiency virus type 1 (HIV-1), the representative lentivirus, differs from gammaretroviruses in that it can be incorporated during a non-mitotic phase. This is one advantage of lentiviral vectors in HSC gene therapy.

Both lentiviruses and gammaretroviruses have *gag, pol,* and *env* genes sandwiched between LTRs with promoter activity at both ends. In addition, lentiviruses have accessory genes (*vif, vpr, vpu, nef*) and regulatory genes (*tat, rev*). Double-stranded cDNA produced from the viral genome enters the cell, and a pre-integration complex is formed with a host protein. This complex can pass through the pores of the nuclear membrane during non-mitotic phases, allowing the viral genome to be inserted into the host cell chromosome.

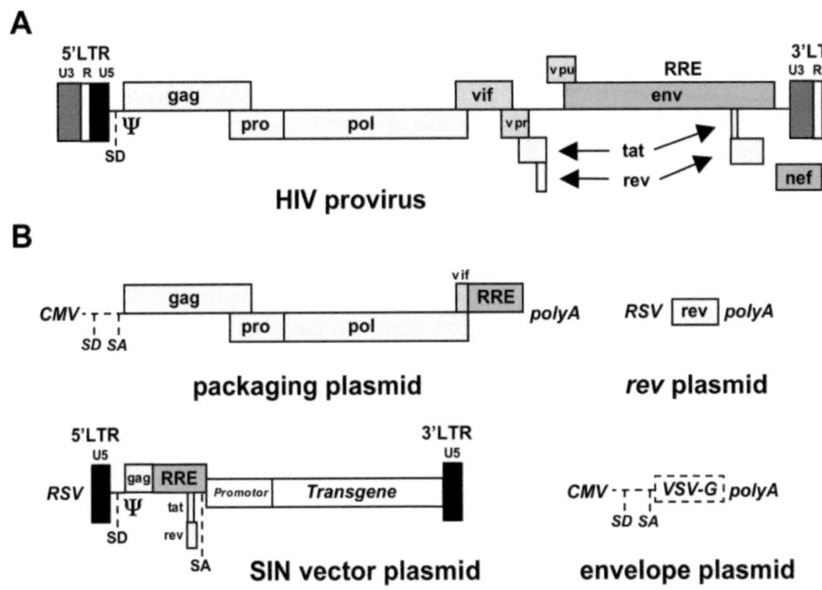

Figure 2. HIV provirus (A) and the four plasmids of a third-generation lentiviral vector (B). The viral long terminal repeats (LTRs), reading frames of the viral genes, splice donor site (SD), splicing acceptor site (SA), packaging signal (Ψ), and rev-responsive element (RRE) are indicated. The packaging plasmid contains the *gag* and *pol* genes under the influence of the CMV promoter, intervening sequences, and the polyadenylation site (polyA) of the human β-globin gene. As the transcripts of the *gag* and *pol* genes contain cis-repressive sequences, they are expressed only if rev promotes their nuclear export by binding to the RRE. All *tat* and *rev* exons have been deleted, and the viral sequences upstream of the *gag* gene have been replaced. The rev plasmid expresses *rev* cDNA. The SIN vector plasmid contains HIV-1 cis-acting sequences and an expression cassette for the transgene. It is the only portion transferred to the target cells and does not contain wild-type copies of the HIV LTR. The 5′ LTR is chimeric, with the RSV enhancer and promoter replacing the U3 region to rescue transcriptional dependence on *tat*. The 3′ LTR has an almost completely deleted U3 region, which includes the TATA box. As the latter is the template used to generate both copies of the LTR in the integrated provirus, transduction of this vector results in transcriptional inactivation of both LTRs; thus, it is a self-inactivating (SIN) vector. The envelope plasmid encodes a heterologous envelope to pseudotype the vector, here shown coding for vesicular stomatitis virus (VSV)-G. Only the relevant parts of the constructs are shown (Reproduced with modifications from [26]).

Although first-generation lentiviral vectors included modification genes, they were removed in the second generation because it was discovered that the modification genes are not required for infection during non-mitotic phases. In the third generation, further modifications included the deletion of *tat*, use of multiple vector plasmids, and introduction of self-inactivating (SIN) vectors. The structure of HIV-1 and a typical third-generation lentiviral vector system are shown in Fig. 2 [26]. Approximately one-third of the HIV-1 genome has been deleted, and the vector system has been divided into four plasmids, namely, the packaging plasmid, *rev* plasmid, SIN vector plasmid and envelope plasmid. To prevent production of wild type HIV-1, *tat,* a regulatory gene indispensable to viral reproduction was deleted, and the *rev* gene was moved to a separate plasmid. Moreover, since the HIV-1 LTR promoter is weak in the absence of *tat*, it was replaced with the cytomegalovirus (CMV) promoter in the packaging plasmid. Since an envelope plasmid can only infect CD4 positive cells with a HIV-1 envelope, the envelope gene was replaced with the vesicular stomatitis virus G glycoprotein (VSV-G) envelope. The SIN vector further improved safety by replacing the enhancer / promoter portion of the LTR, suppressing the activation of unnecessary genes with the integrated gene (Fig. 3) [27].

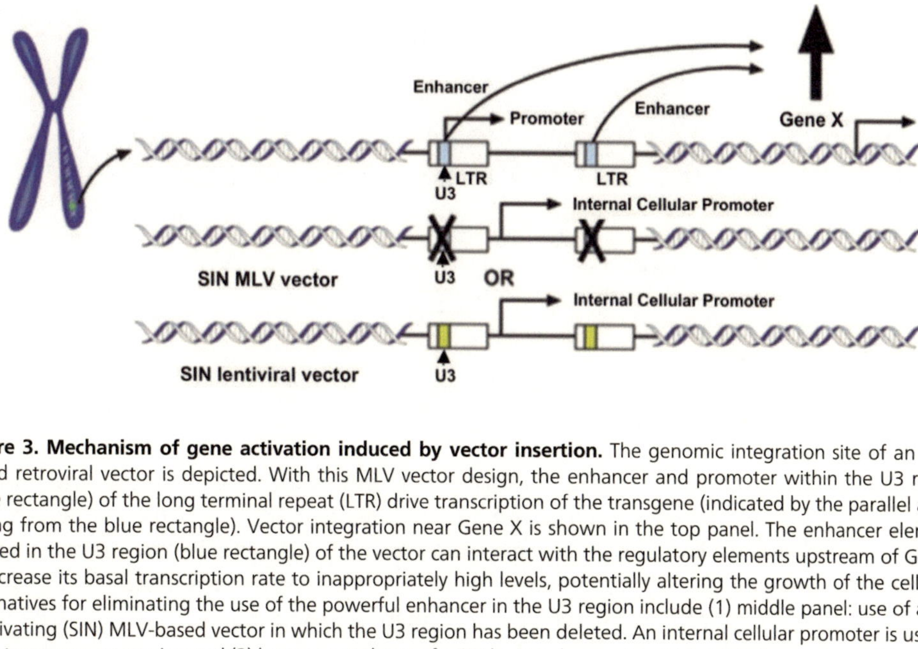

Figure 3. Mechanism of gene activation induced by vector insertion. The genomic integration site of an MLV-based retroviral vector is depicted. With this MLV vector design, the enhancer and promoter within the U3 region (blue rectangle) of the long terminal repeat (LTR) drive transcription of the transgene (indicated by the parallel arrow arising from the blue rectangle). Vector integration near Gene X is shown in the top panel. The enhancer elements located in the U3 region (blue rectangle) of the vector can interact with the regulatory elements upstream of Gene X to increase its basal transcription rate to inappropriately high levels, potentially altering the growth of the cell. Two alternatives for eliminating the use of the powerful enhancer in the U3 region include (1) middle panel: use of a self-inactivating (SIN) MLV-based vector in which the U3 region has been deleted. An internal cellular promoter is used to drive transgene expression and (2) bottom panel: use of a SIN lentiviral vector in which U3 (yellow rectangle) has been eliminated. This system also uses an internal cellular promoter to drive transgene expression (Reproduced with modification from [27]).

To improve the gene transfer into HSCs, Verhoeyen and colleagues designed lentiviral vectors displaying "early-acting cytokines" such as TPO and SCF. This vector can promote survival of CD34 positive HSCs and achieve selective transduction of long-term repopulating cells in a humanized mouse model (Fig. 4) [28, 29].

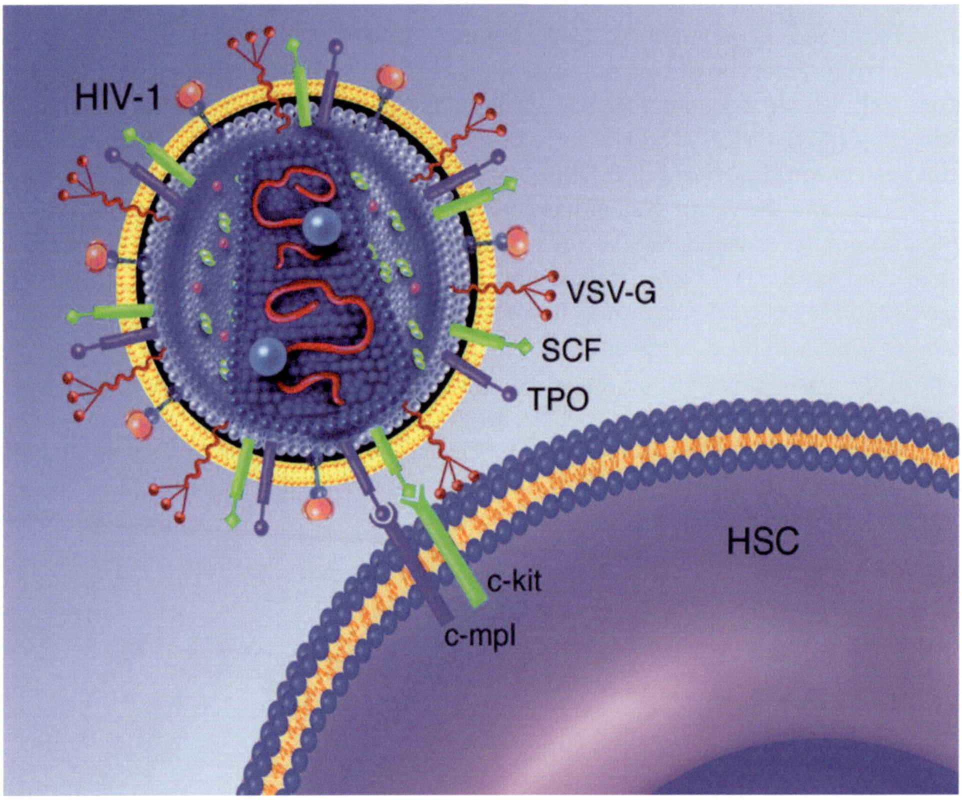

Figure 4. Lentiviral vector particles (HIV-1) display recombinant membrane envelope proteins such as stem cell factor (SCF), thrombopoietin (TPO), and vesicular stomatitis virus G glycoprotein (VSV-G). This vector can specifically target vector particles to hematopoietic stem cells (HSCs) expressing c-kit and c-mpl receptors for SCF and TPO, respectively. VSV-G envelope protein can bind to phospholipids in the HSC cell membrane. (Karlsson S, Gene therapy: efficient targeting of hematopoietic stem cells. Blood. 2005;106(10):3333)

3.3. Genotoxicity of viral vectors

The most serious problem with using viral vectors to incorporate a gene into a chromosome is the potential development of clonal proliferative diseases such as leukemia, which was observed in clinical trials involving gene therapy for SCID-X1 and chronic granulomatous disease (CGD). Although this problem of genotoxicity represents a great hurdle in the development of clinical applications for gene therapy, there is promising ongoing research on the mechanisms underlying genotoxicity and how to avoid it.

The mechanisms of retrovirus-induced oncogenesis are shown in Fig. 5 [30]. In oncogene capture, an acute transforming replication-competent retrovirus captures a cellular proto-oncogene and mediates transformation. This mechanism does not occur in replication-incompetent vectors. Second, the provirus 3′ LTR can trigger increased transcription of a cellular proto-oncogene. Third, enhancers in the provirus LTRs can activate transcription from nearby cellular proto-oncogene promoters. Fourth, a novel isoform can be expressed when transcription from the provirus 5′ LTR creates a novel truncated isoform of a cellular proto-oncogene via splicing. Fifth, an inserted provirus can disrupt transcription by causing premature polyadenylation. The same mechanisms can occur in cellular oncogenesis when a gene is inserted by a retroviral vector [30].

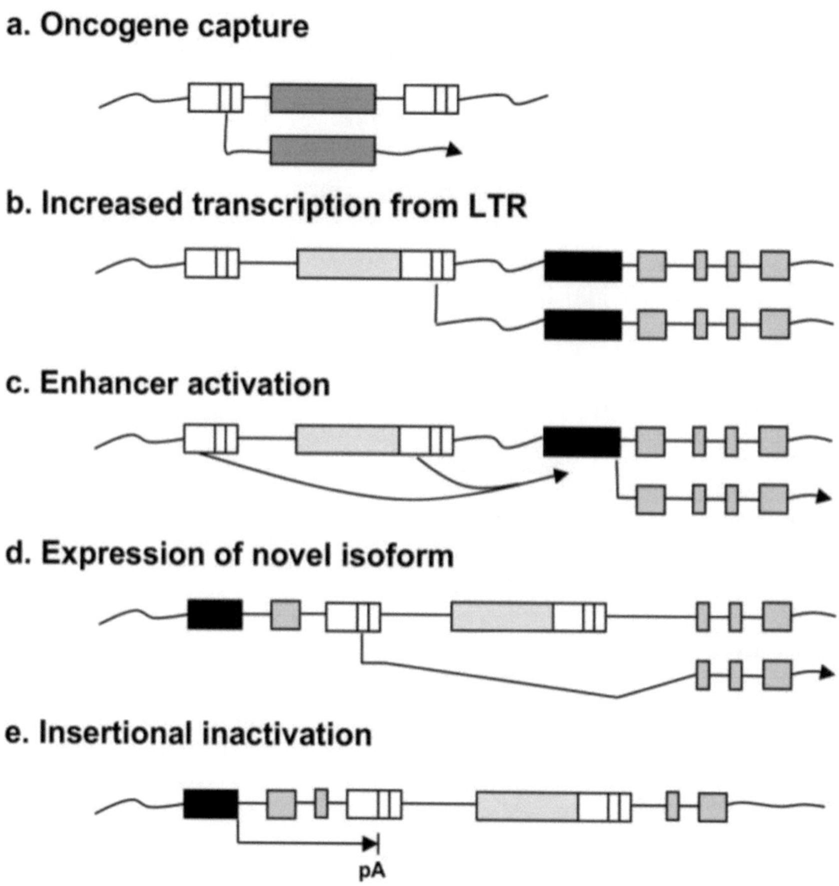

Figure 5. Retroviral mechanisms of oncogenesis. The detailed mechanisms are shown in the text. The integrated provirus is indicated by two LTRs. Cellular proto-oncogene promoter and exons are indicated by black and grey boxes respectively (Reproduced from [30]).

Even if a gene is inserted into a HSC similarly, it is also known that there are diseases which may develop a tumor, and diseases a tumor is not accepted to be. Each type of virus has a unique integration profile, and the following observations have been made [30]: (a) Different retroviral vectors have distinct integration profiles. (b) The route of entry does not appear to strongly affect distribution of integration sites. VSV-G–pseudotyped HIV vectors have an integration profile similar to HIV virions with the native HIV envelope despite differences in the route of entry. (c) The integration profile is largely independent of the target cell type, although the transcriptional program and epigenetic status of the target cell can influence integration site selection. (d) For lentiviruses, which can integrate independently of mitosis, the cell-cycle status of the target cell has only a modest effect on the distribution of integration sites.

In order to avoid genotoxicity, various SIN vectors have been developed and improved. In general, lentiviral vectors are considered to have a lower risk of oncogenesis than retroviral vectors [31]. However, when a HSC is the target cell, more attention should be required because tumorigenesis can occur when the cell with the inserted gene undergoes differentiation.

4. Clinical applications of HSC gene therapy

Diseases in which gene therapy using HSCs are being studied are shown in Table 1. They are roughly divided into hematological disorders, immunodeficiencies, and metabolic diseases. Most are congenital or hereditary diseases. The characteristic clinical features and recent basic science or clinical studies on HSC gene therapy for each disease are discussed below.

Congenital hematopoietic disorders
β-thalassaemia
Fanconi anemia
Hemophilia
Primary immunodeficiencies
X-linked severe combined immunodeficiency (SCID-X1)
Adenosine deaminase deficiency (ADA-SCID)
Chronic granulomatous disease (CGD)
Wiskott-Aldrich syndrome (WAS)
Janus kinase 3 (JAK3) deficiency
Purine nucleoside phosphorylase (PNP) deficiency
Leukocyte adhesion deficiency type 1 (LAD-1)
Congenital metabolic diseases
Mucopolysaccharidosis (MPS) types I, II, III, VII
Gaucher disease
X-linked adrenoleukodystrophy (X-ALD)

Table 1. Clinical applications of hematopoietic stem cell gene therapy.

4.1. β-thalassemia

Hemoglobin A (HbA), comprising 98% of adult human hemoglobin, is a tetramer with two α-globin and two β-globin chains combined with a heme group. β-thalassemia is an

http://dx.doi.org/10.5772/53587

autosomal hemoglobin disorder caused by decreased β-globin chain synthesis. Although individuals with β-thalassemia minor (heterozygote) may be asymptomatic or have mild to moderate microcytic anemia, β-thalassemia major (homozygote) progresses to serious anemia by one or two years of age, and hemosiderosis, iron overload caused by transfusion or increased iron absorption, develops. Since most patients develop life-threatening complications such as heart failure by adolescence, HSCT has been performed in patients with advanced disease [32]. In recent years, gene therapy using a lentiviral vector containing a functional β-globin gene has been performed in an HbE/ β-thalassemia (β^E/ β^0) transfusion-dependent adult male, who subsequently did not require transfusions for over 21 months [33].

The human β-globin locus is located in a large 70kb area which also contains some β-like globulin genes (ε, Gγ, Aγ, δ, β). Gene switching takes place according to the development stage, and the β-globin gene is transcribed and expressed specifically after birth. A powerful enhancer called the LCR (locus control region) exists on the 5′ side of the promoter. The LCR contains five DNase I hypersensitive sites, referred to as HS5 to HS1 starting from the 5′ side. Furthermore, HS5 contains CCCTC-binding factor (CTCF)-dependent insulator.

The structure of the lentiviral SIN vector used in gene therapy for β-thalassemia is shown in Fig. 6. To improve safety, two stop codons were inserted into the packaging signal (ψ) of GAG, the HS5 portion with insulator activity was deleted, and two copies of the 250 base pair (bp) core of the cHS4 chromatin insulators (chicken β-globin insulators) were inserted in the U3 region of the HIV 3′ LTR. Furthermore, the amino acid at the 87th position of β-globin was changed from threonine to glutamine. This altered β-globin can be distinguished from normal adult β-globin by high performance liquid chromatography (HPLC) analysis in individuals receiving red blood cell transfusion and β^+-thalassemia patients [33].

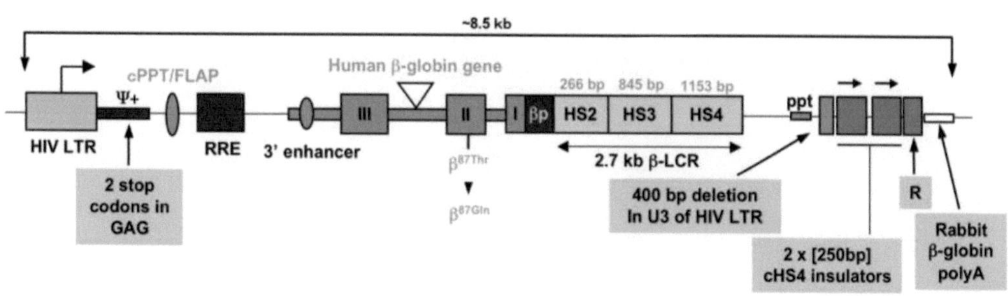

Figure 6. Diagram of the human β-globin gene in a lentiviral vector. HIV LTR, human immunodeficiency type-1 virus long terminal repeat; Ψ+, packaging signal; cPPT/flap, central polypurine tract/DNA flap; RRE, rev-responsive element; βp, human β-globin promoter; ppt, polypurine tract; HS, DNase I Hypersensitive Sites (Reproduced with color modification from [33])

A clinical study using this vector was performed in two β-thalassemia patients. As with autologous bone marrow transplantation, some of the patients' marrow cells were cryopreserved as a backup. The lentiviral vector particles containing a functional β-globin were introduced into the remaining cells. After the transfected cells were cultured for one week *ex vivo*, some were also cryopreserved. The patients were conditioned with intravenous busulfan (3.2 mg/kg/day for four days) without the addition of cyclophosphamide, before transplantation using the autologous gene-modified cryopreserved cells (Fig. 7) [34].

The first patient failed to engraft because the HSCs had been compromised by how they were handled, not because of any issues with the gene therapy vector, and ultimately used backup bone marrow. The second patient, as described previously, achieved long-term β-globin production; one-third of the patient's hemoglobin was produced by the genetically modified cells [33].

Furthermore, the detailed examination of the transgenic cells showed significantly increased expression of high mobility group AT-hook 2 (HMGA2), which interacts with transcription factors to regulate gene expression, in the clones where gene insertion occurred in the *HMGA2* gene. The proportion of the HMGA2 overexpressing clones increased with time, to over 50% of transgenic cells at 20 months after gene therapy. In this patient, the HMGA2 overexpressing cells were only 5% of all circulating hematopoietic cells and there was no evidence of malignant transformation. However, researchers point out that there was expressive production of a truncated form of the HMGA2 protein. Since truncated or overexpressed HMGA2 is observed with some blood cancers and non-malignant expansions of blood cells, caution is recommended with this therapy [34].

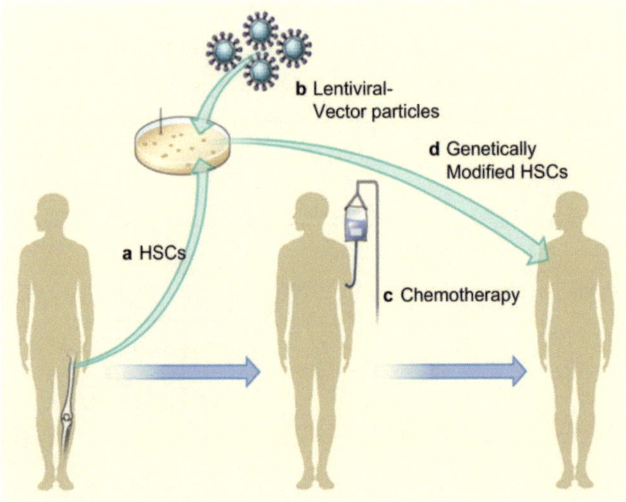

Figure 7. Gene-therapy procedure for patient with b-thalassemia. a. Hematopoietic stem cells (HSCs) are collected from the bone marrow of a patient with β-thalassemia and maintained them in culture. **b**, Lentiviral-vector particles containing a functional β-globin gene were then introduced into the cells and allowed them to expand further in culture. **c.** To eradicate the patient's remaining HSCs and make room for the geneticaaly modified cells, the patient underwent chemotherapy. **d.** The genetically modified HSCs were then transplanted into the patient (Reproduced from [34]).

Recently, researchers generated a LCR-free SIN lentiviral vector that combines two hereditary persistence of fetal hemoglobin (HPFH)-activating elements, resulting in therapeutic lev-

http://dx.doi.org/10.5772/53587

els of Aγ-globin protein produced by erythroid progenitors derived from thalassemic HSCs [35]. Both lentiviral-mediated γ-globin gene addition and genetic reactivation of endogenous γ-globin genes are considered potentially capable of providing therapeutic levels of hemoglobin F to patients with β-globin deficiency [36]. In addition, a trial of γ-globin induction with β-globin production using mithramycin, an inducer of γ-globin expression, to remove excess α-globin proteins in β-thalassemic erythroid progenitor cells was reported [37].

4.2. Fanconi anemia

Fanconi anemia is a hereditary disease characterized by cellular hypersensitivity to DNA crosslinking agents. It leads to bone marrow failure, such as aplastic anemia, by approximately eight years of age. Since there is a high risk of developing malignancy, HSCT has been performed as a curative treatment for bone marrow insufficiency. Although the ten-year probability of survival after transplant from an Human leukocyte antigen (HLA) -identical donor is over 80%, results with other donors are not satisfactory. HSC gene therapy is considered an alternative in cases where there is no HLA-identical donor available [38-40].

There are currently 13 discovered Fanconi anemia complement groups and 13 distinct genes (*FANCA, FANCB, FANCC, FANCD1, FANCD2, FANCE, FANCF, FANCG, FANCI, FANCJ, FANCL, FANCM, FANCN*) have been cloned. Mutations in *FANCB* are associated with an X-linked form of Fanconi anemia; mutations in the other genes are associated with autosomal recessive transmission. Although frequencies vary by geographical region, *FANCA* gene abnormalities are found in more than half of all Fanconi anemia patients [41]. Although one of the major hurdles in the development of gene therapy for Fanconi anemia is the increased sensitivity of Fanconi anemia stem cells to free radical-induced DNA damage during *ex vivo* culture and manipulation, retroviral and lentiviral vectors have been successfully employed to deliver complementing Fanconi anemia cDNA to HSCs with targeted disruptions of the *FANCA* and *FANCC* genes [20, 42-44]. In a phase I trial of *FANCA* gene therapy, gene transfer was performed with patient bone marrow-derived $CD34^+$ cells and the MSCV retroviral vector [38]. Whether sufficient HSCs can be obtained is a potential problem in Fanconi anemia patients due to possible bone marrow insufficiency, but in this study, sufficient target $CD34^+$ cells were obtained from most patients. Two patients had *FANCA*-transduced cells successfully infused. The procedure was safe, well tolerated, and resulted in transient improvements in hemoglobin and platelet counts [39]. However, transduced cell products were not obtained in one patient who required cryopreserved bone marrow. The first clinical study of *FANCC* gene therapy using a retroviral vector involved four patients. Although functional *FANCC* gene expression was observed in peripheral blood and bone marrow cells, the results were transient [43].

Engraftment efficiency of *FANCA*-modified cells using a lentiviral vector was studied in a mouse model. Rapid transduction with four hours of culture using only SCF and megakaryocyte growth and development factor and minimal differentiation of gene-induced cells is better than standard 96-hour culture using a variety of cytokines, including SCF, interleukin-11, Flt-3 ligand, and IL-3 [44]. Moreover, a recent trial demonstrated enhanced viability and engraftment of gene-corrected cells in patients with *FANCA* abnormalities with short transduction (overnight), low oxidative stress (5% oxygen), and the anti-oxidant N-acetyl-L-cysteine [20]. Lentiviral transduction of unselected Fanconi anemia bone marrow cells mediates efficient phenotypic correction of hematopoietic progenitor cells and CD34$^-$ mesenchymal stromal cells, with increased efficacy in hematopoietic engraftment [45]. In *Fancg -/-* mice, the wild-type mesenchymal stem and progenitor cells play important roles in the reconstitution of exogenous HSCs *in vitro* [46]. Recently, a new approach that directly injects lentiviral vector particles into the femur for *FANCC* gene transfer in mice was able to successfully introduce the *FANCC* gene to HSCs. This result provides evidence that targeting the HSCs directly in their native environment enables efficient and long-term correction of bone marrow defects in Fanconi anemia [47].

In recent years, the design of lentiviral vectors used for gene therapy in Fanconi anemia has improved. Although the vav and phosphoglycerate kinase (PGK) promoters are relatively weak, physiological levels of *FANCA* gene expression can be obtained in lymphoblastoid cells. CMV and spleen focus-forming virus (SFFV) promoters result in overexpression of *FANCA*. The PGK-*FANCA* lentiviral vectors with either a wild-type woodchuck hepatitis virus posttranscriptional regulatory element (WPRE) or a mutated WPRE in the 3′ region have higher levels of *FANCA* gene expression. In conclusion, lentiviral vectors with a mutated WPRE and a PGK promoter are considered the most suitable with respect to safety and efficiency for Fanconi anemia gene therapy [48].

There was a recent interesting report on the use of induced pluripotent stem cells (iPS cell). Instead of introducing a repaired gene into the HSCs of a patient with a *FANCA* gene abnormality, the modified gene was introduced into more stable somatic cells, e.g. fibroblasts, and iPS cells were derived from the genetically modified somatic cells. If HSCs can be produced from genetically modified iPS cells, hematological function can be efficiently reconstructed in patients with hematologic disorders [49].

4.3. Hemophilia

Hemophilia is a common congenital coagulopathy caused by coagulation factor VIII (hemophilia A) or IX (hemophilia B) deficiency. Although the genes encoding both factor VIII (Xq28) and factor IX (Xq27) are located on the X chromosome and most cases are X-linked, many sporadic variations have been reported. Factor substitution therapies have been used to treat hemophilia for many years. However, there is great hope for gene therapy with hemophilia because coagulation factors have short half-lives (factor VIII, 8 to 12 hours; factor IX, 18 to 24 hours), and an inhibitor is produced in many cases. Furthermore, it is possible for gene therapy to suppress immunogenicity by introducing a mutant protein that lacks the domain with which the inhibitor interacts. Since both coagulation factors are usually produced in the liver, there are few studies involving HSCs. In addition to hepatocytes, trials introducing the modified gene directly into splenic cells, endothelial cells, myoblasts, fibroblasts, etc. have been reported [50-52]. Since the factor IX gene (34 kb) is smaller than the factor VIII gene (186 kb), hemophilia B gene therapy can be possible with an adenovirus vector or an AAV vector. Therefore, hemophilia B is progressing more as a field of gene therapy research even through there are five times more patients with hemophilia A [51-53].

Recently, human factor VIII variant genes were successfully introduced into the HSCs of a mouse with hemophilia A resulting in therapeutic levels of factor VIII variant protein expression. This variant factor VIII has changes in the B and A2 domain in addition to the A1 domain for improved secretion and reduced immunogenicity (wild-type factor VIII has six domains, A1, A2, B, A3, C1, and C2) [54]. To ameliorate the symptoms of hemophilia A, partial replacement of the mutated liver cells by healthy cells in hemophilia A mice was challenged with allogeneic bone marrow progenitor cell transplantation. In this study, the bone marrow progenitor cell-derived hepatocytes and sinusoidal endothelial cells synthesized factor VIII, showing that autologous gene-modified bone marrow progenitor cells have the potential to treat hemophilia [55].

4.4. Primary immunodeficiencies

Although HSCT has been widely performed as curative treatment for primary immunodeficiencies, gene therapy has been considered when there is no HLA-identical donor available. As previously shown, the first successful gene therapy was performed in a patient with ADA deficiency in the U.S. in 1990. Since the gene was introduced into T lymphocytes, frequent treatment was required. However, this treatment was associated with an unacceptable level of toxicity. Since transfected vector and normal ADA gene expression in T lymphocytes continued for two years after the cessation of treatment [1], gene therapy attracted attention. With advances in HSC gene-transfer technology, gene therapy for many primary immunodeficiencies can now be considered [56].

4.4.1. SCID-X1

SCID-X1 is an X-linked disease caused by deficiency of the common γ (γc) chain in the IL-2 receptor. Because the γc chain is common to the IL-4, IL-7, IL-9, IL-15, and IL-21 receptors, in SCID-X1 patients, there are defects in T and natural killer (NK) cells, and B cell dysfunction are usually observed [57]. Patients begin suffering from various infections starting several weeks after birth. Without curative treatment, such as HSCT, patients die in infancy.

In SCID-X1, since T cells are lacking, engraftment of the gene-transduced cells can be achieved without pre-conditioning therapy. In the clinical studies of SCID-X1 patients in France and the U.K., the MFG retroviral vector was used with HSCs obtained from the patient. After gene therapy, many patients had improvements in immune function. However, since the genes regulating lymphocyte proliferation, such as *LIM domain only 2 (LMO2), Bmi1, cyclin D2 (CCND2)* are near the gene insertion region, there was a high frequency of T-cell leukemia after treatment. Furthermore, in the patients who developed leukemia, additional chromosomal changes, including activating mutations of *Notch1*, changes in the T cell receptor β region, and deletion of tumor suppressor genes, e.g. cyclin-dependent kinase-2A (*CDKN2A*) were observed [58]. Almost gene integration sites by the retroviral vector were inside or near genes that are highly expressed in CD34 positive stem cells. Furthermore, the activity of protein kinases or transferases coded by these activated genes was stronger in CD3 positive T cells than CD34 positive cells [59]. Thus, gene integration mediated by a retrovirus influences the target cell's dormant capacity for survival, engraftment, and proliferation.

Although continuous T cell production was founded in many cases, there was little reconstruction of myeloid cells and B cells, and some patients required continuous immunoglobulin substitution therapy. The use of conditioning therapy is also related to immunological reconstruction after γc chain gene therapy. There is decreased NK cell reconstruction without conditioning therapy, so conditioning chemotherapy is required for the engraftment of undifferentiated stem cells [58]. A trial of SCID-X1 gene therapy in the U.S. involved three patients ranging from 10 to 14 years of age. They had poor immunological recovery after allergenic HSCT and T cell recovery was only observed in the youngest patient, suggesting there is a limit to the recovery of the function of the thymus in older children [60].

To study whether activation of genes near the region of gene insertion or inserted γc chain gene expression itself induces oncogenicity during SCID-X1 gene therapy, a study of the human γc chain gene being expressed under the control of the human CD2 promoter and LTR (CD2- γc chain gene) was performed in mice. When the CD2- γc chain gene was expressed in transgenic mice, a few abnormalities involving T cells were observed, but tumorigenesis was not observed and T and B cell functions were recovered in γc chain-gene deficient mice. This study demonstrated that when the γc ▪ c chain gene is expressed externally, SCID-X1 may be treated safely [61].

Although SIN vectors were developed from earlier retroviral [62] or lentiviral vectors [63] to reduce the risk of oncogenicity in SCID-X1 gene therapy, genotoxicity unrelated to mutations in gene insertion regions or γc chain gene overexpression have been reported with lentiviral vectors in recent years, and it seems that more sophisticated vector development is required [64].

4.4.2. ADA-SCID

ADA is an enzyme that catalyzes the conversion of purine metabolism products adenosine and deoxyadenosine into inosine or deoxyinosine. ADA-SCID is an autosomal recessive disease that results in the accumulation of adenosine, deoxyadenosine, and deoxyadenosinetriphosphate (dATP). Accumulated phosphorylated purine metabolism products act on the thymus and cause the maturational or functional disorder of lymphocytes. Because ADA-SCID patients have both T and B cell production fail, patients have a severe combined immunodeficiency disease with a clinical presentation similar to SCID-X1 results, but unlike SCID-X1, many patients have a low level of T cells. Although enzyme replacement therapy with polyethylene glycol–modified bovine ADA (PEG-ADA) was developed to treat ADA-SCID, it is limited by the development of neutralizing antibodies and the cost of lifelong treatment.

In ADA-SCID, since T cell counts are increased by PEG-ADA, gene therapy to increase peripheral T cell counts was attempted during the early stages of gene therapy. Although adverse events were not observed and continuous expression of ADA was achieved in many patients, reconstruction of immune function was not obtained and substitution therapy with PEG-ADA remained necessary. Therefore, HSCs were no longer the target of gene therapy for ADA-SCID. Since ADA-SCID patients have T cells, nonmyeloablative conditioning was performed to achieve gene-transduced HSC engraftment [25, 65].

In a joint Italian-Israeli study started in 2000, ten ADA-SCID children were infused with CD34 positive cells transduced with a MoMLV retroviral vector containing the *ADA* gene after nonmyeloablative conditioning with busulfan (2mg/kg/day for two days). T cell counts or function were improved in nine out of the ten patients, and PEG-ADA was discontinued in eight. Many patients also had improvements in B or NK cell function, and immunoglobulin substitution therapy was discontinued in five patients. Although some patients had serious adverse events including prolonged neutropenia, hypertension, Epstein-Barr virus infection, and autoimmune hepatitis, there were no cases of treatment-induced leukemia [25].

As with SCID-X1, the retroviral vector gene insertion region is also near genes that control cell proliferation or self-duplication, such as *LMO2*, or proto-oncogenes [66]. In clinical studies performed in France, the U.S., and the U.K., none of the ADA-SCID patients had adverse events related to insertional mutagenesis, such as leukemia [67, 68]. Thus, HSC gene therapy for ADA-SCID using a lentiviral vector [69] is expected to become the alternative therapy in cases without a suitable donor for HSCT [70]. As an alternative to HSC-based gene therapy, a study using an AAV vector has reported ADA gene expression in various tissues, including heart, skeletal muscle, and kidney [71].

4.4.3. CGD

CGD is a disease caused by an abnormality in nicotinamide dinucleotide phosphate (NADPH) oxidase expressed in phagocytes, resulting in failure to produce reactive oxygen species and decreased ability to kill bacteria or fungi after phagocytosis. NADPH oxidase consists of $gp91^{phox}$ (Nox2) and $p22^{phox}$ which together constitute the membrane-spanning component flavocytochrome b558 (CYBB), and the cytosolic components $p47^{phox}$, $p67^{phox}$, $p40^{phox}$, and Rac. CGD is caused by a functional abnormality in any of these components. Mutations in *$gp91^{phox}$* on the X chromosome account for approximately 70% of CGD cases. CGD patients are afflicted with recurrent opportunistic bacterial and fungal infections, leading to the formation of chronic granulomas. Although lifelong antibiotic prophylaxis reduces the incidence of infections, the overall annual mortality rate remains high (2%–5%) and the success rate of HSCT is limited by graft-versus-host-disease and inflammatory flare-ups at infected sites [56].

In the initial trials of CGD gene therapy without any conditioning therapy, *$p47^{phox}$* or *$gp91^{phox}$* gene was inserted using a retroviral vector. The inserted gene was expressed in peripheral blood granulocytes three to six weeks after re-infusion and mobilization by granulocyte colony-stimulating factor (G-CSF), but there was no clinical effect within six months [72-74].

In a German study where *gp91phox* was inserted with busulfan conditioning (8mg/kg), there were fewer infections after gene therapy. Gene expression was observed in 20% of leukocytes in the first month, rising to 80% at one year. However, in the gene insertion region there are genes related to myeloid cell proliferation, such as *myelodysplastic syndrome 1-ecotropic virus integration site 1 (MDS1/EVI1), PR domain containing protein 16 (PRDM16), SET binding protein 1 (SETBP1).* Two patients developed myelodysplasia [75]. These two patients had monosomy 7, considered to be related to *EVI1* activation. One died of severe sepsis 27 months after gene therapy. Although the gene-inserted cells remained expressed in this patient, methylation of the CpG site in the LTR of the viral vector was observed and the expression of the inserted *gp91*^{phox} gene was decreased. Interestingly, methylation was restricted to the promoter region of the LTR; the enhancer region was not methylated. Therefore, although *gp91*^{phox} gene expression was decreased, the activation of *EVI1* near the inserted region occurred, leading to clonal proliferation [76]. Since there is a possibility that the transcription activity of genes related to myeloid cell proliferation near the gene insertion site will be increased, there remains a concern about tumorigenesis with peripheral stem cells mobilization by G-CSF in CGD patients, as with X-SCID [74].

Recently, next-generation gene therapy for CGD using lineage- and stage-restricted lentiviral vectors to avoid tumorigenesis [77] and novel approaches involving iPSs derived from CGD patients using zinc finger nuclease (ZFN)-mediated gene targeting were studied [78]. Specific gene targeting can be performed in human iPSs using ZFNs to induce sequence-specific double-strand DNA breaks that enhance site-specific homologous recombination. A single-copy of *gp91*^{phox} was targeted into one allele of the "safe harbor" AAVS1 locus in iPSs [79].

4.4.4. Wiskott-Aldrich Syndrome (WAS)

WAS is a severe X-linked immunodeficiency caused by mutations in the gene encoding the WAS protein (WASP), a key regulator of signaling and cytoskeletal reorganization in hematopoietic cells. Mutations in *WAS* gene result in a wide spectrum of clinical manifestations ranging from relatively mild X-linked thrombocytopenia to the classic WAS phenotype characterized by thrombocytopenia, immunodeficiency, eczema, high susceptibility to developing tumors, and autoimmune manifestations [80]. Preclinical and clinical evidence suggest that WASP-expressing cells have a proliferative or survival advantage over WASP-deficient cells, supporting the development of gene therapy [56]. Furthermore, up to 11% of WAS patients have somatic mosaicism due to spontaneous *in vivo* reversion to the normal genotype, and in WAS patients, accumulation of normal T-cell precursors are sometimes seen [81].

In one preclinical study introducing the *WAS* gene into human T and B cells or mouse HSCs using a retroviral vector, recovery of T cell function and immune reactions to infection were observed [82, 83]. The first clinical study of WAS using HSCs involved two young boys in Germany. The WASP-expressing retroviral vector was transfected into CD34 positive cells obtained by apheresis of peripheral blood. Busulfan was used for conditioning therapy (4mg/kg/day for two days). Over two years, WASP gene expression by HSCs, lymphoid and myeloid cells, and platelets was sustained, and the number and function of monocytes, T, B, and NK cells normalized. Clinically, hemorrhagic diathesis, eczema, autoimmunity, and the predisposition to severe infections were diminished. Since comprehensive insertion-site analysis showed vector integration near multiple genes controlling growth and immunologic responses in a persistently polyclonal hematopoiesis, careful monitoring for tumorigenesis is necessary, as with SCID-X1 and CGD [84, 85].

SIN lentiviral vectors using the minimal domain of the WAS promoter or other ubiquitous promoters, such as the PGK promoter, are currently being developed for WAS gene therapy. Preclinical studies using the HSCs obtained from mice or human patients have yield good results in terms of gene expression and genotoxicity [86-90].

Since a study using human embryonic stem cells (hESCs) and WAS-promoter–driven lentiviral vectors labeled by green fluorescent protein (GFP) showed highly specific gene expression in hESCs-derived HSCs, the WAS promoter will be used specifically in the generation of hESC-derived HSCs [91].

4.4.5. Janus Kinase 3 (JAK3) deficiency

JAK3 deficiency is characterized by the absence of T and NK cells and impaired function of B cells, similar to SCID-X1. Treatment consists of HSCT with an HLA-identical or HLA-haplo-identical donor, often the parents of the patient, with T cell depletion. Engraftment is successful in most cases.

Although the recovery of T cell function is usually observed after HSCT, there are usually no improvements in B or NK cell function [92]. One case report involved introduction of *JAK3* into the patient's bone marrow CD34 positive cells using the MSCV retroviral vector. In this study, immunological recovery was not achieved although gene expression was observed for seven months [93]. Since JAK activation can cause T-cell lymphoma, tumorigenesis remains a concern with JAK gene therapy [92].

4.4.6. Purine Nucleoside Phosphorylase (PNP) deficiency

PNP metabolizes adenosine into adenine, inosine into hypoxanthine, and guanosine into guanine. PNP deficiency is an autosomal recessive metabolic disorder characterized by lethal T cell defects resulting from the accumulation of products from purine metabolism.

In PNP-deficient mice, transplantation of bone marrow cells transduced with a lentiviral vector containing human *PNP* resulted in human PNP expression, improved thymocyte maturation, increased weight gain, and extended survival. However, 12 weeks after transplant, the benefit of *PNP*-transduced cells and the percentage of engrafted cells decreased [94].

4.4.7. Leukocyte Adhesion Deficiency type 1 (LAD-1)

LAD-1 is a primary immunodeficiency disease caused by abnormalities in the leukocyte integrin CD11/CD18 heterodimer due to mutations in the *CD18* gene. It is similar to canine leukocyte adhesion deficiency (CLAD). LAD-1 patients begin experiencing repeated serious bacterial infections immediately after birth.

In order to suppress gene activation near the gene insertion region in CLAD and to obtain the sufficient expression of the *CD18* gene, researches have used various promoters with a lentiviral vector or foamy virus, a retroviral vector. *In vivo* animal experiments using a PGK or an elongation factor 1α promoter did not lead to symptom improvement [95-97], but improvement was seen with CD11b and CD18 promoters, respectively, with a SIN lentiviral vector in one animal study [98].

4.5. Mucopolysaccharidosis (MPS)

MPS is a general term for diseases characterized by glycosaminoglycan (GAG) accumulation into lysosomes as a result of deficiencies in lysosomal enzymes that degrade GAG. Although there are more than ten enzymes that are known to degrade GAG, MPS is divided into seven types: type I (α-L-iduronidase deficiency, Hurler syndrome, Sheie syndrome, Hurler-Sheie syndrome), type II (iduronate sulfatase deficiency, Hunter syndrome), type III (heparan N-sulfatase deficiency, α-N-acetylglucosaminidase deficiency, α-glucosaminidase acetyltransferase deficiency, N-acetylglucosamine 6-sulfatase deficiency, Sanfilippo syndrome), type IV (galactose 6-sulfatase deficiency, Morquio syndrome), type VI (N-acetylgalactosamine 4-sulfatase deficiency, Maroteaux-Lamy syndrome), type VII (β-glucuronidase deficiency, Sly syndrome), and type IX (hyaluronidase deficiency). Type II is X-linked; the other types are autosomal recessive. Although lysosomes are found in almost all cells, MPS mainly affects internal organs such as the brain, heart, bones, joints, eyes, liver, and spleen. The extent of disease, including mental retardation, varies with MPS type.

In types I, II, and VI, enzyme replacement therapy is performed. HSCT is performed in types I, II, IV, and VII. Gene therapy for types I, II, III, and VII type have been investigated. There are trials using an AAV or adenovirus vector to insert the modified gene into various cell types, including hepatocytes, muscle cells, myoblasts, and fibroblasts [99].

The first study of HSC gene therapy for MPS using a retroviral vector was performed on type VII mice in 1992, resulting in decreased accumulation of GAG in the liver and spleen but not in the brain and eyes [100]. Subsequent studies in type I and III animal models showed decreases in GAG accumulation in the kidneys and brain. Introductory efficiency and immunological reactions are considered challenges in HSC gene therapy for MPS [99].

Restoring or preserving central nervous system (CNS) function is one of the major challenges in the treatment of MPS. Since replaced enzymes easily cannot pass the blood-brain barrier (BBB), a high dose of enzyme is needed to improve CNS function. Gene therapy faces the same challenge. Even with high expression of enzyme by, for example, hepatocytes, the BBB prevents efficient delivery into the CNS. When a lentiviral vector is directly injected into the body, gene expression in brain tissue is observed, although the underlying mechanism is unknown. There are also trials where AAV vectors are directly injected into the CNS of mice or dogs and gene expression was observed in brain tissue [99].

Recently, a lentiviral vector using an ankyrin-1-based erythroid-specific hybrid promoter/enhancer (IHK) was used with HSCs to obtain gene expression only in erythroblasts for type I MPS. This approach resulted in decreased accumulation of GAG in the liver, spleen, heart, and CNS via enzyme expression in erythroblasts [101].

4.6. Gaucher disease

Gaucher disease is the most common lysosomal storage disorder. It is caused by deficiency of glucocerebroside-cleaving enzyme (β-glucocerebrosidase), resulting in the accumulation of glucocerebroside in the reticuloendothelial system [102]. This autosomal recessive disease presents with hepatosplenomegaly, anemia, thrombocytopenia, and convulsions with or without mental retardation. It is classified into three types based on the clinical course or existence of neurological symptoms: type I (non-neuropathic, adult type), type II (acute neuropathic, infantile type), and type III (chronic neuropathic, juvenile type). Enzyme replacement therapy has been established in type I. As with MPS, since it is difficult to improve CNS symptoms with enzyme replacement therapy, HSCT is used, especially with type III. Gene therapy is considered in cases with little improvement with enzyme replacement therapy [103].

For Gaucher disease without CNS symptoms, a animal model using an AAV vector to produce enzyme in hepatocytes yielded good results [103]. HSC gene therapy using a retroviral vector was attempted in type I mice. The treated cells had higher β-glucocerebrosidase activity than the HSCs from wild-type mice. Glucocerebroside levels normalized five to six months after treatment and no infiltration of Gaucher cells could be observed in the bone marrow, spleen, and liver [104]. In recent years, development of lentiviral vectors including the human glucocerebrosidase gene [105] and low-risk HSCT with nonmyeloablative doses of busulfan (25mg/kg) and no radiation therapy have been attempted in mice [106].

4.7. X-ALD

X-ALD is a peroxisomal disease in which a lipid metabolism abnormality causes demyelination of CNS tissues and dysfunction of the adrenal gland. It results from mutations in the ATP-binding cassette sub-family D (*ABCD1*) gene that codes for the adrenoleukodystrophy (ALD) protein. Behavioral disorders, mental retardation, or both occur by the age of five or six. Once symptoms appear, they progress to gait disorder and visual impairment within several months and the prognosis is poor. Increased levels of very long chain fatty acids (VLCFA), such as C25:0 or C26:0, are observed in the CNS, plasma, erythrocytes, leucocytes, etc. If the neurological defects are not severe, arrest of or improvement in symptoms can be obtained with HSCT [107].

One study has reported the introduction of wild-type *ABCD1* using a lentiviral vector into peripheral blood CD34 positive cells of two patients with no HLA-identical donor. The patients received a transfusion of autologous gene-modified cells after myeloablative conditioning therapy. At three years of follow-up, ALD proteins were expressed in approximately 7–14% of neutrophils, monocytes, and T cells. Clinically, cerebral demyelination stopped 14 and 16 months after gene therapy, respectively, similar to results with allergenic HSCT [108, 109].

5. Conclusion

Gene therapy using HSCs was outlined. HSCT with HSCs can replace all of the patient's original HSCs with donor HSCs. Therefore, gene therapy using HSCs is an alternative if the patient does not have an HLA-identical donor or cannot tolerate the conditioning regimen or other HSCT-related side effects. Fully myeloablative or nonmyeloablative conditioning regimens are still necessary to eliminate potential competition within the bone marrow compartment, in an attempt to increase the number of gene-modified HSCs or progenitors that produce the therapeutic enzyme or protein. With gene therapy, eliminating the risk of immune reactions against the transgene is necessary. Lentiviral vectors in clinical use must not be contaminated by replication-competent recombinant vectors related to the parent HIV-1 virus. The main risk of retrovirus- or lentivirus-mediated gene therapy may prove to be insertional mutagenesis caused by random retroviral integration leading to activation of proto-oncogenes or inactivation of tumor-suppressor genes, ultimately leading to malignancy [107]. However, with advances in gene introduction technology, such as the development of the SIN vector and advances in cell or gene-region targeting, gene therapy can be done more safely and efficiently. Furthermore, since cells more immature than HSCs, i.e., iPS cells, are available, further advances in HSC gene therapy are expected in the future.

Author details

Toshihisa Tsuruta

Department of Clinical Examination, National Hospital Organization Kumamoto Medical Center, Japan

References

[1] Blaese RM, Culver KW, Miller AD, Carter CS, Fleisher T, Clerici M, et al. T lymphocyte-directed gene therapy for ADA- SCID: initial trial results after 4 years. Science. 1995;270(5235):475-80. Epub 1995/10/20.

[2] Bordignon C, Notarangelo LD, Nobili N, Ferrari G, Casorati G, Panina P, et al. Gene therapy in peripheral blood lymphocytes and bone marrow for ADA- immunodeficient patients. Science. 1995;270(5235):470-5. Epub 1995/10/20.

[3] Hoogerbrugge PM, van Beusechem VW, Fischer A, Debree M, le Deist F, Perignon JL, et al. Bone marrow gene transfer in three patients with adenosine deaminase deficiency. Gene therapy. 1996;3(2):179-83. Epub 1996/02/01.

[4] Hacein-Bey-Abina S, von Kalle C, Schmidt M, Le Deist F, Wulffraat N, McIntyre E, et al. A serious adverse event after successful gene therapy for X-linked severe combined immunodeficiency. N Engl J Med. 2003;348(3):255-6. Epub 2003/01/17.

[5] Till JE, Mc CE. A direct measurement of the radiation sensitivity of normal mouse bone marrow cells. Radiat Res. 1961;14:213-22. Epub 1961/02/01.

[6] Abe T, Masuya M, Ogawa M. An efficient method for single hematopoietic stem cell engraftment in mice based on cell-cycle dormancy of hematopoietic stem cells. Exp Hematol. 2010;38(7):603-8.

[7] Osawa M, Hanada K, Hamada H, Nakauchi H. Long-term lymphohematopoietic reconstitution by a single CD34-low/negative hematopoietic stem cell. Science. 1996;273(5272):242-5.

[8] Wilson A, Laurenti E, Oser G, van der Wath RC, Blanco-Bose W, Jaworski M, et al. Hematopoietic stem cells reversibly switch from dormancy to self-renewal during homeostasis and repair. Cell. 2008;135(6):1118-29.

[9] Ikuta K, Weissman IL. Evidence that hematopoietic stem cells express mouse c-kit but do not depend on steel factor for their generation. Proceedings of the National Academy of Sciences of the United States of America. 1992;89(4):1502-6.

[10] Solar GP, Kerr WG, Zeigler FC, Hess D, Donahue C, de Sauvage FJ, et al. Role of c-mpl in early hematopoiesis. Blood. 1998;92(1):4-10.

[11] Kimura S, Roberts AW, Metcalf D, Alexander WS. Hematopoietic stem cell deficiencies in mice lacking c-Mpl, the receptor for thrombopoietin. Proceedings of the National Academy of Sciences of the United States of America. 1998;95(3):1195-200.

[12] Buza-Vidas N, Antonchuk J, Qian H, Mansson R, Luc S, Zandi S, et al. Cytokines regulate postnatal hematopoietic stem cell expansion: opposing roles of thrombopoietin and LNK. Genes Dev. 2006;20(15):2018-23.

[13] Seita J, Ema H, Ooehara J, Yamazaki S, Tadokoro Y, Yamasaki A, et al. Lnk negatively regulates self-renewal of hematopoietic stem cells by modifying thrombopoietin-mediated signal transduction. Proceedings of the National Academy of Sciences of the United States of America. 2007;104(7):2349-54.

[14] Arai F, Hirao A, Ohmura M, Sato H, Matsuoka S, Takubo K, et al. Tie2/angiopoietin-1 signaling regulates hematopoietic stem cell quiescence in the bone marrow niche. Cell. 2004;118(2):149-61.

[15] Yamazaki S, Iwama A, Morita Y, Eto K, Ema H, Nakauchi H. Cytokine signaling, lipid raft clustering, and HSC hibernation. Ann N Y Acad Sci. 2007;1106:54-63.

[16] Yamazaki S, Iwama A, Takayanagi S, Eto K, Ema H, Nakauchi H. TGF-beta as a candidate bone marrow niche signal to induce hematopoietic stem cell hibernation. Blood. 2009;113(6):1250-6. Epub 2008/10/24.

[17] Hosokawa K, Arai F, Yoshihara H, Nakamura Y, Gomei Y, Iwasaki H, et al. Function of oxidative stress in the regulation of hematopoietic stem cell-niche interaction. Biochemical and biophysical research communications. 2007;363(3):578-83. Epub 2007/09/28.

[18] Arai F, Suda T. Quiescent stem cells in the niche. StemBook. Cambridge (MA)2008.

[19] Suda T, Takubo K, Semenza GL. Metabolic regulation of hematopoietic stem cells in the hypoxic niche. Cell stem cell. 2011;9(4):298-310. Epub 2011/10/11.

[20] Becker PS, Taylor JA, Trobridge GD, Zhao X, Beard BC, Chien S, et al. Preclinical correction of human Fanconi anemia complementation group A bone marrow cells using a safety-modified lentiviral vector. Gene therapy. 2010;17(10):1244-52. Epub 2010/05/21.

[21] Hawley RG, Lieu FH, Fong AZ, Hawley TS. Versatile retroviral vectors for potential use in gene therapy. Gene therapy. 1994;1(2):136-8. Epub 1994/03/01.

[22] Halene S, Wang L, Cooper RM, Bockstoce DC, Robbins PB, Kohn DB. Improved expression in hematopoietic and lymphoid cells in mice after transplantation of bone marrow transduced with a modified retroviral vector. Blood. 1999;94(10):3349-57. Epub 1999/11/24.

[23] Riviere I, Brose K, Mulligan RC. Effects of retroviral vector design on expression of human adenosine deaminase in murine bone marrow transplant recipients engrafted with genetically modified cells. Proceedings of the National Academy of Sciences of the United States of America. 1995;92(15):6733-7. Epub 1995/07/18.

[24] Hacein-Bey-Abina S, Le Deist F, Carlier F, Bouneaud C, Hue C, De Villartay JP, et al. Sustained correction of X-linked severe combined immunodeficiency by ex vivo gene therapy. N Engl J Med. 2002;346(16):1185-93.

[25] Aiuti A, Cattaneo F, Galimberti S, Benninghoff U, Cassani B, Callegaro L, et al. Gene therapy for immunodeficiency due to adenosine deaminase deficiency. N Engl J Med. 2009;360(5):447-58.

[26] Dull T, Zufferey R, Kelly M, Mandel RJ, Nguyen M, Trono D, et al. A third-generation lentivirus vector with a conditional packaging system. J Virol. 1998;72(11): 8463-71.

[27] Persons DA. Hematopoietic stem cell gene transfer for the treatment of hemoglobin disorders. Hematology Am Soc Hematol Educ Program. 2009:690-7.

[28] Verhoeyen E, Wiznerowicz M, Olivier D, Izac B, Trono D, Dubart-Kupperschmitt A, et al. Novel lentiviral vectors displaying "early-acting cytokines" selectively promote survival and transduction of NOD/SCID repopulating human hematopoietic stem cells. Blood. 2005;106(10):3386-95. Epub 2005/08/04.

[29] Frecha C, Fusil F, Cosset FL, Verhoeyen E. In vivo gene delivery into hCD34+ cells in a humanized mouse model. Methods Mol Biol. 2011;737:367-90. Epub 2011/05/19.

[30] Trobridge GD. Genotoxicity of retroviral hematopoietic stem cell gene therapy. Expert Opin Biol Ther. 2011;11(5):581-93.

[31] Modlich U, Navarro S, Zychlinski D, Maetzig T, Knoess S, Brugman MH, et al. Insertional transformation of hematopoietic cells by self-inactivating lentiviral and gammaretroviral vectors. Mol Ther. 2009;17(11):1919-28.

[32] Angelucci E. Hematopoietic stem cell transplantation in thalassemia. Hematology Am Soc Hematol Educ Program.2010:456-62.

[33] Cavazzana-Calvo M, Payen E, Negre O, Wang G, Hehir K, Fusil F, et al. Transfusion independence and HMGA2 activation after gene therapy of human beta-thalassaemia. Nature. 2010;467(7313):318-22.

[34] Persons DA. Gene therapy: Targeting beta-thalassaemia. Nature. 2010;467(7313): 277-8.

[35] Papanikolaou E, Georgomanoli M, Stamateris E, Panetsos F, Karagiorga M, Tsaftaridis P, et al. The new self-inactivating lentiviral vector for thalassemia gene therapy combining two HPFH activating elements corrects human thalassemic hematopoietic stem cells. Hum Gene Ther. 2012;23(1):15-31. Epub 2011/08/31.

[36] Wilber A, Hargrove PW, Kim YS, Riberdy JM, Sankaran VG, Papanikolaou E, et al. Therapeutic levels of fetal hemoglobin in erythroid progeny of beta-thalassemic CD34+ cells after lentiviral vector-mediated gene transfer. Blood. 2011;117(10): 2817-26. Epub 2010/12/16.

[37] Zuccato C, Breda L, Salvatori F, Breveglieri G, Gardenghi S, Bianchi N, et al. A combined approach for beta-thalassemia based on gene therapy-mediated adult hemoglobin (HbA) production and fetal hemoglobin (HbF) induction. Annals of hematology. 2012;91(8):1201-13. Epub 2012/03/31.

[38] Cheng L, Du C, Lavau C, Chen S, Tong J, Chen BP, et al. Sustained gene expression in retrovirally transduced, engrafting human hematopoietic stem cells and their lympho-myeloid progeny. Blood. 1998;92(1):83-92.

[39] Kelly PF, Radtke S, von Kalle C, Balcik B, Bohn K, Mueller R, et al. Stem cell collection and gene transfer in Fanconi anemia. Mol Ther. 2007;15(1):211-9.

[40] Tolar J, Adair JE, Antoniou M, Bartholomae CC, Becker PS, Blazar BR, et al. Stem cell gene therapy for fanconi anemia: report from the 1st international Fanconi anemia gene therapy working group meeting. Mol Ther. 2011;19(7):1193-8. Epub 2011/05/05.

[41] Kutler DI, Singh B, Satagopan J, Batish SD, Berwick M, Giampietro PF, et al. A 20-year perspective on the International Fanconi Anemia Registry (IFAR). Blood. 2003;101(4):1249-56. Epub 2002/10/24.

[42] Liu JM, Young NS, Walsh CE, Cottler-Fox M, Carter C, Dunbar C, et al. Retroviral mediated gene transfer of the Fanconi anemia complementation group C gene to hematopoietic progenitors of group C patients. Hum Gene Ther. 1997;8(14):1715-30. Epub 1997/10/10.

[43] Liu JM, Kim S, Read EJ, Futaki M, Dokal I, Carter CS, et al. Engraftment of hematopoietic progenitor cells transduced with the Fanconi anemia group C gene (FANCC). Hum Gene Ther. 1999;10(14):2337-46.

[44] Muller LU, Milsom MD, Kim MO, Schambach A, Schuesler T, Williams DA. Rapid lentiviral transduction preserves the engraftment potential of Fanca(-/-) hematopoietic stem cells. Mol Ther. 2008;16(6):1154-60.

[45] Jacome A, Navarro S, Rio P, Yanez RM, Gonzalez-Murillo A, Lozano ML, et al. Lentiviral-mediated genetic correction of hematopoietic and mesenchymal progenitor cells from Fanconi anemia patients. Mol Ther. 2009;17(6):1083-92.

[46] Li Y, Chen S, Yuan J, Yang Y, Li J, Ma J, et al. Mesenchymal stem/progenitor cells promote the reconstitution of exogenous hematopoietic stem cells in Fancg-/- mice in vivo. Blood. 2009;113(10):2342-51.

[47] Habi O, Girard J, Bourdages V, Delisle MC, Carreau M. Correction of Fanconi Anemia Group C Hematopoietic Stem Cells Following Intrafemoral Gene Transfer. Anemia. 2010;2010.

[48] Gonzalez-Murillo A, Lozano ML, Alvarez L, Jacome A, Almarza E, Navarro S, et al. Development of lentiviral vectors with optimized transcriptional activity for the gene therapy of patients with Fanconi anemia. Hum Gene Ther. 2010;21(5):623-30.

[49] Raya A, Rodriguez-Piza I, Guenechea G, Vassena R, Navarro S, Barrero MJ, et al. Disease-corrected haematopoietic progenitors from Fanconi anaemia induced pluripotent stem cells. Nature. 2009;460(7251):53-9.

[50] Matsui H. Endothelial progenitor cell-based therapy for hemophilia A. International journal of hematology. 2012;95(2):119-24. Epub 2012/02/09.

[51] Chuah MK, Nair N, Vandendriessche T. Recent progress in gene therapy for hemophilia. Hum Gene Ther. 2012;23(6):557-65. Epub 2012/06/08.

[52] Orlova NA, Kovnir SV, Vorobiev, II, Gabibov AG. Coagulation Factor IX for Hemophilia B Therapy. Acta naturae. 2012;4(2):62-73. Epub 2012/08/09.

[53] High KA. The gene therapy journey for hemophilia: are we there yet? Blood. 2012. Epub 2012/07/26.

[54] Ramezani A, Hawley RG. Correction of murine hemophilia A following nonmyeloablative transplantation of hematopoietic stem cells engineered to encode an enhanced human factor VIII variant using a safety-augmented retroviral vector. Blood. 2009;114(3):526-34.

[55] Yadav N, Kanjirakkuzhiyil S, Ramakrishnan M, Das TK, Mukhopadhyay A. Factor VIII can be synthesized in hemophilia A mice liver by bone marrow progenitor cell-derived hepatocytes and sinusoidal endothelial cells. Stem cells and development. 2012;21(1):110-20. Epub 2011/04/13.

[56] Aiuti A, Roncarolo MG. Ten years of gene therapy for primary immune deficiencies. Hematology Am Soc Hematol Educ Program. 2009:682-9. Epub 2009/12/17.

[57] Uribe L, Weinberg KI. X-linked SCID and other defects of cytokine pathways. Seminars in hematology. 1998;35(4):299-309. Epub 1998/11/04.

[58] Qasim W, Gaspar HB, Thrasher AJ. Progress and prospects: gene therapy for inherited immunodeficiencies. Gene therapy. 2009;16(11):1285-91.

[59] Schwarzwaelder K, Howe SJ, Schmidt M, Brugman MH, Deichmann A, Glimm H, et al. Gammaretrovirus-mediated correction of SCID-X1 is associated with skewed vector integration site distribution in vivo. J Clin Invest. 2007;117(8):2241-9.

[60] Chinen J, Davis J, De Ravin SS, Hay BN, Hsu AP, Linton GF, et al. Gene therapy improves immune function in preadolescents with X-linked severe combined immunodeficiency. Blood. 2007;110(1):67-73.

[61] Scobie L, Hector RD, Grant L, Bell M, Nielsen AA, Meikle S, et al. A novel model of SCID-X1 reconstitution reveals predisposition to retrovirus-induced lymphoma but no evidence of gammaC gene oncogenicity. Mol Ther. 2009;17(6):1031-8.

[62] Thornhill SI, Schambach A, Howe SJ, Ulaganathan M, Grassman E, Williams D, et al. Self-inactivating gammaretroviral vectors for gene therapy of X-linked severe combined immunodeficiency. Mol Ther. 2008;16(3):590-8.

[63] Zhou S, Mody D, DeRavin SS, Hauer J, Lu T, Ma Z, et al. A self-inactivating lentiviral vector for SCID-X1 gene therapy that does not activate LMO2 expression in human T cells. Blood. 2010;116(6):900-8.

[64] Ginn SL, Liao SH, Dane AP, Hu M, Hyman J, Finnie JW, et al. Lymphomagenesis in SCID-X1 mice following lentivirus-mediated phenotype correction independent of insertional mutagenesis and gammac overexpression. Mol Ther. 2010;18(5):965-76.

[65] Aiuti A, Slavin S, Aker M, Ficara F, Deola S, Mortellaro A, et al. Correction of ADA-SCID by stem cell gene therapy combined with nonmyeloablative conditioning. Science. 2002;296(5577):2410-3.

[66] Aiuti A, Cassani B, Andolfi G, Mirolo M, Biasco L, Recchia A, et al. Multilineage hematopoietic reconstitution without clonal selection in ADA-SCID patients treated with stem cell gene therapy. J Clin Invest. 2007;117(8):2233-40.

[67] Gaspar HB, Cooray S, Gilmour KC, Parsley KL, Zhang F, Adams S, et al. Hematopoietic stem cell gene therapy for adenosine deaminase-deficient severe combined immunodeficiency leads to long-term immunological recovery and metabolic correction. Science translational medicine. 2011;3(97):97ra80. Epub 2011/08/26.

[68] Shaw KL, Kohn DB. A tale of two SCIDs. Science translational medicine. 2011;3(97):97ps36. Epub 2011/08/26.

[69] Mortellaro A, Hernandez RJ, Guerrini MM, Carlucci F, Tabucchi A, Ponzoni M, et al. Ex vivo gene therapy with lentiviral vectors rescues adenosine deaminase (ADA)-deficient mice and corrects their immune and metabolic defects. Blood. 2006;108(9):2979-88. Epub 2006/07/13.

[70] Ferrua F, Brigida I, Aiuti A. Update on gene therapy for adenosine deaminase-deficient severe combined immunodeficiency. Current opinion in allergy and clinical immunology. 2010;10(6):551-6. Epub 2010/10/23.

[71] Silver JN, Elder M, Conlon T, Cruz P, Wright AJ, Srivastava A, et al. Recombinant Adeno-Associated Virus-Mediated Gene Transfer for the Potential Therapy of Adenosine Deaminase-Deficient Severe Combined Immune Deficiency. Hum Gene Ther. 2011.

[72] Malech HL, Maples PB, Whiting-Theobald N, Linton GF, Sekhsaria S, Vowells SJ, et al. Prolonged production of NADPH oxidase-corrected granulocytes after gene therapy of chronic granulomatous disease. Proceedings of the National Academy of Sciences of the United States of America. 1997;94(22):12133-8. Epub 1997/10/29.

[73] Barese CN, Goebel WS, Dinauer MC. Gene therapy for chronic granulomatous disease. Expert Opin Biol Ther. 2004;4(9):1423-34. Epub 2004/09/01.

[74] Kohn DB. Update on gene therapy for immunodeficiencies. Clin Immunol. 2010;135(2):247-54.

[75] Ott MG, Schmidt M, Schwarzwaelder K, Stein S, Siler U, Koehl U, et al. Correction of X-linked chronic granulomatous disease by gene therapy, augmented by insertional activation of MDS1-EVI1, PRDM16 or SETBP1. Nat Med. 2006;12(4):401-9.

[76] Stein S, Ott MG, Schultze-Strasser S, Jauch A, Burwinkel B, Kinner A, et al. Genomic instability and myelodysplasia with monosomy 7 consequent to EVI1 activation after gene therapy for chronic granulomatous disease. Nat Med. 2010;16(2):198-204.

[77] Barde I, Laurenti E, Verp S, Wiznerowicz M, Offner S, Viornery A, et al. Lineage- and stage-restricted lentiviral vectors for the gene therapy of chronic granulomatous disease. Gene therapy. 2011;18(11):1087-97. Epub 2011/05/06.

[78] Kang EM, Malech HL. Gene therapy for chronic granulomatous disease. Methods in enzymology. 2012;507:125-54. Epub 2012/03/01.

[79] Zou J, Sweeney CL, Chou BK, Choi U, Pan J, Wang H, et al. Oxidase-deficient neutrophils from X-linked chronic granulomatous disease iPS cells: functional correction by zinc finger nuclease-mediated safe harbor targeting. Blood. 2011;117(21):5561-72. Epub 2011/03/18.

[80] Bosticardo M, Marangoni F, Aiuti A, Villa A, Grazia Roncarolo M. Recent advances in understanding the pathophysiology of Wiskott-Aldrich syndrome. Blood. 2009;113(25):6288-95.

[81] Davis BR, Candotti F. Revertant somatic mosaicism in the Wiskott-Aldrich syndrome. Immunol Res. 2009;44(1-3):127-31.

[82] Wada T, Jagadeesh GJ, Nelson DL, Candotti F. Retrovirus-mediated WASP gene transfer corrects Wiskott-Aldrich syndrome T-cell dysfunction. Hum Gene Ther. 2002;13(9):1039-46. Epub 2002/06/18.

[83] Klein C, Nguyen D, Liu CH, Mizoguchi A, Bhan AK, Miki H, et al. Gene therapy for Wiskott-Aldrich syndrome: rescue of T-cell signaling and amelioration of colitis upon transplantation of retrovirally transduced hematopoietic stem cells in mice. Blood. 2003;101(6):2159-66. Epub 2002/11/16.

[84] Boztug K, Dewey RA, Klein C. Development of hematopoietic stem cell gene therapy for Wiskott-Aldrich syndrome. Curr Opin Mol Ther. 2006;8(5):390-5.

[85] Boztug K, Schmidt M, Schwarzer A, Banerjee PP, Diez IA, Dewey RA, et al. Stem-cell gene therapy for the Wiskott-Aldrich syndrome. N Engl J Med. 2010;363(20):1918-27.

[86] Dupre L, Marangoni F, Scaramuzza S, Trifari S, Hernandez RJ, Aiuti A, et al. Efficacy of gene therapy for Wiskott-Aldrich syndrome using a WAS promoter/cDNA-containing lentiviral vector and nonlethal irradiation. Hum Gene Ther. 2006;17(3):303-13. Epub 2006/03/21.

[87] Galy A, Roncarolo MG, Thrasher AJ. Development of lentiviral gene therapy for Wiskott Aldrich syndrome. Expert Opin Biol Ther. 2008;8(2):181-90. Epub 2008/01/16.

[88] Astrakhan A, Sather BD, Ryu BY, Khim S, Singh S, Humblet-Baron S, et al. Ubiquitous high-level gene expression in hematopoietic lineages provides effective lentiviral gene therapy of murine Wiskott-Aldrich syndrome. Blood. 2012;119(19):4395-407. Epub 2012/03/21.

[89] Bosticardo M, Draghici E, Schena F, Sauer AV, Fontana E, Castiello MC, et al. Lentiviral-mediated gene therapy leads to improvement of B-cell functionality in a murine model of Wiskott-Aldrich syndrome. The Journal of allergy and clinical immunology. 2011;127(6):1376-84 e5. Epub 2011/05/03.

[90] Scaramuzza S, Biasco L, Ripamonti A, Castiello MC, Loperfido M, Draghici E, et al. Preclinical Safety and Efficacy of Human CD34(+) Cells Transduced With Lentiviral Vector for the Treatment of Wiskott-Aldrich Syndrome. Mol Ther. 2012. Epub 2012/03/01.

[91] Munoz P, Toscano MG, Real PJ, Benabdellah K, Cobo M, Bueno C, et al. Specific marking of hESCs-derived hematopoietic lineage by WAS-promoter driven lentiviral vectors. PloS one. 2012;7(6):e39091. Epub 2012/06/22.

[92] Roberts JL, Lengi A, Brown SM, Chen M, Zhou YJ, O'Shea JJ, et al. Janus kinase 3 (JAK3) deficiency: clinical, immunologic, and molecular analyses of 10 patients and outcomes of stem cell transplantation. Blood. 2004;103(6):2009-18.

[93] Sorrentino BP, Lu T, Ihle J, Buckley RH, Cunningham JM. A clinical attempt to treat JAK3-deficient SCID using retroviral-mediated gene tranfer to bone marrow CD34+ cells. Mol Ther. 2003;7:S449.

[94] Liao P, Toro A, Min W, Lee S, Roifman CM, Grunebaum E. Lentivirus gene therapy for purine nucleoside phosphorylase deficiency. The journal of gene medicine. 2008;10(12):1282-93. Epub 2008/10/17.

[95] Nelson EJ, Tuschong LM, Hunter MJ, Bauer TR, Jr., Burkholder TH, Hickstein DD. Lentiviral vectors incorporating a human elongation factor 1alpha promoter for the treatment of canine leukocyte adhesion deficiency. Gene therapy. 2010;17(5):672-7.

[96] Bauer TR, Jr., Olson EM, Huo Y, Tuschong LM, Allen JM, Li Y, et al. Treatment of canine leukocyte adhesion deficiency by foamy virus vectors expressing CD18 from a PGK promoter. Gene therapy. 2011.

[97] Hunter MJ, Zhao H, Tuschong LM, Bauer Jr TR, Burkholder TH, Persons D, et al. Gene therapy for canine leukocyte adhesion deficiency with lentiviral vectors using the MSCV and human PGK promoters. Hum Gene Ther. 2011.

[98] Hunter MJ, Tuschong LM, Fowler CJ, Bauer TR, Jr., Burkholder TH, Hickstein DD. Gene therapy of canine leukocyte adhesion deficiency using lentiviral vectors with human CD11b and CD18 promoters driving canine CD18 expression. Mol Ther. 2011;19(1):113-21.

[99] Ponder KP, Haskins ME. Gene therapy for mucopolysaccharidosis. Expert Opin Biol Ther. 2007;7(9):1333-45.

[100] Wolfe JH, Sands MS, Barker JE, Gwynn B, Rowe LB, Vogler CA, et al. Reversal of pathology in murine mucopolysaccharidosis type VII by somatic cell gene transfer. Nature. 1992;360(6406):749-53.

[101] Wang D, Zhang W, Kalfa TA, Grabowski G, Davies S, Malik P, et al. Reprogramming erythroid cells for lysosomal enzyme production leads to visceral and CNS cross-correction in mice with Hurler syndrome. Proceedings of the National Academy of Sciences of the United States of America. 2009;106(47):19958-63.

[102] Brady RO, Kanfer JN, Bradley RM, Shapiro D. Demonstration of a deficiency of glucocerebroside-cleaving enzyme in Gaucher's disease. J Clin Invest. 1966;45(7):1112-5. Epub 1966/07/01.

[103] Cox TM. Gaucher disease: clinical profile and therapeutic developments. Biologics. 2010;4:299-313.

[104] Enquist IB, Nilsson E, Ooka A, Mansson JE, Olsson K, Ehinger M, et al. Effective cell and gene therapy in a murine model of Gaucher disease. Proceedings of the National Academy of Sciences of the United States of America. 2006;103(37):13819-24.

[105] Kim EY, Hong YB, Lai Z, Cho YH, Brady RO, Jung SC. Long-term expression of the human glucocerebrosidase gene in vivo after transplantation of bone-marrow-derived cells transformed with a lentivirus vector. The journal of gene medicine. 2005;7(7):878-87. Epub 2005/02/16.

[106] Enquist IB, Nilsson E, Mansson JE, Ehinger M, Richter J, Karlsson S. Successful low-risk hematopoietic cell therapy in a mouse model of type 1 Gaucher disease. Stem Cells. 2009;27(3):744-52. Epub 2008/12/06.

[107] Cartier N, Aubourg P. Hematopoietic stem cell transplantation and hematopoietic stem cell gene therapy in X-linked adrenoleukodystrophy. Brain Pathol. 2010;20(4): 857-62. Epub 2010/07/16.

[108] Cartier N, Hacein-Bey-Abina S, Bartholomae CC, Veres G, Schmidt M, Kutschera I, et al. Hematopoietic stem cell gene therapy with a lentiviral vector in X-linked adreno-leukodystrophy. Science. 2009;326(5954):818-23.

[109] Cartier N, Hacein-Bey-Abina S, Bartholomae CC, Bougneres P, Schmidt M, Kalle CV, et al. Lentiviral hematopoietic cell gene therapy for X-linked adrenoleukodystrophy. Methods in enzymology. 2012;507:187-98. Epub 2012/03/01.

Section 2

Clinical Aspects of Stem Cell Transplantation

Chapter 6

Progress in Hematopoietic Stem Cell Transplantation

Miroslaw Markiewicz,
Malgorzata Sobczyk-Kruszelnicka,
Monika Dzierzak Mietla, Anna Koclega,
Patrycja Zielinska and Slawomira Kyrcz-Krzemien

Additional information is available at the end of the chapter

http://dx.doi.org/10.5772/53086

1. Introduction

Transplantation of autologous or allogeneic hematopoietic stem cells is a method currently used to treat many malignant and nonmalignant hematological diseases. The indications, methods, goals of therapy have evolved since the introduction of transplantation to the clinical practice. Progress that has been achieved allowed for the improvement of results. Thanks to the availability of various conditioning regimens, various hematopoietic cells sources as well as variable possibilities of anti-GvHD prophylaxis the individualization of the transplantation procedure has been more and more widely used in the recent years. This chapter summarizes current clinical practices and presents major clinical problems that have to be optimally managed in order to improve the outcomes of transplantation.

2. Autologous hematopoietic stem cells transplantation

Autologous peripheral hematopoietic stem cells transplantation (auto-HSCT) was for the first time performed at Hammersmith Hospital in London in 1981 to treat the patient in accelerated phase of CML. Although auto-HSCT does not play any role in the treatment of CML nowadays, indications for this valuable therapeutic method have evolved for many years. In acute leukemia auto-HSCT should be recommended only in the context of clinical studies. Auto-HSCT after myeloablative chemotherapy or radiotherapy has originally been developed as an alternative to allogeneic hematopoietic stem cell transplantation for patients with AML with no suitable donor. Several randomized studies in patients with AML in first complete remission (CR1) subsequently suggested reduced relapse rates after auto-HSCT [1]. Auto-HSCT is also widely used to consolidate first remission in AML. The novel molecular and cytogenetic stratification methods may allow the identification of AML entities which could benefit from autografting. The overall survival of patients receiving auto-HSCT in ALL in first remission is around 40%. The high-dose therapy followed by auto-HSCT can be an alternative treatment in patients in whom allo-HSCT is precluded.

The results of a large European study showed that auto-HSCT can be recommended in patients with good-risk cytogenetic characteristics of myelodysplastic syndrome [2]. Auto-HSCT can be recommended as post-remission therapy to reduce the risk of relapse. The longer remission was observed in patients who undergo auto-HSCT.

In myeloproliferative disorders auto-HSCT can induce responses in patients with primary myelofibrosis, but this procedure cannot be recommended out of clinical protocols.

In chronic lymphocytic leukemia auto-HSCT can be considered for patients with poor-risk disease in complete or good partial remission able to withstand high-dose therapy, but it should be performed preferably in the context of clinical protocols.

Auto-HSCT is the standard therapy for patients with Hodgkin's lymphoma (HL) in first chemosensitive relapse or second complete remission as shown by two prospective randomized clinical trials [3,4]. There is no indication for auto-HSCT in first remission, even in patients with poor prognosis at diagnosis [5,6]. Patients refractory to first-line therapy but sensitive to salvage therapy might benefit from auto-HSCT [7]. Auto-HCT might be considered as a part of a clinical protocol for patients with resistant Hodgkin's lymphoma, as an initial debulking therapy to be followed by an allo-HSCT as consolidation therapy [8].

In many non-Hodgkin's lymphomas auto-HSCT is a standard therapy. In diffuse large B-cell lymphoma (DLBCL) auto-HSCT is a standard therapy for patients with chemosensitive relapse [9]. The role of auto-HSCT is being re-evaluated with the advance of monoclonal antibodies and use of chemo-immunotherapy as first-line treatment. Auto-HSCT remains also the standard approach for early relapsing patients with follicular lymphoma (FL) [10]. In both DLBCL and FL, auto-HSCT does not provide any clinical benefit in patients with refractory disease. Otherwise, most patients with mantle cell lymphoma are being offered an early intensification with an auto-HSCT, owing it to the inherent poor prognosis of the disease. The retrospective analysis indicates that the results of auto-HSCT performed beyond the first remission are inferior [11]. Few studies showed an improved survival in patients with T-cell non-Hodgkin's lymphoma (NHL) who received auto-HSCT as a first line treatment, compared to those who did not.

Patients with multiple myeloma form a large group of patients being transplanted. Auto-HSCT is clearly indicated for patients <70 years of age with satisfactory general health and fitness who respond to the first-line treatment. Although new agents change the place of auto-HSCT in MM, this procedure still has an established position in treatment. Best results are observed in patients achieving good response before the auto-HSCT, but some non-responding patients also may benefit from this approach. Double auto-HSCT (or tandem auto-HSCT) has been shown to be superior to consolidation and maintenance with agents such as thalidomide in patients not achieving the remission or a very good partial response after the first transplant [12].

Auto-HSCT constitutes an important treatment option for patients with solid tumors. Selected subgroups of oncological patients may benefit from high-dose chemotherapy supported by auto-HSCT. High-dose chemotherapy for refractory germ cell tumors is considered a standard therapy. Conditioning regimen in this case incorporates carboplatin and etoposide.

Auto-HSCT after conditioning regimen aimed to increase the immunosuppression is being considered in clinical protocols for selected patients with severe multiple sclerosis [13], rheumatoid arthritis [14], systemic lupus erythromatous [15], systemic sclerosis [16], immune cytopenias and Crohn's disease [17]. Auto-HSCT for other autoimmune disorders is being considered on a developmental basis. Steroid dependency with Cushing threshold and skeletal damage could be an indication.

3. Allogeneic hematopoietic stem cell transplantation

Allogeneic hematopoietic stem cell transplantation (allo-HSCT) constitutes a standard treatment of hematological malignant and nonmalignant disorders. The possibility of finding a donor has been increased by use of unrelated donors, with similar results of transplantation when compared to results of sibling donor transplants. The use of peripheral blood stem cells, instead of bone marrow, results in faster engraftment, but also in the increased risk of chronic GvHD (Graft versus Host Disease). Reduced-intensity conditioning is used instead of high–dose myeloablative conditioning for older patients and those with comorbidities. Disease relapse is a major problem and thus it should be detected as early as possible, at the stage of the minimal residual disease or recurrent recipient chimerism and managed by immunotherapy with donor lymphocyte infusions. Novel diagnostic tools and anti-microbial drugs have reduced the morbidity and mortality from infections.

Allogeneic hematopoietic stem cell transplantation connected with application of high-dose chemo- and radiotherapy was first carried out by Thomas et al. in 1957 to treat leukemia patient in advanced stage [18]. The concept of treatment at that time was based on the previous observations conducted during the second world war, referring to destructive activity of radiation on the function of bone marrow, as well as further research conducted in the 1950's, which showed that it was possible to avoid irreversible pancytopenia thanks to the bone marrow cells transplantation in the irradiated animals. The discovery of Human Leucocyte Antigen (HLA) enabled to match appropriately the donor and the recipient, what contributed to the significant increase in overall survival after transplantation which has been observed since 1968 [19]. The improvement of the results was undoubtedly also influenced by other factors: performing the transplantation in the optimal phase- remission of the disease, GvHD prevention, the improvement of adjunctive treatment. Nowadays more that 25.000 allo-HSCTs are being performed each year.

The main indication for allo-HSCT is acute myeloblastic leukemia (AML) and acute lymphoblastic leukemia (ALL). In high risk ALL and AML, when favorable prognostic genetic changes are lacking, the allo-HSCT is recommended in the first remission of the disease. The transplantation in more advanced stages of the disease leads to the higher relapse rate, as well as to the increased incidence of transplantation complications.

Despite the introduction of tyrosine kinase inhibitors (TKIs) into the treatment of chronic myeloid leukemia (CML) over ten years ago, allo-HSCT still remains the only way of treatment capable to provide the complete recovery. The standard indication for allo-HSCT is resistance to TKIs treatment, especially in young patients. Other indications for allo-HSCT are myelodysplastic syndrome, high-risk chronic lymphocytic leukemia, selected patients with high-risk lymphoma, patients with myelofibrosis and other myeloproliferative neoplasms of unfavorable prognosis. The results of multiple myeloma treatment with the use alloHSCT are encouraging. AlloHSCT with the reduced conditioning regimen after previous auto-HSCT constitutes an interesting alternative in patients with multiple myeloma patients, who undergo single or tandem autologous transplantation [20].

Allo-HSCT is also the standard treatment in nonmalignant diseases of hematopoietic system such as severe aplastic anemia (SAA), paroxysmal nocturnal hemoglobinuria (PNH) and hemoglobinopathies. In some cases of inborn metabolic defects, allogeneic transplantation of donor's cells can restore the production of the deficient or lacking enzyme and eliminate the disease [21].

4. Hematopoietic stem cell donors

The optimal donors are siblings possessing both haplotypes identical with the recipient. The syngeneic transplantation, i.e. from monozygotic twins, is the safest from the immunological point of view, however it is connected with the increased risk of the relapse of the disease resulting from the lack of immunological interaction between the donor cells and the recipient cells [22].

Probability of possessing matched sibling donor is defined by the formula: $1-(0.75)^n$, were n indicates the number of siblings. The observed decrease in the number of newly born children causes problems in finding matched family donors for many patients. In rare cases with no matched sibling donor, matching donor could be found among other members of the family. In the vast majority of patients without matched sibling donor, transplantation from unrelated donor is the most frequently chosen option. The number of such transplantations has increased considerably in the last 20 years [23]. It has been made possible thanks to dynamic development of bone marrow donors' registries, whose number of potential donors exceeded 20 million in the current year 2012. Alternatively, for those patients who are unlikely to find a matched donor, partial incompatibility could be accepted.

The most desirable model of the donors' registry organization is the development of national ones which, for many reasons (safety of donors, clearness of procedures and financial reasons), according to WMDA's (World Marrow Donors Association) recommendations should control and supervise the recruitment of the donors within the country. The chance to find

http://dx.doi.org/10.5772/53086

the matched donor depends on the frequency of occurrence of HLA-haplotypes in the whole population and the race of recipient – most donors recruited by registries worldwide belong to Caucasian race. The efforts are being made, especially in the USA, aiming to recruit higher number of donors of other races.

Phenotypic HLA-matching involving the testing of HLA-antigens by means of specific sera has been replaced by more precise molecular testing enabling the precise identification of HLA allelic determinants.

The question of accepting a mismatched donor for a patients, who didn't find a fully matched donor has not been finally solved. With the increasing number of observations there are recommendations concerning optimal matching not only in HLA-A, B and DR, but also in C, DQ and even DP. Many centers aim to transplant patients only from donors fully matched in 10/10 alleles of HLA-A, B, C, DR and DQ. The improved methods of typing enabling more precise molecular donor matching has improved the results of allo-HSCT from unrelated donors, which are now similar to those of allo-HSCT from siblings [24]. As the allo-HSCT from an unrelated donor has to be preceded by often time-consuming search for a donor, it is important to plan the transplantation carefully in advance.

5. The sources of hematopoietic stem cells

The choice of the cells source depends of diagnosis and the type of conditioning treatment applied. Collection of bone marrow is preferred in nonmalignant diseases in order to avoid chronic GvHD. Transplantation of hematopoietic cells from peripheral blood is preferred when reduced intensity conditioning regimen is used, with regard to the fact that transplantation of larger number of hematopoietic cells is able to break the resistance of the recipient and to result in the engraftment.

The bone marrow aspirated in general anesthesia from iliac spine was for many years the main source of cells for transplantation. Except of hematopoietic cells, the bone marrow also consists of multipotential mesenchymal cells which, although are not hematopoietic cells, have a potential to differentiate in vitro and in vivo into various mesenchymal tissues, such as bone, cartilage, fat tissue, tendons and bone matrix. Mesenchymal stem cells can reduce the alloreactivity, they inhibit lymphocytes T proliferation and act immunosuppressively, what has been implemented in the form of unrelated or haploidentical mesenchymal cells infusion into the treatment of acute GvHD.

During the 1990's the cells collected in apheresis from peripheral blood after their previous stimulation with granulocytic stimulating factor (G-CSF) completely superseded the bone marrow aspiration in autologous transplantations. In the beginning of 21st century, the similar trend occurred also in allo-HSCT. The apheresis of cells from peripheral blood usually results in collection of higher number of nucleated cells, CD34+ cells, lymphocytes CD3+ and NK cells when compared with cells aspiration from the bone marrow; it enables faster regeneration of granulocytes and platelets. It translates into the smaller risk of infections and smaller demand for transfusions of blood derivatives.

In the beginning allo-HSCT in the form of PBSCT (Peripheral Blood Stem Cells Transplantation) was applied only in sibling transplantations due to the anxiety of acute GvHD occurrence, however, the frequency of GvHD is similar to the one after bone marrow transplantation, despite greater number of T-lymphocytes in the transplantation material collected from peripheral blood, as it was shown in the number of studies. Thus allo-PBSCT has been successfully applied also in allo-HSCT from unrelated donors. However, the frequency of chronic GvHD is higher, and thus allo-PBSCT is applied seldom in patients with nonmalignant disease, who do not benefit from Graft versus Leukemia (GvL) effect, which is usually connected with chronic GvHD [25].

The important source of hematopoietic stem cells for allo-HSCT is a cord blood (CB), usually intended to be discarded. In many countries there are banks of frozen CB units where there are over 0,5 million units ready to be transplanted. The advantage of applying CB cells is their immediate availability and a reduced risk of GvHD, related to a relative shortage of mature T-lymphocytes in the CB. Therefore the higher level of HLA-mismatching between the donor and the the recipient is more acceptable in CB transplantation than in traditional transplantations. The unfavorable factors are a more frequent occurrence of graft failure and a slower regeneration responsible for higher risk of infections. The number of necessary nucleated cells and CD34+ cells calculated per kilogram of the recipient's body mass is lower by about one logarithm when compared to the bone marrow. A number of studies showed the importance of sufficient number of cord blood nucleated cells, for this reason it is recommended to transplant more than 2x10^7 nucleated cells per kilogram of recipient's body mass. It constitutes limitation in CB application in adults due to the small volume of cord blood and small total number of cells. Simultaneous transplantation of two CB units is successfully applied to solve this problem [26,27]. In vitro cells expansion to increase the number of CB cells has not been widely used. Because of the limited, usually small number of cord blood cells, it is most often applied as the source of cells for transplantation in children.

6. Preparative treatment before transplantation

The preparative treatment before transplantation (or conditioning regimen) aims to eradicate the remains of the disease and to make immunological system of recipient weaker in order to enable the acceptance of the graft by the recipient. The preparative treatment is connected with toxicities which turned out to be impossible to eliminate so far.

The choice of conditioning treatment depends of the patient's age, the main disease and co-existing diseases. Myeloablative conditioning regimens are characterized by strong cytotoxicity as well as strong immunosuppressive potential, while reduced intensity conditioning regimens differ in cytotoxic activity and immunosuppressive potential. They are chosen depending on the main disease and evaluation of the risk of graft failure.

The combination of radiotherapy (TBI- total body irradiation- at total dose of 12 Gy, delivered in fractions) and cyclophosphamide (Cy, at total dose of 120 mg/kg administered within 2 days) has been used for over 40 years for conditioning [28]. TBI treatment is recommended as a standard in ALL. In order to avoid potential TBI consequences, such as bronchiolitis obliterans, cataract, secondary malignancy, endocrinological disorders, inhibition of the growing process in children, TBI in AML and MDS has been replaced by busulfan given at 16 mg/kg dose within 4 consecutive days before Cy [29]. The BuCy treatment has higher risk of SOS (sinusoidal obstruction syndrome), hemorrhagic cystitis and chronic GvHD. The high serum concentration of Bu (Busulfan) occurring during its oral treatment has influenced considerably its toxic complications. It is difficult to avoid it because of various degree of absorption from digestive tract. Thus the intravenous use of busulfan is more favorable. The reduction of SOS incidence and decrease of transplant related mortality (TRM) after intravenous use of Bu has been reported [30]. In order to further limit the toxicity, treosulfan is used instead of Bu in modern treatment programs nowadays, and additional immunosuppressive effect is obtained by parallel application of purine analoque, e.g. fludarabine.

The standard preparative treatment applied in SAA comprises of Cy 200 mg/kg and antithymocyte globulin (ATG).

Although the intensive conditioning treatment decreases the risk of relapse after transplantation, it does not prolong the overall survival because greater toxicity leads to increased transplant related mortality [31].

The concept of so-called RIC (reduced intensity conditioning) incorporates the advantage of anti-leukemic effect of donor T-lymphocytes while cytotoxic effect of conditioning regimen is decreased. The main result of RIC treatment is immunosuppressive therapy aiming to enable the acceptance of the transplant by braking the immunological defence of the recipient. The anti-leukemic effect can be escalated after transplantation by means of DLI (Donor Lymphocyte Infusion), whenever it is required. DLI was first used with success in CML patients, in whom the disease relapsed after conventional allo-HSCT [32]. Since then it has been used in many other diseases, including many clonal diseases of hematopoietic system, most often lymphomas and chronic lymphocytic leukemia (CLL). RIC treatment has lower toxicity when compared to conventional conditioning treatment, thus it is suitable for transplantation in older patients and in patients with coexisting diseases in whom the application of myeloablative treatment is contraindicated. RIC treatment consists most often of purine analoque. The example of RIC treatment reduced to the minimum, after which graft occurs, is the combination of TBI dose 2Gy with fludarabine. Other exemplary RIC protocols are the combination of fludarabine with Bu at dose 8 mg/kg and ATG with Cy or with melphalan. The important element of RIC treatment is the use of immunosuppressive therapy after transplantation e.g. cyclosporine A and mycophenolate mofetil. The reduced intensity of conditioning enables the immunocompetent recipient cells to survive until the moment of transplantation, what leads to the higher risk of graft failure or incomplete graft. In some centers transplantation with RIC are performed in ambulatory, however patients often require further hospitalization due to infections or GvHD [33].

Allo-HSCT with use of RIC can be applied when autologous transplantation is ineffective. Other possibility is to apply the tandem transplantation: at first autologous one and then the allogeneic one, with use of RIC in order to reduce the TRM by separation of high-dosed cytotoxic treatment from immunotherapy related to allogeneic HSCT, which has been applied for the first time in patients with multiple myeloma (MM).

7. Adjunctive treatment

During the phase of pancytopenia after myeloablative conditioning patients are usually susceptible to infections and thus they have to stay in a sterile environment, e.g. in HEPA-filtered rooms with reversed isolation. They routinely receive preventive treatment against bacteria, viruses, fungi. Moreover, the substitution treatment is applied with the use of irradiated, CMV-negative red blood cells and single donor platelets concentrates. Analgetic drugs and parenteral nutrition are applied when needed. Ursodeoxycholic acid is used in order to avoid hepatic complications. G-CSF is applied to accelerate the regeneration of granulocytes, however it can delay the recovery of platelets and can increase the risk of GvHD. Erythropoietin accelerates the recovery of red blood cells system and thus it reduces the need for transfusions, but it increases the cost of the transplant procedure and it is not used on a regular basis.

8. Post-transplant complications

8.1. Graft versus host disease

Acute and chronic graft versus host disease are the main complications of allo-HSCT. In pathophysiology of acute and chronic GvHD, T-lymphocytes of the donor recognize HLA-molecules of the recipient presented by the antigen presenting cells. It results in the release of interleukin-2 and activation of cytotoxic T-lymphocytes, NK-cells and macrophages. The main targets of the attack are skin, gut and liver. The most important risk factor is the HLA-incompatibility between the donor and the recipient, but also minor histocompatibility antigens are responsible for the risk of GvHD, especially HY mismatch in case when the donor is female and the recipient is male [34]. Chronic GvHD occurs most often from 100 days to one year after allo-HSCT. It resembles autoimmunological diseases, e.g. systemic sclerodermia and Sjoegren syndrome. Symptoms such as lichen and sclerodermic skin changes, mucositis, kserostomia, keratoconjunctivitis sicca, stricture of esophagus and vagina, cholestatic liver failure, bronchiolitis obliterans and musculitis also occur. Cachexia, immunological deficiency, additionally increasing the risk of infections especially caused by gram-plus bacteria can be also observed. The initial stage of chronic GvHD is usually more progressive when it is preceded by the acute form of the disease. It can also occurr after nonsymptomatic (quiescent) period or de-novo, without any preceding symptoms of acute GvHD. The chronic progressive GvHD has the worst prognosis.

In order to decrease the risk of GvHD a preventive immunosuppression, usually with the use of cyclosporine A (CsA) and methotrexate is applied. The removal of T-lymphocytes from the transplanted cells (T-depletion) constitutes the effective form of prevention, however, it is connected with the higher risk of the graft failure and relapse of the disease. In cord blood transplantations, instead of methotrexate which prolonges the regeneration period, prednisolon is used. New immunosuppressive protocols include calcineurin-inhibitors other than cyclosporine A – tacrolimus, macrolid immunosuppressant– syrolimus and mycophenolan mofetil. The administration of ATG before transplantation is an important immunosuppressive element used in allo-HSCT from unrelated donors. As the effective serum concentration of ATG is maintained for many weeks after infusion, it effects not only T-lymphocytes of the recipient but also those of the donor [35]. The increased risk of infections is an undesirable side effect of ATG.

The type of GvHD prevention depends of the diagnosis, the type of conditioning treatment and the applied cell source. The GvHD prevention should be more effective in nonmalignant disease and less intensive when lower number of cells have been transplanted.

When symptoms of acute GvHD develop despite its prophylaxis, methylprednisolone at the dose of 2-5 mg per kilogram of body weight per day is used on the standard basis, usually effectively. In case of steroid resistance the risk of failure is high. The second line treatment consists of ATG, anti-IL-2 antibodies, anti-IL-2 receptor antibodies and antibodies against TNF-alpha. Photosensitizing psolarens and ultraviolet radiation in a form of extracorporal photopheresis and transplantation of mesenchymal stem cells can be also applied, but are not everywhere available. Mesenchymal stem cells have strong immunosuppressive effect, they can be obtained from the primitive connective tissue of the umbilical cord, called the Wharton's jelly, and they do not require any matching due to low levels of HLA-ABC and lack of HLA-DR antigens.

The treatment of chronic GvHD consists of CsA and steroids. In patients not responding to the treatment tacrolimus, thalidomide, mycophenolan mofetil, sirolimus and irradiation of lymphatic system with dose of 1 Gy can be applied.

8.2. Infections

Immunological reconstitution is of a primary importance to avoid infections after allo-HSCT. The highest risk of the infection occurs in patients with GvHD, but also in the remaining patients with no GvHD it is 20 times higher than in the whole population. From 20% up to 50% of patients still require immunosuppressive treatment after 3 years from allo-HSCT, what considerably increases the risk of infectious complications in this group of patients [36].

Normal endogenous Gram-negative flora from the gastrointestinal tract and exogenous catheter-related Gram-positive bacteria constitute the most frequent cause of infections in the early stage after allo-HSCT. In this stage fungal infections are also the problem, especially other than Candida albicans, which are usually recognized with the delay. Although mycological diagnosis based on PCR method is available, it has not been introduced into practice yet. Galactomannan testing and detection of specific fungal antigens in the blood are sometimes helpful. In the treatment we already administer not only conventional amphotericine B with considerable side effects, but also its lipid-based preparations (Abelcet,

AmBisome, Amphocil) being better tolerated, but unfortunately expensive. New antifungal drugs such as echinocandins (caspofungin, anidulafungin, micafungin) and newer azole drugs (voriconazole, posaconazole) are also currently available.

After resolution of pancytopenia cytomegalovirus infection is a most frequent problem. Thanks to a modern diagnostic approach based on early CMV antigen detection by means of PCR methods, CMV reactivation can be detected and cured before CMV disease is developed. The most common cause of CMV infection is latent virus reactivation in CMV-seropositive patient or CMV-transmission from a seropositive donor to a seronegative recipient. Therefore the optimal situation is when the serological status of the donor and the recipient is identical. The antiviral prevention includes the substitution of blood products from CMV-seronegative donors, transfusion of immunoglobulines and administration of antiviral drugs such as gancyclovir, foscavir, cydofovir and oral valgancyclovir. Polyoma- BK virus and adenovirus are common causes of dysuria, urinary tract infections and haemorrhagic cystitis in immunocompromised patients. Epstein-Barr virus can cause post-transplant lymphoproliferative disease (PTLD). The risk factors are the use of anti-thymocyte globulin and transplantation from unrelated donors. Monitoring of EBV-viremia by means of PCR methods enables to start the treatment early – to reduce the immunosuppression and to use rituximab (anti-CD20 antibody) and donor lymphocyte infusion (DLI).

8.3. Relapse of the disease

Having better adjunctive treatment and more effective GvHD prevention, the relapse of basic disease constitutes the main cause of allo-HSCT failure. The risk of the relapse depends on the type of the disease, its stage at the moment of transplantation and the GvHD prevention applied (the more effective immunosuppression, the higher risk of the relapse). The longest survival time is observed in patients with moderate acute or limited chronic GvHD, because of the lowest risk of the relapse.

Although the relapse after allo-HSCT can be treated by means by DLI, good prognosis refers usually to the patients with CML. In acute leukemia relapsing after allo-HSCT the temporary response can be also achieved, but it is usually not stable.

Patients with molecular CML relapse, i.e. with reappearance of bcr/abl transcript in PCR tests, have better prognosis than those with hematological relapse. The prognosis in patients with more advanced stages of CML- relapse, acceleration phase or blastic transformation- is much worse. The relapse should by detected as soon as possible, when there is still a chance for effective immunotherapy after allo-HSCT.

The alternative to specific disease markers determination is post-transplant chimerism testing. The PCR short tandem repeats (STR) method is used. The goal of allo-HSCT is to obtain the full donor's chimerism. Detection of returning or increasing recipient's chimerism can be a sign of the relapse of the disease, similarly to the re-occurrence of minimal residual disease [37]. In such case it is recommended to use adoptive immunotherapy by reduction of immunosuppressive treatment and DLI application. The chimerism testing is important also for

http://dx.doi.org/10.5772/53086

prediction and analysis of the graft failure and GvHD risks. GvHD and pancytopenia can develop as the side effects of DLI. The use of T-lymphocytes in escalating doses is equally effective as high DLI dose, but it decreases the risk of GvHD [38].

9. New indications for transplantation of hematopoietic stem cells

Allo-HSCT with subsequent immunotherapy can be applied in patients with metastatic solid tumors. The presence of graft versus tumor effect has been shown in kidney cancer, colon adenoma, metastatic breast, ovarian, prostate and pancreas cancers. RIC treatment has been used in these conditions in order to reduce TRM while enabling to achieve the response, which was complete in some cases [39]. The presence of fewer than 3 metastases and Karnofsky scale ≥70 constitutes beneficial prognostic factors [40]. The survival is longer in patients who develop chronic GvHD after DLI.

Allo-HSCT is currently tested in animal models and in experimental clinical applications. Hematopoietic stem cells are characterized by plasticity, which means that they can form not only blood cells. Hematopoietic stem cells can be forced to transform into the cells of various tissues such as heart muscle, bone or blood vessels in suitable conditions [41]. The science dealing with plasticity of stem cells is just developing, but it arises hope for revolution in the way of thinking about transplantation and organ regeneration.

10. Conclusion

Allo-HSCT procedure has transformed from the experimental method of treatment of leukemia in its final stage into routine procedure applied in patients with various hematological diseases. The ability to collect and to transplant hematopoietic cells makes it possible to cure many patients with malignant and nonmalignant diseases incurable with other methods. Thanks to development of unrelated donor registries the treatment with allo-HSCT can be currently offered not only to the patients having HLA-matched sibling donor but to almost every patient in need. The observed increase of transplantations of peripheral hematopoietic stem cells results from observed faster regeneration of hematopoietic system than after bone marrow transplantation and from beneficial GvL effect in clonal diseases, although it coincides with more frequent occurrence of GvHD.

The patients in the older age group and those with comorbidities can be treated with allo-HSCT after preparation with RIC. Still, the main problem is the relapse of the disease, however when it is detected early basing on chimerism analysis and minimal residual disease evaluation, it can be successfully treated with immunological intervention with DLI. Recent and current studies indicate that hematopoietic stem cells will be used for new clinical applications in the near future.

Author details

Miroslaw Markiewicz*, Malgorzata Sobczyk-Kruszelnicka, Monika Dzierzak Mietla, Anna Koclega, Patrycja Zielinska and Slawomira Kyrcz-Krzemien

*Address all correspondence to: mir.markiewicz@wp.pl

Medical University of Silesia, Department of Hematology and Bone Marrow Transplantation, Katowice, Poland

References

[1] Woods WG, Neudorf S, Gold S, Sanders J, Buckley JD, Barnard DR, Dusenbery K, DeSwarte J, Arthur DC, Lange BJ, Kobrinsky NL; Children's Cancer Group. A comparison of allogeneic bone marrow transplantation, autologous bone marrow transplantation, and aggressive chemotherapy in children with acute myeloid leukemia in remission. Blood. 2001; 97(1):56-62.

[2] Oosterveld M, Suciu S, Verhoef G, Labar B, Belhabri A, Aul C, Selleslag D,Ferrant A, Wijermans P, Mandelli F, Amadori S, Jehn U, Muus P, Zittoun R, Hess U,Anak O, Beeldens F, Willemze R, de Witte T. The presence of an HLA-identical sibling donor has no impact on outcome of patients with high-risk MDS or secondary AML (sAML) treated with intensive chemotherapy followed by transplantation: results of a prospective study of the EORTC, EBMT, SAKK and GIMEMA Leukemia Groups (EORTC study 06921). Leukemia. 2003; 17(5):859-68.

[3] Schmitz N, Pfistner B, Sextro M, Sieber M, Carella AM, Haenel M, Boissevain F, Zschaber R, Müller P, Kirchner H, Lohri A, Decker S, Koch B, Hasenclever D, Goldstone AH, Diehl V; German Hodgkin's Lymphoma Study Group; Lymphoma Working Party of the European Group for Blood and Marrow Transplantation. Aggressive conventional chemotherapy compared with high-dose chemotherapy with autologous haemopoietic stem-cell transplantation for relapsed chemosensitive Hodgkin's disease: a randomised trial. Lancet. 2002; 359(9323):2065-71.

[4] Carella AM, Bellei M, Brice P, Gisselbrecht C, Visani G, Colombat P, Fabbiano F, Donelli A, Luminari S, Feugier P, Browett P, Hagberg H, Federico M. High-dose therapy and autologous stem cell transplantation versus conventional therapy for patients with advanced Hodgkin's lymphoma responding to front-line therapy: long-term results. Haematologica. 2009; 94(1):146-8.

[5] Federico M, Bellei M, Brice P, Brugiatelli M, Nagler A, Gisselbrecht C, Moretti L, Colombat P, Luminari S, Fabbiano F, Di Renzo N, Goldstone A, Carella AM; EBMT/GISL/ANZLG/SFGM/GELA Intergroup HD01 Trial. High-dose therapy and autologous stem-cell transplantation versus conventional therapy for patients with ad-

vanced Hodgkin's lymphoma responding to front-line therapy. J Clin Oncol. 2003; 21(12):2320-5.

[6] Proctor SJ, Mackie M, Dawson A, White J, Prescott RJ, Lucraft HL, Angus B, Jackson GH, Lennard AL, Hepplestone A, Taylor PR. A population-based study of intensive multi-agent chemotherapy with or without autotransplant for the highest risk Hodgkin's disease patients identified by the Scotland and Newcastle Lymphoma Group (SNLG) prognostic index. A Scotland and Newcastle Lymphoma Group study (SNLG HD III). Eur J Cancer. 2002; 38(6):795-806.

[7] Moskowitz CH, Kewalramani T, Nimer SD, Gonzalez M, Zelenetz AD, Yahalom J. Effectiveness of high dose chemoradiotherapy and autologous stem cell transplantation for patients with biopsy-proven primary refractory Hodgkin's disease. Br J Haematol. 2004; 124(5):645-52.

[8] Carella AM, Cavaliere M, Lerma E, Ferrara R, Tedeschi L, Romanelli A, Vinci M, Pinotti G, Lambelet P, Loni C, Verdiani S, De Stefano F, Valbonesi M, Corsetti MT. Autografting followed by nonmyeloablative immunosuppressive chemotherapy and allogeneic peripheral-blood hematopoietic stem-cell transplantation as treatment of resistant Hodgkin's disease and non-Hodgkin's lymphoma. J Clin Oncol. 2000; 18(23):3918-24.

[9] Philip T, Guglielmi C, Hagenbeek A, Somers R, Van der Lelie H, Bron D, Sonneveld P, Gisselbrecht C, Cahn JY, Harousseau JL, et al. Autologous bone marrow transplantation as compared with salvage chemotherapy in relapses of chemotherapy-sensitive non-Hodgkin's lymphoma. N Engl J Med. 1995; 333(23):1540-5.

[10] Schouten HC, Qian W, Kvaloy S, Porcellini A, Hagberg H, Johnson HE, Doorduijn JK, Sydes MR, Kvalheim G. High-dose therapy improves progression-free survival and survival in relapsed follicular non-Hodgkin's lymphoma: results from the randomized European CUP trial. J Clin Oncol. 2003; 21(21):3918-27.

[11] Ljungman P, Bregni M, Brune M, Cornelissen J, de Witte T, Dini G, Einsele H, Gaspar HB, Gratwohl A, Passweg J, Peters C, Rocha V, Saccardi R, Schouten H, Sureda A, Tichelli A, Velardi A, Niederwieser D; European Group for Blood and Marrow Transplantation. Allogeneic and autologous transplantation for haematological diseases, solid tumours and immune disorders: current practice in Europe 2009. Bone Marrow Transplant. 2010; 45(2):219-34.

[12] Rosiñol L, Pérez-Simón JA, Sureda A, de la Rubia J, de Arriba F, Lahuerta JJ, González JD, Díaz-Mediavilla J, Hernández B, García-Frade J, Carrera D, León A, Hernández M, Abellán PF, Bergua JM, San Miguel J, Bladé J; Programa para el Estudio y la Terapéutica de las Hemopatías Malignas y Grupo Español de Mieloma (PETHEMA/GEM). A prospective PETHEMA study of tandem autologous transplantation versus autograft followed by reduced-intensity conditioning allogeneic transplantation in newly diagnosed multiple myeloma. Blood. 2008 ; 112(9):3591-3.

[13] Saccardi R, Freedman MS, Sormani MP, Atkins H, Farge D, Griffith LM, Kraft G, Mancardi GL, Nash R, Pasquini M, Martin R, Muraro PA; European Blood and Marrow Transplantation Group; Center for International Blood and Marrow Research; HSCT in MS International Study Group. A prospective, randomized, controlled trial of autologous haematopoietic stem cell transplantation for aggressive multiple sclerosis: a position paper. Mult Scler. 2012; 18(6):825-34.

[14] Snowden JA, Kapoor S, Wilson AG. Stem cell transplantation in rheumatoid arthritis. Autoimmunity. 2008; 41(8): 625-31.

[15] Marmont AM, Burt RK. Hematopoietic stem cell transplantation for systemic lupus erythematosus, the antiphospholipid syndrome and bullous skin diseases. Autoimmunity. 2008; 41(8):639-47.

[16] Farge D, Nash R, Laar JM. Autologous stem cell transplantation for systemic sclerosis. Autoimmunity. 2008; 41(8):616-24.

[17] Cassinotti A, Annaloro C, Ardizzone S, Onida F, Della Volpe A, Clerici M,Usardi P, Greco S, Maconi G, Porro GB, Deliliers GL. Autologous haematopoietic stem cell transplantation without CD34+ cell selection in refractory Crohn's disease. Gut. 2008; 57(2):211-7.

[18] Thomas ED, Lochte HL Jr, Lu Wc, Ferrebbe JW. Intravenous infusion of bone marrow in patients receiving radiation and chemotherapy. N Engl J Med. 1957;257(11): 491-6.

[19] Bach FH, Albertini RJ, Joo P, Anderson JL, Bortin MM. Bone-marrow transplantation in a patient with the Wiskott-Aldrich syndrome. Lancet. 1968;2(7583):1364-6.

[20] Rotta M, Storer BE, Sahebi F, Shizuru JA, Bruno B, Lange T, Agura ED, McSweeney PA, Pulsipher MA, Hari P, Maziarz RT, Chauncey TR, Appelbaum FR, Sorror ML, Bensinger W, Sandmaier BM, Storb RF, Maloney DG. Long-term outcome of patients with multiple myeloma after autologous hematopoietic cell transplantation and nonmyeloablative allografting. Blood. 2009; 113(14):3383-91.

[21] Peters C, Steward CG; National Marrow Donor Program; International Bone Marrow Transplant Registry; Working Party on Inborn Errors, European Bone Marrow Transplant Group. Hematopoietic cell transplantation for inherited metabolic diseases: an overview of outcomes and practice guidelines. Bone Marrow Transplant. 2003; 31(4): 229-39.

[22] Horowitz MM, Gale RP, Sondel PM, Goldman JM, Kersey J, Kolb HJ, Rimm AA, Ringdén O, Rozman C, Speck B, et al. Graft-versus-leukemia reactions after bone marrow transplantation. Blood. 1990; 75(3):555-62.

[23] Hows J, Bradley BA, Gore S, Downie T, Howard M, Gluckman E. Prospective evaluation of unrelated donor bone marrow transplantation. The International Marrow Unrelated Search and Transplant (IMUST) Study. Bone Marrow Transplant. 1993; 12(4): 371-80.

[24] Ringdén O, Schaffer M, Le Blanc K, Persson U, Hauzenberger D, Abedi MR, Olerup O, Ljungman P, Remberger M. Which donor should be chosen for hematopoietic stem cell transplantation among unrelated HLA-A, -B, and -DRB1 genomically identical volunteers? Biol Blood Marrow Transplant. 2004; 10(2):128-34.

[25] Bensinger WI, Storb R. Allogeneic peripheral blood stem cell transplantation. Rev Clin Exp Hematol. 2001; 5(2):67-86.

[26] de Lima M, Shpall E. Strategies for widening the use of cord blood in hematopoietic stem cell transplantation. Haematologica. 2006; 91(5):584-7.

[27] Ballen KK. New trends in umbilical cord blood transplantation. Blood. 2005; 105(10): 3786-92.

[28] Ferry C, Socié G. Busulfan-cyclophosphamide versus total body irradiation-cyclophosphamide as preparative regimen before allogeneic hematopoietic stem cell transplantation for acute myeloid leukemia: what have we learned? Exp Hematol. 2003; 31(12):1182-6.

[29] Tutschka PJ, Copelan EA, Klein JP. Bone marrow transplantation for leukemia following a new busulfan and cyclophosphamide regimen. Blood. 1987; 70(5):1382-8.

[30] Kashyap A, Wingard J, Cagnoni P, Roy J, Tarantolo S, Hu W, Blume K, Niland J, Palmer JM, Vaughan W, Fernandez H, Champlin R, Forman S, Andersson BS. Intravenous versus oral busulfan as part of a busulfan/cyclophosphamide preparative regimen for allogeneic hematopoietic stem cell transplantation: decreased incidence of hepatic venoocclusive disease (HVOD), HVOD-related mortality, and overall 100-day mortality. Biol Blood Marrow Transplant. 2002; 8(9):493-500.

[31] Clift RA, Buckner CD, Appelbaum FR, Bearman SI, Petersen FB, Fisher LD, Anasetti C, Beatty P, Bensinger WI, Doney K, et al. Allogeneic marrow transplantation in patients with acute myeloid leukemia in first remission: a randomized trial of two irradiation regimens. Blood. 1990; 76(9):1867-71.

[32] Kolb HJ, Mittermüller J, Clemm C, Holler E, Ledderose G, Brehm G, Heim M, Wilmanns W. Donor leukocyte transfusions for treatment of recurrent chronic myelogenous leukemia in marrow transplant patients. Blood. 1990; 76(12):2462-5.

[33] McSweeney PA, Niederwieser D, Shizuru JA, Sandmaier BM, Molina AJ, Maloney DG, Chauncey TR, Gooley TA, Hegenbart U, Nash RA, Radich J, Wagner JL, Minor S, Appelbaum FR, Bensinger WI, Bryant E, Flowers ME, Georges GE, Grumet FC, Kiem HP, Torok-Storb B, Yu C, Blume KG, Storb RF. Hematopoietic cell transplantation in older patients with hematologic malignancies: replacing high-dose cytotoxic therapy with graft-versus-tumor effects. Blood. 2001; 97(11):3390-400.

[34] Markiewicz M, Siekiera U , Dzierzak-Mietla M, Zielinska P, Kyrcz-Krzemien S. The impact of H-Y mismatches on results of HLA-matched unrelated allogeneic hematopoietic stem cell transplantation. Transplant Proc 2010; 42(8):3297-300.

[35] Remberger M, Storer B, Ringdén O, Anasetti C. Association between pretransplant Thymoglobulin and reduced non-relapse mortality rate after marrow transplantation from unrelated donors. Bone Marrow Transplant. 2002; 29(5):391-7.

[36] Savani BN, Griffith ML, Jagasia S, Lee SJ. How I treat late effects in adults after allogeneic stem cell transplantation. Blood. 2011; 117(11):3002-9.

[37] Mattsson J, Uzunel M, Tammik L, Aschan J, Ringdén O. Leukemia lineage-specific chimerism analysis is a sensitive predictor of relapse in patients with acute myeloid leukemia and myelodysplastic syndrome after allogeneic stem cell transplantation. Leukemia. 2001; 15(12):1976-85.

[38] Dazzi F, Szydlo RM, Craddock C, Cross NC, Kaeda J, Chase A, Olavarria E, van Rhee F, Kanfer E, Apperley JF, Goldman JM. Comparison of single-dose and escalating-dose regimens of donor lymphocyte infusion for relapse after allografting for chronic myeloid leukemia. Blood. 2000; 95(1):67-71.

[39] Childs R, Chernoff A, Contentin N, Bahceci E, Schrump D, Leitman S, Read EJ, Tisdale J, Dunbar C, Linehan WM, Young NS, Barrett AJ. Regression of metastatic renal-cell carcinoma after nonmyeloablative allogeneic peripheral-blood stem-cell transplantation. N Engl J Med. 2000; 343(11):750-8.

[40] Barkholt L, Bregni M, Remberger M, Blaise D, Peccatori J, Massenkeil G, Pedrazzoli P, Zambelli A, Bay JO, Francois S, Martino R, Bengala C, Brune M, Lenhoff S, Porcellini A, Falda M, Siena S, Demirer T, Niederwieser D, Ringdén O; French ITAC group and the EBMT Solid Tumour Working Party. Allogeneic haematopoietic stem cell transplantation for metastatic renal carcinoma in Europe. Ann Oncol.; 17(7):1134-40.

[41] Li ZH, Liao W, Cui XL, Zhao Q, Liu M, Chen YH, Liu TS, Liu NL, Wang F, Yi Y, Shao NS. Intravenous transplantation of allogeneic bone marrow mesenchymal stem cells and its directional migration to the necrotic femoral head. Int J Med Sci. 2011; 8(1): 74-83.

Chapter 7

Current Approach to Allogeneic Hematopoietic Stem Cell Transplantation

Hugo F. Fernandez and Lia Perez

Additional information is available at the end of the chapter

http://dx.doi.org/10.5772/53488

1. Introduction

1.1. Aims of chapter

In this Chapter we will discuss the indications for allogeneic hematopoietic stem cell transplantation (HCT). We will focus on the appropriate timing of this procedure for the different hematologic malignancies. We reviewed past approaches using myeloablative conditioning and present some of the newer reduced intensity therapies. Allogeneic transplantation is one of the first known uses of stem cells. Born from the need to rescue damaged bone marrow, it was first used in the setting of aplastic anemia and acute leukemia. Over the years, the technique has changed steadily and support for this procedure has improved immensely. Today this procedure is used to treat multiple malignant blood disorders, bone marrow failure syndromes, immune deficiency syndromes, and hemoglobinopathies. This chapter will focus on the malignant hematopathies. Another aspect of this Chapter will be to review the conditioning regimens used in allogeneic HCT.

2. Indications for transplantation

2.1. Acute myeloid leukemia

Acute myeloid leukemia (AML) Is heterogeneous group of clonal disorders. The disease can present at all ages, but this disorder is most commonly seen in older patients, with a median age at presentation of 67 years. [1] AML can present in a de novo fashion or can progress from antecedent hematological disorders, including myelodysplasia and myeloproliferative neoplasms (secondary AML), or after prior exposure to chemotherapy and/or radiation therapy (treatment-related AML). Patients who are deemed fit enough to receive therapy can be given various combinations of chemotherapy to induce a remission of the disease. The most common induction therapy is that of cytarabine given as a continuous infusion for 7 days in combination with an anthracycline for 3 days (the 7+3 regimen). This approach has been used for over 40 years with very good results [2-6]. Attempts to improve on this by adding other therapies have not resulted in improved outcomes. More recently, dose intensification of the anthracycline has resulted in improved complete remission (CR) rates and more importantly overall survival (OS) for patients below the age of 65 years [7-10]. Although current induction chemotherapy regimens are successful in obtaining a CR with rates approaching 70-80%; without consolidation chemotherapy, most patients will relapse and die of the disease. Because of the high risk of relapse, AML is the leading indication for allogeneic transplant.

There are several significant prognostic factors that will affect the patient's ability to achieve a CR. The most important is that of age. Other recognized factors are cytogenetic risk profile, molecular mutations, prior exposures to chemotherapy and radiation therapy, and antecedent hematological disorders. [11] These factors also impact on the patient's ability to maintain long-term remission and be cured of the disease. More recently, molecular mutations have come to the forefront in determining overall prognosis. These mutations include nucleophosmin-1 (*NPM1*), fms-like tyrosine kinase-3 (*FLT3), CAAT* enhancer binding protein alpha (*CEBPA*), and c-KIT. Retrospective analyses have shown that, in cytogenetically normal individuals, *NPM-1*and *CEBPA* have improved survival in comparison to those with other mutations [12]. *FLT3-ITD* negatively impacts all cytogenetic and molecular risk groups [12-14]. The European Leukemia Network proposed a new prognostic designation based on both accepted cytogenetic and molecular abnormalities [15]. More recently, newer molecular mutations have been described which in the future may help further delineate the prognostic risk [14]. A recent retrospective study from the Center for International Blood and Marrow Transplantation Research has also re-classified the cytogenetic risk for those patients proceeding to transplantation. [16]

The potential for relapse and the patient's clinical status are factors that determine the consolidation approach. Currently, prognostic factors are used to decide on the most appropriate consolidation therapy for patients with this disease. Multiple studies have demonstrated that patients with the core binding factor AML (*AML/ETO* and *RUNX/RUNX1*] have an excellent response to induction and consolidation chemotherapy. [17] For these patients, allogeneic hematopoietic cell transplantation (HCT) should be reserved for relapse of the disease. Contrary to this, an unfavorable risk profile usually portends a very poor prognosis. Patients with unfavorable cytogenetics (complex cytogenetics, single or multiple monosomal karyotype, *MLL* (11q23) [18]) respond very poorly to induction chemotherapy, and remissions are usually shorter. In patients with cytogenetically normal AML, the presence of *FLT3, MLL, DNMT3A*, and others have also demonstrated shorter disease-free survival (DFS) and OS [12, 19-21].

For more than 15 years, the standard of consolidation therapy for patients with AML in first CR (CR1) has been intensive chemotherapy using high-dose cytarabine. However, this approach is only effective in patients who are below the age of 60 years and have favorable risk cytogenetics [22]. Initially, allogeneic HCT was used as salvage therapy for patients who failed conventional chemotherapy. The sentinel paper was published by Thomas et al., who used allogeneic HCT as salvage therapy for 100 patients who had relapsed or refractory AML. The 13% OS gave great hope to the use of this modality [23] Subsequent reports from the same group promoted the use of allogeneic HCT as front-line consolidation therapy [24-27]. Randomized trials using genetic randomization demonstrated an improved DFS in patients receiving allogeneic transplantation [28]. Although the US Intergroup trial demonstrated there was no advantage to allogeneic transplantation compared to intensive chemotherapy in patients with de novo AML below the age of 60in CR1[29], more recent studies have demonstrated effectiveness of this approach. The US Intergroup trial had a significant flaw in that a large number of patients allocated to transplantation did not receive the intended therapy. However, retrospective subset analysis did note a significant improvement in patients with unfavorable-risk cytogenetics [30]. A meta-analysis of five trials performed by Yanada et al. (3100 patients) demonstrated an improved OS for patients with unfavorable-risk cytogenetic profiles. Until recently, there was no consensus as to how to treat patients with intermediate risk AML in CR1. Meta-analyses by the HOVON-SAKK group (925 patients) and a systematic review by Koreth et al. (6007 patients) all showed an improved OS for patients with intermediate- and unfavorable-risk cytogenetic profiles. These analyses were limited to related donor transplantations and to younger patients. [31-33] A Markov analysis of 2090 Japanese patients with *de novo* AML in CR1 confirmed the appropriateness of a related or alternative donor HCT over chemotherapy in this setting but not for patients without a matched donor [34]. A recent evaluation of patients with AML with a monosomal karyotype also demonstrated a benefit of allogeneic HCT in this group. [35] The appropriate intensity of the conditioning regimen for patients with myeloid malignancies in first CR is currently being evaluated by the Bone Marrow Transplant Clinical Trials Network (BMT-CTN) in a prospective randomized multi-center trial (0901).

About two-thirds of the patients with AML will not have a matched related donor (MRD). For these patients, matched unrelated donor (MUD) transplantation is an option particularly for those patients with unfavorable-risk profiles. A retrospective study from the CIBMTR reviewed MRD, MUD and partial MUD transplantation in patients with unfavorable-risk cytogenetics. Here the investigators found that MRD and MSD had similar leukemia-free survival and OS. The benefit was not seen in partially MUD or those over the age of 50 years. Other studies have demonstrated the similarities in outcomes compared to sibling transplants. [36-39] The trade-off is an increase in graft versus host disease (GVHD) and its associated mortality for increased disease control (graft versus leukemia effect). The only randomized trial using MUD was a German AML 01/99 trial. Here patients < 60 years of age with high-risk features (non CBF AML and > 5% blasts on the day 15 bone marrow biopsy) who did not have a MRD were randomized to a MUD allogeneic versus autologous HCT. The patients who had a MUD HCT had a superior OS to those treated with an autograft. [36]

Improvements in human leukocyte antigen (HLA) sequencing and selection of donors have reduced the effect of GVHD in this setting. [40] Better treatment options for the conditioning regimen and preventing and treating acute GVHD have provided more confidence in the procedure. [41] Tacrolimus and methotrexate are widely used as GVHD prophylaxis with or without anti-thymocyte globulin (ATG). Newer GVHD prophylaxis combinations such as sirolimus and tacrolimus [42-44], and ATG-Fresenius have reduced the incidence of both acute and chronic GVHD without impacting relapse or OS. [45]

A major challenge which remains is the older patient conventionally described as older than 60 years of age. [46] Interestingly in the case of allogeneic transplantation the threshold for the older patient is closer to 50 years. These patients are affected by worse prognostic factors, comorbidities, and intolerance to therapy. [47] However, multiple reports have demonstrated that transplant is possible with the appropriate conditioning regimen utilizing a non-myeloablative or reduced intensity dosing of therapy. [48]Although no randomized trial between conventional therapy and HCT has been reported to date, results suggest that outcomes are better than conventional chemotherapy for this group of patients. [49, 50] More on this will be discussed later in this chapter.

3. Chronic myeloid leukemia

Translocation between chromosomes 9 and 22 (t(9;22) or Philadelphia chromosome (Ph+) leads to an abnormal fusion protein (BCR-ABL) with dysregulated tyrosine kinase activity resulting in a myeloproliferative disorder characterized by abnormal white cell production known as chronic myeloid leukemia (CML). Without therapy, CML has a predictable progression from a chronic phase (CP) to the more advanced accelerated (AP) and/or blast (BP) phases.Since the introduction of tyrosine kinase inhibitors (TKIs) in October 2001, allogeneic hematopoietic stem cell transplant (HCT) has shifted from a first-line treatment option and to a second-, third-, or even a fourth-line option [51, 52]. The number of allogeneic transplantations in the post-TKI era has significantly decreased in CP CML patients; however, the number of patients transplanted in AP or BP remains the same [53].

Given the excellent results of studies using TKIs as upfront treatment for CP CML, a randomized trial to compare HCT to TKIs has not been performed and has not been justified. The use of TKIs as standard front-line therapy has been supported by few retrospective and/or genetically randomized studies [54, 55]. Imatinib mesylate has activity against progenitors and mature cells but has limited activity against leukemia stem cells [56, 57]. Unfortunately, the majority of patients achieving remission with imatinib mesylate continue to have molecular evidence of persistent disease [58]. Even in those patients who are treated for over 4 years with imatinib mesylate and in remission, BCR-ABL + stem cells are still detected in bone marrow [59].

Allogeneic HCT remains a curative approach with long-term molecular remissions, seen only rarely with TKIs, as the mechanism of the graft versus leukemia effect relies on the presence of antigens on leukemia stem cells [60]. Current indications of transplant are reserved, according to the European leukemia net [61], to the following CML subjects:

- At diagnosis for patients presenting in AP or BP
- Imatinib failure (after second-generation TKI pretreatment) progressing to AP or BP
- Patients with TKI resistant mutations such us T315I
- All patients failing second-generation TKI treatment.

Definitions of imatinib mesylate failure are: 1) a lack to achieve complete hematological remission at 3 months; 2) failure to achieve any cytogenetic response at 6 months; 3) persistence of more than 35% Ph+ metaphases at 12 months; or 4) less than complete cytogenetic response at 18 months. Resistance to imatinib mesylate is defined as loss of complete hematological response or complete cytogenetic response or development during imatinib mesylate treatment of an ABL kinase mutation leading to its resistance.

In summary, the present use of allogeneic HCT is reserved for patients with poor response to TKIs and/or those with advanced disease. Saussele et al. reported an interim analysis from the German CML Study group IV in patients who underwent a 5-arm randomization where 84 patients underwent allogeneic HCT as second-line therapy after imatinib mesylate failure [62].The 3-year survival in CP was 91%, with 59% in AP. The majority of patients (88%) achieved a molecular remission and reported a very low treatment-related mortality (TRM) (8%).The authors at that time concluded that allogeneic HCT could become the preferred second-line option after imatinib mesylate failure for suitable patients with a donor.

Because most patients are treated with TKI before transplant, it is important to understand whether this strategy could potentially jeopardize HCT results. Retrospective comparison of patients treated with imatinib mesylate pre-HCT compared with historical controls showed no effect on OS, progression-free survival, and non-relapse mortality [63]. Based on a Center for International Blood and Marrow Transplant Research (CIBMTR) study reported by Lee et al., imatinib mesylate before HCT in patients with CP CML leads to a better survival but no statistically significant difference in TRM, relapse, and leukemia-free survival and no differences reported in advanced CML. These results are re-assuring for the majority of patients that today are treated with TKIs prior to allogeneic HCT [64]. In summary, imatinib mesylate use before HCT has been shown to not increase toxicity and/or engraftment of subsequent allogeneic HCT [65-68]. Interestingly, risk of chronic GVHD may be decreased with the use of imatinib mesylate pre-HCT [67] and may potentially target GVHD-related fibrotic features if they developed post-HCT[69, 70]. In addition, the use of TKIs before HCT has been shown to improve outcomes if a patient achieves major cytogenetic remission compared to those who do not [67].

Imatinib mesylate as frontline for CP patients leads to a major cytogenetic response rate of 89% and OS of 86% at 7 years. Unfortunately, secondary resistance develops at a rate of 4% per year for CP [71] and 70-90% in AP/BP phases [72-74].With the development of second-generation TKIs (dasatinib and nilotinib) and the compelling results shown of a major cytogenetic response of up to 45% for imatinib mesylate failure patients [75, 76], recommendations for HCT are reserved for patients who have failed not only imatinib but also second-generation TKIs[61]. Front-line therapy with second-generation TKIs for CP CML it is now warranted [77, 78].

The majority of mutations are susceptible to second-generation TKIs, but some are resistant not only to first-generation but also to all second-generation TKIs. Threonine-to-isoleucine substitution at position 315 of Bcr-Abl fusion protein (T315I mutation) is well established to confer resistance to most TKIs [61]. Multiple reports have shown encouraging results with allogeneic HCT in patients for whom allogeneic HCT is recommended earlier in the disease course [79-82]. The results from efforts to develop third-line TKIs to target resistant mutations are encouraging. On September 4, 2012, the U.S. Food and Drug Administration approved bosutinib tablets (Bosulif®, Pfizer, Inc.) for the treatment of CP, AP, and BP Ph+ CML in adult patients with imatinib-resistant mutants of Abl or intolerance to prior therapy (http://www.fda.gov/Drugs/InformationOnDrugs/ApprovedDrugs/ucm318203.htm). The pivotal PACE trial data have shown robust anti-leukemic activity of ponatinib in patients with CML at all stages, who are either resistant or intolerant to dasatinib or nilotinib or who have the T315I mutation [83].

For advanced patients, TKIs have facilitated a bridge to the HCT procedure. Long-term outcomes with imatinib mesylate for AP CML are only up to 47 months and 7 months for BP CML[84-86] with a 2-year OS of only 47% and 16% for patients in AP and BP, respectively [87]. The goal for advanced disease patients is to achieve a second CP in order to proceed with allogeneic HCT. Because the rate of mutations is highly increased for these patients, assessment of mutation profile is quite vital to guide TKI selection. Allogeneic HCT represents the best chance for long-term success or even cure in AP/BP CML [88]. Given selection bias, only unfavorable risk CML patients should proceed to allogeneic HCT these days. Reduced intensity conditioning (RIC) regimens have facilitated transplant access to more frail populations; unfortunately a higher relapse risk remains due to aggressive disease and reduced chemotherapy [89-92]. Therefore, there is a need for strategies to improve current leukemia-free survival post-allogeneic HCT.Measurement of minimal residual disease has become particularly important as it has been shown that patients who have increased BCR-ABL expression levels (more than 10^{-4}) experience higher relapses rates [93-95]. Serial BCR-ABL RT-PCR is considered a standard practice and can be used to guide clinical interventions. It is not unusual to decect low level molecular disease; however treatment should be reserved for those patients whose markers increase over time or remain persistently positive. Maintenance therapy with TKIs post-transplant has proven to be tolerable [96]. Carpenter et al. reported that prophylactic use of imatinib mesylate for 1 year in Ph+ acute lymphoblastic leukemia (ALL) and CML lead to a low risk of relapse (18%) [97]. Other groups have also shown that use of TKIs post-HCT can help to minimize relapse risk [98, 99] and/or effectively control relapse post-HCT [100]. Experience of second-generation TKIs in the post-HCT setting are currently being explored in clinical trials (http://clinicaltrials.gov/ct2/show/NCT00702403). An early approach is to consider maintenance with a TKI in those who have shown activity prior to transplant, and BCR-ABL mutation analysis should guide TKI selection. Role of TKIs in the post-HCT setting should also be studied in the context of donor lymphocyte infusions (DLI) as immunotherapy, as it has been shown to be effective for management of early relapse in the pre-TKI era. The synergistic role of TKI with DLI should be further explored [101].

In conclusion, several effective drugs are available today to treat CML upfront during the chronic phase of the disease. Careful monitoring for BCR/ABL and mutation analysis are warranted to determine which patients will be in need of second- or third-line therapies. For patients with advanced-phase disease, HCT remains the option of choice, using a TKI to bridge to allogeneic HCT.

http://dx.doi.org/10.5772/53488

4. Myelodysplastic syndrome

Myelodysplastic syndrome (MDS) is a clonal stem cell disorder that results in a heterogeneous group of disorders characterized by excessive apoptosis of bone marrow cells. It is characterized by low peripheral counts, marrow dysplasia, proliferation and loss of differentiation of hematopoietic progenitors with a median age of 60-70 years at presentation. Mortality is related to bone marrow failure and evolution to secondary AML [102]. Despite development of novel therapeutic agents over the past decades, allogeneic HCT remains the only curative option in this disease. To date, HCT indications, timing, and incorporation of novel drugs before and/or after HCT remains a challenge. Additionally, whether novel treatment agents for elderly MDS patients should be pursued instead of allogeneic HCT remains unanswered. A recent retrospective cohort analysis suggested a survival advantage for allogeneic HCT (39%) compared with azacytidine (23%) therapy in medically fit patients with high-risk MDS of 60-70 years of age [103]. The German MDS study group is testing 5-azacytidine compared to allogeneic HCT in a prospective study for patients with International Prognostic Scoring System (IPSS) intermediate II or high-risk up to age 70 years (NCT01404741).

The IPSS system is based on peripheral blood cytopenias, cytogenetics, and marrow myeloblast percentages and is generally used to identify HCT candidates [104]. A limitation of the IPSS score is that it does not take into account patient age; therefore, development of other scoring system has been proposed. The World Health Organization classification and the World Health Organization classification–based Prognostic Scoring System have both shown relevant prognostic values in post-HCT MDS outcome for OS and relapse [105, 106]. In a recent analysis of 1915 patients with MDS, only 26% had primary MDS without prior therapy that could be classified with the IPSS system. A multivariate analysis of prognostic factors determined worst outcome for poor performance, older age, thrombocytopenia, anemia, increased bone marrow blasts, leukocytosis, chromosome 7 or complex (≥3) abnormalities, and prior transfusions. This new MDS prognostic model divided patients into 4 prognostic groups with significantly different outcomes with the advantage that it accounts for duration of MDS and prior therapy and is applicable to any patient with MDS at any time during the course of MDS [107].

A Markov decision analysis model designed by Cutler et al. showed that for low and intermediate-1 IPSS groups, delayed transplantation maximized OS; for intermediate-2 and high IPSS groups, HCT at diagnosis maximized OS and was associated with maximal life expectancy [108]. In contrast, other studies have suggested that younger patients with less advanced disease have a better transplantation outcome [105, 109]. An evidence-based review consensus by the American Society of Blood and Marrow Transplantation recommended early HCT for patients with IPSS intermediate-2 or high-risk at diagnosis and selected patients with lower risk disease at diagnosis who have poor prognostic features (such as older age, refractory cytopenias, and/or transfusion dependence) [110]. The American Society of Blood and Marrow Transplantation recommendations are limited as they are based on studies using IPSS score instead of more comprehensive ones; in addition, it only applies to newly diagnosed patients and excludes MDS subjects with treatment-related MDS/t-AML and chronic myelomonocytic leukemia subtype [111].

Factors that determine risk of progression from MDS to t-AML and that more accurately predict disease progression and HCT indication have been studied in the context of MDS phenotype and/or disease biology. With a patient group of 692 MDS patients, a European group analyzed outcome and reported worse OS and relapse rates based on poor cytogenetics [112]. In a multivariate analysis by Chang et al. comparing patients with secondary MDS or transformed to AML(t-AML) to de novo MDS, no significant differences in outcome were shown between the 2 cohorts and overall inferior outcome was shown in patients with secondary MDS/tAML, as the majority of advanced patients has increased frequency of high-risk cytogenetics [113]. Flow cytometric scoring system is predictive of post-HCT outcomes even after adjusting for risk factors such as marrow myeloblast percentage and IPSS score [114]. Cases of MDS classified as AML by microarray-based GEP assays had more aggressive disease and more rapid progression to AML, whereas MDS cases classified as "none-of-the-targets" had a more indolent clinical course [115]. Tumor necrosis factor-α polymorphisms affect HCT outcome in a disease-dependent manner [116]. There are many others risk categorization factors in MDS like FISH, spectral karyotyping, and mutation or deletion analyses [117-119], although clinical significance remains controversial [120]. Development of a revised scoring system is warranted to guide the decision-making process to recommend HCT for such a diverse and heterogeneous clonal condition.

Clinical evolution of disease such us increased transfusion, recurrent infections or bleeding may also precipitate the decision to proceed with HCT. Elevated serum ferritin levels, as reflection of increased body iron storage, have been showed to be associated with decreased OS and DFS, acute GVHD, and infections with myeloablative HCT [121, 122]. Ferritin levels should guide the need of chelation therapy prior to HCT and/or may guide conditioning regimen selection [123]. Co-morbidity as a determinant of HCT outcomes has been elegantly studied by Sorror et al. [124] and applied in the context of AML-MDS [125]. This group investigated the role of comorbidities, among other risk factors, in stratifying and comparing patients conditioned with non-myeloablative or myeloablative regimens. Patients with low HCT-CI scores and either low or high disease risks had probabilities of OS at 2 years of 70% and 57% after nonmyeloablative conditioning compared to 78% and 50% after myeloablative conditioning, respectively. Patients with higher HCT-CI scores (≥ 3) and either low or high disease risks had probabilities of OS of 41% and 29% with nonmyeloablative conditioning compared with 45% and 24% with myeloablative regimens, respectively. After adjusting for pretransplantation differences, stratified outcomes were not significantly different among patients receiving nonmyeloablative compared with myeloablative conditioning, with the exception of lessened nonrelapse mortality (hazard ratio, 0.50; P =.05) in the highest risk group. This group concluded that patients with low comorbidity scores could be candidates for prospective randomized trials comparing nonmyeloablative and myeloablative conditioning regardless of disease status [125]. An additional scoring system has also emphasized the negative influence of comorbidities on HCT outcomes [126].

Based on published literature, patients up to 70 years of age can tolerate allogeneic HCT and age per se should not be a criterion for patient selection and/or intensity of the conditioning regimen rather than performance status, comorbidity, and disease status [127].Results from a

European Group for Blood and Marrow Transplantation (EBMT) report suggested that age is not a contraindication to HCT; the cumulative incidence of non-relapse mortality at 4 years was 36% in the 50- to 60-year-old patient group and 39% for the group 60 years or older (P =. 39], with OS not differing between the groups (34% versus 27%, P =.2). In a multivariate analysis for OS, only advanced stage of the disease at time of transplantation (hazard ratio = 1.55] was associated with inferior survival [128]. Similar results were reported by the CIBMTR; in a multivariate analysis, they showed that OS was inferior with low performance status, mismatched unrelated donors, and unfavorable cytogenetic, but age had no impact [129].

To facilitate HCT access to the majority of MDS patients, a RIC regimen has been developed. The rationale for RIC is to promote graft-versus-leukemia effect without excessive toxicity to minimize TRM. Many RIC regimens have been developed using combinations of busulfan with cyclophosphamide or fludarabine, fludarabine with cyclophosphamide, or low-dose total body irradiation (TBI) (200cG) among others versus the more intense or conventional regimens based on TBI or busulfan/cyclophosphomide-based regimens. Unfortunately, due to the lack randomized prospective trials, it remains unknown which conditioning regimen should be chosen and how "intense and/or reduced" the conditioning should be. In general, the highest tolerable regimen should be chosen since reduced intensity is associated with a higher relapse rate, as suggested in multiple retrospective studies [130-136]. RIC HCT with fludarabine/melphalan and tacrolimus/sirolimus-based GVHD prophylaxis resulted in a relapse incidence of 20.9% with low-grade acute GVHD [137]. An ongoing prospective randomized trial comparing RIC versus myeloablative conditioning has been developed to address selection bias for allogeneic HCT by the EBMT group (NCT00682396).

Disease relapse post-HCT remains a critical issue as long-term outcome is compromised. Approaches to tackle this issue include pre-HCT induction chemotherapy and/or novel agents for high-risk patients or drug maintenance to prevent relapse pre-emptively post-HCT, as opposed to strategies for relapse treatment. Still debatable to date is whether pre-HCT induction chemotherapy has a role to minimize relapse post-HCT for patients with advanced MDS. Unfortunately, this remains unanswered due to lack of randomized and/or definitive data [138-141]. Introduction of novel agents in the pre-HCT setting seems feasible, associated with less toxicity, and may allow for similar post-HCT outcomes when compared to chemotherapy [142].Another approach is to use low-dose 5-azacytidine as maintenance post-HCT. De Lima et al. determined that the optimal combination was 32 mg/m^2 given for at least 4 cycles, with reversible thrombocytopenia as the dose-limiting toxicity. The authors suggested that this treatment prolonged event-free survival (EFS) and OS [143]. In the event of disease relapse post-HCT, azacytidine administration is feasible and may induce durable remissions [144]. DLIs can result in complete remission in some patients, but long-term survival is infrequent [145]. The Azarela trial, a prospective multicenter phase II trial, was developed to test whether a combination of 5-azacytidine and DLI would benefit patients with relapsed MDS post-HCT. Overall response rate was 64% with 20% achieving and staying in CR, 12% achieved partial response, and 32% showed stable disease with low incidence of acute GVHD occurring (24%). These data suggest that salvage therapy with combination azacytidine + DLI is feasible and has significant anti-leukemic activity in relapsed MDS post-HCT [146].

In conclusion, several factors influence HCT indication and timing for MDS patients. Incorporation of evolving prognostic indicators might help to develop treatment algorithms to decide the appropriate timing for allogeneic HCT. The ultimate objective is to proceed with HCT when non-transplantation approaches would result in outcomes lower than those that would result with allogeneic HCT. Currently, novel HCT approaches are allowing the consideration of older patients and/or the use of alternative donors to treat MDS. A remaining question is how to incorporate HCT for those patients that are achieving a CR with hypomethylating agents and/or other novel agents. Development of prospective clinical trial may help to elucidate these questions within a fast evolving field.

5. Acute lymphoblastic leukemia

Acute lymphoblastic leukemia (ALL) is a bone marrow clonal disease characterized by the rapid proliferation of immature lymphoblasts. Despite initial control of the disease, the majority of adult patients will relapse with poor long-term outcomes. Allogeneic HCT has been used as a salvage therapy for both relapsed patient and high-risk patients with ALL early in the disease process. The availability of unrelated donors and/or alternative stem cell sources and the development of RIC transplants have resulted in far more allogeneic transplants being performed for this rare disease. For adults with ALL, indication and timing of allogeneic HCT remains debatable as defining the optimal role for allogeneic HCT has been limited by the lack of prospective data that can only be gained by large multicenter-national trials.

Historically, allogeneic HCT was reserved for high-risk patients, especially for those with Ph + ALL. Patients with high-risk features benefit from upfront HCT, including those with increased white blood count at presentation (>25,000/μL), chromosomal translocations [t(9;22), t(4;11), t(8;14)], older age (≥30 years), extra-medullary disease at diagnosis, and/or requiring more than 4 weeks to achieve CR [147]. Strategy to take ALL patients in CR1 for t(9;22) and t(1;19) have been supported by a trial by the French Group of Therapy for adult ALL (LALA-94) in a subgroup analysis [148]. Improvement in detection of minimal residual disease has also helped to assess disease risk, as 10% of patients with a rapid MRD decline to lower than 10(-4) or below detection limits at day 11 and day 24 were classified as low risk as their 3-year relapse rate was 0% [149]. Testing MRD with flow cytometry and/or molecular analysis for gene rearrangements may help to guide transplant decisions.

The largest prospective study of HCT in adult ALL was conducted by the Medical Research Council in Great Britain (UKALL XII) and the Eastern Cooperative Oncology Group in the United States (ECOG 2993). In this trial, allogeneic HCT resulted in improved disease control in all adult patients with ALL, with younger patients with low-risk disease benefiting the most with allogeneic HCT [150]. This international collaboration prospectively evaluated the role of allogeneic HCT for adults with ALL and compared autologous HCT with standard chemotherapy. Patients received 2 phases of induction and, if in remission, were assigned to allogeneic HCT if they had a compatible sibling donor. Patients without a donor were randomized to chemotherapy for 2.5 years versus an autologous HCT. A donor versus no-donor analysis showed that Ph- ALL patients (standard risk) with a donor had a 5-year improved OS of 53% versus 45% for no donor (P =.01). The relapse rate was significantly lower (P ≤.001) with HCT in the standard-risk ALL patients. The survival difference was significant only in standard-risk patients, but not in high-risk patients, who had an impressive reduction in relapse rate but increased non-relapse mortality that abrogated the OS benefit of allogeneic HCT. For the no donor group, patients randomized to chemotherapy had a higher 5-year OS (46%) than those randomized to autologous transplantation (37%; P =.03). In conclusion, MRD allogeneic HCT for ALL in CR1 provide the most potent anti-leukemic therapy and considerable survival benefit for standard-risk patients. We may also conclude that there is no role for a single autologous HCT to replace consolidation/maintenance in any risk group.

For high-risk patients, results are conflicting with a recent large meta-analysis from seven studies of adult high-risk ALL (n=1274) using natural randomization based on donor availability combined with intent-to-treat analyses. This study demonstrated that patients in the donor groups had significantly better survival than patients in the no-donor groups (hazard ratio, 1.29; 95% confidence interval [95% CI], 1.02-1.63 [P =.037]). When only high-risk patients were included in the analysis, the superiority of the survival advantage was even greater (hazard ratio, 1.42; 95% CI, 1.06-1.90 [P =.019]) [151]. In addition, a recent systematic review and meta-analysis supported MRD HCT as the optimal post-remission therapy in ALL patients aged 15 years or over, resulting in improved OS and DFS with a significant reduction of disease relapse but with increased non-relapse mortality[152]. Interpretation of the results of the multicenter international trial has led to advocating early allogeneic HCT for patients with standard risk for some transplantation teams while others have preferred a more personalized approach as reports from various study groups differ and are often contradictory, leading to difficulty in interpreting the data [153, 154].

Historically, allogeneic HCT has been the standard of care for patients with high-risk Ph+ ALL in CR1. With the introduction of TKIs over the past decade, a treatment algorithm introducing TKIs in combination with allogeneic HCT for adult patients with Ph+ ALL is mandated. TKIs have been used in the upfront induction/maintenance chemotherapy setting and as maintenance post-HCT to prevent disease relapse in Ph+ ALL patients. Whether use of TKIs has an impact on OS when combined with HCT or whether TKIs will replace the use of allogeneic HCT remains unanswered to date. Multiple studies have shown the advantage of using imatinib mesylate in the induction/consolidation phase, allowing better remission rates and durable response with minimal toxicity as well as facilitating access and planning for an allogeneic HCT [154-159]. Review of these trials has suggested that over 90% of patients achieved a complete response as previously reviewed [154, 160]. Dasatinib, a multi-target kinase inhibitor of BCR-ABL and SRC family kinases, has been shown to induce responses in patients with imatinib-resistant or intolerant Ph+ ALL. In the START-L trial, major hematologic responses were achieved in 42%(15/36) of patients, 67% of whom remained progression-free when used at a dose of 140 mg. Complete cytogenetic responses were attained by 58% (21/36) of patients. The presence of BCR-ABL mutations conferring imatinib resistance did not preclude a response to dasatinib in this trial [161], suggesting a role for dasatinib to manage Ph+ ALL upfront [161]. Ravandi et al. examined the efficacy and safety of combining chemotherapy with dasatinib in patients with Ph+ ALL and determined that 94% achieved CR with an estimated 2-year survival of 64%. The combination of chemotherapy with dasatinib is effective in achieving long-term remissions in patients with newly diagnosed Ph+ ALL[162]. Nilotinib has also been tested for the management of relapsed/refractory Ph+ ALL with encouraging results[163].

TKI treatment is also a promising strategy when used as a consolidation strategy to induce and/or maintain molecular responses to decrease relapse rate after allogeneic HCT. Carpenter el al. reported safety data in 15 patients with Ph+ ALL who were enrolled in a prospective study and given imatinib from the time of engraftment until day 265 after HCT [97]. A clinical trial is currently ongoing to determine the safety of the administration of nilotinib between day 81 and day 365 after HCT in patients with Ph+ leukemia (http://clinicaltrials.gov/show/NCT00702403). Lastly, TKIs have been shown to be effective for management of relapse in Ph+ ALL in the post-HCT setting, although these data are based on few reports [160]. In summary, TKIs should be incorporated as a pre-HCT strategy to facilitate higher response rate and to improve both quality and durability of responses prior to allografting. TKIs are also a reasonable and promising strategy after allogeneic HCT to consolidate and maintain molecular responses that may ultimately improve survival for patients with Ph+ ALL. The optimal duration of therapy post-HCT, particularly in patients with sustained molecular response, remains to be determined. Whether TKI incorporation in the treatment strategy would impact OS is still unclear. In the absence of large prospective randomized trials comparing imatinib-chemotherapy regimens versus allo-HCT as a consolidative strategy, allo-HCT remains the best therapeutic approach that offers a possibility of cure in Ph+ ALL [160].

There is increased interest in developing strategies to minimize toxicity associated with allogeneic HCT, especially after the results of the UK ALL XII ECOG 2993 study, which showed a significant TRM in patients over the age of 35 years despite better control of disease [150]. Several groups have sought to minimize morbidity and mortality in this group of patents through reduced intensity approaches, allowing for access to HCT for majority of Ph+ ALL subjects [164]. Unfortunately, there is no prospective trials using RIC for this disease published in the literature. Few recent retrospective series have been reported with 2-year OS and DFS between 50 and 61.5% [165]. We previously published our initial experience with FLU and BU in adult ALL patients, which showed a 2-year cumulative incidence of relapse of 19% (95% CI 8%-41%) for those transplanted in CR1 and 48% (29%-80%) in those with more advanced disease, with a 2-year OS of 54% (95% CI 39%-69%). Relapse-free survival at 2 years was 63% (95% CI 45%-81%) for patients transplanted in CR1 and 34% (95% CI 11%-57%) for patients transplanted in more advanced disease. We concluded that, compared to irradiation-containing regimens, FLU and PK-targeted BU appear safer and similarly effective in controlling ALL, providing a treatment option for adult patients with ALL [166]. Nonmyeloablative allogeneic HCT approach is promising but its role for management of Ph+ ALL requires further investigations [154].

http://dx.doi.org/10.5772/53488

6. Lymphoma

Both Hodgkin lymphoma (HL) and non-Hodgkin lymphoma (NHL) represent a large group of diverse diseases. They are characterized by enlarged lymph nodes, splenomegaly, and constitutional symptoms. These disorders can present with bone marrow and extramedullary consequences. As a whole, they respond to combination chemotherapy. For patients who have relapsed or are refractory to initial therapy autologous HCT is the treatment of choice. The Parma group study, established the superiority of high-dose chemotherapy and autologous HCT over conventional salvage chemotherapy in a randomized multi-center trial for relapsed aggressive NHL [167]. Based on this study, autologous HCT became the standard of care for chemotherapy-sensitive relapsed or primary refractory aggressive NHL. There are instances where allogeneic HCT is the preferred approach for lymphoma.

6.1. Non-hodgkin lymphoma

6.1.1. Diffuse large B cell lymphoma (DLBCL)

The number of published studies using allogeneic HCT in DLBCL are limited and do not allow definitive conclusions. Allogeneic HCT has generally been used as treatment for patients who have relapsed after autologous HCT and on occasion for relapsed high-risk or refractory disease. No prospective comparative studies are available in this setting. A retrospective study by the CIBMTR compared the outcomes of DLBCL patients undergoing first autologous HCT (n = 837) or HLA-identical MSD allogeneic HCT with myeloablative conditioning (n =79). Allogeneic HCT was associated with higher TRM but with a similar risk of disease progression compared with lower-risk patients who received autologus HCT. [168] The European Group for Blood and Marrow Transplantation (EBMT) registry published a retrospective analysis of 101 patients. Approximately two-thirds of the patients received a reduced-intensity conditioning (RIC) regimen and 70% had an MSD. Non relapse mortality (NRM) was low with a rate of 28.2%, a relapse rate of 30% and an OS rate of 53%. Patients with a long remission after autologous HCT and with sensitive disease at allogeneic HCT appear to be the best candidates for this approach. [169] Thus, the use of allogeneic transplantation should be reserved for relapsed and refractory DLBCL that is responsive to the last line of therapy.

6.2. Follicular lymphoma (FL)

FL comprises approximately 25% of all newly diagnosed NHL cases. As an indolent lymphoma, the disease course is one of remissions and relapses with chemotherapy, followed inevitably by resistance and transformation to a more aggressive NHL histology. Trials from the several European Groups compared consolidative autologous HCT to chemotherapy ± interferon alfa (IFN-α) maintenance therapy or rituximab. [170-173] As autologous HCT provides no benefit in OS in FL it is currently not recommended as consolidation therapy.

The graft-vs-lymphoma effect afforded by allogeneic HCT is appealing as a potential curative approach in FL. Myeloablative conditioning allogeneic HCT, due to high TRM has not resulted in an improved OS in this disease. [174, 175] RIC allogeneic HCT is associated with a lower

TRM and the graft-vs-lymphoma effect may be beneficial in this indolent disease. Several studies have been published using this approach. The MD Anderson BMT program published results of their single institution trial of 43 patients with relapsed/refractory FL receiving a RIC allogeneic HCT with high doses of rituximab during and after conditioning. The PFS and OS rates were robust at 83% and 85%, respectively. [176] Currently, the BMT-CTN (0701) is confirming these results in a multi-institution trial.

6.3. Mantle cell lymphoma (MCL)

MCL is an aggressive NHL that often is responsive to initial chemotherapy but has a very high relapse rate and is incurable with conventional chemotherapy. With intensified induction regimens and the addition of rituximab, a higher proportion of patients achieve complete remission; however, long term cures are rare. [177] Autologous HCT provides very good control of the disease particularly in patients who received transplants in CR1. [178, 179] The Mantle Cell Lymphoma International Prognostic Index (MIPI) predicted good outcomes for patients in the good- and intermediate-risk. Unfortunately the poor-risk group had a disappointing survival, suggesting that these patients may be better suited for allogeneic HCT. [180]

To reduce toxicity and mortality in these heavily pretreated and older patients, RIC allogeneic HCT has been proposed with promising results. Treatment with a nonmyeloablative conditioning regimen and allogeneic HCT in 33 patients with relapsed and refractory MCL resulted in an OS rate of 65%. None of the patients transplanted in CR had relapsed after a median follow-up of 2 years. [181] Long term follow upof RIC allogeneic HCT in 35 patients with relapsed or refractory MCL demonstrated a low TRM rate and outcomes in which median OS had not been reached. [182] Finally, The British Society for Blood and Marrow Transplantation published the results of a retrospective analysis of 70 heavily pretreated patients with relapsed/refractory MCL who received RIC allogeneic HCT with or without alemtuzumab with or without DLI to boost the graft vs-lymphoma effect. The 3-year OS rate for patients who received donor lymphocyte infusions for relapse was 79%. [183] All of these studies demonstrated a plateau on the survival curves. Based on these reports, allogeneic HCT appears to be effective therapy for relapsed and refractory MCL and the only one associated with long-term remission. It will be necessary to complete a prospective, randomized study to define the role of upfront allogeneic HCT in MCL patients.

6.4. T-cell lymphoma

T-cell NHL (Peripheral T-cell lymphoma-not otherwise specified, angioimmunoblastic T-cell lymphoma (AITL), and anaplastic large-cell lymphoma (ALCL)) are a heterogeneous group of lymphomas which for the most part have an inferior prognosis when compared to B-cell NHL after CHOP therapy. With the exception of anaplastic large-cell kinase-positive (ALK) positive anaplastic large-cell lymphoma, T-cell NHL Carries a poor prognosis with low DFS and OS with standard chemotherapy. Several studies have demonstrated the use of autologous HCT in T-cell lymphoma has similar results to DLBCL. [184-189].

Allogeneic HCT has been proposed for the treatment of T-cell Lymphoma because of the potential graft-vs-lymphoma effect. There are limited studies in this field but the results have been promising. A retrospective analysis from France on 77 patients who underwent allogeneic HCT for PTCL resulted in a 5-year OS rates of 57%. Myeloablative conditioning was used in the majority of the patients. Patients with AITL had the best outcome, with a 5-year OS rate of 80%. Risk of relapse was low; however, the high TRM limited the benefit of the myeloablative approach. [190] RIC allogeneic HCT was published a prospective phase II trial using a reduced intensity regimen in 17 patients with PTCL. As expected TRM was low and the estimated 3-year OS was 81%. [191] In summary, the use of RIC allogeneic HCT through a lower TRM and allows transplant in older and heavily pretreated patients with reasonable OS. Certain T-cell entities such as hepatosplenic T-cell lymphoma, adult T-cell leukemia/ lymphoma, and systemic extranodal NK/T-cell lymphoma carry such a poor prognosis that allogeneic HCT is justified as part of the initial treatment. The use of prognostic indexes such help identify patients with extremely high risk of relapse who may also benefit from an allograft. Only prospective multicenter trials will define the role of allogeneic HCT in these aggressive lymphomas.

6.5. Hodgkin lymphoma

Combination chemotherapy with or without radiation therapy results in long-term DFS and OS for about 80% of newly diagnosed patients with HL. [192] As in NHL autologous HCT is well established for the treatment of disease. [193] An approach to minimize relapse after autologous HCT for high-risk patients using the anti-CD30 antibody (brentuximab) conjugated to an anti-tubulin drug (vedotin) [SGN-35][194] is currently being studied in a randomized phase III placebo-controlled trial as maintenance therapy following autologous HCT.

Because of prior intensive therapy, RIC allogeneic HCT is an appropriate option in candidates for patients with HL. [195-198] Recent retrospective analyses demonstrate improved PFS and OS compared to additional salvage therapy for patients treated with this approach after relapse following autologous HCT. [197, 199] More importantly, outcomes with MRD vs MUD do not appear to be different. [196, 198]

7. Conditioning regimens

7.1. Myeloablative conditioning

Allogeneic bone marrow transplantation is the most intensive post-remission therapy used for management of malignant disorders over the past decades. Toxicity of a conditioning regimen can impact on overall morbidity, including interstitial pneumonitis, sinusoidal obstruction syndrome/veno-occlusive disease, and may lead to an increased incidence of GVHD. Despite current understanding of the transplantation process, the optimal chemotherapy and/or radiation conditioning regimen remains unknown. Few data from comparative or randomized studies are available to address this issue. Allogeneic hematopoietic cells serve a dual purpose, not only to restore hematopoiesis but also to impose immunologic effects against malignant clones, a process known as graft versus leukemia. This has led to the development of a conditioning regimen that will minimize toxicity with preservation of graft versus leukemia effect as the main mechanism of action to eradicate disease.

The spectrum of conditioning intensity has been defined in three categories: 1) myeloablative, which causes irreversible marrow aplasia if transplantation is not performed; 2) nonmyeloablative, which cause minimal marrow suppression; and 3) RIC, which causes cytopenias of intermediate duration [200]. Assignment to these categories is based on the duration of cytopenia and on the requirement for stem cell support. Myeloablative regimens cause irreversible cytopenia, and stem cell support is mandatory. Nonmyeloablative regimens cause minimal cytopenia and can be given also without stem cell support. RIC causes cytopenias of variable duration and should be given with stem cell support, although cytopenia may not be irreversible. Compared with high-dose MA preparative regimens, NMA or RIC regimens are associated with shorter inpatient hospital stays, reduced need for transfusions [201], and a shorter duration of neutropenia with fewer bacterial infections [202-204]. There is current trend to adopt less-toxic conditioning regimens to allow access for patients to undergo HCT who has been previously been excluded because of age or comorbidities. Standardized classification of conditioning regimen intensities will allow comparisons across studies and interpretation of study results [200].

Myeloablative regimens, a combination of agents expected to produce profound pancytopenia and myeloablation within 1-3 weeks from administration, have caused pancytopenia that is long lasting, usually irreversible, and in most instances fatal, unless hematopoiesis is restored by hemopoietic stem cell infusion [200]. Early use of this approached invested on the theory of dose intensity to eradicate disease. [205]. The two most commonly used myeloablative conditioning regimens for allografts for leukemia/lymphoma use a combination of high-dose busulfan combined with cyclophosphamide and cyclophosphamide in combination with TBI. The Cyclophosphamide-TBI regimen uses a cyclophosphamide dose of 120 mg/kg and 10–15 Gy TBI [23] and busulfan-cyclophosphamide uses a busulfan dose of 16 mg/kg orally and Cy 120 mg/kg [206]. From the available data, there are no significant differences in survival with these two regimens. There is also no evidence that intensified conditioning improves survival, as a higher dose of TBI is associated with increased toxicity [205]. Cyclophosphamide or TBI has also been tested in addition to other chemotherapy agents like melphalan, thiotepa, etoposide, and dimethylbusulfan. The problem with myeloablative conditioning is the high TRM that ultimately jeopardizes overall success. The risk of TRM after a myeloablative regimen has decreased over time, attributed to improved HLA-typing and better supportive care [207]. Neither regimen explored in the myeloablative setting is suitable for all the situations and a particular regimen should be selected depending on the clinical situations if myeloablative approaches are still an option nowadays [208] with the introduction of less toxic transplantation approaches.

Several attempts have been made in the past 30 years to limit early transplant toxicity, by reducing the intensity of the conditioning regimen as previously reviewed [200]. Within the past 20 years, the introduction of fludarabine (Flu) [209, 210] and further dose reductions of alkylating agents [211, 212] or TBI has led to minimized toxicity.

http://dx.doi.org/10.5772/53488

These regimens were designed to allow access to HCT for older patients or because of comorbidities that would preclude HCT. Enthusiasm in the transplant community has led to adoption of these reduced toxicity modalities [213]. A workshop convened by the CIBMTR addressed the dose spectrum, which defines a RIC regimen [214]. A total of 56 participants were surveyed, and 67% agreed that a RIC regimen should cause reversible myelosuppression when administered without stem cell support, result in low nonhematologic toxicity, and, after transplantation, result in mixed donor–recipient chimerism at the time of first assessment in most patients. Likewise, the majority (71%) agreed or strongly agreed that regimens including <500 cGy of TBI as a single fraction or 800 cGy in fractionated doses, busulfan dose <9 mg/kg, melphalan dose <140 mg/m^2, and thiotepa dose < 10 mg/kg should be considered RIC regimens. However, only 32% agreed or strongly agreed that the combination of carmustine, etoposide, cytarabine, and melphalan (BEAM) should be considered a RIC regimen. These results demonstrate that, although HCT professionals have not reached a consensus on what constitutes a RIC regimen, most accept currently used criteria and operational definitions [214].

RIC is an intermediate category of regimens that causes pancytopenia and requires stem cell support if prolonged and autologous recovery is possible. An improved rate of toxicity is achieved by reducing the dose of alkylating agents or TBI by at least 30%. Most often, these regimens combine Flu with an alkylating agent, melphalan [215],Bu [211],thiotepa[212] in reduced doses, or Flu with reduced-dose TBI [216]. Decreased TRM has been successfully achieved with this approach [217, 218] Among the published phase II trials, leukemia relapse remained consistently the main cause of treatment failure after RIC or nonmyeloablative conditioning, with 2- to 4-year relapse rates ranging from 30% to 61%. Mohty et al.recently updated results of the first prospective trial directly comparing RIC allogeneic HCT versus consolidation chemotherapy in patients with AML using "genetic allocation." In an intent-to-treat analysis, leukemia-free survival was superior in the donor group (60% versus 23% at 7 years; P =.003) but with a significant relapse risk [219]. Recent retrospective analysis demonstrated that RIC has similar outcomes to MAC in patients with AML or MDS. [217, 220] Because of prior therapy and older age, as described above in the Lymphoma section RIC allogeneic HCT is appropriate for most those patients. Allogeneic transplantation has evolved significantly in the last 40 plus years of use as stem cell therapy. To further improve its outcomes patients should be selected early and the appropriate regimen should be used to optimize the anti-malignancy effect.

Author details

Hugo F. Fernandez and Lia Perez

Moffitt Cancer Center and Research Institute, Tampa, Florida, USA

References

[1] Howlader N NA, Krapcho M, Neyman N, Aminou R, Altekruse SF, Kosary CL, Ruhl J, Tatalovich Z, Cho H, Mariotto A, Eisner MP, Lewis DR, Chen HS, Feuer EJ, Cronin KA (eds). , National Cancer Institute. Bethesda, MD, , . *SEER Cancer Statistics Review, 1975-2009 (Vintage 2009 Populations).*based on November 2011 SEER data submission, posted to the SEER web site, 2012.

[2] Rai KR, Holland JF, Glidewell OJ, Weinberg V, Brunner K, Obrecht JP, et al. Treatment of acute myelocytic leukemia: a study by cancer and leukemia group B. Blood. 1981 Dec;58(6):1203-12.

[3] A systematic collaborative overview of randomized trials comparing idarubicin with daunorubicin (or other anthracyclines) as induction therapy for acute myeloid leukaemia. AML Collaborative Group. British journal of haematology. 1998 Oct;103(1): 100-9.

[4] Weick JK, Kopecky KJ, Appelbaum FR, Head DR, Kingsbury LL, Balcerzak SP, et al. A randomized investigation of high-dose versus standard-dose cytosine arabinoside with daunorubicin in patients with previously untreated acute myeloid leukemia: a Southwest Oncology Group study. Blood. 1996 Oct 15;88(8):2841-51.

[5] Vogler WR, Velez-Garcia E, Weiner RS, Flaum MA, Bartolucci AA, Omura GA, et al. A phase III trial comparing idarubicin and daunorubicin in combination with cytarabine in acute myelogenous leukemia: a Southeastern Cancer Study Group Study. J Clin Oncol. 1992 Jul;10(7):1103-11.

[6] Pautas C, Thomas X, Merabet F, Raffoux E, Bourhis JH, de Botton S, et al. Randomized Comparison of Standard Induction with Daunorubicin (DNR) for 3 Days vs Idarubicin (IDA) for 3 or 4 Days in AML pts Aged 50 to 70 and of Maintenance with Interleukin 2. Final Analysis of the ALFA 9801 Study. ASH Annual Meeting Abstracts. 2007 November 16, 2007;110(11):162-.

[7] Ohtake S, Miyawaki S, Fujita H, Kiyoi H, Shinagawa K, Usui N, et al. Randomized study of induction therapy comparing standard-dose idarubicin with high-dose daunorubicin in adult patients with previously untreated acute myeloid leukemia: the JALSG AML201 Study. Blood. 2011 Feb 24;117(8):2358-65.

[8] Fernandez HF, Sun Z, Yao X, Litzow MR, Luger SM, Paietta EM, et al. Anthracycline dose intensification in acute myeloid leukemia. The New England journal of medicine. 2009 Sep 24;361(13):1249-59.

[9] Lowenberg B, Ossenkoppele GJ, van Putten W, Schouten HC, Graux C, Ferrant A, et al. High-dose daunorubicin in older patients with acute myeloid leukemia. The New England journal of medicine. 2009 Sep 24;361(13):1235-48.

[10] Lee JH, Joo YD, Kim H, Bae SH, Kim MK, Zang DY, et al. A randomized trial comparing standard versus high-dose daunorubicin induction in patients with acute myeloid leukemia. Blood. 2011 Oct 6; 118(14):3832-41.

[11] Ganzel C, Rowe JM. Prognostic factors in adult acute leukemia. Hematology/oncology clinics of North America. 2011 Dec;25(6):1163-87.

[12] Schlenk RF, Dohner K, Krauter J, Frohling S, Corbacioglu A, Bullinger L, et al. Mutations and treatment outcome in cytogenetically normal acute myeloid leukemia. The New England journal of medicine. 2008 May 1;358(18):1909-18.

[13] Kottaridis PD, Gale RE, Frew ME, Harrison G, Langabeer SE, Belton AA, et al. The presence of a FLT3 internal tandem duplication in patients with acute myeloid leukemia (AML) adds important prognostic information to cytogenetic risk group and response to the first cycle of chemotherapy: analysis of 854 patients from the United Kingdom Medical Research Council AML 10 and 12 trials. Blood. 2001 Sep 15;98(6): 1752-9.

[14] Patel JP, Gonen M, Figueroa ME, Fernandez H, Sun Z, Racevskis J, et al. Prognostic relevance of integrated genetic profiling in acute myeloid leukemia. The New England journal of medicine. 2012 Mar 22;366(12):1079-89.

[15] Dohner H, Estey EH, Amadori S, Appelbaum FR, Buchner T, Burnett AK, et al. Diagnosis and management of acute myeloid leukemia in adults: recommendations from an international expert panel, on behalf of the European LeukemiaNet. Blood. 2010 Jan 21;115(3):453-74.

[16] Armand P, Kim HT, Zhang MJ, Perez WS, Dal Cin PS, Klumpp TR, et al. Classifying cytogenetics in patients with acute myelogenous leukemia in complete remission undergoing allogeneic transplantation: a Center for International Blood and Marrow Transplant Research study. Biol Blood Marrow Transplant. 2012 Feb;18(2):280-8.

[17] Byrd JC, Dodge RK, Carroll A, Baer MR, Edwards C, Stamberg J, et al. Patients with t(8;21)(q22;q22) and acute myeloid leukemia have superior failure-free and overall survival when repetitive cycles of high-dose cytarabine are administered. J Clin Oncol. 1999 Dec;17(12):3767-75.

[18] Lugthart S, Groschel S, Beverloo HB, Kayser S, Valk PJ, van Zelderen-Bhola SL, et al. Clinical, Molecular, and Prognostic Significance of WHO Type inv(3)(q21q26.2)/t(3;3) (q21;q26.2) and Various Other 3q Abnormalities in Acute Myeloid Leukemia. J Clin Oncol. 2010 Aug 9.

[19] Schnittger S, Kinkelin U, Schoch C, Heinecke A, Haase D, Haferlach T, et al. Screening for MLL tandem duplication in 387 unselected patients with AML identify a prognostically unfavorable subset of AML. Leukemia. 2000 May;14(5):796-804.

[20] Ley TJ, Ding L, Walter MJ, McLellan MD, Lamprecht T, Larson DE, et al. DNMT3A mutations in acute myeloid leukemia. N Engl J Med. 2010 Dec 16;363(25):2424-33.

[21] Yanada M, Matsuo K, Suzuki T, Kiyoi H, Naoe T. Prognostic significance of FLT3 internal tandem duplication and tyrosine kinase domain mutations for acute myeloid leukemia: a meta-analysis. Leukemia. 2005 Aug;19(8):1345-9.

[22] Mayer RJ, Davis RB, Schiffer CA, Berg DT, Powell BL, Schulman P, et al. Intensive postremission chemotherapy in adults with acute myeloid leukemia. Cancer and Leukemia Group B. The New England journal of medicine. 1994 Oct 6;331(14): 896-903.

[23] Thomas ED, Buckner CD, Banaji M, Clift RA, Fefer A, Flournoy N, et al. One hundred patients with acute leukemia treated by chemotherapy, total body irradiation, and allogeneic marrow transplantation. Blood. 1977 Apr;49(4):511-33.

[24] Thomas ED, Buckner CD, Clift RA, Fefer A, Johnson FL, Neiman PE, et al. Marrow transplantation for acute nonlymphoblastic leukemia in first remission. N Engl J Med. 1979 Sep 13;301(11):597-9.

[25] Clift R, Buckner CD, Bianco J, Petersen F, Appelbaum F. Marrow transplantation in patients with acute myeloid leukemia. Leukemia. 1992;6 Suppl 2:104-9.

[26] Clift RA, Buckner CD. Marrow transplantation for acute myeloid leukemia. Cancer Invest. 1998;16(1):53-61.

[27] Clift RA, Buckner CD, Appelbaum FR, Bearman SI, Petersen FB, Fisher LD, et al. Allogeneic marrow transplantation in patients with acute myeloid leukemia in first remission: a randomized trial of two irradiation regimens. Blood. 1990 Nov 1;76(9): 1867-71.

[28] Zittoun RA, Mandelli F, Willemze R, de Witte T, Labar B, Resegotti L, et al. Autologous or allogeneic bone marrow transplantation compared with intensive chemotherapy in acute myelogenous leukemia. European Organization for Research and Treatment of Cancer (EORTC) and the Gruppo Italiano Malattie Ematologiche Maligne dell'Adulto (GIMEMA) Leukemia Cooperative Groups. The New England journal of medicine. 1995 Jan 26;332(4):217-23.

[29] Cassileth PA, Harrington DP, Appelbaum FR, Lazarus HM, Rowe JM, Paietta E, et al. Chemotherapy compared with autologous or allogeneic bone marrow transplantation in the management of acute myeloid leukemia in first remission. The New England journal of medicine. 1998 Dec 3;339(23):1649-56.

[30] Slovak ML, Kopecky KJ, Cassileth PA, Harrington DH, Theil KS, Mohamed A, et al. Karyotypic analysis predicts outcome of preremission and postremission therapy in adult acute myeloid leukemia: a Southwest Oncology Group/Eastern Cooperative Oncology Group Study. Blood. 2000 Dec 15;96(13):4075-83.

[31] Koreth J, Schlenk R, Kopecky KJ, Honda S, Sierra J, Djulbegovic BJ, et al. Allogeneic stem cell transplantation for acute myeloid leukemia in first complete remission: sys-

tematic review and meta-analysis of prospective clinical trials. Jama. 2009 Jun 10;301(22):2349-61.

[32] Cornelissen JJ, van Putten WL, Verdonck LF, Theobald M, Jacky E, Daenen SM, et al. Results of a HOVON/SAKK donor versus no-donor analysis of myeloablative HLA-identical sibling stem cell transplantation in first remission acute myeloid leukemia in young and middle-aged adults: benefits for whom? Blood. 2007 May 1;109(9): 3658-66.

[33] Yanada M, Matsuo K, Emi N, Naoe T. Efficacy of allogeneic hematopoietic stem cell transplantation depends on cytogenetic risk for acute myeloid leukemia in first disease remission: a metaanalysis. Cancer. 2005 Apr 15;103(8):1652-8.

[34] Kurosawa S, Yamaguchi T, Miyawaki S, Uchida N, Kanamori H, Usuki K, et al. A Markov decision analysis of allogeneic hematopoietic cell transplantation versus chemotherapy in patients with acute myeloid leukemia in first remission. Blood. 2011 Feb 17;117(7):2113-20.

[35] Cornelissen JJ, Breems D, van Putten WL, Gratwohl AA, Passweg JR, Pabst T, et al. Comparative analysis of the value of allogeneic hematopoietic stem-cell transplantation in acute myeloid leukemia with monosomal karyotype versus other cytogenetic risk categories. J Clin Oncol. 2012 Jun 10;30(17):2140-6.

[36] Krauter J, Heil G, Hoelzer D, Ottmann OG, Martin H, Lubbert M, et al. Treatment of Patients up to 60 Years with High Risk AML: Final Results of the AML SHG-Hannover 01/99 Trial. ASH Annual Meeting Abstracts. 2006 November 16, 2006;108(11): 433-.

[37] Ho VT, Kim HT, Aldridge J, Liney D, Kao G, Armand P, et al. Use of matched unrelated donors compared with matched related donors is associated with lower relapse and superior progression-free survival after reduced-intensity conditioning hematopoietic stem cell transplantation. Biol Blood Marrow Transplant. 2011 Aug;17(8): 1196-204.

[38] Lee SJ, Kang BW, Moon JH, Chae YS, Kim JG, Jung JS, et al. Comparable analysis of outcomes for allogeneic peripheral blood stem cell transplantation from matched related and matched unrelated donors in acute myeloid leukemia. Acta haematologica. 2012;127(2):81-9.

[39] Sierra J, Bjerke J, Hansen J, Martin P, Petersdorf E, Woolfrey A, et al. Marrow transplants from unrelated donors as treatment for acute leukemia. Leukemia & lymphoma. 2000 Nov;39(5-6):495-507.

[40] Horowitz MM. High-resolution typing for unrelated donor transplantation: how far do we go? Best Pract Res Clin Haematol. 2009 Dec;22(4):537-41.

[41] Gooley TA, Chien JW, Pergam SA, Hingorani S, Sorror ML, Boeckh M, et al. Reduced mortality after allogeneic hematopoietic-cell transplantation. The New England journal of medicine. 2010 Nov 25;363(22):2091-101.

[42] Cutler C, Antin JH. Sirolimus for GVHD prophylaxis in allogeneic stem cell transplantation. Bone marrow transplantation. 2004 Sep;34(6):471-6.

[43] Cutler C, Li S, Ho VT, Koreth J, Alyea E, Soiffer RJ, et al. Extended follow-up of methotrexate-free immunosuppression using sirolimus and tacrolimus in related and unrelated donor peripheral blood stem cell transplantation. Blood. 2007 Apr 1;109(7): 3108-14.

[44] Pidala J, Kim J, Jim H, Kharfan-Dabaja MA, Nishihori T, Fernandez H, et al. A randomized phase II study to evaluate tacrolimus in combination with sirolimus or methotrexate after allogeneic hematopoietic cell transplantation. Haematologica. 2012 Jun 11.

[45] Finke J, Bethge WA, Schmoor C, Ottinger HD, Stelljes M, Zander AR, et al. Standard graft-versus-host disease prophylaxis with or without anti-T-cell globulin in haematopoietic cell transplantation from matched unrelated donors: a randomised, open-label, multicentre phase 3 trial. Lancet Oncol. 2009 Sep;10(9):855-64.

[46] Juliusson G, Antunovic P, Derolf A, Lehmann S, Mollgard L, Stockelberg D, et al. Age and acute myeloid leukemia: real world data on decision to treat and outcomes from the Swedish Acute Leukemia Registry. Blood. 2009 Apr 30;113(18):4179-87.

[47] Frohling S, Schlenk RF, Kayser S, Morhardt M, Benner A, Dohner K, et al. Cytogenetics and age are major determinants of outcome in intensively treated acute myeloid leukemia patients older than 60 years: results from AMLSG trial AML HD98-B. Blood. 2006 Nov 15;108(10):3280-8.

[48] Herr AL, Labopin M, Blaise D, Milpied N, Potter M, Michallet M, et al. HLA-identical sibling allogeneic peripheral blood stem cell transplantation with reduced intensity conditioning compared to autologous peripheral blood stem cell transplantation for elderly patients with de novo acute myeloid leukemia. Leukemia. 2007 Jan;21(1): 129-35.

[49] Schetelig J, Bornhauser M, Schmid C, Hertenstein B, Schwerdtfeger R, Martin H, et al. Matched unrelated or matched sibling donors result in comparable survival after allogeneic stem-cell transplantation in elderly patients with acute myeloid leukemia: a report from the cooperative German Transplant Study Group. J Clin Oncol. 2008 Nov 10;26(32):5183-91.

[50] Eom KS, Kim HJ, Cho BS, Choi SM, Lee DG, Lee SE, et al. Hematopoietic stem cell transplant following remission induction chemotherapy including gemtuzumab ozogamicin is a feasible and effective treatment option in elderly patients with acute myeloid leukemia. Leukemia & lymphoma. 2011 Dec;52(12):2321-8.

[51] Druker BJ, Talpaz M, Resta DJ, Peng B, Buchdunger E, Ford JM, et al. Efficacy and safety of a specific inhibitor of the BCR-ABL tyrosine kinase in chronic myeloid leukemia. The New England journal of medicine. [Clinical Trial Clinical Trial, Phase I Research Support, Non-U.S. Gov't Research Support, U.S. Gov't, P.H.S.]. 2001 Apr 5;344(14):1031-7.

[52] Baccarani M, Saglio G, Goldman J, Hochhaus A, Simonsson B, Appelbaum F, et al. Evolving concepts in the management of chronic myeloid leukemia: recommendations from an expert panel on behalf of the European LeukemiaNet. Blood. [Practice Guideline Research Support, Non-U.S. Gov't Review]. 2006 Sep 15;108(6):1809-20.

[53] Giralt SA, Arora M, Goldman JM, Lee SJ, Maziarz RT, McCarthy PL, et al. Impact of imatinib therapy on the use of allogeneic haematopoietic progenitor cell transplantation for the treatment of chronic myeloid leukaemia. British journal of haematology. [Comparative Study Research Support, N.I.H., Extramural Research Support, Non-U.S. Gov't Research Support, U.S. Gov't, Non-P.H.S.]. 2007 Jun;137(5):461-7.

[54] Hehlmann R, Berger U, Pfirrmann M, Heimpel H, Hochhaus A, Hasford J, et al. Drug treatment is superior to allografting as first-line therapy in chronic myeloid leukemia. Blood. 2007 Jun 1;109(11):4686-92.

[55] Bittencourt H, Funke V, Fogliatto L, Magalhaes S, Setubal D, Paz A, et al. Imatinib mesylate versus allogeneic BMT for patients with chronic myeloid leukemia in first chronic phase. Bone Marrow Transplant. 2008 Nov;42(9):597-600.

[56] Michor F, Hughes TP, Iwasa Y, Branford S, Shah NP, Sawyers CL, et al. Dynamics of chronic myeloid leukaemia. Nature. 2005 Jun 30;435(7046):1267-70.

[57] Roeder I, Horn M, Glauche I, Hochhaus A, Mueller MC, Loeffler M. Dynamic modeling of imatinib-treated chronic myeloid leukemia: functional insights and clinical implications. Nat Med. 2006 Oct;12(10):1181-4.

[58] Bhatia R, Holtz M, Niu N, Gray R, Snyder DS, Sawyers CL, et al. Persistence of malignant hematopoietic progenitors in chronic myelogenous leukemia patients in complete cytogenetic remission following imatinib mesylate treatment. Blood. 2003 June 15, 2003;101(12):4701-7.

[59] Chu S, McDonald T, Lin A, Chakraborty S, Huang Q, Snyder DS, et al. Persistence of leukemia stem cells in chronic myelogenous leukemia patients in prolonged remission with imatinib treatment. Blood. 2011 Nov 17;118(20):5565-72.

[60] Pavlu J, Szydlo RM, Goldman JM, Apperley JF. Three decades of transplantation for chronic myeloid leukemia: what have we learned? Blood. 2011 Jan 20;117(3):755-63.

[61] Baccarani M, Cortes J, Pane F, Niederwieser D, Saglio G, Apperley J, et al. Chronic myeloid leukemia: an update of concepts and management recommendations of European LeukemiaNet. Journal of clinical oncology : official journal of the American

Society of Clinical Oncology. [Practice Guideline Review]. 2009 Dec 10;27(35): 6041-51.

[62] Saussele S, Lauseker M, Gratwohl A, Beelen DW, Bunjes D, Schwerdtfeger R, et al. Allogeneic hematopoietic stem cell transplantation (allo SCT) for chronic myeloid leukemia in the imatinib era: evaluation of its impact within a subgroup of the randomized German CML Study IV. Blood. [Clinical Trial, Phase IV Multicenter Study Randomized Controlled Trial Research Support, Non-U.S. Gov't]. 2010 Mar 11;115(10):1880-5.

[63] Deininger M, Schleuning M, Greinix H, Sayer HG, Fischer T, Martinez J, et al. The effect of prior exposure to imatinib on transplant-related mortality. Haematologica. 2006 Apr;91(4):452-9.

[64] Lee SJ, Kukreja M, Wang T, Giralt SA, Szer J, Arora M, et al. Impact of prior imatinib mesylate on the outcome of hematopoietic cell transplantation for chronic myeloid leukemia. Blood. 2008 Oct 15;112(8):3500-7.

[65] Zaucha JM, Prejzner W, Giebel S, Gooley TA, Szatkowski D, Kalwak K, et al. Imatinib therapy prior to myeloablative allogeneic stem cell transplantation. Bone marrow transplantation. [Clinical Trial Comparative Study]. 2005 Sep;36(5):417-24.

[66] Perz JB, Khorashad JS, Marin D, Apperley JF, Olavarria E. Imatinib preceding allogeneic stem cell transplantation in chronic myeloid leukemia. Haematologica. [Letter]. 2006 Aug;91(8):1145-6.

[67] Oehler VG, Gooley T, Snyder DS, Johnston L, Lin A, Cummings CC, et al. The effects of imatinib mesylate treatment before allogeneic transplantation for chronic myeloid leukemia. Blood. [Comparative Study Research Support, N.I.H., Extramural Research Support, Non-U.S. Gov't]. 2007 Feb 15;109(4):1782-9.

[68] Shimoni A, Kroger N, Zander AR, Rowe JM, Hardan I, Avigdor A, et al. Imatinib mesylate (STI571) in preparation for allogeneic hematopoietic stem cell transplantation and donor lymphocyte infusions in patients with Philadelphia-positive acute leukemias. Leukemia. 2003 Feb;17(2):290-7.

[69] Magro L, Mohty M, Catteau B, Coiteux V, Chevallier P, Terriou L, et al. Imatinib mesylate as salvage therapy for refractory sclerotic chronic graft-versus-host disease. Blood. [Multicenter Study Research Support, Non-U.S. Gov't]. 2009 Jul 16;114(3): 719-22.

[70] Olivieri A, Locatelli F, Zecca M, Sanna A, Cimminiello M, Raimondi R, et al. Imatinib for refractory chronic graft-versus-host disease with fibrotic features. Blood. [Clinical Trial Research Support, Non-U.S. Gov't]. 2009 Jul 16;114(3):709-18.

[71] Druker BJ, Guilhot F, O'Brien SG, Gathmann I, Kantarjian H, Gattermann N, et al. Five-year follow-up of patients receiving imatinib for chronic myeloid leukemia. N Engl J Med. 2006 Dec 7;355(23):2408-17.

[72] Druker BJ, Sawyers CL, Kantarjian H, Resta DJ, Reese SF, Ford JM, et al. Activity of a specific inhibitor of the BCR-ABL tyrosine kinase in the blast crisis of chronic myeloid leukemia and acute lymphoblastic leukemia with the Philadelphia chromosome. N Engl J Med. 2001 Apr 5;344(14):1038-42.

[73] Kantarjian HM, Cortes J, O'Brien S, Giles FJ, Albitar M, Rios MB, et al. Imatinib mesylate (STI571) therapy for Philadelphia chromosome-positive chronic myelogenous leukemia in blast phase. Blood. 2002 May 15;99(10):3547-53.

[74] Kantarjian H, Sawyers C, Hochhaus A, Guilhot F, Schiffer C, Gambacorti-Passerini C, et al. Hematologic and cytogenetic responses to imatinib mesylate in chronic myelogenous leukemia. N Engl J Med. 2002 Feb 28;346(9):645-52.

[75] Brave M, Goodman V, Kaminskas E, Farrell A, Timmer W, Pope S, et al. Sprycel for chronic myeloid leukemia and Philadelphia chromosome-positive acute lymphoblastic leukemia resistant to or intolerant of imatinib mesylate. Clin Cancer Res. 2008 Jan 15;14(2):352-9.

[76] Hazarika M, Jiang X, Liu Q, Lee SL, Ramchandani R, Garnett C, et al. Tasigna for chronic and accelerated phase Philadelphia chromosome--positive chronic myelogenous leukemia resistant to or intolerant of imatinib. Clin Cancer Res. 2008 Sep 1;14(17):5325-31.

[77] Kantarjian H, Shah NP, Hochhaus A, Cortes J, Shah S, Ayala M, et al. Dasatinib versus imatinib in newly diagnosed chronic-phase chronic myeloid leukemia. N Engl J Med. 2010 Jun 17;362(24):2260-70.

[78] Saglio G, Kim DW, Issaragrisil S, le Coutre P, Etienne G, Lobo C, et al. Nilotinib versus imatinib for newly diagnosed chronic myeloid leukemia. N Engl J Med. 2010 Jun 17;362(24):2251-9.

[79] Velev N, Cortes J, Champlin R, Jones D, Rondon G, Giralt S, et al. Stem cell transplantation for patients with chronic myeloid leukemia resistant to tyrosine kinase inhibitors with BCR-ABL kinase domain mutation T315I. Cancer. 2010 Aug 1;116(15): 3631-7.

[80] Basak G, Torosian T, Snarski E, Niesiobedzka J, Majewski M, Gronkowska A, et al. Hematopoietic stem cell transplantation for T315I-mutated chronic myelogenous leukemia. Ann Transplant. 2010 Apr-Jun;15(2):68-70.

[81] Sanchez-Guijo FM, Lopez-Jimenez J, Gonzalez T, Santamaria C, Gonzalez M, Del Canizo MC. Multitargeted sequential therapy with MK-0457 and dasatinib followed by stem cell transplantation for T315I mutated chronic myeloid leukemia. Leuk Res. 2009 Jun;33(6):e20-2.

[82] Jabbour E, Cortes J, Santos FP, Jones D, O'Brien S, Rondon G, et al. Results of allogeneic hematopoietic stem cell transplantation for chronic myelogenous leukemia pa-

tients who failed tyrosine kinase inhibitors after developing BCR-ABL1 kinase domain mutations. Blood. 2011 Mar 31;117(13):3641-7.

[83] Cortes J, Talpaz M, Bixby D, Deininger M, Shah N, Flinn IW, et al. A Phase 1 Trial of Oral Ponatinib (AP24534) In Patients with Refractory Chronic Myelogenous Leukemia (CML) and Other Hematologic Malignancies: Emerging Safety and Clinical Response Findings. American Society of Haematology Annual meeting. 2010(210).

[84] Palandri F, Castagnetti F, Testoni N, Luatti S, Marzocchi G, Bassi S, et al. Chronic myeloid leukemia in blast crisis treated with imatinib 600 mg: outcome of the patients alive after a 6-year follow-up. Haematologica. 2008 December 1, 2008;93(12): 1792-6.

[85] Palandri F, Castagnetti F, Alimena G, Testoni N, Breccia M, Luatti S, et al. The long-term durability of cytogenetic responses in patients with accelerated phase chronic myeloid leukemia treated with imatinib 600 mg: the GIMEMA CML Working Party experience after a 7-year follow-up. Haematologica. 2009 February 1, 2009;94(2): 205-12.

[86] Silver RT, Cortes J, Waltzman R, Mone M, Kantarjian H. Sustained durability of responses and improved progression-free and overall survival with imatinib treatment for accelerated phase and blast crisis chronic myeloid leukemia: long-term follow-up of the STI571 0102 and 0109 trials. Haematologica. 2009 May 1, 2009;94(5):743-4.

[87] Gratwohl A, Brand R, Apperley J, Crawley C, Ruutu T, Corradini P, et al. Allogeneic hematopoietic stem cell transplantation for chronic myeloid leukemia in Europe 2006: transplant activity, long-term data and current results. An analysis by the Chronic Leukemia Working Party of the European Group for Blood and Marrow Transplantation (EBMT). Haematologica. 2006 2006 Apr;91(4):513-21.

[88] Hehlmann R. How I treat CML blast crisis. Blood. 2012 Jul 26;120(4):737-47.

[89] Bornhauser M, Kiehl M, Siegert W, Schetelig J, Hertenstein B, Martin H, et al. Dose-reduced conditioning for allografting in 44 patients with chronic myeloid leukaemia: a retrospective analysis. British journal of haematology. [Multicenter Study]. 2001 Oct;115(1):119-24.

[90] Or R, Shapira MY, Resnick I, Amar A, Ackerstein A, Samuel S, et al. Nonmyeloablative allogeneic stem cell transplantation for the treatment of chronic myeloid leukemia in first chronic phase. Blood. 2003 January 15, 2003;101(2):441-5.

[91] Baron F, Storb R. Current roles for allogeneic hematopoietic cell transplantation following nonmyeloablative or reduced-intensity conditioning. Clinical advances in hematology & oncology : H&O. [Research Support, N.I.H., Extramural Research Support, Non-U.S. Gov't Review]. 2005 Oct;3(10):799-819.

[92] Warlick E, Ahn KW, Pedersen TL, Artz A, de Lima M, Pulsipher M, et al. Reduced intensity conditioning is superior to nonmyeloablative conditioning for older chronic

myelogenous leukemia patients undergoing hematopoietic cell transplant during the tyrosine kinase inhibitor era. Blood. 2012 Apr 26;119(17):4083-90.

[93] Asnafi V, Rubio MT, Delabesse E, Villar E, Davi F, Damaj G, et al. Prediction of relapse by day 100 BCR-ABL quantification after allogeneic stem cell transplantation for chronic myeloid leukemia. Leukemia : official journal of the Leukemia Society of America, Leukemia Research Fund, UK. [Clinical Trial Comparative Study Research Support, Non-U.S. Gov't]. 2006 May;20(5):793-9.

[94] Ballestrero A, Cirmena G, Dominietto A, Garuti A, Rocco I, Cea M, et al. Peripheral blood vs. bone marrow for molecular monitoring of BCR-ABL1 levels in chronic myelogenous leukemia, a retrospective analysis in allogeneic bone marrow recipients. International Journal of Laboratory Hematology. 2010;32(4):387-91.

[95] Lange T, Deininger M, Brand R, Hegenbart U, Al-Ali H, Krahl R, et al. BCR-ABL transcripts are early predictors for hematological relapse in chronic myeloid leukemia after hematopoietic cell transplantation with reduced intensity conditioning. Leukemia. 2004 Sep;18(9):1468-75.

[96] Klyuchnikov E, Schafhausen P, Kroger N, Brummendorf TH, Osanmaz O, Asenova S, et al. Second-generation tyrosine kinase inhibitors in the post-transplant period in patients with chronic myeloid leukemia or Philadelphia-positive acute lymphoblastic leukemia. Acta haematologica. 2009;122(1):6-10.

[97] Carpenter PA, Snyder DS, Flowers ME, Sanders JE, Gooley TA, Martin PJ, et al. Prophylactic administration of imatinib after hematopoietic cell transplantation for high-risk Philadelphia chromosome-positive leukemia. Blood. [Clinical Trial Research Support, N.I.H., Extramural]. 2007 Apr 1;109(7):2791-3.

[98] Olavarria E, Siddique S, Griffiths MJ, Avery S, Byrne JL, Piper KP, et al. Posttransplantation imatinib as a strategy to postpone the requirement for immunotherapy in patients undergoing reduced-intensity allografts for chronic myeloid leukemia. Blood. [Clinical Trial Research Support, Non-U.S. Gov't]. 2007 Dec 15;110(13):4614-7.

[99] DeAngelo DJ, Hochberg EP, Alyea EP, Longtine J, Lee S, Galinsky I, et al. Extended follow-up of patients treated with imatinib mesylate (gleevec) for chronic myelogenous leukemia relapse after allogeneic transplantation: durable cytogenetic remission and conversion to complete donor chimerism without graft-versus-host disease. Clinical cancer research : an official journal of the American Association for Cancer Research. 2004 Aug 1;10(15):5065-71.

[100] Kantarjian HM, O'Brien S, Cortes JE, Giralt SA, Rios MB, Shan J, et al. Imatinib mesylate therapy for relapse after allogeneic stem cell transplantation for chronic myelogenous leukemia. Blood. [Clinical Trial Research Support, Non-U.S. Gov't]. 2002 Sep 1;100(5):1590-5.

[101] Savani BN, Montero A, Kurlander R, Childs R, Hensel N, Barrett AJ. Imatinib synergizes with donor lymphocyte infusions to achieve rapid molecular remission of CML

relapsing after allogeneic stem cell transplantation. Bone marrow transplantation. 2005 Dec;36(11):1009-15.

[102] Garcia-Manero G. Myelodysplastic syndromes: 2012 update on diagnosis, risk-stratification, and management. American Journal of Hematology. 2012;87(7):692-701.

[103] Platzbecker U, Schetelig J, Finke J, Trenschel R, Scott BL, Kobbe G, et al. Allogeneic Hematopoietic Cell Transplantation in Patients Age 60-70 Years with De Novo High-Risk Myelodysplastic Syndrome or Secondary Acute Myelogenous Leukemia: Comparison with Patients Lacking Donors Who Received Azacitidine. Biology of blood and marrow transplantation : journal of the American Society for Blood and Marrow Transplantation. 2012;18(9):1415-21.

[104] Greenberg P, Cox C, LeBeau MM, Fenaux P, Morel P, Sanz G, et al. International Scoring System for Evaluating Prognosis in Myelodysplastic Syndromes. Blood. 1997 March 15, 1997;89(6):2079-88.

[105] Alessandrino EP, Della Porta MG, Bacigalupo A, Van Lint MT, Falda M, Onida F, et al. WHO classification and WPSS predict posttransplantation outcome in patients with myelodysplastic syndrome: a study from the Gruppo Italiano Trapianto di Midollo Osseo (GITMO). Blood. 2008 August 1, 2008;112(3):895-902.

[106] Malcovati L, Germing U, Kuendgen A, Della Porta MG, Pascutto C, Invernizzi R, et al. Time-Dependent Prognostic Scoring System for Predicting Survival and Leukemic Evolution in Myelodysplastic Syndromes. J Clin Oncol. 2007 August 10, 2007;25(23): 3503-10.

[107] Kantarjian H, O'Brien S, Ravandi F, Cortes J, Shan J, Bennett JM, et al. Proposal for a new risk model in myelodysplastic syndrome that accounts for events not considered in the original International Prognostic Scoring System. Cancer. 2008;113(6):1351-61.

[108] Cutler CS, Lee SJ, Greenberg P, Deeg HJ, Pérez WS, Anasetti C, et al. A decision analysis of allogeneic bone marrow transplantation for the myelodysplastic syndromes: delayed transplantation for low-risk myelodysplasia is associated with improved outcome. Blood. 2004 July 15, 2004;104(2):579-85.

[109] Mandelli F, Vignetti M, Suciu S, Stasi R, Petti MC, Meloni G, et al. Daunorubicin versus mitoxantrone versus idarubicin as induction and consolidation chemotherapy for adults with acute myeloid leukemia: the EORTC and GIMEMA Groups Study AML-10. J Clin Oncol. 2009 Nov 10;27(32):5397-403.

[110] Oliansky DM, Antin JH, Bennett JM, Deeg HJ, Engelhardt C, Heptinstall KV, et al. The Role of Cytotoxic Therapy with Hematopoietic Stem Cell Transplantation in the Therapy of Myelodysplastic Syndromes: An Evidence-Based Review. Biology of blood and marrow transplantation : journal of the American Society for Blood and Marrow Transplantation. 2009;15(2):137-72.

[111] Appelbaum FR. The role of hematopoietic cell transplantation as therapy for myelodysplasia. Best Practice & Research Clinical Haematology. 2011;24(4):541-7.

[112] Onida F, Brand R, Van Biezen A, Barge R, Verdonck L, Finke J, et al. Impact of cytogenetics on outcome of patients with MDS or secondary AML undergoing allogeneic HSCT from HLA-identical siblings: a retrospective analysis of the EBMT-CLWP. Blood (ASH Annual Meeting Abstracts). 2006;108(11):A750-A.

[113] Chang C, Storer BE, Scott BL, Bryant EM, Shulman HM, Flowers ME, et al. Hematopoietic cell transplantation in patients with myelodysplastic syndrome or acute myeloid leukemia arising from myelodysplastic syndrome: similar outcomes in patients with de novo disease and disease following prior therapy or antecedent hematologic disorders. Blood. 2007 August 15, 2007;110(4):1379-87.

[114] Scott BL, Wells DA, Loken MR, Myerson D, Leisenring WM, Deeg HJ. Validation of a flow cytometric scoring system as a prognostic indicator for posttransplantation outcome in patients with myelodysplastic syndrome. Blood. 2008 October 1, 2008;112(7):2681-6.

[115] Mills KI, Kohlmann A, Williams PM, Wieczorek L, Liu W-m, Li R, et al. Microarray-based classifiers and prognosis models identify subgroups with distinct clinical outcomes and high risk of AML transformation of myelodysplastic syndrome. Blood. 2009 July 30, 2009;114(5):1063-72.

[116] Newell LF, Gooley T, Hansen JA, Stirewalt DL, Petersdorf EW, Deeg HJ. Tumor Necrosis Factor Polymorphism Affects Transplantation Outcome in Patients with Myelodysplastic Syndrome but Not in Those with Chronic Myelogenous Leukemia, Independent of the Presence of HLA-DR15. Biology of blood and marrow transplantation : journal of the American Society for Blood and Marrow Transplantation. 2010;16(12):1700-6.

[117] Bejar R, Stevenson K, Abdel-Wahab O, Galili N, Nilsson B, Garcia-Manero G, et al. Clinical Effect of Point Mutations in Myelodysplastic Syndromes. New England Journal of Medicine. 2011;364(26):2496-506.

[118] Yoshida K, Sanada M, Shiraishi Y, Nowak D, Nagata Y, Yamamoto R, et al. Frequent pathway mutations of splicing machinery in myelodysplasia. Nature. [10.1038/nature10496]. 2011;478(7367):64-9.

[119] Malcovati L, Papaemmanuil E, Bowen DT, Boultwood J, Della Porta MG, Pascutto C, et al. Clinical significance of SF3B1 mutations in myelodysplastic syndromes and myelodysplastic/myeloproliferative neoplasms. Blood. 2011 December 8, 2011;118(24):6239-46.

[120] Bacher U, Haferlach C, Kroger N, Schnittger S, Kern W, Wiedemann B, et al. Diagnostic tools in the indications for allogeneic stem cell transplantation in myelodysplastic syndromes. Biol Blood Marrow Transplant. 2010 Jan;16(1):1-11.

[121] Armand P, Kim HT, Cutler CS, Ho VT, Koreth J, Alyea EP, et al. Prognostic impact of elevated pretransplantation serum ferritin in patients undergoing myeloablative stem cell transplantation. Blood. 2007 15 May 2007;109(10):4586-8.

[122] Pullarkat V, Blanchard S, Tegtmeier B, Dagis A, Patane K, Ito J, et al. Iron overload adversely affects outcome of allogeneic hematopoietic cell transplantation. Bone Marrow Transplant. 2008;42(12):799-805.

[123] Cutler C. Allogeneic Hematopoietic Stem-Cell Transplantation for Myelodysplastic Syndrome. ASH Education Program Book. 2010 December 4, 2010;2010(1):325-9.

[124] Sorror ML, Maris MB, Storb R, Baron F, Sandmaier BM, Maloney DG, et al. Hematopoietic cell transplantation (HCT)-specific comorbidity index: a new tool for risk assessment before allogeneic HCT. Blood. 2005 October 15, 2005;106(8):2912-9.

[125] Sorror ML, Sandmaier BM, Storer BE, Maris MB, Baron F, Maloney DG, et al. Comorbidity and Disease Status–Based Risk Stratification of Outcomes Among Patients With Acute Myeloid Leukemia or Myelodysplasia Receiving Allogeneic Hematopoietic Cell Transplantation. Journal of Clinical Oncology. 2007 September 20, 2007;25(27):4246-54.

[126] Boehm A, Sperr WR, Leitner G, Worel N, Oehler L, Jaeger E, et al. Comorbidity predicts survival in myelodysplastic syndromes or secondary acute myeloid leukaemia after allogeneic stem cell transplantation. European Journal of Clinical Investigation. 2008;38(12):945-52.

[127] Kröger N. Allogeneic stem cell transplantation for elderly patients with myelodysplastic syndrome. Blood. 2012 June 14, 2012;119(24):5632-9.

[128] Lim Z, Brand R, Martino R, van Biezen A, Finke J, Bacigalupo A, et al. Allogeneic Hematopoietic Stem-Cell Transplantation for Patients 50 Years or Older With Myelodysplastic Syndromes or Secondary Acute Myeloid Leukemia. Journal of Clinical Oncology. 2010 January 20, 2010;28(3):405-11.

[129] McClune BL, Weisdorf DJ, Pedersen TL, Tunes da Silva G, Tallman MS, Sierra J, et al. Effect of Age on Outcome of Reduced-Intensity Hematopoietic Cell Transplantation for Older Patients With Acute Myeloid Leukemia in First Complete Remission or With Myelodysplastic Syndrome. Journal of Clinical Oncology. 2010 April 10, 2010;28(11):1878-87.

[130] Rodrigo M, Simona I, Ronald B, Thekla J, Anja van B, Jürgen F, et al. Retrospective comparison of reduced-intensity conditioning and conventional high-dose conditioning for allogeneic hematopoietic stem cell transplantation using HLA-identical sibling donors in myelodysplastic syndromes. Blood. 2006;108(3):836-46.

[131] Scott BL, Sandmaier BM, Storer B, Maris MB, Sorror ML, Maloney DG, et al. Myeloablative vs nonmyeloablative allogeneic transplantation for patients with myelodys-

plastic syndrome or acute myelogenous leukemia with multilineage dysplasia: a retrospective analysis. Leukemia. 2006;20(1):128-35.

[132] Kojima R, Kami M, Kanda Y, Kusumi E, Kishi Y, Tanaka Y, et al. Comparison between reduced intensity and conventional myeloablative allogeneic stem-cell transplantation in patients with hematologic malignancies aged between 50 and 59 years. Bone Marrow Transplant. 2005;36(8):667-74.

[133] Martino R, Iacobelli S, Brand R, Jansen T, van Biezen A, Finke J, et al. Retrospective comparison of reduced-intensity conditioning and conventional high-dose conditioning for allogeneic hematopoietic stem cell transplantation using HLA-identical sibling donors in myelodysplastic syndromes. Blood. 2006 August 1, 2006;108(3):836-46.

[134] de Lima M, Anagnostopoulos A, Munsell M, Shahjahan M, Ueno N, Ippoliti C, et al. Nonablative versus reduced-intensity conditioning regimens in the treatment of acute myeloid leukemia and high-risk myelodysplastic syndrome: dose is relevant for long-term disease control after allogeneic hematopoietic stem cell transplantation. Blood. 2004 August 1, 2004;104(3):865-72.

[135] Luger SM, Ringden O, Zhang MJ, Perez WS, Bishop MR, Bornhauser M, et al. Similar outcomes using myeloablative vs reduced-intensity allogeneic transplant preparative regimens for AML or MDS. Bone Marrow Transplant. 2012;47(2):203-11.

[136] Shimoni A, Hardan I, Shem-Tov N, Yeshurun M, Yerushalmi R, Avigdor A, et al. Allogeneic hematopoietic stem-cell transplantation in AML and MDS using myeloablative versus reduced-intensity conditioning: the role of dose intensity. Leukemia. 2005;20(2):322-8.

[137] Nakamura R, Palmer JM, O'Donnell MR, Stiller T, Thomas SH, Chao J, et al. Reduced intensity allogeneic hematopoietic stem cell transplantation for MDS using tacrolimus/sirolimus-based GVHD prophylaxis. Leukemia Research. 2012;36(9):1152-6.

[138] Nakai K, Kanda Y, Fukuhara S, Sakamaki H, Okamoto S, Kodera Y, et al. Value of chemotherapy before allogeneic hematopoietic stem cell transplantation from an HLA-identical sibling donor for myelodysplastic syndrome. Leukemia. 2005;19(3): 396-401.

[139] Field T, Perkins J, Huang Y, Kharfan-Dabaja MA, Alsina M, Ayala E, et al. 5-Azacitidine for myelodysplasia before allogeneic hematopoietic cell transplantation. Bone Marrow Transplant. 2009;45(2):255-60.

[140] Buchholz S, Dammann E, Stadler M, Krauter J, Beutel G, Trummer A, et al. Cytoreductive treatment with clofarabine/ara-C combined with reduced-intensity conditioning and allogeneic stem cell transplantation in patients with high-risk, relapsed, or refractory acute myeloid leukemia and advanced myelodysplastic syndrome. European Journal of Haematology. 2012;88(1):52-60.

[141] De Padua Silva L, de Lima M, Kantarjian H, Faderl S, Kebriaei P, Giralt S, et al. Feasibility of allo-SCT after hypomethylating therapy with decitabine for myelodysplastic syndrome. Bone Marrow Transplant. 2009;43(11):839-43.

[142] Gerds AT, Gooley TA, Estey EH, Appelbaum FR, Deeg HJ, Scott BL. Pretransplantation Therapy with Azacitidine vs Induction Chemotherapy and Posttransplantation Outcome in Patients with MDS. Biology of blood and marrow transplantation : journal of the American Society for Blood and Marrow Transplantation. 2012;18(8): 1211-8.

[143] de Lima M, Giralt S, Thall PF, de Padua Silva L, Jones RB, Komanduri K, et al. Maintenance therapy with low-dose azacitidine after allogeneic hematopoietic stem cell transplantation for recurrent acute myelogenous leukemia or myelodysplastic syndrome. Cancer. 2010;116(23):5420-31.

[144] Jabbour E, Giralt S, Kantarjian H, Garcia-Manero G, Jagasia M, Kebriaei P, et al. Low-dose azacitidine after allogeneic stem cell transplantation for acute leukemia. Cancer. 2009;115(9):1899-905.

[145] Campregher PV, Gooley T, Scott BL, Moravec C, Sandmaier B, Martin PJ, et al. Results of donor lymphocyte infusions for relapsed myelodysplastic syndrome after hematopoietic cell transplantation. Bone Marrow Transplant. 2007;40(10):965-71.

[146] Schroeder T, Czibere AG, Kröger N, Platzbecker U, Bug G, Uharek L, et al. Combining Azacitidine (Vidaza®, Aza) and Donor Lymphocyte Infusions (DLI) as First Salvage Treatment in Patients (PTS) With Acute Myeloid Leukemia (AML) or Myelodysplastic Syndromes (MDS) Relapsing After Allogeneic Hematopoietic Stem Cell Transplantation (allo-SCT): Interim-Analysis From the Azarela-Trial (NCT-00795548). Biology of blood and marrow transplantation : journal of the American Society for Blood and Marrow Transplantation. 2011;17(2):S159.

[147] Jamieson CHM, Amylon MD, Wong RM, Blume KG. Allogeneic hematopoietic cell transplantation for patients with high-risk acute lymphoblastic leukemia in first or second complete remission using fractionated total-body irradiation and high-dose etoposide: A 15-year experience. Experimental Hematology. 2003;31(10):981-6.

[148] Thomas X, Boiron JM, Huguet F, Dombret H, Bradstock K, Vey N, et al. Outcome of treatment in adults with acute lymphoblastic leukemia: analysis of the LALA-94 trial. J Clin Oncol. 2004 Oct 15;22(20):4075-86.

[149] Bruggemann M, Raff T, Flohr T, Gokbuget N, Nakao M, Droese J, et al. Clinical significance of minimal residual disease quantification in adult patients with standard-risk acute lymphoblastic leukemia. Blood. 2006 Feb 1;107(3):1116-23.

[150] Goldstone AH, Richards SM, Lazarus HM, Tallman MS, Buck G, Fielding AK, et al. In adults with standard-risk acute lymphoblastic leukemia, the greatest benefit is achieved from a matched sibling allogeneic transplantation in first complete remission, and an autologous transplantation is less effective than conventional consolida-

tion/maintenance chemotherapy in all patients: final results of the International ALL Trial (MRC UKALL XII/ECOG E2993). Blood. 2008 February 15, 2008;111(4):1827-33.

[151] Yanada M, Matsuo K, Suzuki T, Naoe T. Allogeneic hematopoietic stem cell transplantation as part of postremission therapy improves survival for adult patients with high-risk acute lymphoblastic leukemia: a metaanalysis. Cancer. 2006 Jun 15;106(12): 2657-63.

[152] Ram R, Gafter-Gvili A, Vidal L, Paul M, Ben-Bassat I, Shpilberg O, et al. Management of adult patients with acute lymphoblastic leukemia in first complete remission: systematic review and meta-analysis. Cancer. 2010 Jul 15;116(14):3447-57.

[153] Rowe JM. Interpreting Data on Transplant Selection and Outcome in Adult Acute Lymphoblastic Leukemia (ALL). Biology of blood and marrow transplantation : journal of the American Society for Blood and Marrow Transplantation. 2011;17(1):S76-S83.

[154] Fielding AK. Philadelphia-Positive Acute Lymphoblastic Leukemia—Is Bone Marrow Transplant Still Necessary? Biology of blood and marrow transplantation : journal of the American Society for Blood and Marrow Transplantation. 2011;17(1):S84-S8.

[155] Thomas DA, Faderl S, Cortes J, O'Brien S, Giles FJ, Kornblau SM, et al. Treatment of Philadelphia chromosome–positive acute lymphocytic leukemia with hyper-CVAD and imatinib mesylate. Blood. 2004 June 15, 2004;103(12):4396-407.

[156] Wassmann B, Pfeifer H, Goekbuget N, Beelen DW, Beck J, Stelljes M, et al. Alternating versus concurrent schedules of imatinib and chemotherapy as front-line therapy for Philadelphia-positive acute lymphoblastic leukemia (Ph+ALL). Blood. 2006 September 1, 2006;108(5):1469-77.

[157] de Labarthe A, Rousselot P, Huguet-Rigal F, Delabesse E, Witz F, Maury S, et al. Imatinib combined with induction or consolidation chemotherapy in patients with de novo Philadelphia chromosome–positive acute lymphoblastic leukemia: results of the GRAAPH-2003 study. Blood. 2007 February 15, 2007;109(4):1408-13.

[158] Yanada M, Sugiura I, Takeuchi J, Akiyama H, Maruta A, Ueda Y, et al. Prospective monitoring of BCR-ABL1 transcript levels in patients with Philadelphia chromosome-positive acute lymphoblastic leukaemia undergoing imatinib-combined chemotherapy. Br J Haematol. 2008 Nov;143(4):503-10.

[159] Lee S, Kim YJ, Min CK, Kim HJ, Eom KS, Kim DW, et al. The effect of first-line imatinib interim therapy on the outcome of allogeneic stem cell transplantation in adults with newly diagnosed Philadelphia chromosome-positive acute lymphoblastic leukemia. Blood. 2005 May 1;105(9):3449-57.

[160] Abou Mourad YR, Fernandez HF, Kharfan-Dabaja MA. Allogeneic hematopoietic cell transplantation for adult Philadelphia-positive acute lymphoblastic leukemia in

the era of tyrosine kinase inhibitors. Biol Blood Marrow Transplant. 2008 Sep;14(9): 949-58.

[161] Ottmann O, Dombret H, Martinelli G, Simonsson B, Guilhot F, Larson RA, et al. Dasatinib induces rapid hematologic and cytogenetic responses in adult patients with Philadelphia chromosome–positive acute lymphoblastic leukemia with resistance or intolerance to imatinib: interim results of a phase 2 study. Blood. 2007 October 1, 2007;110(7):2309-15.

[162] Ravandi F, O'Brien S, Thomas D, Faderl S, Jones D, Garris R, et al. First report of phase 2 study of dasatinib with hyper-CVAD for the frontline treatment of patients with Philadelphia chromosome–positive (Ph+) acute lymphoblastic leukemia. Blood. 2010 September 23, 2010;116(12):2070-7.

[163] Ottmann OG, Larson RA, Kantarjian HM, le Coutre P, Baccarani M, Haque A, et al. Nilotinib in Patients (pts) with Relapsed/Refractory Philadelphia Chromosome-Positive Acute Lymphoblastic Leukemia (Ph+ ALL) Who Are Resistant or Intolerant to Imatinib. ASH Annual Meeting Abstracts. 2007 November 16, 2007;110(11):2815-.

[164] Mohty M, Labopin M, Volin L, Gratwohl A, Socie G, Esteve J, et al. Reduced-intensity versus conventional myeloablative conditioning allogeneic stem cell transplantation for patients with acute lymphoblastic leukemia: a retrospective study from the European Group for Blood and Marrow Transplantation. Blood. 2010 Nov 25;116(22): 4439-43.

[165] Bachanova V, Verneris MR, DeFor T, Brunstein CG, Weisdorf DJ. Prolonged survival in adults with acute lymphoblastic leukemia after reduced-intensity conditioning with cord blood or sibling donor transplantation. Blood. 2009 March 26, 2009;113(13): 2902-5.

[166] Santarone S, Pidala J, Di Nicola M, Field T, Alsina M, Ayala E, et al. Fludarabine and Pharmacokinetic-Targeted Busulfan before Allografting for Adults with Acute Lymphoid Leukemia. Biology of blood and marrow transplantation : journal of the American Society for Blood and Marrow Transplantation. 2011;17(10):1505-11.

[167] Philip T, Guglielmi C, Hagenbeek A, Somers R, Van der Lelie H, Bron D, et al. Autologous bone marrow transplantation as compared with salvage chemotherapy in relapses of chemotherapy-sensitive non-Hodgkin's lymphoma. N Engl J Med. 1995 Dec 7;333(23):1540-5.

[168] Lazarus HM, Zhang MJ, Carreras J, Hayes-Lattin BM, Ataergin AS, Bitran JD, et al. A comparison of HLA-identical sibling allogeneic versus autologous transplantation for diffuse large B cell lymphoma: a report from the CIBMTR. Biol Blood Marrow Transplant. 2010 Jan;16(1):35-45.

[169] van Kampen RJ, Canals C, Schouten HC, Nagler A, Thomson KJ, Vernant JP, et al. Allogeneic stem-cell transplantation as salvage therapy for patients with diffuse large B-cell non-Hodgkin's lymphoma relapsing after an autologous stem-cell trans-

plantation: an analysis of the European Group for Blood and Marrow Transplantation Registry. J Clin Oncol. 2011 Apr 1;29(10):1342-8.

[170] Gyan E, Foussard C, Bertrand P, Michenet P, Le Gouill S, Berthou C, et al. High-dose therapy followed by autologous purged stem cell transplantation and doxorubicin-based chemotherapy in patients with advanced follicular lymphoma: a randomized multicenter study by the GOELAMS with final results after a median follow-up of 9 years. Blood. 2009 Jan 29;113(5):995-1001.

[171] Lenz G, Dreyling M, Schiegnitz E, Forstpointner R, Wandt H, Freund M, et al. Myeloablative radiochemotherapy followed by autologous stem cell transplantation in first remission prolongs progression-free survival in follicular lymphoma: results of a prospective, randomized trial of the German Low-Grade Lymphoma Study Group. Blood. 2004 Nov 1;104(9):2667-74.

[172] Sebban C, Mounier N, Brousse N, Belanger C, Brice P, Haioun C, et al. Standard chemotherapy with interferon compared with CHOP followed by high-dose therapy with autologous stem cell transplantation in untreated patients with advanced follicular lymphoma: the GELF-94 randomized study from the Groupe d'Etude des Lymphomes de l'Adulte (GELA). Blood. 2006 Oct 15;108(8):2540-4.

[173] Ladetto M, De Marco F, Benedetti F, Vitolo U, Patti C, Rambaldi A, et al. Prospective, multicenter randomized GITMO/IIL trial comparing intensive (R-HDS) versus conventional (CHOP-R) chemoimmunotherapy in high-risk follicular lymphoma at diagnosis: the superior disease control of R-HDS does not translate into an overall survival advantage. Blood. 2008 Apr 15;111(8):4004-13.

[174] van Besien K, Loberiza FR, Jr., Bajorunaite R, Armitage JO, Bashey A, Burns LJ, et al. Comparison of autologous and allogeneic hematopoietic stem cell transplantation for follicular lymphoma. Blood. 2003 November 15, 2003;102(10):3521-9.

[175] Hosing C, Saliba RM, McLaughlin P, Andersson B, Rodriguez MA, Fayad L, et al. Long-term results favor allogeneic over autologous hematopoietic stem cell transplantation in patients with refractory or recurrent indolent non-Hodgkin's lymphoma. Ann Oncol. 2003 May 1, 2003;14(5):737-44.

[176] Khouri IF, McLaughlin P, Saliba RM, Hosing C, Korbling M, Lee MS, et al. Eight-year experience with allogeneic stem cell transplantation for relapsed follicular lymphoma after nonmyeloablative conditioning with fludarabine, cyclophosphamide, and rituximab. Blood. 2008 Jun 15;111(12):5530-6.

[177] Romaguera JE, Fayad LE, Feng L, Hartig K, Weaver P, Rodriguez MA, et al. Ten-year follow-up after intense chemoimmunotherapy with Rituximab-HyperCVAD alternating with Rituximab-high dose methotrexate/cytarabine (R-MA) and without stem cell transplantation in patients with untreated aggressive mantle cell lymphoma. Br J Haematol. 2010 Jul;150(2):200-8.

[178] Vandenberghe E, Ruiz de Elvira C, Loberiza FR, Conde E, Lopez-Guillermo A, Gisselbrecht C, et al. Outcome of autologous transplantation for mantle cell lymphoma:

a study by the European Blood and Bone Marrow Transplant and Autologous Blood and Marrow Transplant Registries. Br J Haematol. 2003 Mar;120(5):793-800.

[179] Geisler CH, Kolstad A, Laurell A, Andersen NS, Pedersen LB, Jerkeman M, et al. Long-term progression-free survival of mantle cell lymphoma after intensive front-line immunochemotherapy with in vivo-purged stem cell rescue: a nonrandomized phase 2 multicenter study by the Nordic Lymphoma Group. Blood. 2008 Oct 1;112(7): 2687-93.

[180] Geisler CH, Kolstad A, Laurell A, Raty R, Jerkeman M, Eriksson M, et al. The Mantle Cell Lymphoma International Prognostic Index (MIPI) is superior to the International Prognostic Index (IPI) in predicting survival following intensive first-line immuno-chemotherapy and autologous stem cell transplantation (ASCT). Blood. 2010 Feb 25;115(8):1530-3.

[181] Maris MB, Sandmaier BM, Storer BE, Chauncey T, Stuart MJ, Maziarz RT, et al. Allogeneic hematopoietic cell transplantation after fludarabine and 2 Gy total body irradiation for relapsed and refractory mantle cell lymphoma. Blood. 2004 Dec 1;104(12): 3535-42.

[182] Tam CS, Bassett R, Ledesma C, Korbling M, Alousi A, Hosing C, et al. Mature results of the M. D. Anderson Cancer Center risk-adapted transplantation strategy in mantle cell lymphoma. Blood. 2009 Apr 30;113(18):4144-52.

[183] Cook G, Smith GM, Kirkland K, Lee J, Pearce R, Thomson K, et al. Outcome following Reduced-Intensity Allogeneic Stem Cell Transplantation (RIC AlloSCT) for relapsed and refractory mantle cell lymphoma (MCL): a study of the British Society for Blood and Marrow Transplantation. Biol Blood Marrow Transplant. 2010 Oct;16(10): 1419-27.

[184] Blystad AK, Enblad G, Kvaloy S, Berglund A, Delabie J, Holte H, et al. High-dose therapy with autologous stem cell transplantation in patients with peripheral T cell lymphomas. Bone marrow transplantation. 2001 Apr;27(7):711-6.

[185] Rodriguez J, Caballero MD, Gutierrez A, Gandarillas M, Sierra J, Lopez-Guillermo A, et al. High dose chemotherapy and autologous stem cell transplantation in patients with peripheral T-cell lymphoma not achieving complete response after induction chemotherapy.The GEL-TAMO experience. Haematologica. 2003 Dec;88(12):1372-7.

[186] Rodriguez J, Caballero MD, Gutierrez A, Marin J, Lahuerta JJ, Sureda A, et al. High-dose chemotherapy and autologous stem cell transplantation in peripheral T-cell lymphoma: the GEL-TAMO experience. Ann Oncol. 2003 December 1, 2003;14(12): 1768-75.

[187] Corradini P, Tarella C, Zallio F, Dodero A, Zanni M, Valagussa P, et al. Long-term follow-up of patients with peripheral T-cell lymphomas treated up-front with high-dose chemotherapy followed by autologous stem cell transplantation. Leukemia. 2006 Sep;20(9):1533-8.

[188] Mercadal S, Briones J, Xicoy B, Pedro C, Escoda L, Estany C, et al. Intensive chemotherapy (high-dose CHOP/ESHAP regimen) followed by autologous stem-cell transplantation in previously untreated patients with peripheral T-cell lymphoma. Ann Oncol. 2008 May;19(5):958-63.

[189] Reimer P, Rudiger T, Geissinger E, Weissinger F, Nerl C, Schmitz N, et al. Autologous stem-cell transplantation as first-line therapy in peripheral T-cell lymphomas: results of a prospective multicenter study. J Clin Oncol. 2009 Jan 1;27(1):106-13.

[190] Le Gouill S, Milpied N, Buzyn A, De Latour RP, Vernant JP, Mohty M, et al. Graft-versus-lymphoma effect for aggressive T-cell lymphomas in adults: a study by the Societe Francaise de Greffe de Moelle et de Therapie Cellulaire. J Clin Oncol. 2008 May 10;26(14):2264-71.

[191] Corradini P, Dodero A, Zallio F, Caracciolo D, Casini M, Bregni M, et al. Graft-versus-lymphoma effect in relapsed peripheral T-cell non-Hodgkin's lymphomas after reduced-intensity conditioning followed by allogeneic transplantation of hematopoietic cells. J Clin Oncol. 2004 Jun 1;22(11):2172-6.

[192] Advani R. Optimal therapy of advanced Hodgkin lymphoma. Hematology Am Soc Hematol Educ Program. 2011;2011:310-6.

[193] Bierman PJ, Anderson JR, Freeman MB, Vose JM, Kessinger A, Bishop MR, et al. High-dose chemotherapy followed by autologous hematopoietic rescue for Hodgkin's disease patients following first relapse after chemotherapy. Ann Oncol. 1996 Feb;7(2):151-6.

[194] Younes A, Gopal AK, Smith SE, Ansell SM, Rosenblatt JD, Savage KJ, et al. Results of a pivotal phase II study of brentuximab vedotin for patients with relapsed or refractory Hodgkin's lymphoma. J Clin Oncol. 2012 Jun 20;30(18):2183-9.

[195] Sarina B, Castagna L, Farina L, Patriarca F, Benedetti F, Carella AM, et al. Allogeneic transplantation improves the overall and progression-free survival of Hodgkin lymphoma patients relapsing after autologous transplantation: a retrospective study based on the time of HLA typing and donor availability. Blood. 2010 May 6;115(18): 3671-7.

[196] Robinson SP, Sureda A, Canals C, Russell N, Caballero D, Bacigalupo A, et al. Reduced intensity conditioning allogeneic stem cell transplantation for Hodgkin's lymphoma: identification of prognostic factors predicting outcome. Haematologica. 2009 Feb;94(2):230-8.

[197] Thomson KJ, Peggs KS, Smith P, Cavet J, Hunter A, Parker A, et al. Superiority of reduced-intensity allogeneic transplantation over conventional treatment for relapse of Hodgkin's lymphoma following autologous stem cell transplantation. Bone marrow transplantation. 2008 May;41(9):765-70.

[198] Devetten MP, Hari PN, Carreras J, Logan BR, van Besien K, Bredeson CN, et al. Unrelated donor reduced-intensity allogeneic hematopoietic stem cell transplantation

for relapsed and refractory Hodgkin lymphoma. Biol Blood Marrow Transplant. 2009 Jan;15(1):109-17.

[199] Sureda A, Canals C, Arranz R, Caballero D, Ribera JM, Brune M, et al. Allogeneic stem cell transplantation after reduced intensity conditioning in patients with relapsed or refractory Hodgkin's lymphoma. Results of the HDR-ALLO study - a prospective clinical trial by the Grupo Espanol de Linfomas/Trasplante de Medula Osea (GEL/TAMO) and the Lymphoma Working Party of the European Group for Blood and Marrow Transplantation. Haematologica. 2012 Feb;97(2):310-7.

[200] Bacigalupo A, Ballen K, Rizzo D, Giralt S, Lazarus H, Ho V, et al. Defining the Intensity of Conditioning Regimens: Working Definitions. Biology of blood and marrow transplantation : journal of the American Society for Blood and Marrow Transplantation. 2009;15(12):1628-33.

[201] Weissinger F, Sandmaier BM, Maloney DG, Bensinger WI, Gooley T, Storb R. Decreased transfusion requirements for patients receiving nonmyeloablative compared with conventional peripheral blood stem cell transplants from HLA-identical siblings. Blood. 2001 December 15, 2001;98(13):3584-8.

[202] Fukuda T, Hackman RC, Guthrie KA, Sandmaier BM, Boeckh M, Maris MB, et al. Risks and outcomes of idiopathic pneumonia syndrome after nonmyeloablative and conventional conditioning regimens for allogeneic hematopoietic stem cell transplantation. Blood. 2003 October 15, 2003;102(8):2777-85.

[203] Hogan WJ, Maris M, Storer B, Sandmaier BM, Maloney DG, Schoch HG, et al. Hepatic injury after nonmyeloablative conditioning followed by allogeneic hematopoietic cell transplantation: a study of 193 patients. Blood. 2004 January 1, 2004;103(1):78-84.

[204] Junghanss C, Marr KA, Carter RA, Sandmaier BM, Maris MB, Maloney DG, et al. Incidence and outcome of bacterial and fungal infections following nonmyeloablative compared with myeloablative allogeneic hematopoietic stem cell transplantation: A matched control study. Biology of blood and marrow transplantation : journal of the American Society for Blood and Marrow Transplantation. 2002;8(9):512-20.

[205] Clift RA, Buckner CD, Appelbaum FR, Sullivan KM, Storb R, Thomas ED. Long-term follow-Up of a randomized trial of two irradiation regimens for patients receiving allogeneic marrow transplants during first remission of acute myeloid leukemia. Blood. 1998;92(4):1455-6.

[206] Santos GW, Tutschka PJ, Brookmeyer R, Saral R, Beschorner WE, Bias WB, et al. Marrow transplantation for acute nonlymphocytic leukemia after treatment with busulfan and cyclophosphamide. N Engl J Med. 1983;309(22):1347-53.

[207] Bacigalupo A, Sormani M, Lamparelli T, Gualandi F, Occhini D, Bregante S, et al. Reducing transplant-related mortality after allogeneic hematopoietic stem cell transplantation. Haematologica. 2004 2004 Oct;89(10):1238-47.

[208] Gupta V, Lazarus HM, Keating A. Myeloablative conditioning regimens for AML allografts: 30 years later. Bone Marrow Transplant. 0000;32(10):969-78.

[209] Giralt S, Estey E, Albitar M, Besien Kv, Rondón G, Anderlini P, et al. Engraftment of allogeneic hematopoietic progenitor cells with purine analog-containing chemotherapy: harnessing graft-versus-leukemia without myeloablative therapy. Blood. 1997;89(12):4531-6.

[210] Giralt S, Thall PF, Khouri I, Wang X, Braunschweig I, Ippolitti C, et al. Melphalan and purine analog-containing preparative regimens: reduced-intensity conditioning for patients with hematologic malignancies undergoing allogeneic progenitor cell transplantation. Blood. 2001;97(3):631-7.

[211] Slavin S, Nagler A, Naparstek E, Kapelushnik Y, Aker M, Cividalli G, et al. Nonmyeloablative stem cell transplantation and cell therapy as an alternative to conventional bone marrow transplantation with lethal cytoreduction for the treatment of malignant and nonmalignant hematologic diseases. Blood. 1998;91(3):756-63.

[212] Corradini P, Zallio F, Mariotti J, Farina L, Bregni M, Valagussa P, et al. Effect of age and previous autologous transplantation on nonrelapse mortality and survival in patients treated with reduced-intensity conditioning and allografting for advanced hematologic malignancies. J Clin Oncol. 2005;23(27):6690-8.

[213] Bacigalupo A. Third EBMT/AMGEN Workshop on reduced-intensity conditioning allogeneic haemopoietic stem cell transplants (RIC-HSCT), and panel consensus. Bone Marrow Transplant. 2004;33(7):691-6.

[214] Giralt S, Ballen K, Rizzo D, Bacigalupo A, Horowitz M, Pasquini M, et al. Reduced-Intensity Conditioning Regimen Workshop: Defining the Dose Spectrum. Report of a Workshop Convened by the Center for International Blood and Marrow Transplant Research. Biology of blood and marrow transplantation : journal of the American Society for Blood and Marrow Transplantation. 2009;15(3):367-9.

[215] Sudhir T, Charles C, Karl P, Gulnaz B, Premini M, Gordon C, et al. Allogeneic stem-cell transplantation using a reduced-intensity conditioning regimen has the capacity to produce durable remissions and long-term disease-free survival in patients with high-risk acute myeloid leukemia and myelodysplasia. J Clin Oncol. 2005;23(36): 9387-93.

[216] Christoph S, Michael S, Rainer S, Bernd H, Eva M-W, Donald B, et al. Long-term survival in refractory acute myeloid leukemia after sequential treatment with chemotherapy and reduced-intensity conditioning for allogeneic stem cell transplantation. Blood. 2006;108(3):1092-9.

[217] Aoudjhane M, Labopin M, Gorin NC, Shimoni A, Ruutu T, Kolb HJ, et al. Comparative outcome of reduced intensity and myeloablative conditioning regimen in HLA identical sibling allogeneic haematopoietic stem cell transplantation for patients older than 50 years of age with acute myeloblastic leukaemia: a retrospective survey

from the Acute Leukemia Working Party (ALWP) of the European group for Blood and Marrow Transplantation (EBMT). Leukemia. 2005;19(12):2304-12.

[218] Charles C, Richard S, Marc L, Andrea B, Andrzej L, Mats B, et al. Outcomes of reduced-intensity transplantation for chronic myeloid leukemia: an analysis of prognostic factors from the Chronic Leukemia Working Party of the EBMT. Blood. 2005;106(9):2969-76.

[219] Mohty M, de Lavallade H, El-Cheikh J, Ladaique P, Faucher C, Furst S, et al. Reduced intensity conditioning allogeneic stem cell transplantation for patients with acute myeloid leukemia: long term results of a /`donor/' versus /`no donor/' comparison. Leukemia. 2008;23(1):194-6.

[220] Luger SM, Ringden O, Zhang MJ, Perez WS, Bishop MR, Bornhauser M, et al. Similar outcomes using myeloablative vs reduced-intensity allogeneic transplant preparative regimens for AML or MDS. Bone marrow transplantation. 2012 Feb;47(2):203-11.

Chapter 8

Controversies in Autologous Stem Cell Transplantation for the Treatment of Multiple Myeloma

Bhavana Bhatnagar and Ashraf Z. Badros

Additional information is available at the end of the chapter

http://dx.doi.org/10.5772/54115

1. Introduction

The treatment paradigm for multiple myeloma has evolved considerably over the past three decades with the incorporation of autologous stem cell transplantation (ASCT) in upfront therapy for eligible patients, and the use of novel agents. As a result, although multiple myeloma remains an incurable disease, clinical outcomes have significantly improved. In this chapter we will review the seminal studies that established the role of ASCT in multiple myeloma and as well as the current controversies with regard to the role of ASCT in the management of myeloma in the era of novel agents. We will review conditioning regimens, post-transplant maintenance strategies with novel agents and immune modulation. We will summarize the current data on early versus late ASCT, single versus tandem transplant and the role of ASCT in patients with relapsed or progressive disease.

2. The role of autologous stem cell transplantation in multiple myeloma

The advent of autologous stem cell transplantation has changed the therapeutic landscape for the management of multiple myeloma and has been the standard frontline therapy for younger patients with normal renal function since the 1990's. The standard of care for multiple myeloma patients prior to the incorporation of ASCT was conventional chemotherapy using melphalan and prednisone with the primary goals of treatment being achievement of partial response or disease stabilization. Treatment complications and later resistance were associated with poor outcomes with median overall survival ranged between two and three years.

High-dose chemotherapy was initially explored as a therapeutic approach in the 1980's after a landmark study demonstrated its effectiveness in inducing 100-percent complete remission rates in nine high-risk multiple myeloma and plasma cell leukemia patients after preconditioning with high-dose melphalan. The observation that high-dose melphalan had significant anti-tumor activity and could overcome primary drug resistance was confirmed in a later study.

Since its initial description, there have been seven randomized clinical trials comparing high-dose ASCT to conventional chemotherapy (Table 1). The first of these trials was conducted by the Intergroupe Français du Myélome (IFM) in which 200 untreated multiple myeloma patients under 65 years of age were randomized to receive either conventional chemotherapy or high dose chemotherapy in combination with ASCT. Response rates were significantly higher in patients receiving high-dose chemotherapy and ASCT compared to those who received conventional chemotherapy alone (81% vs 57%, $p<0.001$). Furthermore, patients who received high-dose therapy had a higher probability of 5-year event-free survival (28% vs 10%, $p = 0.01$) and estimated 5-year rate of overall survival (52% vs 12%, $p = 0.03$). Seven years later, the findings from the IFM study were corroborated by the British Medical Research Council Myeloma VII Trial (MRCM-VII) in a larger 407 patient multicenter study.

These findings prompted modifications to the disease response criteria as proposed by the International Myeloma Working Group as the achievement of complete responses (CRs), which were rare using conventional chemotherapy, became more achievable and, most importantly, were found to correlate with survival endpoints.

An additional five prospective randomized trials comparing ASCT to conventional chemotherapy followed. Most, but not all, demonstrated superiority of ASCT to conventional chemotherapy with respect to higher rates of CR and very good partial responses (VGPR) which ultimately translated into longer progression-free survival (PFS). An overall survival (OS) benefit was reported in three of the seven studies [5,6,12]. Differences in methodology and trial design between studies may account for some of the discordance in results. A systematic review and meta-analysis of these randomized trials reported improved overall median PFS with no significant improvement in OS following ASCT when compared to conventional chemotherapy .

In summary, high dose chemotherapy and ASCT has markedly improved the depth of response, overall response rates, and length of progression-free survival in multiple myeloma patients. Most importantly, ASCT has improved overall survival from a median of 36 months to 50-55 months, thereby establishing it as the standard of care for multiple myeloma patients under the age of 65 with normal renal function. However, there remains considerable heterogeneity between myeloma patients with regard to underlying disease characteristics and post-ASCT clinical responses. A number of prognostic markers have been identified that influence disease response to chemotherapy, ASCT and survival, specifically age, elevated β-2-microglobulin levels, LDH and serum free light chain ratio. Additionally, the recognition of recurrent chromosomal abnormalities, which have been reported in as many as 90% of patients has allowed myeloma patients to be categorized into low, in-

http://dx.doi.org/10.5772/54115

termediate and high risk groups on the basis of these aberrations. Translocation (4;14), t(14;20), deletion 17p and gain of 1q have been well associated with poor disease responses and negatively impact overall survival. A recent update from the IFM group have demonstrated a 75% 8-year survival rate in patients who did not have these chromosomal abnormalities and β-2-microglobulin values less than 5.5mg/L.

Trial/group (Year of publication)	No. Patients	Age, years	Median Follow-up	Response Rates (%) (CCT vs ASCT)	EFS, mos (CCT vs ASCT)	OS, mos (CCT vs ASCT)
IFM, (1996)	200	<65	7 years	ORR: 57 vs 81 CR: 5 vs 22 VGPR: 9 vs 16	18 vs 28	44 vs 57
MAG91 (1998)	185	55-65	58 mos	ORR: 62 vs 86 CR: 5 vs 19	19 vs 24	50 vs 55
BMRC VII, (2003)	407	<65	42 mos	ORR: 67 vs 90 CR: 8 vs 44 PR: 40 vs 42	19 vs 31	42 vs 54
Italian MMSG M97G(2004)	194	50-70	39 mos	ORR: 42 vs 73 nCR: 6 vs 25	15.6 vs 28	42 vs 58+
MAG95 (2005)	190	55-65	120 mos	ORR: 77 vs 70 CR +MRD: 20 vs 36 PR: 38.5 vs 26 MR: 18 vs 7	19 vs 31	42 vs 54
PETHEMA, (2005)	164	<65	44 mos	CR: 11 vs 30	33 vs 42	66 vs 61
US Intergroup 9321 (2006)	516	≤70	76 mos	CR: 15 vs 17	14% vs 17%**	38% vs 38%**

Abbreviations: IFM:,Intergroupe Français du Myélome, MAG: Myélome Autogreffe, BMRC: British Medical Research Council, MMSG: Multiple Myeloma Study Group, PETHEMA: Programa para el Estudio de la Terapéutica en Hemaopatia Maligna, ORR: Overall response rate, CR: Complete remission, nCR: Near complete remission, VGPR: Very good partial response, PR: Partial response, MRD: Minimal residual disease, MR: Minimal response

** 7-year estimated EFS and OS rate

Table 1. Randomized trials comparing ASCT to conventional chemotherapy (CCT)

While clinical outcomes have improved significantly since the widespread implementation of ASCT, there are several unanswered questions relating to the use of ASCT in multiple myeloma, particularly in the era of novel therapies, which remain as areas of active investigation. However, before these controversies can be fully addressed, it is important to understand the role of novel agents and their impact on myeloma management before discussing their current use in the context of ASCT.

3. Immune modulation and the advent of novel agents

The concept of immune modulation was formulated and developed after a greater understanding of the complex interaction between myeloma cells and their microenvironment as well as the discovery that myeloma cells, through a variety of mechanisms, are inherently able to evade host natural immune defenses, thereby potentiating their own survival. The immune dysregulation that is known to accompany multiple myeloma is believed to be the result of multiple biological pathways and mechanisms including excess production of myeloma–derived cytokines, inadequate antigen presentation, resistance to NK-cell lysis and impaired activity of B, T and NK cells. Additionally, multiple myeloma is also associated with defective humoral and cellular immunity leading to abnormal B-cell differentiation and function. Reduced numbers of CD4+ T cells, abnormal Th1/Th2 CD4+ T-cell ratios, impaired cytotoxic T-cell responses, dysfunction of NK and NK T-cells and abnormal dendritic cell function further compound the immune dysfunction associated with multiple myeloma.

The immunomodulatory drugs (IMiDs), lenalidomide and pomalidomide are thalidomide analogs that were specifically developed in response to the resurgence of interest in thalidomide after it was incidentally discovered to be an effective treatment in patients with cutaneous leprosy presumably through inhibition of TNFα. Subsequent preclinical trials revealed that thalidomide, in fact, had several favorable properties that would optimize its use as an anti-cancer agent.

The IMiDs, were created with the intent to maximize the pleiotropic activity directed against myeloma cells that was demonstrated by thalidomide, and, in fact are 50,000 times more potent than thalidomide in their immunomodulatory properties, including CD4+ and CD8+ T-cell costimulation, Th1 cytokine production, NK and NK T-cell activation, and antibody-dependent cellular cytotoxicity. Furthermore, they also disrupt the interaction between myeloma cells and the tumor microenvironment through potent inhibition of angiogenesis and downregulation of inflammatory cytokines, specifically TNFα, from peripheral blood mononuclear cells. The IMiDs also directly exert anti-tumor proliferation effects. Additionally the IMiDs are more capable of stimulating T-cells with without incurring the same degree of toxicity as thalidomide. The manipulation of the immune system by IMiDs has established their efficacy in the management of multiple myeloma. Lenalidomide and thalidomide, in addition to the proteasome inhibitor, bortezomib, are considered the main novel agents, and, in light of their significant disease activity, are now routinely integrated into multiple myeloma management in ASCT eligible and ineligible patients.

http://dx.doi.org/10.5772/54115

4. The impact of novel agents on induction and stem cell mobilization

Prior to the widespread use and incorporation of novel agents, the standard induction regimen was vincristine, doxorubicin and dexamethasone (VAD). Dexamethasone was the most active drug in this regimen and has long since remained the cornerstone of upfront treatment for multiple myeloma,. The investigation and incorporation of novel agents into induction chemotherapy regimens was prompted by the discovery that the quality of disease response following induction therapy, preceding ASCT, corresponded to better clinical outcomes, including subsequent response to ASCT, PFS and OS. Novel agents were initially investigated to determine whether the rates of these responses could be improved. Table 3 summarizes the results of published studies using novel agents as part of induction therapy prior to ASCT.

Thalidomide-based induction regimens were initially compared to VAD and were found to produce higher VGPR, but not CR, rates prior to transplant.. However, the increased incidence of thromboembolic complications and drug toxicity rendered the overall benefit of thalidomide containing regimens somewhat modest. A 10-year clinical follow-up study of 169 myeloma with advanced or refractory disease who were initially treated with thalidomide demonstrated remarkably improved event-free survival and OS in patients with normal cytogenetics and non-lambda light chain isotype.

Lenalidomide and high-dose dexamethasone (RD) was compared to lenalidomide and low-dose dexamethasone (Rd) as initial therapy in transplant eligible and ineligible patients and, while improved response rates (≥ VGPR) were significantly improved in patients receiving RD, increased toxicities and mortality were also more pronounced with this regimen, especially in patients older than 65 years of age. Furthermore, ASCT in combination with RD or Rd improved 3-year OS rates compared to patients who did not undergo ASCT [92% vs 79%]. Three drug-combinations using lenalidomide, bortezomib and dexamethasone (RVD) have also been investigated in a few phase I/II studies and have shown even greater improvements in response rates pre and post-transplant.

The proteasome inhibitor, bortezomib, in combination with dexamethasone was initially discovered to significantly improve near complete remission (nCR) and CR rates in the landmark IFM2005-1 trial when it was compared to VAD, VAD and dexamethasone, cyclophosphamide, etoposide and cisplatin (DCEP) consolidation, and bortezomib and dexamethasone followed by DCEP consolidation followed by ASCT. Bortezomib-containing regimens resulted in higher CR/nCR rates irrespective of disease stage or cytogenetic risk. Post-transplant, these improved response rates were associated with improved CR, nCR and VGPR rates as well as improved PFS after a median follow-up of 32 months compared to patients treated with VAD alone (36 mos vs 30 mos). In the VISTA trial, the addition of bortezomib to melphalan and prednisone also produced longer OS, and was not found to incur more resistant relapses in a long term follow-up study. The IFM 2005-1 and VISTA trials were critical in establishing the role of bortezomib in induction therapy for myeloma. To further improve the depth of disease response several phase II and III clinical trials have evaluated the efficacy of adding a third novel agent, either lenalidomide or thalidomide, to the bortezomib and dexamethasone backbone, and have demonstrated improved responses following the addition of a third agent.

Although novel agents have vastly improved the quality of disease response as well as overall response rates in the pre- and post-transplant settings, the use of these agents as part of induction therapy has resulted in greater difficulties with stem cell collection prior to autologous transplant, particularly with the use of lenalidomide and, to a lesser extent, bortezomib although the exact mechanisms by which stem cell collection is hindered has not yet been fully elucidated. To address this issue, the International Myeloma Working Group has recommended early stem cell mobilization, following 3-4 cycles of induction therapy. Mobilization using G-CSF alone or in combination with cyclophosphamide is typically considered adequate; and while a large multi-center randomized phase III trial demonstrated a significant improvement in the number of CD34+ cells/kg collected in patients receiving G-CSF and the CXCR4 inhibitor, plerixafor (AMD3100) compared to G-CSF and placebo, the routine use of plerixafor upfront for mobilization remains controversial.

5. The importance of pre-transplant disease response

Complete remissions in the pre-ASCT era were rare, but have now become a very attainable and desirable treatment goal in the pre and post-transplant settings, especially as they are considered to be strong surrogate markers for progression-free and survival overall survival in several studies. The prognostic impact of CR was not fully appreciated until ASCT was adopted as frontline therapy in the management of multiple myeloma, and this is reflected in the International Myeloma Working Group response criteria by the introduction of stringent CRs to further qualify the depth of response [Table 2]. Furthermore, the duration of CR is also described as a favorable prognostic variable ; however, in several patient subgroups, including those with a prior history of monoclonal gammopathy of undetermined significance and smoldering myeloma or with low-risk disease achievement of CR appears to be of less importance.

sCR: Meets criteria for CR plus normal FLC ratio and no clonal cells bone marrow IHC or immunofluorescence
CR: Absence of M protein in serum and urine by immunofixation, < 5% bone marrow plasma cells, no increase of lytic bone lesions, disappearance of soft tissue plasmacytomas
VGPR: Serum and urine M protein detectable by immunofixation but not on electrophoresis OR ≤ 90% reduction in serum M-protein plus urine M-protein <100mg/24hr
Abbreviations: IMWG: International Myeloma Working Group, sCR: stringent complete remission, CR: Complete Remission, VGPR: Very Good Partial Response, FLC: Free Light Chains, IHC: Immunohistochemistry

Table 2. IMWG Complete Response Criteria (Durie et al, Leukemia 2006)

http://dx.doi.org/10.5772/54115

Author, date of publication	No. of patients	Treatment regimen	Median follow-up	RR after induction (%)	RR after transplant (%)	PFS, median	OS
Rajkumar, 2006	207	TD vs Dex	207	CR: 4 vs 0 ≥ VGPR: --- ≥ PR: 63 vs 41	CR: --- ≥ VGPR: --- ≥ PR: ---	---	---
Lokhorst, 2010	536	VAD vs TAD	52 mos	CR: 2 vs 3 ≥ VGPR: 18 vs 37 ≥ PR: 57 vs 71	CR: 12 vs 14 ≥ VGPR: 44 vs 54 ≥ PR: 76 vs 84	22 vs 34 mos	60 vs 73 mos
Harousseau, 2010	482	VAD vs VD	31.2 mos	CR/nCR: 6.4 vs 14.8 ≥ VGPR: 15.1 vs 37.7 ≥ PR: 62.8 vs 72.5	CR/nCR: 18.4 vs 35 ≥ VGPR: 37.2 vs 54.3 ≥ PR: 77.1 vs 80.3	29.7 vs 36 mos	3 yr OS 77.4% vs 81.4%
Cavo, 2010	480	VTD vs TD	36 mos	CR/nCR: 31 vs 11 ≥ VGPR: 62 vs 28 ≥ PR: 93 vs 79	CR/nCR: 55 vs 41 ≥ VGPR: 82 vs 64 ≥ PR: 93 vs 84	68 % vs 56%*	86% vs 84%*
Rajkumar, 2010	445	RD vs Rd	35.8 mos	CR: 5 vs 4 ≥ VGPR: 71vs 26 ≥ PR: 81 vs 70	---- ---- ----	19 vs 25 mos	2yr OS 87% vs 75%
Moreau, 2011	199	VD vs vtD	32 mos	CR: 22 vs 31 ≥ VGPR: 36 vs 49 ≥ PR: 81 vs 88	CR: 52 vs 61 ≥ VGPR: 58 vs 74 ≥ PR: 86 vs 89	30 vs 26 mos	---
Rosinol, 2012	386	VTD vs TD vs VBMCP/ VBAD/B	35.2 mos	CR: 35 vs 14 vs 21 ≥ VGPR: 25 vs 15 vs 15 ≥ PR: 25 vs 33 vs 39	CR: 46 vs 24 vs 38 ≥ VGPR: --- ≥ PR: ---	56.2 vs 28.2 vs 35.3 mos	4yr OS 74% vs 65% vs 70%
Sonneveld, 2012	833	VAD vs PAD	41 mos	CR/nCR: 15 vs 11 ≥ VGPR: 14 vs 42 ≥ PR: 54 vs 78	CR/nCR: 15 vs 31 ≥ VGPR: 36 vs 62 ≥ PR: 75 vs 88	28 vs 35 mos	5 yr OS, 55% vs 65%

Abbreviations: TD: Thalidomide and Dexamethasone,Dex: Dexamethasone VAD: Vincristine, Adriamycin and Dexamethasone, TAD: Thalidomide, Adriamycin and Dexamethasone, VD: Bortezomib and Dexamethasone, VTD: Bortezomib, Thalidomide and Dexamethasone, RD: Lenalidomide and high-dose dexamethasone, Rd: Lenalidomide and low-dose dexamethasone, PAD: Bortezomib, Adriamycin and Dexamethasone, vtD: reduced dose bortezomib, thalidomide and Dexamethasone

Table 3. Published phase III studies using novel agents as part of induction therapy prior to ASCT

6. Early versus late transplant

Only one randomized trial has compared upfront ASCT to late ASCT at the time of relapse. Upfront ASCT improved event-free survival and quality of life compared to patients treated with conventional chemotherapy and who underwent ASCT as rescue treatment at the time of relapse; interestingly there was no appreciable difference in 5-year overall survival between the arms . However, in the era of novel agents and resultant improvements in complete remission rates, the question as to whether ASCT could potentially be delayed until disease relapse or progression has, again, resurfaced. The International Myeloma Working Group recommends that all eligible patients be offered ASCT at some point in their disease course and while there are no published randomized phase III trials incorporating the use of novel agents in induction therapy to support the use of ASCT upon disease relapse, many clinicians opt to collect stem cells early and preserve them for use following disease relapse. We believe that upfront ASCT should be the standard of care until ongoing trials establish that delayed ASCT after novel agents has a role.

7. Single ASCT versus tandem transplant

Tandem ASCT, as part of a more intensified treatment strategy ("total therapy") was initially shown to improve CR rates, EFS and OS in comparison to standard therapy. The superiority of double ASCT was later appreciated in a landmark randomized clinical trial which demonstrated significantly improved OS, particularly in patients who had not achieved VGPR following transplant. Several other randomized trials have also attempted to compare single versus double ASCT and have reported conflicting results regarding the superiority of one approach over the other. A recent meta-analysis attempted to answer this question and concluded that tandem transplant offered no benefit in terms of disease outcomes and was, in fact, associated with greater morbidity; however, this analysis has received criticism due to the heterogeneity of the selected studies which were evaluated as well as variability in treatment methodology. A more recent analysis suggests that tandem ASCT does offer a survival benefit. Most clinicians speculate that tandem and single transplants are equivocal, however, there have been no definitive trials evaluating this issue and it remains an area of considerable debate.

8. Methods to improve conditioning regimens: The addition of total body irradiation or other agents to high-dose melphalan

The quality of disease response following pre-transplant conditioning is critical to the success of ASCT. High dose melphalan 200mg/m2 is the standard conditioning chemotherapy regimen prior to autologous HSCT in multiple myeloma as this approach has demonstrated superior overall survival rates in comparison to chemotherapy alone. Various approaches to

http://dx.doi.org/10.5772/54115

improve responses to this conditioning regimen while minimizing toxicities have been evaluated in a number of studies.

Total body irradiation (TBI) in combination with melphalan demonstrated improved CR rates, relapse and progression rates and five year OS rates when compared to TBI and cyclophosphamide as a myeloablative conditioning regimen in myeloma patients undergoing allogeneic HSCT. A landmark study, however, in which melphalan and TBI was compared with high-dose melphalan 200mg/m2 demonstrated more rapid hematologic recovery, reduced transfusion requirements, shorter hospitalization and improved survival in patients receiving high-dose melphalan alone. As such, the routine use of TBI in conjunction with melphalan is not widely used.

The alkylating agent, busulfan, has been used in several studies in combination melphalan with promising outcomes, particularly in patients with non-remission disease at the time of transplant. A recent analysis of multiple myeloma patients treated with oral busulfan and melphalan 140mg/m^2 compared to standard melphalan 200mg/m^2 did demonstrate improved median PFS (41 mos vs 31 mos, p=0.009), however, the increased incidence of veno-occlusive disease and transplant related mortality counteracted the benefits; additionally, patients who received busulfan had less access to salvage therapies using novel agents than patients who had relapsed following treatment with melphalan 200mg/m2.

The Intergroupe Francophone du Myelome study group also evaluated the efficacy of adding bortezomib to high-dose melphalan in a recent phase II study and were able to demonstrate, that, in comparison to historical controls, patients treated with the bortezomib and melphalan achieved higher CR rates (35% vs 11%, p=.001) with no increase in hematologic toxicity.

9. Novel agents as post-transplant maintenance therapy

Maintenance with interferon, steroids, and chemotherapy has been tried in many centers for over 30 years with no clear benefit. Maintenance interferon frequently resulted in worsened quality of life; furthermore, future development of therapy-related myelodysplastic syndrome following chemotherapy led to these maintenance treatments to fall out of favor. The availability of novel agents and their tolerable toxicity profiles has renewed interest in post-transplant maintenance treatment. The results of this approach have, thus far, been encouraging, including upgrades in disease responses and improvements in PFS/EFS, and OS in many studies; however, none of these agents are currently approved in the post-transplant setting. The recently released consensus statement from the International Myeloma Working Group does not advocate for or against maintenance therapy and recommends that the decision to use maintenance therapy be made on an individualized basis. In the following paragraph we will review various agents with a brief summary of the randomized trials data.

9.1. Thalidomide

Thalidomide maintenance therapy following ASCT has been evaluated in six randomized clinical trials all of which have reported a significant improvement in progression free survival in patients receiving thalidomide maintenance versus the comparator arm, but only 3 had shown improvement in OS by 6-9 months. Two meta-analyses have confirmed improved OS with thalidomide maintenance. However, most patients (> 50%) eventually discontinued thalidomide, between 7 months and 2 years of treatment, due to side effects, particularly development of peripheral neuropathy. Interestingly, thalidomide maintenance does not benefit patients with poor-risk cytogenetics, and, in fact, this patient subset was shown to have a shorter survival duration. Similar results from the Total Therapy 2 study were reported although a longer follow up showed improvement in long-term survival in high risk disease, although the main impact was most appreciable in patients with favorable cytogenetics.

9.2. Lenalidomide

Lenalidomide has a favorable toxicity profile, and its efficacy extends beyond inhibition of the growth of myeloma cells as the drug also causes alterations within the bone marrow microenvironment leading to an enhancement of immune responses, thereby making it an ideal drug for post-transplant maintenance. Two very recently published trials from the Cancer and Leukemia Group B (CALGB) and IFM evaluated the efficacy of lenalidomide following transplant and demonstrated that lenalidomide maintenance therapy was associated with a significant improvement in PFS compared to placebo (48 vs 30.9 mos, and 41 vs 24 mos in the CALGB and IFM studies, respectively). The benefits of lenalidomide maintenance were appreciated across all patient subgroups, including those with high-risk cytogenetics, although the data is limited to a small number pf patients in the IFM study, β2-microglobulin and response following transplant. In the IFM-2005-02 trial, patients were given two courses of lenalidomide as consolidation treatment which led to an upgrade in the number of disease response with rates of CR increasing from 14% to 20% and responses higher than or equal to VGPR from 58% to 67%. The side effects were tolerable, mostly hematologic, and responded well to dose adjustments, supportive growth factor injections and transfusion support. A meta-analysis by the International Myeloma Working Group, which included a total of 1380 patients demonstrated a 65% reduction in risk of disease progression for patients treated with lenalidomide maintenance therapy. There is a notable increased risk of second cancers in association with this drug as noted by both IFM and CALGB. The IFM reported the incidence of second cancers as 3.1 per 100 patient years, compared to 1.2 per 100 patient years in the placebo group ($p = 0.002$). In the CALGB study, 8% of patients treated with lenalidomide developed second cancers, compared to 3% in the control arm.

9.3. Bortezomib

An interesting study to evaluate the effect of minimal residual disease, by qualitative and real-time quantitative polymerase chain reaction (RQ-PCR) after ASCT showed that a consolidation regimen comprised of bortezomib, thalidomide, and dexamethasone (VTD) increased CRs from 15% after ASCT to 49% after VTD. Most importantly, molecular remissions increased from 3% after ASCT to 18% after VTD. No patients had relapsed at the time of reporting (median follow-up, 42 months). These unprecedented levels of tumor cell reduction are very encouraging and have laid the foundation for a new area of investigation to better evaluate the depth of treatment response in myeloma.

A subsequent randomized phase 3 study specifically assessed the efficacy and safety of consolidation therapy using bortezomib, thalidomide and dexamethasone (VTD) versus thalidomide and dexamethasone (TD). Before starting consolidation, CR/nCR rates were not significantly different in the VTD and TD arms (63.1% 54.7%, respectively). However, after consolidation, CR (60.6% vs 46.6%) and CR/nCR (73.1% vs 60.9%) rates were significantly higher for VTD-treated versus TD-treated patients. With a median follow-up of 30.4 months from start of consolidation, 3-year PFS was significantly longer for the VTD group compared to TD (60% vs 48%). The VTD consolidation therapy was shown to significantly improve clinical outcomes after ASCT.

The evaluation of novel agents in the post-transplant setting has resulted in significant improvements in disease responses and survival endpoints. Moreover combination regimens in the form of consolidation and/or long term maintenance are well tolerated with further improvements and achievement of molecular remissions. Future studies to determine the optimal duration of maintenance therapy are urgently needed.

10. Combined ASCT/Allogeneic Hematopoeitic Stem Cell Transplant approaches

Early trials evaluating myeloablative allogeneic stem cell transplantation in the treatment of multiple myeloma demonstrated improvements in relapse and progression rates attributed to graft versus myeloma effects; however, development of graft versus host disease and infectious complications resulted in high transplant-related mortality [TRM]. A critical advantage of allogeneic transplantation was the development of reduced intensity conditioning (RIC) regimens that were associated with decreased toxicities and profound graft versus tumor effects as demonstrated in early trials evaluating the efficacy of RIC in relapsed and refractory myeloma patients. However, higher rates of disease progression and relapse, were noted and attributed to the late use of this modality underscoring the importance of using effective regimens early before the disease becomes refractory especially since the goals of allogeneic transplant are curative in intent.

Combined sequential therapy utilizing ASCT for cytoreduction followed RIC allogeneic transplant (i.e. the auto-allo approach) to exploit the graft versus myeloma effect has been compared to tandem ASCT in several studies; randomization in these trials was biological; i.e. patients with an HLA-matched sibling received RIC allogeneic transplant and all others underwent tandem ASCT. The first published study from the IFM compared tandem ASCT in 219 patients to auto-allo in 65 patients with high-risk multiple myeloma and reported no significant difference in response rates or event-free survival between groups; however, there was an observed trend toward better overall survival in patients treated with tandem ASCT; these findings remained unchanged in a long-term follow-analysis from the same group. Subsequent comparisons have reported improved CR rates and PFS durations and only one has shown superior OS in auto-allo treated patients. However a recently published large multi-center phase 3 study reported that the auto-allo approach was not superior to auto-auto in terms of progression-free survival (43 % vs 46% at 3 years) or 3-year OS (77% vs 80%). Additionally, there was no significant difference in the development of grade 3-5 adverse events between groups by three years (46% vs 42%). Further modifications to allogeneic transplantation would be needed to offset the graft versus myeloma effect as well as the increase in transplant-related mortality. The Eastern Cooperative Oncology Group conducted a small trial in which 32 patients received non-myeloablative matched sibling donor transplant following ASCT and reported a 78% ORR (30% CR and 48% PR) with low TRM; however over half of patients developed chronic GVHD. A recent Swedish study compared auto-allo approach to single ASCT in 357 previously untreated multiple myeloma patients and was demonstrated that the auto-allo approach was superior in terms of PFS, OS and relapse rate with a 12% nonrelapse mortality rate. The data remain conflicting; however, a meta-analysis reviewing outcomes on 7 published and unpublished studies concluded that the auto-allo approach offers no benefit compared to autologous transplant approaches and is associated with higher TRM. The International Myeloma Working Group does not recommend the routine use of allogeneic transplantation, and, in fact, recommends consideration of RIC transplant only in the setting of a clinical trial.

11. Novel immunotherapy strategies

The post-transplant period is the ideal time point for immunotherapy as the disease burden is, theoretically, low. Immune function remains depressed following high-dose therapy for many months. Ex vivo expansion and subsequent transfer of autologous stimulated T cells may enhance host antitumor immunity and may also allow for enhancement of a post-transplant vaccination strategy against tumor-directed antigens. Early trials focused on the generation of antibodies against myeloma specific antigens. The idiotype [Id] protein has, in a number of pre-clinical studies, demonstrated powerful antibody responses that, in vitro, resulted in apoptosis of myeloma cells. However, durable clinical responses were not seen in subsequent clinical studies. Idiotype-pulsed dendritic cell vaccinations following ASCT have also demonstrated that cellular immune responses can be elicited in the context of minimal residual disease following transplantation; however, again, there is no definitive evidence that these vaccination strategies alter the course of disease. It has been suggested that the immune dysfunction in myeloma patients is the primary barrier to successful vaccination strategies. A low number of T-cells with activity against myeloma have been detected in multiple myeloma patients . Several attempts to expand T cells, collected from the peripheral blood of affected patients, and infused after ASCT, have shown that rapid recovery of T-cell numbers can be achieved but, unfortunately, with no clear anti-myeloma benefits. The results of one interesting study in which myeloma patients received the conjugated pneumococcal vaccine before T-cell collection and after ASCT showed profound antibody responses, suggesting that T cells may improve immune responses to vaccination. A subsequent study in which adoptive transfer of vaccine primed autologous T-cells to the htert/survivin multipeptide vaccine, a target in myeloma cells, corroborated these findings and demonstrated that vaccination was associated with robust antibody responses in most patients; however, again, there was no definitive activity directed against myeloma cells specifically. Clinical trials building on the expansion of T cells and targeting various myeloma antigen such as MAGE A3 and NYESO1 are ongoing. Of important note, several studies focusing on expansion of marrow infiltrating lymphocytes (MILs) had yielded interesting results with regards to antimyeloma activities, but, again, the clinical benefit was quite limited.

Several antibody trials in myeloma are ongoing. A recently published phase 1 study has provided encouraging evidence that elotuzumab in combination with lenalidomide yields impressive responses in relapsed and refractory myeloma; whether the responses seen in the relapsed setting can be confirmed and implemented in patients with minimal disease states would require further investigation. In light of these findings, it is suggested that enhancement of T-cell function could potentially lay the groundwork for subsequent trials aimed to improve immune function, and by extension, clinical outcomes following ASCT in myeloma patients.

12. Salvage ASCT for relapsed disease

At present, the optimal treatment approach for patients with relapsed disease following initial ASCT has not yet been defined. Potential options include treatment with novel agents, conventional chemotherapy or a second salvage ASCT. While the data evaluating the role of a second ASCT are limited, several small retrospective analyses have demonstrated that it is an effective and well tolerated treatment option with overall response rates reported between 55-90%. Overall survival and progression free survival is significantly improved for patients who have received fewer lines of therapy prior to transplant and for those who have experienced a late disease relapse. However, the length of time which constitutes a late relapse has varied between studies, ranging between 12 months and > 36 months. A recently published retrospective review suggested a time-dependent association between remission duration following initial ASCT and PFS following transplant. Patients who relapsed within 18 months of initial ASCT had significantly shorter PFS compared to those who relapsed between 18 and 36 months and those who relapsed 36 months or more (4.2 mos vs 13.8 mos vs 49.1 mos) [111]. Although larger studies would provide greater insight regarding the optimal timing of a second transplant, consideration of salvage ASCT is generally regarded as feasible approach which offers the greatest benefit in select patients who have relapsed at least more than 12 months after their initial ASCT. Salvage allogeneic transplant following failure of initial autografting has also been compared to salvage ASCT in a limited number of studies and has been reported to have comparable PFS due to lower rates of disease progression following allogeneic transplant, but superior OS in autografted patients; furthermore, the increased incidence of graft versus host disease in allografted patients has rendered this approach less preferable. Refinements in allogeneic transplant techniques may potentially generate renewed interest in this treatment approach.

13. Conclusion

The widespread implementation of autologous stem cell transplantation, in conjunction with novel agents, has revolutionized the management of multiple myeloma and has markedly altered the natural history of the disease by improving disease responses and response duration, which, by extension, have led to significant improvements in overall survival. While treatment options for multiple myeloma have expanded considerably over the past several decades, long-term survivorship remains low. Continued investigative efforts are targeted towards refining our current treatment modalities with the hope of ultimately developing a treatment approach which results in cure.

Author details

Bhavana Bhatnagar[1] and Ashraf Z. Badros[2*]

*Address all correspondence to: abadros@umm.edu

1 University of Maryland Marlene and Stewart Greenebaum Cancer Center. Division of Hematology and Medical Oncology, Baltimore, United States of America

2 Department of Medicine, University of Maryland School of Medicine, Stem Cell Transplant Program, Marelene and Stewart Greenebaum Cancer CenterUniversity of Maryland Medical Center, Maryland, United States of America

References

[1] Combination chemotherapy versus melphalan plus prednisone as treatment for multiple myeloma: an overview of 6,633 patients from 27 randomized trials. Myeloma Trialists' Collaborative Group. J Clin Oncol. 1998 Dec;16[12]:3832-42. PubMed PMID: 9850028.

[2] Corso A, Nozza A, Lazzarino M, Klersy C, Zappasodi P, Arcaini L, et al. Plateau phase in multiple myeloma: an end-point of conventional-dose chemotherapy. Haematologica. 1999;84[4]:336-41.

[3] McElwain TJ, Powles RL. High-dose intravenous melphalan for plasma-cell leukaemia and myeloma. Lancet. 1983;2[8354]:822-4.

[4] Barlogie B, Hall R, Zander A, Dicke K, Alexanian R. High-dose melphalan with autologous bone marrow transplantation for multiple myeloma. Blood. 1986;67[5]:1298-301.

[5] Attal M, Harousseau JL, Stoppa AM, Sotto JJ, Fuzibet JG, Rossi JF, et al. A prospective, randomized trial of autologous bone marrow transplantation and chemotherapy in multiple myeloma. Intergroupe Francais du Myelome. The New England journal of medicine. 1996;335[2]:91-7.

[6] Child JA, Morgan GJ, Davies FE, Owen RG, Bell SE, Hawkins K, et al. High-dose chemotherapy with hematopoietic stem-cell rescue for multiple myeloma. The New England journal of medicine. 2003;348[19]:1875-83.

[7] Durie BG, Harousseau JL, Miguel JS, Blade J, Barlogie B, Anderson K, et al. International uniform response criteria for multiple myeloma. Leukemia : official journal of the Leukemia Society of America, Leukemia Research Fund, UK. 2006;20[9]:1467-73.

[8] Barlogie B, Kyle RA, Anderson KC, Greipp PR, Lazarus HM, Hurd DD, et al. Standard chemotherapy compared with high-dose chemoradiotherapy for multiple myeloma: final results of phase III US Intergroup Trial S9321. Journal of clinical oncology : official journal of the American Society of Clinical Oncology. 2006;24[6]:929-36.

[9] Blade J, Rosinol L, Sureda A, Ribera JM, Diaz-Mediavilla J, Garcia-Larana J, et al. High-dose therapy intensification compared with continued standard chemotherapy in multiple myeloma patients responding to the initial chemotherapy: long-term results from a prospective randomized trial from the Spanish cooperative group PETHEMA. Blood. 2005;106[12]:3755-9.

[10] Fermand JP, Katsahian S, Divine M, Leblond V, Dreyfus F, Macro M, et al. High-dose therapy and autologous blood stem-cell transplantation compared with conventional treatment in myeloma patients aged 55 to 65 years: long-term results of a randomized control trial from the Group Myelome-Autogreffe. Journal of clinical oncology : official journal of the American Society of Clinical Oncology. 2005;23[36]:9227-33.

[11] Fermand JP, Ravaud P, Chevret S, Divine M, Leblond V, Belanger C, et al. High-dose therapy and autologous peripheral blood stem cell transplantation in multiple myeloma: up-front or rescue treatment? Results of a multicenter sequential randomized clinical trial. Blood. 1998;92[9]:3131-6.

[12] Palumbo A, Bringhen S, Petrucci MT, Musto P, Rossini F, Nunzi M, et al. Intermediate-dose melphalan improves survival of myeloma patients aged 50 to 70: results of a randomized controlled trial. Blood. 2004;104[10]:3052-7.

[13] Koreth J, Cutler CS, Djulbegovic B, Behl R, Schlossman RL, Munshi NC, et al. High-dose therapy with single autologous transplantation versus chemotherapy for newly diagnosed multiple myeloma: A systematic review and meta-analysis of randomized

controlled trials. Biol Blood Marrow Transplant. 2007 Feb;13[2]:183-96. PubMed PMID: 17241924.

[14] Harousseau JL, Moreau P. Autologous hematopoietic stem-cell transplantation for multiple myeloma. The New England journal of medicine. 2009;360[25]:2645-54.

[15] Ludwig H, Durie BG, Bolejack V, Turesson I, Kyle RA, Blade J, et al. Myeloma in patients younger than age 50 years presents with more favorable features and shows better survival: an analysis of 10 549 patients from the International Myeloma Working Group. Blood. 2008 Apr 15;111[8]:4039-47. PubMed PMID: 18268097.

[16] Avet-Loiseau H, Attal M, Moreau P, Charbonnel C, Garban F, Hulin C, et al. Genetic abnormalities and survival in multiple myeloma: the experience of the Intergroupe Francophone du Myelome. Blood. 2007;109[8]:3489-95.

[17] Rossi D, Fangazio M, De Paoli L, Puma A, Riccomagno P, Pinto V, et al. Beta-2-microglobulin is an independent predictor of progression in asymptomatic multiple myeloma. Cancer. 2010;116[9]:2188-200.

[18] Stella-Holowiecka B, Czerw T, Holowiecka-Goral A, Giebel S, Wojnar J, Holowiecki J. Beta-2-microglobulin level predicts outcome following autologous hematopoietic stem cell transplantation in patients with multiple myeloma. Transplantation proceedings. 2007 Nov;39[9]:2893-7. PubMed PMID: 18022010.

[19] Maltezas D, Dimopoulos MA, Katodritou I, Repousis P, Pouli A, Terpos E, et al. Re-evaluation of prognostic markers including staging, serum free light chains or their ratio and serum lactate dehydrogenase in multiple myeloma patients receiving novel agents. Hematological oncology. 2012.

[20] Fonseca R, Blood E, Rue M, Harrington D, Oken MM, Kyle RA, et al. Clinical and biologic implications of recurrent genomic aberrations in myeloma. Blood. 2003;101[11]:4569-75.

[21] Gertz MA, Lacy MQ, Dispenzieri A, Greipp PR, Litzow MR, Henderson KJ, et al. Clinical implications of t[11;14][q13;q32], t[4;14][p16.3;q32], and -17p13 in myeloma patients treated with high-dose therapy. Blood. 2005;106[8]:2837-40.

[22] Avet-Loiseau H, Attal M, Campion L, Caillot D, Hulin C, Marit G, et al. Long-term analysis of the IFM 99 trials for myeloma: cytogenetic abnormalities [t[4;14], del[17p], 1q gains] play a major role in defining long-term survival. Journal of clinical oncology : official journal of the American Society of Clinical Oncology. 2012;30[16]:1949-52.

[23] Brown RD, Pope B, Murray A, Esdale W, Sze DM, Gibson J, et al. Dendritic cells from patients with myeloma are numerically normal but functionally defective as they fail to up-regulate CD80 [B7-1] expression after huCD40LT stimulation because of inhibition by transforming growth factor-beta1 and interleukin-10. Blood. 2001;98[10]: 2992-8.

[24] Quach H, Ritchie D, Stewart AK, Neeson P, Harrison S, Smyth MJ, et al. Mechanism of action of immunomodulatory drugs [IMiDS] in multiple myeloma. Leukemia. 2010 Jan;24[1]:22-32. PubMed PMID: 19907437.

[25] Sampaio EP, Sarno EN, Galilly R, Cohn ZA, Kaplan G. Thalidomide selectively inhibits tumor necrosis factor alpha production by stimulated human monocytes. The Journal of experimental medicine. 1991;173[3]:699-703.

[26] D'Amato RJ, Loughnan MS, Flynn E, Folkman J. Thalidomide is an inhibitor of angiogenesis. Proceedings of the National Academy of Sciences of the United States of America. 1994;91[9]:4082-5.

[27] Davies FE, Raje N, Hideshima T, Lentzsch S, Young G, Tai YT, et al. Thalidomide and immunomodulatory derivatives augment natural killer cell cytotoxicity in multiple myeloma. Blood. 2001;98[1]:210-6.

[28] Haslett PA, Corral LG, Albert M, Kaplan G. Thalidomide costimulates primary human T lymphocytes, preferentially inducing proliferation, cytokine production, and cytotoxic responses in the CD8+ subset. The Journal of experimental medicine. 1998;187[11]:1885-92.

[29] Gorgun G, Calabrese E, Soydan E, Hideshima T, Perrone G, Bandi M, et al. Immunomodulatory effects of lenalidomide and pomalidomide on interaction of tumor and bone marrow accessory cells in multiple myeloma. Blood. 2010;116[17]:3227-37.

[30] Corral LG, Haslett PA, Muller GW, Chen R, Wong LM, Ocampo CJ, et al. Differential cytokine modulation and T cell activation by two distinct classes of thalidomide analogues that are potent inhibitors of TNF-alpha. Journal of immunology [Baltimore, Md: 1950]. 1999;163[1]:380-6.

[31] Muller GW, Corral LG, Shire MG, Wang H, Moreira A, Kaplan G, et al. Structural modifications of thalidomide produce analogs with enhanced tumor necrosis factor inhibitory activity. Journal of medicinal chemistry. 1996;39[17]:3238-40.

[32] Barlogie B, Smith L, Alexanian R. Effective treatment of advanced multiple myeloma refractory to alkylating agents. N Engl J Med. 1984 May 24;310[21]:1353-6. PubMed PMID: 6546971.

[33] Alexanian R, Barlogie B, Tucker S. VAD-based regimens as primary treatment for multiple myeloma. Am J Hematol. 1990 Feb;33[2]:86-9. PubMed PMID: 2301376.

[34] Alexanian R, Dimopoulos MA, Delasalle K, Barlogie B. Primary dexamethasone treatment of multiple myeloma. Blood. 1992 Aug 15;80[4]:887-90. PubMed PMID: 1498331.

[35] Harousseau JL, Attal M, Avet-Loiseau H. The role of complete response in multiple myeloma. Blood. 2009;114[15]:3139-46.

[36] Harousseau JL, Avet-Loiseau H, Attal M, Charbonnel C, Garban F, Hulin C, et al. Achievement of at least very good partial response is a simple and robust prognostic

factor in patients with multiple myeloma treated with high-dose therapy: long-term analysis of the IFM 99-02 and 99-04 Trials. Journal of clinical oncology : official journal of the American Society of Clinical Oncology. 2009;27[34]:5720-6.

[37] Cavo M, Tacchetti P, Patriarca F, Petrucci MT, Pantani L, Galli M, et al. Bortezomib with thalidomide plus dexamethasone compared with thalidomide plus dexamethasone as induction therapy before, and consolidation therapy after, double autologous stem-cell transplantation in newly diagnosed multiple myeloma: a randomised phase 3 study. Lancet. 2010 Dec 18;376[9758]:2075-85. PubMed PMID: 21146205.

[38] Harousseau JL, Attal M, Avet-Loiseau H, Marit G, Caillot D, Mohty M, et al. Bortezomib plus dexamethasone is superior to vincristine plus doxorubicin plus dexamethasone as induction treatment prior to autologous stem-cell transplantation in newly diagnosed multiple myeloma: results of the IFM 2005-01 phase III trial. Journal of clinical oncology : official journal of the American Society of Clinical Oncology. 2010;28[30]:4621-9.

[39] Lokhorst HM, van der Holt B, Zweegman S, Vellenga E, Croockewit S, van Oers MH, et al. A randomized phase 3 study on the effect of thalidomide combined with adriamycin, dexamethasone, and high-dose melphalan, followed by thalidomide maintenance in patients with multiple myeloma. Blood. 2010 Feb 11;115[6]:1113-20. PubMed PMID: 19880501.

[40] Moreau P, Avet-Loiseau H, Facon T, Attal M, Tiab M, Hulin C, et al. Bortezomib plus dexamethasone versus reduced-dose bortezomib, thalidomide plus dexamethasone as induction treatment before autologous stem cell transplantation in newly diagnosed multiple myeloma. Blood. 2011;118[22]:5752-8; quiz 982.

[41] Rajkumar SV, Blood E, Vesole D, Fonseca R, Greipp PR. Phase III clinical trial of thalidomide plus dexamethasone compared with dexamethasone alone in newly diagnosed multiple myeloma: a clinical trial coordinated by the Eastern Cooperative Oncology Group. J Clin Oncol. 2006 Jan 20;24[3]:431-6. PubMed PMID: 16365178. Epub 2005/12/21. eng.

[42] Rajkumar SV, Jacobus S, Callander NS, Fonseca R, Vesole DH, Williams ME, et al. Lenalidomide plus high-dose dexamethasone versus lenalidomide plus low-dose dexamethasone as initial therapy for newly diagnosed multiple myeloma: an open-label randomised controlled trial. The lancet oncology. 2010;11[1]:29-37.

[43] Rosinol L, Oriol A, Teruel AI, Hernandez D, Lopez-Jimenez J, de la Rubia J, et al. Superiority of bortezomib, thalidomide, and dexamethasone [VTD] as induction pretransplantation therapy in multiple myeloma: a randomized phase 3 PETHEMA/GEM study. Blood. 2012;120[8]:1589-96.

[44] Sonneveld P, Schmidt-Wolf IG, van der Holt B, El Jarari L, Bertsch U, Salwender H, et al. Bortezomib Induction and Maintenance Treatment in Patients With Newly Diagnosed Multiple Myeloma: Results of the Randomized Phase III HOVON-65/

GMMG-HD4 Trial. Journal of clinical oncology : official journal of the American Society of Clinical Oncology. 2012;30[24]:2946-55.

[45] Cavo M, Zamagni E, Tosi P, Tacchetti P, Cellini C, Cangini D, et al. Superiority of thalidomide and dexamethasone over vincristine-doxorubicindexamethasone [VAD] as primary therapy in preparation for autologous transplantation for multiple myeloma. Blood. 2005;106[1]:35-9.

[46] Goldschmidt H, Sonneveld P, Cremer FW, van der Holt B, Westveer P, Breitkreutz I, et al. Joint HOVON-50/GMMG-HD3 randomized trial on the effect of thalidomide as part of a high-dose therapy regimen and as maintenance treatment for newly diagnosed myeloma patients. Annals of hematology. 2003 Oct;82[10]:654-9. PubMed PMID: 12845480.

[47] van Rhee F, Dhodapkar M, Shaughnessy JD, Jr., Anaissie E, Siegel D, Hoering A, et al. First thalidomide clinical trial in multiple myeloma: a decade. Blood. 2008;112[4]: 1035-8.

[48] Richardson PG, Weller E, Lonial S, Jakubowiak AJ, Jagannath S, Raje NS, et al. Lenalidomide, bortezomib, and dexamethasone combination therapy in patients with newly diagnosed multiple myeloma. Blood. 2010;116[5]:679-86.

[49] Wang M, Delasalle K, Giralt S, Alexanian R. Rapid control of previously untreated multiple myeloma with bortezomib-lenalidomide-dexamethasone [BLD]. Hematology [Amsterdam, Netherlands]. 2010;15[2]:70-3.

[50] Mateos MV, Richardson PG, Schlag R, Khuageva NK, Dimopoulos MA, Shpilberg O, et al. Bortezomib plus melphalan and prednisone compared with melphalan and prednisone in previously untreated multiple myeloma: updated follow-up and impact of subsequent therapy in the phase III VISTA trial. Journal of clinical oncology : official journal of the American Society of Clinical Oncology. 2010;28[13]:2259-66.

[51] Cavo M, Tacchetti P, Patriarca F, Petrucci MT, Pantani L, Galli M, et al. Bortezomib with thalidomide plus dexamethasone compared with thalidomide plus dexamethasone as induction therapy before, and consolidation therapy after, double autologous stem-cell transplantation in newly diagnosed multiple myeloma: a randomised phase 3 study. Lancet. 2010;376[9758]:2075-85.

[52] Popat U, Saliba R, Thandi R, Hosing C, Qazilbash M, Anderlini P, et al. Impairment of filgrastim-induced stem cell mobilization after prior lenalidomide in patients with multiple myeloma. Biol Blood Marrow Transplant. 2009 Jun;15[6]:718-23. PubMed PMID: 19450756.

[53] Kumar S, Dispenzieri A, Lacy MQ, Hayman SR, Buadi FK, Gastineau DA, et al. Impact of lenalidomide therapy on stem cell mobilization and engraftment post-peripheral blood stem cell transplantation in patients with newly diagnosed myeloma. Leukemia. 2007 Sep;21[9]:2035-42. PubMed PMID: 17581613.

[54] DiPersio JF, Micallef IN, Stiff PJ, Bolwell BJ, Maziarz RT, Jacobsen E, et al. Phase III prospective randomized double-blind placebo-controlled trial of plerixafor plus granulocyte colony-stimulating factor compared with placebo plus granulocyte colony-stimulating factor for autologous stem-cell mobilization and transplantation for patients with non-Hodgkin's lymphoma. Journal of clinical oncology : official journal of the American Society of Clinical Oncology. 2009;27[28]:4767-73.

[55] Kumar S, Giralt S, Stadtmauer EA, Harousseau JL, Palumbo A, Bensinger W, et al. Mobilization in myeloma revisited: IMWG consensus perspectives on stem cell collection following initial therapy with thalidomide-, lenalidomide-, or bortezomib-containing regimens. Blood. 2009 Aug 27;114[9]:1729-35. PubMed PMID: 19561323.

[56] Lahuerta JJ, Martinez-Lopez J, Serna JD, Blade J, Grande C, Alegre A, et al. Remission status defined by immunofixation vs. electrophoresis after autologous transplantation has a major impact on the outcome of multiple myeloma patients. British journal of haematology. 2000;109[2]:438-46.

[57] van de Velde HJ, Liu X, Chen G, Cakana A, Deraedt W, Bayssas M. Complete response correlates with long-term survival and progression-free survival in high-dose therapy in multiple myeloma. Haematologica. 2007;92[10]:1399-406.

[58] Barlogie B, Anaissie E, Haessler J, van Rhee F, Pineda-Roman M, Hollmig K, et al. Complete remission sustained 3 years from treatment initiation is a powerful surrogate for extended survival in multiple myeloma. Cancer. 2008;113[2]:355-9.

[59] Pineda-Roman M, Bolejack V, Arzoumanian V, Anaissie E, van Rhee F, Zangari M, et al. Complete response in myeloma extends survival without, but not with history of prior monoclonal gammopathy of undetermined significance or smouldering disease. British journal of haematology. 2007;136[3]:393-9.

[60] Cavo M, Rajkumar SV, Palumbo A, Moreau P, Orlowski R, Blade J, et al. International Myeloma Working Group consensus approach to the treatment of multiple myeloma patients who are candidates for autologous stem cell transplantation. Blood. 2011;117[23]:6063-73.

[61] Barlogie B, Jagannath S, Vesole DH, Naucke S, Cheson B, Mattox S, et al. Superiority of tandem autologous transplantation over standard therapy for previously untreated multiple myeloma. Blood. 1997;89[3]:789-93.

[62] Attal M, Harousseau JL, Facon T, Guilhot F, Doyen C, Fuzibet JG, et al. Single versus double autologous stem-cell transplantation for multiple myeloma. The New England journal of medicine. 2003;349[26]:2495-502.

[63] Cavo M, Tosi P, Zamagni E, Cellini C, Tacchetti P, Patriarca F, et al. Prospective, randomized study of single compared with double autologous stem-cell transplantation for multiple myeloma: Bologna 96 clinical study. Journal of clinical oncology : official journal of the American Society of Clinical Oncology. 2007;25[17]:2434-41.

[64] Goldschmidt H. Single vs double high-dose therapy in multiple myeloma: second analysis of the GMMG-HD2 trial. Haematologica2005.

[65] Segeren CM, Sonneveld P, van der Holt B, Vellenga E, Croockewit AJ, Verhoef GE, et al. Overall and event-free survival are not improved by the use of myeloablative therapy following intensified chemotherapy in previously untreated patients with multiple myeloma: a prospective randomized phase 3 study. Blood. 2003;101[6]: 2144-51.

[66] Kumar A, Kharfan-Dabaja MA, Glasmacher A, Djulbegovic B. Tandem versus single autologous hematopoietic cell transplantation for the treatment of multiple myeloma: a systematic review and meta-analysis. J Natl Cancer Inst. 2009 Jan 21;101[2]: 100-6. PubMed PMID: 19141779. Epub 2009/01/15. eng.

[67] Barlogie B, Attal M, Crowley J, van Rhee F, Szymonifka J, Moreau P, et al. Long-term follow-up of autotransplantation trials for multiple myeloma: update of protocols conducted by the intergroupe francophone du myelome, southwest oncology group, and university of arkansas for medical sciences. Journal of clinical oncology : official journal of the American Society of Clinical Oncology. 2010;28[7]:1209-14.

[68] Palumbo A, Bringhen S, Bruno B, Falcone AP, Liberati AM, Grasso M, et al. Melphalan 200 mg/m[2] versus melphalan 100 mg/m[2] in newly diagnosed myeloma patients: a prospective, multicenter phase 3 study. Blood. 2010;115[10]:1873-9.

[69] Hunter HM, Peggs K, Powles R, Rahemtulla A, Mahendra P, Cavenagh J, et al. Analysis of outcome following allogeneic haemopoietic stem cell transplantation for myeloma using myeloablative conditioning--evidence for a superior outcome using melphalan combined with total body irradiation. British journal of haematology. 2005;128[4]:496-502.

[70] Moreau P, Facon T, Attal M, Hulin C, Michallet M, Maloisel F, et al. Comparison of 200 mg/m[2] melphalan and 8 Gy total body irradiation plus 140 mg/m[2] melphalan as conditioning regimens for peripheral blood stem cell transplantation in patients with newly diagnosed multiple myeloma: final analysis of the Intergroupe Francophone du Myelome 9502 randomized trial. Blood. 2002;99[3]:731-5.

[71] Lahuerta JJ, Grande C, Blade J, Martinez-Lopez J, de la Serna J, Alegre A, et al. Myeloablative treatments for multiple myeloma: update of a comparative study of different regimens used in patients from the Spanish registry for transplantation in multiple myeloma. Leukemia & lymphoma. 2002;43[1]:67-74.

[72] Ria R, Falzetti F, Ballanti S, Minelli O, Di Ianni M, Cimminiello M, et al. Melphalan versus melphalan plus busulphan in conditioning to autologous stem cell transplantation for low-risk multiple myeloma. The hematology journal : the official journal of the European Haematology Association / EHA. 2004;5[2]:118-22.

[73] Blanes M, Lahuerta JJ, Gonzalez JD, Ribas P, Solano C, Alegre A, et al. Intravenous Busulfan and Melphalan as Conditioning Regimen for Autologous Stem Cell Trans-

plantation in Patients with Newly Diagnosed Multiple Myeloma: A Matched Comparison to a Melphalan-only Approach. Biology of blood and marrow transplantation : journal of the American Society for Blood and Marrow Transplantation. 2012.

[74] Carreras E, Rosinol L, Terol MJ, Alegre A, de Arriba F, Garcia-Larana J, et al. Veno-occlusive disease of the liver after high-dose cytoreductive therapy with busulfan and melphalan for autologous blood stem cell transplantation in multiple myeloma patients. Biology of blood and marrow transplantation : journal of the American Society for Blood and Marrow Transplantation. 2007;13[12]:1448-54.

[75] Lahuerta JJ, Mateos MV, Martinez-Lopez J, Grande C, de la Rubia J, Rosinol L, et al. Busulfan 12 mg/kg plus melphalan 140 mg/m2 versus melphalan 200 mg/m2 as conditioning regimens for autologous transplantation in newly diagnosed multiple myeloma patients included in the PETHEMA/GEM2000 study. Haematologica. 2010;95[11]:1913-20.

[76] Roussel M, Moreau P, Huynh A, Mary JY, Danho C, Caillot D, et al. Bortezomib and high-dose melphalan as conditioning regimen before autologous stem cell transplantation in patients with de novo multiple myeloma: a phase 2 study of the Intergroupe Francophone du Myelome [IFM]. Blood. 2010 Jan 7;115[1]:32-7. PubMed PMID: 19884643.

[77] Ludwig H, Durie BG, McCarthy P, Palumbo A, San Miguel J, Barlogie B, et al. IMWG consensus on maintenance therapy in multiple myeloma. Blood. 2012;119[13]: 3003-15.

[78] Attal M, Harousseau JL, Leyvraz S, Doyen C, Hulin C, Benboubker L, et al. Maintenance therapy with thalidomide improves survival in patients with multiple myeloma. Blood. 2006;108[10]:3289-94.

[79] Barlogie B, Pineda-Roman M, van Rhee F, Haessler J, Anaissie E, Hollmig K, et al. Thalidomide arm of Total Therapy 2 improves complete remission duration and survival in myeloma patients with metaphase cytogenetic abnormalities. Blood. 2008;112[8]:3115-21.

[80] Morgan GJ, Gregory WM, Davies FE, Bell SE, Szubert AJ, Brown JM, et al. The role of maintenance thalidomide therapy in multiple myeloma: MRC Myeloma IX results and meta-analysis. Blood. 2012;119[1]:7-15.

[81] Spencer A, Prince HM, Roberts AW, Prosser IW, Bradstock KF, Coyle L, et al. Consolidation therapy with low-dose thalidomide and prednisolone prolongs the survival of multiple myeloma patients undergoing a single autologous stem-cell transplantation procedure. Journal of clinical oncology : official journal of the American Society of Clinical Oncology. 2009;27[11]:1788-93.

[82] Hahn-Ast C. Improved progression-free and overall survival with thalidomide maintenance therapy after autologous transplantion in multiple myeloma: a meta-analysis of five randomized trials. Haematologica. 2011;96:suppl 2, abstract 884.

[83] Barlogie B, Tricot G, Anaissie E, Shaughnessy J, Rasmussen E, van Rhee F, et al. Thalidomide and hematopoietic-cell transplantation for multiple myeloma. The New England journal of medicine. 2006;354[10]:1021-30.

[84] Attal M, Lauwers-Cances V, Marit G, Caillot D, Moreau P, Facon T, et al. Lenalidomide maintenance after stem-cell transplantation for multiple myeloma. The New England journal of medicine. 2012;366[19]:1782-91.

[85] McCarthy PL, Owzar K, Hofmeister CC, Hurd DD, Hassoun H, Richardson PG, et al. Lenalidomide after stem-cell transplantation for multiple myeloma. The New England journal of medicine. 2012;366[19]:1770-81.

[86] Ladetto M, Pagliano G, Ferrero S, Cavallo F, Drandi D, Santo L, et al. Major tumor shrinking and persistent molecular remissions after consolidation with bortezomib, thalidomide, and dexamethasone in patients with autografted myeloma. J Clin Oncol. 2010 Apr 20;28[12]:2077-84. PubMed PMID: 20308672. Epub 2010/03/24. eng.

[87] Cavo M, Pantani L, Petrucci MT, Patriarca F, Zamagni E, Donnarumma D, et al. Bortezomib-thalidomide-dexamethasone is superior to thalidomide-dexamethasone as consolidation therapy after autologous hematopoietic stem cell transplantation in patients with newly diagnosed multiple myeloma. Blood. 2012 Jul 5;120[1]:9-19. PubMed PMID: 22498745.

[88] Gahrton G, Tura S, Flesch M, Gratwohl A, Gravett P, Lucarelli G, et al. Bone marrow transplantation in multiple myeloma: report from the European Cooperative Group for Bone Marrow Transplantation. Blood. 1987;69[4]:1262-4.

[89] Aschan J, Lonnqvist B, Ringden O, Kumlien G, Gahrton G. Graft-versus-myeloma effect. Lancet. 1996;348[9023]:346.

[90] Tricot G, Vesole DH, Jagannath S, Hilton J, Munshi N, Barlogie B. Graft-versus-myeloma effect: proof of principle. Blood. 1996;87[3]:1196-8.

[91] Badros A, Barlogie B, Morris C, Desikan R, Martin SR, Munshi N, et al. High response rate in refractory and poor-risk multiple myeloma after allotransplantation using a nonmyeloablative conditioning regimen and donor lymphocyte infusions. Blood. 2001 May 1;97[9]:2574-9. PubMed PMID: 11313244.

[92] Crawley C, Iacobelli S, Bjorkstrand B, Apperley JF, Niederwieser D, Gahrton G. Reduced-intensity conditioning for myeloma: lower nonrelapse mortality but higher relapse rates compared with myeloablative conditioning. Blood. 2007;109[8]:3588-94.

[93] Garban F, Attal M, Michallet M, Hulin C, Bourhis JH, Yakoub-Agha I, et al. Prospective comparison of autologous stem cell transplantation followed by dose-reduced allograft [IFM99-03 trial] with tandem autologous stem cell transplantation [IFM99-04 trial] in high-risk de novo multiple myeloma. Blood. 2006;107[9]:3474-80.

[94] Moreau P, Garban F, Attal M, Michallet M, Marit G, Hulin C, et al. Long-term follow-up results of IFM99-03 and IFM99-04 trials comparing nonmyeloablative allotrans-

plantation with autologous transplantation in high-risk de novo multiple myeloma. Blood. 2008 Nov 1;112[9]:3914-5. PubMed PMID: 18948589.

[95] Bruno B, Rotta M, Patriarca F, Mordini N, Allione B, Carnevale-Schianca F, et al. A comparison of allografting with autografting for newly diagnosed myeloma. The New England journal of medicine. 2007;356[11]:1110-20.

[96] Rosinol L, Perez-Simon JA, Sureda A, de la Rubia J, de Arriba F, Lahuerta JJ, et al. A prospective PETHEMA study of tandem autologous transplantation versus autograft followed by reduced-intensity conditioning allogeneic transplantation in newly diagnosed multiple myeloma. Blood. 2008;112[9]:3591-3.

[97] Krishnan A, Pasquini MC, Logan B, Stadtmauer EA, Vesole DH, Alyea E, 3rd, et al. Autologous haemopoietic stem-cell transplantation followed by allogeneic or autologous haemopoietic stem-cell transplantation in patients with multiple myeloma [BMT CTN 0102]: a phase 3 biological assignment trial. The lancet oncology. 2011;12[13]:1195-203.

[98] Vesole DH, Zhang L, Flomenberg N, Greipp PR, Lazarus HM, Huff CA, et al. A Phase II trial of autologous stem cell transplantation followed by mini-allogeneic stem cell transplantation for the treatment of multiple myeloma: an analysis of Eastern Cooperative Oncology Group ECOG E4A98 and E1A97. Biology of blood and marrow transplantation : journal of the American Society for Blood and Marrow Transplantation. 2009;15[1]:83-91.

[99] Bjorkstrand B, Iacobelli S, Hegenbart U, Gruber A, Greinix H, Volin L, et al. Tandem autologous/reduced-intensity conditioning allogeneic stem-cell transplantation versus autologous transplantation in myeloma: long-term follow-up. Journal of clinical oncology : official journal of the American Society of Clinical Oncology. 2011;29[22]: 3016-22.

[100] Armeson KE, Hill EG, Costa LJ. Tandem autologous vs autologous plus reduced intensity allogeneic transplantation in the upfront management of multiple myeloma: meta-analysis of trials with biological assignment. Bone marrow transplantation. 2012.

[101] Lokhorst H, Einsele H, Vesole D, Bruno B, San Miguel J, Perez-Simon JA, et al. International Myeloma Working Group consensus statement regarding the current status of allogeneic stem-cell transplantation for multiple myeloma. J Clin Oncol. 2010 Oct 10;28[29]:4521-30. PubMed PMID: 20697091.

[102] Froyland M, Ruffini PA, Thompson KM, Gedde-Dahl T, Fredriksen AB, Bogen B. Targeted idiotype-fusion DNA vaccines for human multiple myeloma: preclinical testing. European journal of haematology. 2011;86[5]:385-95.

[103] Houet L, Veelken H. Active immunotherapy of multiple myeloma. European journal of cancer [Oxford, England : 1990]. 2006;42[11]:1653-60.

[104] Massaia M, Borrione P, Battaglio S, Mariani S, Beggiato E, Napoli P, et al. Idiotype vaccination in human myeloma: generation of tumor-specific immune responses after high-dose chemotherapy. Blood. 1999 Jul 15;94[2]:673-83. PubMed PMID: 10397734.

[105] Osterborg A, Yi Q, Henriksson L, Fagerberg J, Bergenbrant S, Jeddi-Tehrani M, et al. Idiotype immunization combined with granulocyte-macrophage colony-stimulating factor in myeloma patients induced type I, major histocompatibility complex-restricted, CD8- and CD4-specific T-cell responses. Blood. 1998;91[7]:2459-66.

[106] Lim SH, Bailey-Wood R. Idiotypic protein-pulsed dendritic cell vaccination in multiple myeloma. International journal of cancerJournal international du cancer. 1999;83[2]:215-22.

[107] Liso A, Stockerl-Goldstein KE, Auffermann-Gretzinger S, Benike CJ, Reichardt V, van Beckhoven A, et al. Idiotype vaccination using dendritic cells after autologous peripheral blood progenitor cell transplantation for multiple myeloma. Biology of blood and marrow transplantation : journal of the American Society for Blood and Marrow Transplantation. 2000;6[6]:621-7.

[108] Reichardt VL, Okada CY, Liso A, Benike CJ, Stockerl-Goldstein KE, Engleman EG, et al. Idiotype vaccination using dendritic cells after autologous peripheral blood stem cell transplantation for multiple myeloma--a feasibility study. Blood. 1999;93[7]:2411-9.

[109] Titzer S, Christensen O, Manzke O, Tesch H, Wolf J, Emmerich B, et al. Vaccination of multiple myeloma patients with idiotype-pulsed dendritic cells: immunological and clinical aspects. British journal of haematology. 2000;108[4]:805-16.

[110] Dhodapkar MV, Krasovsky J, Olson K. T cells from the tumor microenvironment of patients with progressive myeloma can generate strong, tumor-specific cytolytic responses to autologous, tumor-loaded dendritic cells. Proceedings of the National Academy of Sciences of the United States of America. 2002;99[20]:13009-13.

[111] Rapoport AP, Stadtmauer EA, Aqui N, Badros A, Cotte J, Chrisley L, et al. Restoration of immunity in lymphopenic individuals with cancer by vaccination and adoptive T-cell transfer. Nature medicine. 2005;11[11]:1230-7.

[112] Rapoport AP, Aqui NA, Stadtmauer EA, Vogl DT, Fang HB, Cai L, et al. Combination immunotherapy using adoptive T-cell transfer and tumor antigen vaccination on the basis of hTERT and survivin after ASCT for myeloma. Blood. 2011;117[3]:788-97.

[113] Noonan K, Huff CA, Sproul JMD, Lemas MVM, Rudraraju L, Luznik L, et al. Phase I/II Study of Marrow Infiltrating Lymphocytes [MILs] Generates Measurable Myeloma-Specific Immunity in the Autologous Stem Cell Transplant [SCT] Setting. ASH Annual Meeting Abstracts. 2011 November 18, 2011;118[21]:997-.

[114] Noonan KA, Mgebroff S, Serafini P, Meckel K, Bonyhadi M, Borrello I. CD4+CD25+ Marrow Infiltrating Lymphocytes in Myeloma Patients Display an Activated Pheno-

type and Lack Suppressive Function. ASH Annual Meeting Abstracts. 2006 November 16, 2006;108[11]:1741-.

[115] Lonial S, Vij R, Harousseau JL, Facon T, Moreau P, Mazumder A, et al. Elotuzumab in combination with lenalidomide and low-dose dexamethasone in relapsed or refractory multiple myeloma. Journal of clinical oncology : official journal of the American Society of Clinical Oncology. 2012;30[16]:1953-9.

[116] Chow AW, Lee CH, Hiwase DK, To LB, Horvath N. Relapsed Multiple Myeloma: Who Benefits from Salvage Autografts? Internal Medicine Journal. 2012.

[117] Jimenez-Zepeda VH, Mikhael J, Winter A, Franke N, Masih-Khan E, Trudel S, et al. Second autologous stem cell transplantation as salvage therapy for multiple myeloma: impact on progression-free and overall survival. Biology of blood and marrow transplantation : journal of the American Society for Blood and Marrow Transplantation. 2012;18[5]:773-9.

[118] Olin RL, Vogl DT, Porter DL, Luger SM, Schuster SJ, Tsai DE, et al. Second auto-SCT is safe and effective salvage therapy for relapsed multiple myeloma. Bone marrow transplantation. 2009;43[5]:417-22.

[119] Qazilbash MH, Saliba R, De Lima M, Hosing C, Couriel D, Aleman A, et al. Second autologous or allogeneic transplantation after the failure of first autograft in patients with multiple myeloma. Cancer. 2006;106[5]:1084-9.

[120] Shah N, Ahmed F, Bashir Q, Qureshi S, Dinh Y, Rondon G, et al. Durable remission with salvage second autotransplants in patients with multiple myeloma. Cancer. 2012;118[14]:3549-55.

[121] Gonsalves WI, Gertz MA, Lacy MQ, Dispenzieri A, Hayman SR, Buadi FK, et al. Second auto-SCT for treatment of relapsed multiple myeloma. Bone marrow transplantation. 2012.

[122] Fenk R, Liese V, Neubauer F, Bruns I, Kondakci M, Balleisen S, et al. Predictive factors for successful salvage high-dose therapy in patients with multiple myeloma relapsing after autologous blood stem cell transplantation. Leuk Lymphoma. 2011 Aug;52[8]:1455-62. PubMed PMID: 21657961.

[123] Mehta J, Tricot G, Jagannath S, Ayers D, Singhal S, Siegel D, et al. Salvage autologous or allogeneic transplantation for multiple myeloma refractory to or relapsing after a first-line autograft? Bone marrow transplantation. 1998;21[9]:887-92.

Chapter 9

Proteasome Inhibition and Hematopoietic Stem Cell Transplantation in Multiple Myeloma

Helgi van de Velde and Andrew Cakana

Additional information is available at the end of the chapter

http://dx.doi.org/10.5772/54117

1. Introduction

Multiple myeloma is a malignant plasma cell disorder in which the proliferation of the malignant plasma cells leads to anemia, infections, bone fractures, hypercalcemia and renal dysfunction [1]. Affecting approximately 32,000 people each year worldwide, with a median age of onset of approximately 68 years, it is the second most common hematological malignancy after non-Hodgkin's lymphoma (NHL). Two major advances have occurred in the treatment of multiple myeloma in the last two decades: the introduction of high-dose chemotherapy with autologous stem cell transplantation (ASCT), and the development of active drugs with a novel mechanism of action (proteasome inhibition and immunomodulation). Both advances have led to significant improvements in overall survival in this disease.

The superiority of ASCT over conventional chemotherapy treatment in younger subjects with newly diagnosed multiple myeloma was first established in a French IFM Phase 3 study in the 1990s [2]. The ASCT approach led to higher tumor response rates, better event-free survival (EFS) and overall survival (OS). This superiority of ASCT over conventional treatment was later confirmed in a British phase 3 study [3]. Both EFS and OS appear directly related to the depth of the tumor response to treatment [4]. Due to this correlation achieving complete response (CR) or at least very good partial response (VGPR) has become an important goal of the ASCT approach. ASCT can be complicated by severe myelosuppression and infections and has therefore been reserved for patients who are <65-70 years old without significant comorbidities [5].

ASCT is preceded by an *induction regimen* the primary objective of which is to debulk the tumor without causing damage to the hematopoietic progenitor cells. Until recently, the standard induction regimens in myeloma were vincristine, doxorubicin and dexamethasone (VAD), which has 5-7% CR rate post induction [6], or Thalidomide-Dexamethasone (Thal-Dex), which has a 4% CR rate post induction [7]. After induction therapy autologous CD34+ hematopoietic stem cells are harvested from peripheral blood, and less often collected as part of bone marrow cells, and reinfused after conditioning with high-dose chemotherapy regimen with high-dose melphalan (*conditioning regimen*). The conditioning and reinfusion of CD34+ stem cells can be done once or twice (single or double ASCT). The addition of a limited number of cycles of standard dose chemotherapy (*consolidation treatment*) or of a prolonged exposure to low dose therapy (*maintenance treatment*) after the ASCT is increasingly being used to further improve EFS and OS, although their value has not been fully established [5]

If the ASCT approach is not used in the treatment of a newly diagnosed patient with multiple myeloma, it can still be applied upon relapse. A randomized study by the French GMA group indicated that a second-line rescue with high dose therapy and ASCT resulted in similar overall survival as compared to initial treatment with this approach [8].

The value of allogeneic bone marrow transplantation (Allo-SCT) in multiple myeloma is controversial [9]. The high treatment related mortality associated with myeloablative conditioning in allo-SCT has led to the development of reduced-intensity conditioning (Allo-RIC). Convincing evidence is sofar lacking that Allo-RIC can improve the survival compared with autologous stem-cell transplantation. For this reason, allo-RIC in myeloma is currently only recommended in the context of clinical trials.

It is still unclear whether any of these treatment approaches can be curative, even in a subset of patients, although they have extended the median overall survival of patients with newly diagnosed multiple myeloma beyond 5-6 years [10]. Improvement of outcomes by incorporation of the proteasome inhibitor bortezomib into autologous stem cell transplantation approaches has been an area of intense clinical research over the last decade and is the topic of this review.

Bortezomib is a first-in-class proteasome inhibitor which was originally approved for the treatment of relapsed or refractory multiple myeloma by the US Food and Drug Administration (FDA) in 2003 and by the European Agency for the Evaluation of Medicinal Products (EMEA) in 2004 [11,12]. Bortezomib is a reversible inhibitor of the 26S proteasome which is a large protein complex that degrades ubiquitinated proteins. Regulatory proteins relevant to the initiation and progression of cancers including multiple myeloma are known to be degraded during the cell cycle by the ubiquitin-proteasome pathway [13]. Binding of bortezomib to the 20S β5 subunit of the proteasome results in a reversible inhibition of the chymotrypsin-like protease in the proteasome. In multiple myeloma cells, this results in inhibition of NF-kB activation, in attenuation of interleukin-6-mediated cell growth, and direct apoptotic and anti-angiogenic effects [14,15].

In relapsed multiple myeloma, single agent bortezomib was shown to improve time to progression, response rate and overall survival as compared to high-dose dexamethasone [16]. Median survival of patients treated with bortezomib was 29.9 months as compared to 23.7 months in the dexamethasone control group ($p=0.027$) [17]. In patients with newly diagnosed multiple myeloma who were not candidate for ASCT, the addition of bortezomib to

http://dx.doi.org/10.5772/54117

standard chemotherapy with melphalan-prednisone also resulted in improvement in time to progression, response rate and complete rate, and overall survival. Complete response rate of patients treated with the bortezomib-melphalan-prednisone combination was 30% as compared to 4% in the melphalan-prednisone control group ($p<0.001$) [18]. Median survival of patients treated with the bortezomib-melphalan-prednisone combination was 56.4 months as compared to 43.1 months in the melphalan-prednisone control group ($p=0.0004$) [19]. Based on these significant improvements in outcomes in other settings in multiple myeloma, the introduction of bortezomib in autologous stem cell transplant approaches in myeloma has been an area of intense clinical study activity.

The approved single agent bortezomib dose and schedule in multiple myeloma is 1.3 mg/m^2, on days 1, 4, 8, and 11, followed by a 10-day rest period (21 day cycle). The most clinically significant side-effect is a cumulative dose-related peripheral neuropathy which is managed by treatment interruptions and dose modifications [20]. Other common adverse events include lower grade gastro-intestinal adverse events and thrombocytopenia [16]. In most clinical studies, including those reviewed in this chapter, bortezomib has been given intravenously. Recent data in relapsed multiple myeloma have indicated that the subcutaneous administration of bortezomib is as efficacious and results in less neurotoxicity [21]. However, data on subcutaneous administration of bortezomib as part of transplant regimens in myeloma are currently still lacking.

2. Methodology

We followed published guidelines for medical literature reviews [22,23,24]. The medical literature was searched in the OVID database (Medline; Derwent Drug File; Your journals @ Ovid; Biosis Previews; and Embase). The search was limited to the English language and articles published without a data range limit to August 1, 2012. The following search strategy was used with the following words being entered in the basic search section of OVID:

1. randomized AND phase AND 3 AND bortezomib OR velcade AND stem AND cell AND transplantation.

2. the second phase of searching showed the addition of 'myeloma AND proteosome inhibitor' with 'autologous'

3. the third phase of searching substituted 'bone AND marrow' for the words 'stem AND cell

Further selection of the identified studies included in this review was based on following criteria: (1) prospective study design; (2) publication in peer-reviewed journals; (3) randomized phase 3 or phase 2 design, or single arm phase 2 design with sample size >25 patients. These criteria were chosen to increase likelihood of scientific quality and interpretability of the selected studies. Twenty-nine studies were initially retained in the search and 16 studies were subsequently selected based on these criteria.

VELCADE (Bortezomib for Injection) is a small molecule proteasome inhibitor being codeveloped by Millennium Pharmaceuticals, Inc. (Millennium) and Janssen Research & Development. The selection of the identified was solely based on the criteria indicated above and no studies were de-selected due a conflict-of-interest.

3. Results

3.1. Bortezomib in induction therapy

The search of an optimal induction regimen prior to high-dose therapy and ASCT in multiple myeloma is still ongoing. An ideal induction regimen should, among others, have the following characteristics:

1. Able to give the optimum post induction tumor response, since better response is associated with better long-term outcomes;
2. Able to act quickly to debulk the tumor, as often the patients present with advanced disease and complicated presentations;
3. Able to work even in renal failure, since this is a common feature of multiple myeloma;
4. Allowing collection of an adequate of viable hematopoietic stem cells necessary for successful bone marrow rescue and engraftment.

All randomized studies published in peer reviewed journals and investigating the role of bortezomib in induction therapy have been analyzed in this section. Some of the randomized phase 2 and 3 studies compare a bortezomib containing regimen with a non-bortezomib containing regimen, while other studies investigate various combinations and doses with all treatment groups containing bortezomib. Further, all single arm phase 2 studies with more than 25 patients are also included in this review.

Early studies started soon after the introduction of bortezomib into the treatment of relapsed myeloma compared the role of a bortezomib-containing induction regimen against VAD, the standard regimen at that time.

A first phase 3 study by the French IFM group provided evidence that the combination of bortezomib plus dexamethasone was superior to VAD as induction regimen [25]. In this IFM2005-01 study 481 patients who were eligible for autologous stem cell transplantation (≤65 years) were randomized to receive VAD (n = 121); VAD plus DCEP (n=121); bortezomib plus dexamethasone (n = 121) or bortezomib plus dexamethasone plus DCEP (n = 119). DCEP (dexamethasone, cyclophosphamide, etoposide and cisplatin) were given as a consolidation course, soon after 4 induction cycles and before the high dose melphalan for conditioning. The study allowed a second high-dose therapy and stem cell transplant procedure for patients failing to attain at least a VGPR after first transplant. The primary endpoint was post induction CR/nCR rate. Patients in the VAD group were treated with four 4-week cycles of vincristine 0.4 mg/d and doxorubicin 9 $mg/m^2/d$ by continuous infusion on days 1 to

4, and dexamethasone 40 mg daily po on days 1 to 4 (all cycles) and days 9 to 12 and days 17 to 20 (cycles 1 and 2 only). Bortezomib plus dexamethasone comprised four 3 week cycles of bortezomib 1.3 mg/m^2 iv days 1, 4, 8 and 11 plus dexamethasone po 40 mg/d on days 1 to 4 (all cycles) and days 9 to 12 (cycles 1 and 2 only). DCEP comprised two 4-week cycles of dexamethasone 40 mg/daily on days 1 to 4; plus cyclophosphamide 400 mg/m^2, etoposide 40 mg/m^2 and cisplatin 15 mg/m^2 by continuous iv infusion day 1 to 4. Stem cells were collected after priming with granulocyte stimulating factor alone or cyclophosphamide for those who mobilize poorly. The CR/nCR rate was significantly higher (14.8%) for patients receiving bortezomib plus dexamethasone compared to patients receiving VAD (6.4%, p-value=0.004). ORR was 78.5% vs 62.8% (p=<0.001). Patients with del13, a negative prognostic cytogenetic abnormality in multiple myeloma, also reported higher response rates in the bortezomib containing arm : ORR was 78.2% vs 65.1% (p=0.037); and CR/nCR was 20.8% compared to 5.8%(p=0.002). The study showed that the addition of DCEP did not further improve the outcomes wither either regimen. The PFS for the bortezomib group was 36 months vs 29.7 months in the VAD arm after 32.2 months follow-up (p=0.064 unadjusted). Median OS was not yet reached but the 3 year OS rate was 81.4% compared to 77.4% in favor of the bortezomib-dexamethasone combination. Stem cell collection was adequate in both arms. The safety profile was similar between the groups for most adverse events except for all grade peripheral neuropathy (45.6% for bortezomib and dexamethasone and 28% for VAD) and grade 3/4 neuropathy (7.1% and 2.1%, respectively).

A second phase 3 study by the Dutch-German HOVON-GMMG groups randomized 827 patients to receive 3 cycles of either bortezomib combined with adriamycin and dexamethasone (PAD) or VAD during induction given every 28 days [26,26,27]. This HOVON-65/GMMG-HD4 study had a maintenance part post ASCT in which patients on VAD further received thalidomide 50 mg po daily for a further 2 years while those on PAD received bortezomib 1.3 mg/m^2 iv every two weeks for 2 years. The primary objective of the study was to compare PFS of the two arms. Response rates post induction were analyzed as secondary objectives. The CR/nCR rate post induction was 5% in patients who were randomized to VAD and 11% in patients who received PAD (*P* <.001). The post transplant response rate for nCR/CR was 15% (VAD) versus 31% (PAD), (*P* <.001). Overall nCR/CR rates were 34% versus 49%, (*P* <.001) for patients on VAD and PAD respectively. The median PFS was 28 months for the VAD arm and 35 months for the PAD arm (p=0.002). Median OS was not reached after 66 months of follow-up, with 5-year OS of 55% (VAD) versus 61% (PAD). In patients with del17p, the worst prognostic cytogenetic abnormality in multiple myeloma, both PFS (median PFS, 12 vs 22 months, p=0.01) and OS (median OS, 24 vs > 54 months, p=0.003) were significantly better in the PAD arm. In patients with del13, a negative impact on PFS was observed in both treatment arms. OS in patients with this deletion was similar to the OS in patients with no del13 in the PAD arm and significantly better than OS in the VAD arm (median OS for VAD 49 vs 59 months for the PAD arm, p=0.007). Stem cell collection was adequate in both treatment arms. In patients presenting with a baseline serum creatinine of more than 2 mg/dL, bortezomib significantly improved CR/nCR rates which were 27% (VAD) compared to 53% (PAD) (p=0.02). The PFS in the same population improved from a median of 13 months to 30 months (p= 0.004) and OS from a median of 21 months to

54 months (HR, 0.33; $p < 0.001$) respectively. There was more neuropathy in the PAD arm (40% grades 2 to 4) compared to the VAD arm (18%, $p<0.001$). The contribution of the maintenance regimens in this study is discussed later in this chapter.

The above two large studies showed significant improvement of bortezomib-containing regimens as compared to VAD in terms of post-induction response and PFS, with a positive trend on overall survival. Later studies focused on comparing bortezomib-containing regimens against non-bortezomib containing regimens other than VAD.

The bortezomib-thalidomide-dexamethasone (VTD) regimen was compared to the thalidomide-dexamethasone (TD) regimen in a Phase 3 study randomizing 480 patients over four 21-day cycles [28,29]. The patients received thalidomide 100 mg po daily for the first 14 days and 200 mg daily thereafter, plus dexamethasone (40 mg po daily on 8 of the first 12 days, but not consecutively; total of 320 mg per cycle), either alone or with bortezomib (1 3 mg/m^2 iv on days 1, 4, 8, and 11). Post double ASCT the patients received two 35-day cycles of their assigned drug regimen, VTD or TD, as consolidation therapy (see below). The primary endpoint was the CR/nCR rate to induction therapy. After induction therapy, complete or near complete response was achieved in 31% patients receiving VTD compared to 11% for those on TD ($p<0$ 0001). Rates of complete or near complete response continued to be significantly higher in the VTD group than in the TD group after the first and second autologous stem-cell transplantations (55% vs 41%, $p=0.0024$). Median time to best complete or near complete response was significantly shorter for patients receiving VTD (9 months) than in those on TD (14 months). The contribution of the consolidation therapy is discussed below. The estimated 3-year PFS was 60% in the VTD arm compared to 48% in the TD arm. Overall, PFS was significantly longer with VTD compared to TD (median not reached vs 32 months, $p=0.0061$). The estimated 3-year probability of progression or relapse was 29% in the VTD group versus 39% in the TD group ($p=0$ 0061 by Kaplan-Meier analysis with an HR of 0 61. In the VTD group, the PFS of subjects with or without high-risk cytogenetic abnormalities [del13q, or del17p or t(4;14)] were similar (59% with abnormalities and 60% without). This contrasted with the TD group in which a much lower PFS of 19% for patients with high-risk cytogenetics was observed as compared to the 48% attained by patients without high-risk cytogenetics in the same TD group. Stem cell collection was adequate in both arms. Grade 3 or 4 adverse events were more frequent on VTD (56%) than on TD (33%), with a higher occurrence of grade 3 or higher peripheral neuropathy in patients on VTD (10%) than on TD (2%).

A further phase 3 study (GEM05-MENOS65) performed by the Spanish GIMEMA group randomized 390 patients in a three arm study to receive VTD versus TD versus a regimen called VBMCP/VBAD with bortezomib [30]. Combination chemotherapy with VBMCP/VBAD and bortezomib consisted of a total of 4 cycles of alternating VBMCP (vincristine, BCNU, melphalan, cyclophosphamide, prednisone) and VBAD (vincristine, BCNU, doxorubicine, dexamethasone) followed by 2 cycles of bortezomib (1.3 mg/m^2 iv on days 1, 4, 8 and 11 at 3 weeks intervals), TD consisted of thalidomide 200 mg po daily (escalating doses in the first cycle: 50 mg on days 1 to 14 and 100 mg on days 15 to 28) and dexamethasone 40 mg po on days 1-4 and 9-12 at 4-week intervals for 6 cycles and the VTD arm was identical

http://dx.doi.org/10.5772/54117

to TD plus bortezomib 1.3 mg/m^2 iv on days 1, 4, 8 and 11 of each cycle. The duration of the induction therapy was 24 weeks in all arms. Three months after ASCT patients were randomized to receive maintenance therapy with interferon alfa-2b subcutaneously versus thalidomide 100 mg po daily versus thalidomide 100 mg/day po daily plus one cycle of bortezomib iv on days 1, 4, 8 and 11 every three months (see below). The CR rate after induction was significantly higher with VTD (35%) compared to TD (14%) and VBMCP/VBAD/B (21%) (p=0.0001 and p=0.01, respectively). Of significance in the VBMCP/VBAD/Bortezomib arm, the CR rate increased from 8% after the 4 cycles of VBMCP/VBAD to 21% after the completion of the 2 bortezomib courses. The progressive disease (PD) rate during induction was significantly lower with VTD than with TD (7% vs. 23%, p= 0.0004). In patients with extramedullary soft-tissue plasmacytomas the CR rate after induction was significantly higher with VTD as compared with TD (42% vs. 14%, p=0.02). In all the above analysis the VBMCP/VBAD/Bortezomib arm showed an intermediate efficacy between VTD and TD. In this study VTD also had superior CR rates in the subgroup of patients with high-risk cytogenetic abnormalities as compared to the two other regimens. After a median duration of follow-up of 35.2 months, the median PFS was significantly higher for VTD (56.2m) than with VBMCP/VBAD/Bortezomib (35.3m) or with TD (28.2 m, p=0.01). The difference in the four-year survival rates between VTD (74%), VPMCP/VVBAD/bortezomib (70%) and TD (65%) is not statistically significant at this point. There were two stem cell mobilization failures in the VBMCP/VBAD/Bortezomib group. Peripheral neuropathy grade ≥ 3 with VTD (14%) was significantly higher than with TD (5%) (p=0.01) but not significantly different from VBMCP/VBAD/B (9%). An additional 46% of patients in the VTD arm developed grade 2 peripheral neuropathy compared with 8% and 15% in the TD and VBMCP/VBAD/B arms, respectively (p<0.001). Grade 3 and 4 neutropenia was significantly higher with VBMCP/VBAD/B (22%) than with remaining two arms TD (14%) and VTD (10%). There were no significant difference in incidence of all grade ≥ 3 adverse events between the three treatment groups.

In conclusion, all currently published phase 3 studies indicate that induction regimens containing bortezomib lead to improvements in CR/nCR rates after induction which are maintained after ASCT, and also lead to improved PFS as compared to standard regimens. Where reported, the time to response appear shorter, and the regimens have important activity in poor prognosis situations such as high-risk cytogenetic disease and renal insufficiency. After a relatively short duration of follow-up, a trend towards improved overall survival with the bortezomib regimens has been noted in several studies. All phase 3 studies also provide evidence of good hematopoietic stem cell collection but indicate a higher incidence of neuropathy in patients treated with a bortezomib combination. This phase 3 evidence is further supported by a plethora of randomized and non-randomized phase 2 studies which have incorporated bortezomib in the induction regimens.

In a randomized phase 2 study by the French IFM (IFM2007-02) a lower dose of bortezomib was investigated in combination with thalidomide and dexamethasone in order to reduce the peripheral neuropathy risk. One hundred ninety-nine patients were randomized to receive VD or vTD over four 3 week cycles prior to ASCT [31]. vtD was composed of reduced bortezomib at 1 mg/m^2 iv on days 1, 4, 8, and 11, thalidomide 100 mg/day po, and dexamethasone while VD consisted of bortezomib 1.3 mg/m^2 iv on days 1, 4, 8, and 11 plus the same dexamethasone regimen. In case of less than partial response (PR) after cycle 2, the dose of bortezomib was increased to 1.3 mg/m^2 and the dose of thalidomide to 200 mg/day in the vTD arm. The primary endpoint of this study was post induction CR rate. The CR rate between the groups was the same after 4 cycles, 13% in the vTD arm and 12% in the VD arm. However, both the bortezomib and thalidomide dose had to be increased in 7 patients in the vTD arm. The ORR was 88% in the vTD arm versus 81% in the VD arm, the difference not reaching statistical significance. Further, there was no difference in CR rate post transplant (29% in vTD arm and 31% in VD arm). The target stem cell collection yield of 2 × 10^6 CD34$^+$ cells/kg was achieved in 93% and 80% of VD and vTD patients, respectively (P =.01). While the overall safety profile was similar between the two arms, there was less peripheral neuropathy in the vTD arm (53% all grade vs 70% on VD, 11% grade 3 or higher vs 11% on VD). Results of the VD control group were consistent with prior observations from the IFM group on this VD induction regimen, both in a single arm phase 2 study [32] and in the randomized phase 3 study (see above).

Efficacy of VD in induction was also assessed in another phase 2 study with 57 patients, given over 4 cycles followed by 2 cycles of DCEP consolidation [33]. The median CR34+ cells collected were 7.5 x 10^6/kg and in 86% of these patients the amount was more than twice the minimum required for transplantation. The ORR was 87% and CR 30%. Univariate analysis found no difference between response and cytogenetic abnormalities.

Efficacy of VTD was further confirmed in a single arm phase 2 study of 44 patients treated with bortezomib combined with thalidomide and dexamethasone, administered over eight 3-week cycles [34]. The patient enrolment included both frontline and recurrent disease as long as the patients were eligible for ASCT. Thirty four patients were frontline, 8 with recurrent disease in second line and a further 2 had a third line recurrence. The ORR was 91% with CR/sCR rate of 20%. Post transplant these response rates increased to ORR of 100% and CR/sCR rate of 53%. All 44 patients had successful stem cell collection. Fifty-five percent of the subjects developed neuropathy of all grades, though grade 3 neuropathy was reported in 9%. DVT occurred in 5% of the patients.

Other multidrug combination induction regimens including bortezomib were also investigated in phase 2:

- In a randomized phase 2 study, 140 patients were initially randomized to VDCR, VDR, or VDC to receive eight 3-week cycles of induction therapy followed by four 6-week cycles of bortezomib maintenance therapy [35]. The VDC arm was modified after an interim analysis to add a third dose of cyclophosphamide at 500 mg/m^2 on day 15 (VDC-mod). Bortezomib was given in standard doses. Patients could undergo stem cell mobilization any time after 2 cycles and undergo ASCT any time after 4 cycles. After 4 cycles of induction therapy, the confirmed ORR was 80%, 73%, 63%, and 82% of patients in the VDCR, VDR, VDC, and VDC-mod arms including VGPR or better in 33%, 32%, 13%, and 41%, respectively. After ASCT, the ORR was 88%, 85%, 75%, and 100% for the VDCR, VDR, VDC, and VDC-mod arms including VGPR or better in 58%, 51%, 41%, and 53%, respectively. The 1-year PFS was 100%, 100%, 88%, and 100% for the VDCR, VDR, VDC, and

VDC-mod arms, respectively. The 1-year OS estimate was 100% for all 4 arms. In addition the 1-year PFS for the high-risk patients (n = 24) was 100% and 85% for the standard-risk patients, and was similar across the study arms. The median CD34+ cell yield was 6.8 x 10^6/kg (VDCR); 7.8 (VDR); 7.95 (VDC) and 7.75 (VDC modified). At least one grade ≥ 3 AE was seen in ~ 80% of patients in each arm. AEs leading to discontinuation were seen in 21%, 19%, 12%, and 6% in the VDCR, VDR, VDC, and VDC-mod arms, respectively. The most common adverse event of grade 3 or higher was neutropenia occurring in 44% (VDCR), 10% (VDR), 30% (VDC) and 24% (VDC modified). Neuropathy grade 3 or higher occurred in 13%, 17%, 9% and 18% respectively.

- In two single arm phase 2 studies, bortezomib was combined with cyclophosphamide and dexamethasone (CyBorD) [36,37]. In one study, 33 patients were treated with four 3 weekly cycles with cyclophosphamide 300 mg/m² given orally and once weekly, while bortezomib and dexamethasone were given in standard doses. ORR was 88%, and 39% were CR/nCR. Responses were rapid with a mean 80% decline in the monoclonal protein at the end of two cycles. All patients undergoing stem cell harvest had a successful collection. The most common grade 3-4 adverse events were hematological (anemia in 12%, neutropenia in 13%, thrombocytopenia in 25%) and hyperglycemia (13%). All grade peripheral neuropathy adverse events occurred in 66% of the patients while grade 3 occurred in 7%. In the second study, 30 patients were treated with different IV cyclophosphamide dose levels in combination with bortezomib and dexamethasone for 3 cycles [37]. The recommended dose of IV cyclophosphamide was 900 mg/m² on day 1. The CR rate after induction therapy was 10% and the overall response rate was 90% at the end of the induction therapy. Most frequent adverse events were again hematologic and neuropathy as well as gastro-intestinal.

- The most intense bortezomib-containing induction regimen of VTD-PACE is included in a high-dose therapy approach called Total Therapy 3 and has been investigated in a large cohort of 303 patients [38]. The regimen consists of two cycles of VTD-PACE (bortezomib, thalidomide, dexamethasone and 4-d continuous infusions of cis-platin, doxorubicin, cyclophosphamide, etoposide) during induction and then another two cycles during consolidation after the ASCT. The patients are then maintained for 3 years on monthly cycles of VTD in the first and TD in the remaining years. The response rates of this Total Therapy 3 approach are among the highest reported in multiple myeloma. The 2 year CR rate was 56% whilst the nCR rate was as high as 83%. Two year OS estimates were also high at 82.9% and EFS of 79.9%. Although no randomized comparison was performed, the investigators consider those results better than a similar approach (Total therapy 2) which did not include bortezomib. Stem cell collection was successful. Adverse events grade 3 or higher included thrombo-embolic events in 27% and peripheral neuropathy in 12% of the patients.

In conclusion, these studies have provided evidence of the important role of bortezomib in induction therapy pre-ASCT. Randomized phase 3 studies indicate that induction regimens containing bortezomib lead to improvements in CR/nCR rates after induction which are maintained after ASCT, and also lead to improved PFS as compared to standard non-bortezomib containing regimens. After a relatively short duration of follow-up, a trend towards improved overall survival with the bortezomib regimens has been noted. Particularly in patients with high-risk cytogenetic abnormalities, such as del17p and del13, the addition of bortezomib to induction therapy has improved outcomes. All phase 3 studies also provide evidence of good hematopoietic stem cell collection. While bortezomib can safely be combined with several induction regimens, a higher incidence of neuropathy in patients treated with a bortezomib combination is generally noted. Other toxicities of the induction regimens appear related to the combination partner (such as neutropenia for cyclophosphamide, thrombo-embolic events for thalidomide, hyperglycemia for high-dose dexamethasone) and the optimal combination regimen, as well as the optimal number of induction cycles has not been identified yet. One phase 2 study provided evidence of a lower incidence of neuropathy with a lower dose of bortezomib.

3.2. Bortezomib during conditioning

The high-dose chemotherapy regimen which immediately precedes the autologous stem cell transplantation is referred to as the 'conditioning regimen'. Melphalan is the most frequently used conditioning agent in multiple myeloma and is given at the high-dose of 200 mg/m^2 or at a reduced dose in case of renal function impairment [39].

Two single arm studies have investigated the addition of bortezomib to the high-dose melphalan conditioning regimen. The rationale to combine the two agents in this setting was based on (1) the synergy between bortezomib and melphalan reported both in vitro and in vivo [14,40], as well as on (2) the lack of overlapping toxicities between the two agents (mainly neurologic for bortezomib and hematologic for melphalan).

In a dose and schedule-finding phase ½ study, 39 patients with newly diagnosed multiple myeloma who achieved less than VGPR following induction therapy were randomized to receive a single escalating dose of bortezomib (1 mg, 1.3 mg or 1.6 mg/m^2) either 24 hours before or 24 hours after melphalan (given 100 mg/m^2/d for 2 days) [41]. Stem cells were reinfused 2 days after the last melphalan dose. Median time to neutrophil recovery and platelet recovery was 12 days and 16 days, respectively, for both schedules. Transplant-related toxicities (gastro-intestinal and mucositis) were also similar for the two schedules. No peripheral neuropathy was reported. In the treatment group receiving bortezomib prior to melphalan (n=19) 47% had at least VGPR and 11% had CR post-transplant, while in the treatment group receiving bortezomib after melphalan (n=20) 55% had at least VGPR and 30% had CR. The investigators that the combination was safe with data suggesting improved efficacy and recommend the administration of bortezomib after high-dose melphalan as the preferred schedule.

In a phase 2 study conducted by the French IFM group, 54 patients with newly diagnosed multiple myeloma received melphalan 200 mg/m^2 in combination with four administrations of bortezomib at a dose 1 mg/m^2 (1 and 4 days prior to melphalan, and 3 and 6 days after melphalan) [42]. The autologous peripheral blood stem cells were reinfused 2 days after melphalan administration. While 4% of patients had CR and 28% had PR at the end of the induction therapy, 32% had CR and 68% had at least VGPR 3 months after this conditioning regimen. The median time to neutrophil and platelet recovery was 7 days and 3 days after stem cell reinfusion respectively. No engraftment failure or treatment-related death was reported. Three patients developed de novo neuropathy, while the severity of pre-existing neuropathy was not affected. In a matched control analysis, only 11% of CR post-conditioning were reported.

In a randomized phase 2 study, 60 patients not in CR after induction therapy were randomized to receive an unconventional conditioning regimen with melphalan 200 mg/m^2 in combination with arsenic trioxide and ascorbic acid either without (group 1) or with bortezomib at either $1mg/m^2$ (group 2) or 1.3 mg/m^2 for 3 doses (group 3). Fifty-eight patients were randomized between the 3 treatment groups. Addition of bortezomib to this regimen was found safe with no apparent increase in time to neutrophil or platelet engraftment, in grade ¾ non-hematologic toxicity or in treatment-related mortality. However, there was no significant improvement in the CR rate, PFS and OS rates in the bortezomib groups. The reason for this lack of improvement was interpreted by the authors as related to the high proportion of patients with relapsed disease (25%) and by the concomitant administration of ascorbic acid [43].

In conclusion, these studies have provided evidence that the addition of bortezomib to the conditioning regimen is feasible with no negative impact on hematopoietic recovery or treatment-related mortality after ASCT. From the two phase 2 studies adding bortezomib to high-dose melphalan, high CR rates post-ASCT were noted which appeared superior to historic data. A small randomized phase 2 study was not able to confirm improved efficacy outcomes when bortezomib was added to a multi-drug conditioning evidence.

3.3. Bortezomib in consolidation treatment

At the moment of our search, the results of only one randomized study incorporating bortezomib in consolidation therapy has been published in a peer-reviewed paper.

In the GIMEMA phase 3 study investigating bortezomib-thalidomide-dexamethasone (VTD) vs thalidomide-dexamethasone (TD), the combination regimens were given both in induction therapy pre-ASCT and in consolidation therapy post-ASCT[29]. Patients initially randomized to VTD received 2 post-ASCT consolidation cycles of bortezomib 1.3 mg/m^2 iv on d1,8,15,22 every 5 weeks in combination with thalidomide 100 mg/d po and dexamethasone, patients initially randomized to TD received 2 post-ASCT consolidation cycles without bortezomib. Of the 236 patients initially randomized to VTD induction, 160 patients (68%) continued with VTD consolidation, while of the 238 patients initially randomized to TD induction, 161 patients (68%) continued with TD consolidation. VTD consolidation significantly improved the CR and CR/nCR rates post-ASCT, while the TD consolidation did not. After a median follow-up of 30.4 months from start of consolidation, 3-year PFS was significantly longer for the VTD group (60% vs 48%, p=0.042) but sofar no difference in overall survival from this landmark has been seen (3-year survival rates 90% for VTD and 88% for TD). Grade 2 or 3 peripheral neuropathy (8.1% vs 2.4%) was more frequent with VTD versus TD consolidation. The authors conclude that VTD consolidation therapy significantly contributed to the improved clinical outcomes observed for patients randomly assigned to the VTD arm of the study.

3.4. Bortezomib in maintenance treatment

There were no studies identified which in a randomized fashion have investigated the role of single agent bortezomib as prolonged maintenance therapy post-ASCT. However, a lot of information on single agent bortezomib maintenance therapy can be derived from the HOVON-65/GMMG-HD4 study, the largest phase 3 study ever conducted in ASCT in newly diagnosed multiple myeloma. In addition, preliminary data are available from a randomized phase 3 study investigating the bortezomib-thalidomide combination in maintenance therapy (GEM05-MENOS65) and from a phase 2 study investigating a bortezomib-thalidomide-dexamethasone combination in maintenance therapy post-ASCT [44].

In the HOVON-65/GMMG-HD4 study, as discussed above, patients randomized to the bortezomib-doxorubicin-dexamethasone (PAD) induction treatment group continued bortezomib maintenance 1.3 mg/m^2 iv every 2 weeks for 2 years post-ASCT, whereas the control treatment group of vincristine-doxorubicin-dexamethasone (VAD) induction continued to be treated with thalidomide 50 mg/d po for the same treatment duration [26]. In this study, 833 patients were randomized between the PAD and the VAD induction regimens. After ASCT, 229 patients from the PAD treatment group (55%) continued with bortezomib maintenance, while in the VAD treatment group 270 patients (65%) continued with thalidomide maintenance. Of the 229 patients starting bortezomib maintenance, 109 (48%) completed the 2-year maintenance, while 26 (11%) discontinued because of toxicity and 74 (32%) discontinued earlier because of progression. Of the 270 patients starting thalidomide maintenance, only 73 (27%) completed the 2-year maintenance, while more patients (82 or 30%) discontinued because of toxicity and a similar percentage (86 or 32%) discontinued because of progression. Because of the sequential study design a direct comparison between the two maintenance regimens should be interpreted with caution. However, the main study publication indicates a statistically significantly higher incidence of serious adverse events (34% vs 23%, $p<0.01$) during bortezomib maintenance, mainly related to infection, while on the other hand more peripheral neuropathy was reported during thalidomide maintenance (5% vs 8%, $p<0.001$). The sequential design also limits the interpretation of the efficacy data of the maintenance regimens. Although in the bortezomib maintenance all patients had already been exposed to bortezomib during induction therapy, a similar percentage of patients (23%) had an upgrade of their tumor response post-ASCT as compared to the thalidomide maintenance which introduced a new agent (24%). An analysis of progression-free survival calculated from the last ASCT indicates that bortezomib contributed more to improvement of progression-free survival than thalidomide (31 months vs 26 months). Also, a landmark analysis starting at month 12 shows an improvement in progression-free survival ($p=0.04$) and overall survival ($p=0.05$) in the bortezomib-containing arm.

In phase 3 study GEM05-MENOS65 performed by the Spanish PETHEMA/GEM group, patients initially randomized to answer the induction regimen question of bortezomib-thal-dex vs thal-dex vs VBMCP/VBAD/bortezomib (discussed above) were rerandomized after ASCT between different maintenance regimens: interferon alfa-2b versus thalidomide 100 mg/d vs thalidomide 100 mg/d plus bortezomib (1.3 mg/m^2 d1,4,8,11 q3 month) until progression and for a maximum of 3 years [30]. Three-hundred ninety patients were initially randomized between the three induction arms; the initial study publication does not report how many patients were rerandomized between the 3 maintenance arms nor does it address the toxicities observed. However, the publication states that after a median follow-up of 24 months from initiation of maintenance, the PFS is significantly longer with thalidomide/ bortezomib compared with thalidomide alone and with alfa2-interferon (78% vs 63% vs 49% at 2 years, p=0.01). However, at this early analysis, the overall survival is not significantly different between the 3 maintenance groups.

In a small phase 2 study of 40 patients post-ASCT, a sequential maintenance therapy including bortezomib was investigated [44]. In this study, 6 4-week cycles of weekly bortezomib at a dose of 1.3 mg/m^2 was given in combination with dexamethasone, followed by 6 cycles of thalidomide and dexamethasone and then followed by thalidomide single agent until progression. Of the 40 patients, 32 (80%) completed the bortezomib therapy and in 9 patients the bortezomib-dexamethasone combination upgraded the response from less than CR to CR. The combination regimen was feasible, with peripheral neuropathy grade 1-2 being reported in 27 patients. The authors concluded that this bortezomib maintenance regimen was able to upgrade post-ASCT CR responses with no severe grade ¾ peripheral neuropathy.

In conclusion, currently available data suggest that maintenance therapy with bortezomib, either as single agent or in combination with thalidomide, improves the PFS over thalidomide alone. Prolonged maintenance therapy with bortezomib at lower dose intensity than in the induction setting (either one dose weekly or every 2 weeks, or four doses every 3 months) appears feasible and is able to further improve the CR rate post-ASCT. More follow-up is needed on the impact of these bortezomib maintenance regimens on overall survival.

3.5. Bortezomib during or after ASCT procedure for relapsed myeloma

There were no studies identified which specifically looked into the use of bortezomib as part of an ASCT procedure for relapsed multiple myeloma.

However, one large randomized phase 3 study by the EBMT group (European Group for Blood and Marrow Transplantation) investigated the use of a bortezomib containing regimen to rescue patients with multiple myeloma progressing or relapsing after ASCT [45]. In this study, 269 patients were randomly assigned to receive bortezomib or no bortezomib for one year, in combination with thalidomide (200mg/d) and dexamethasone. Almost half of the patients (47%) had two prior ASCTs. The triplet combination of VTD resulted in a significantly longer time to progression (19.5 m vs 13.8 m, p=0.001) and a significantly better CR/nCR rate (45% vs 25%, p=0.001) with a trend towards improved overall survival (71% vs 65% 24-month survival rate, p=0.093) as compared to the TD control group. On the other hand, the triplet combination had a higher incidence of grade 3 peripheral sensory neuropathy (29% vs 12%, p=0.001) and a higher incidence of grade >=3 thrombocytopenia (17% vs 7%, p=0.016) not associated with serious bleeding complications. The neurotoxicity was attributed by the investigators as due to the combination of the two neurotoxic agents bortezomib and thalidomide given for a prolonged period of time (1 year) and at higher dose levels (200 mg/d thalidomide). The investigators concluded that the VTD combination may be considered as a standard of care for patients relapsing after ASCT, but that the risk for neurotoxicity should be decreased by using lower doses of thalidomide and appropriate dose reductions of bortezomib.

4. Conclusions and future directions

There is an increasing body of literature on the incorporation of bortezomib in the different treatment phases of the autologous stem cell transplantation approach in multiple myeloma. The highest level of evidence on the benefit of bortezomib-containing regimens is available from multiple phase 3 studies in the induction treatment phase. In other treatment phases, the current experimental clinical evidence is more limited. In the conditioning phase, only phase 2 data are available on the addition of bortezomib and comparisons with historic data should be made with caution. In consolidation, only limited phase 3 information is currently available but phase 3 studies comparing bortezomib consolidation versus no consolidation are ongoing or awaiting final publication [46]. In the maintenance phase, randomized phase 3 studies have been published but did not directly test the value of bortezomib maintenance over no maintenance. Despite these limitations some common themes on the incorporation of bortezomib can be observed across the different treatment phases:

- The addition of bortezomib increased the quality of the response (higher complete and near-complete response rates) as compared to control groups or to historic data
- The addition of bortezomb improved the progression-free survival post-ASCT as compared to control groups or to historic data
- Where analyzed, the addition of bortezomib improved the outcomes of patients with poor prognostic features, such as high-risk cytogenetics and renal function impairment
- The addition of bortezomib had no negative impact on hematopoietic stem cell collection or engraftment
- The addition of bortezomib resulted in a higher incidence of peripheral neuropathy

The effect of the addition of bortezomib on the overall survival post-ASCT were variable across studies. While several studies appear to report a favorable survival trend, only in the largest phase 3 study (HOVON-65/GMMG-HD4) the survival improvement reached statistical significance. Potentially contributing to this could be the short duration of follow-up in the initial study publications, the effect of subsequent therapy (and in particular of cross-over use of bortezomib in subsequent therapy lines) and the sample size limitation of the individual studies. An argument for the latter could be found in a recent meta-analysis indicating a survival benefit of bortezomib-containing induction therapy if the different phase 3 study results are combined [47].

Which future research directions can be expected on this topic in the next decade?

First, given the high response rates and complete response rates observed with bortezomib containing regimens, the question will be asked whether in this younger population with newly diagnosed multiple myeloma a non-transplant approach incorporating bortezomib and immunomodulatory agents can delay or prevent the need for a high-dose therapy and autologous transplant approach. Several randomized phase 3 studies are currently underway to test this hypothesis.

Second, if the ultimate goal of the autologous transplant approach is disease eradication and cure, more rigorous definitions of complete response and more sensitive diagnostic techniques will be required to optimize individual therapy decisions. A stringent CR (sCR) category has already been defined by the IMWG criteria [48]. This stringent CR (sCR) category requires a normalization of the free κ/λ ratio in serum and an immunophenotypic normalization of the κ/λ ratio in the bone marrow, but sofar has not been routinely reported in high-dose therapy studies. By the most recent criteria, also an immunophenotypic CR category has been defined to exclude minimal residual disease based on a more extensive immunophenotypic analysis of the bone marrow [49]. Characterization of minimal residual disease by immunophenotyping has only been reported in selected studies [50]. Alternative techniques such as magnetic resonance imaging or positron emission tomography have also been reported but require further characterization before incorporation in routine ASCT procedures. [51,52].

Third, second-generation proteasome inhibitors, such as carfilzomib, marizomib and MLN-9708, are currently in development in multiple myeloma [53]. These agents are also potent inhibitors of proteasome activity in vitro but show differences in enzyme binding kinetics which might affect their pharmacology and result in different efficacy and safety profiles [54]. Most data with the second generation proteasome inhibitors have been generated in the relapsed or refractory myeloma setting. As there are no full publications in peer-reviewed journals available addressing the incorporation of such agents in autologous stem cell transplant approaches, these agents were not included in this review. However, data of early studies combining carfilzomib with either thalidomide-dexamethasone or lenalidomide-dexamethasone as induction treatment prior to ASCT have already been reported at international conferences [55,56]. Further research on the incorporation of second-generation proteasome inhibitors in autologous stem cell transplant approaches in myeloma can therefore definitely be expected.

Acknowledgements

The authors would like to thank Margaret Roll (librarian, Janssen Research & Development, High Wycombe, UK) for her help in performing an extensive literature search and An Van Eyken (Janssen Research & Development, Beerse, Belgium) for excellent editorial assistance.

Author details

Helgi van de Velde[1*] and Andrew Cakana[2]

*Address all correspondence to: hvdvelde@its.jnj.com

1 Oncology R&D, Janssen Research & Development, Beerse, Belgium

2 Oncology R&D, Janssen Research & Development, High Wycombe, UK

References

[1] Kyle RA,Rajkumar SV. Multiple Myeloma. The New England Journal of Medicine 2004; 351(18) 1860-1873.

[2] Attal M et al. A prospective randomized trial of autologous bone marrow transplantation and chemotherapy in multiple myeloma. The New England Journal of Medicine 1996; 335(2) 91-97.

[3] Child JA et al. High-dose chemotherapy with hematopoietic stem-cell rescue for multiple myeloma. The New England Journal of Medicine 2003; 348(19) 1875-1883.

[4] van de Velde H et al. Complete Response correlates with Long-term Survival and Progression-free Survival in High-Dose Therapy in Multiple Myeloma. Haematologica 2007; 92(10) 1399-1406.

[5] Harousseau JL,Moreau P. Autologous Hematopoietic Stem-Cell Transplantation for Multiple Myeloma. The New England Journal of Medicine 2009; 360(25) 2645-2654.

[6] Bladé J,Rosiñol L. Advances in therapy of multiple myeloma. Current Opinion in Oncology 2008; 20(6) 697-704.

[7] Rajkumar SV et al. Phase III Clinical Trial of Thalidomide Plus Dexamethasone Compared With Dexamethasone Alone in Newly Diagnosed Multiple Myeloma: A Clinical Trial Coordinated by the Eastern Cooperative Oncology Group. Journal of Clinical Oncology 2006; 24(3) 431-436.

[8] Fermand J-P et al. High-dose therapy and autologous blood stem-cell transplantation compared with conventional treatment in multiple patients aged 55 to 65 years: long-term results of a randomized control trial form the Group Myelome-Autogreffe. Journal of Clinical Oncology 2005; 23(36) 9227-9233.

[9] Lokhorst H et al. International Myeloma Working Group Consensus Statement Regarding the Current Status of Allogeneic Stem-Cell Transplantation for Multiple Myeloma. Journal of Clinical Oncology 2010; 28(29) 4521-4530.

[10] Barlogie B et al. Long-term follow-up of autotransplantation trials for multiple myeloma: update of protocols conducted by Intergroupe Francophone du Myelome,

Southwest Oncology Group, and University of Arkansas for Medical Sciences. Journal of Clinical Oncology 2010; (28)7 1209-1214.

[11] VELCADE United States Package Insert [USPI].Millennium Pharmaceuticals, Inc. Cambridge MA. December 2010. http://www.accessdata.fda.gov/drugsatfda_docs/label/2010/021602s023lbl.pdf (accessed 21 September 2012).

[12] VELCADE (bortezomib) Summary of Product Characteristics [SmPC].Janssen-Cilag International N.V., Beerse, Belgium. April 2012.http://www.ema.europa.eu/docs/en_GB/document_library/EPAR__Product_Information/human/000539/WC500048471.pdf. (accessed 21 September 2012).

[13] Mitchell BS. The Proteasome - An Emerging Therapeutic Target in Cancer. The New England Journal of Medicine 2003; 348(26) 2597-2598.

[14] Ma MH et al. The proteasome inhibitor PS-341 markedly enhances sensitivity of multiple myeloma tumor cells to chemotherapeutic agents. Clinical Cancer Research 2003; 9(3) 1136-1144.

[15] Hideshima T et al. The Proteasome Inhibitor PS-341 Inhibits Growth, Induces Apoptosis, and Overcomes Drug Resistance in Human Multiple Myeloma Cells. Cancer Research 2001; 61(7) 3071-3076.

[16] Richardson P et al. Bortezomib or High-Dose Dexamethasone for Relapsed Multiple Myeloma. The New England Journal of Medicine 2005; 352(24) 2487-2498.

[17] Richardson P et al. Extended follow-up of a phase 3 trial in relapsed multiple myeloma: final time-to-event results of the APEX trial. Blood 2007; 110(10) 3557-3560.

[18] San Miguel J et al. Bortezomib plus melphalan and prednisone for initial treatment of multiple myeloma. The New England Journal of Medicine 2008; 359(9) 906-917.

[19] San Miguel J et al. Continued Overall Survival Benefit After 5 Years' Follow-Up with Bortezomib-Melphalan-Prednisone (VMP) Versus Melphalan-Prednisone (MP) in Patients with Previously Untreated Multiple Myeloma, and No Increased Risk of Second Primary Malignancies: Final Results of the Phase 3 VISTA Trial. 53rd ASH Annual Meeting and Exposition 2011; abstract 476.

[20] Richardson PG et al. Frequency, characteristics, and reversibility of peripheral neuropathy during treatment of advanced multiple myeloma with bortezomib. Journal of Clinical Oncology 2006; 24(19) 3113-3120.

[21] Moreau P et al. A randomized phase 3 study comparing subcutaneous and intravenous administration of bortezomib in patients with relapsed multiple myeloma. Lancet Oncology 2011; 12(5) 431-440.

[22] Mulrow CD. The medical review article: state of the science. Annals of Internal Medicine 1987; 106(3) 485-488.

[23] Weed DL. Methodologic guidelines for review papers. Journal of the National Cancer Institute 1997; 89(1) 96-97.

[24] Oxman AD. Checklist for review articles. British Medical Journal 1994; 309(6955) 648-651.

[25] Harousseau JL et al. Bortezomib plus dexamethasone is superior to vincristine plus doxorubicin plus dexamethasone as induction treatment prior to autologous stem-cell transplantation in newly diagnosed multiple myeloma: results of the IFM 2005-01 phase III trial. Journal of Clinical Oncology 2010; 28(30) 4621-4629.

[26] Sonneveld P et al. Bortezomib Induction and Maintenance Treatment in Patients With Newly Diagnosed Multiple Myeloma: Results of the Randomized Phase III HOVON-65/ GMMG-HD4 Trial. Journal of Clinical Oncology 2012; 30(24) 2946-2955.

[27] Neben K et al. Administration of bortezomib before and after autologous stem-cell transplantation improves outcome in Multiple Myeloma patients with deletion 17p – a Subgroup Analysis of the HOVON-65/GMMG-HD4 Trial. Blood 2012; 119(4) 940-948.

[28] Cavo M et al. Bortezomib with thalidomide plus dexamethasone compared with thalidomide plus dexamethasone as induction therapy before, and consolidation therapy after, double autologous stem-cell transplantation in newly diagnosed multiple myeloma: a randomised phase 3 study. Lancet 2010; 376(9758) 2075-2085.

[29] Cavo M et al. Bortezomib-thalidomide-dexamethasone is superior to thalidomide-dexamethasone as consolidation therapy after autologous hematopoietic stem cell transplantation in patients with newly diagnosed multiple myeloma. Blood 2012; 120(1) 9-19.

[30] Rosiñol L et al. Superiority of bortezomib, thalidomide, and dexamethasone (VTD) as induction pretransplantation therapy in multiple myeloma: a randomized phase 3 PETHEMA/GEM study. Blood 2012; 120(8) 1589-1596.

[31] Moreau P et al. Bortezomib plus dexamethasone versus reduced-dose bortezomib, thalidomide plus dexamethasone as induction treatment before autologous stem cell transplantation in newly diagnosed multiple myeloma. Blood 2011; 118(22) 5752-5758.

[32] Harousseau JL et al. Bortezomib plus dexamethasone as induction treatment prior to autoloous stem cell transplantation in patients with newly diagnosed multiple myeloma: results of an IFM phase II study. Haematologica 2006; 91(11) 1498-1505.

[33] Corso A et al. Bortezomib plus dexamethasone can improve stem cell collection and overcome the need for additional chemotherapy before autologous transplant in patients with myeloma. Leukemia & Lymphoma 2010; 51(2) 236-242.

[34] Kaufman J et al. Bortezomib, thalidomide, and dexamethasone as induction therapy for patients with symptomatic multiple myeloma: a retrospective study. Cancer 2010; 116(13) 3143-3151.

[35] Kumar S et al. Randomized, multicenter, phase 2 study (Evolution) of combinations of bortezomib, dexamethasone, cyclophosphamide, and lenalidomide in previously untreated multiple myeloma. Blood 2012; 119(19) 4375-4382.

[36] Reeder CB et al. Cyclophosphamide, bortezomib and dexamethasone induction for newly diagnosed multiple myeloma: high response rates in a phase II clinical trial. Leukemia 2009; 23(7) 1337-1341.

[37] Kropff M et al. DSMM XI study: dose definition for intravenous cyclophosphamide in combination with bortezomib/dexamethasone for remission induction in patients with newly diagnosed myeloma. Annals of Hematology 2009; 88(11) 1125-1130.

[38] Barlogie B et al. Incorporating bortezomib into upfront treatment for multiple myeloma: early results of total therapy 3. British Journal of Haematology 2007; 138(2) 176-185.

[39] Moreau P et al. Comparison of 200 mg/m^2 melphalan and 8Gy total body irradiation plus 140 mg/m^2 melphalan as conditioning regimens for peripheral blood stem cell transplantation in patients with newly diagnosed multiple myeloma: final analysis of the Intergroupe Francophone du Myélome 9502 randomized trial. Blood 2002; 99(3) 731-735.

[40] Berenson JR et al. Phase I/II trial assessing bortezomib and melphalan combination therapy for the treatment of patients with relapsed or refractory multiple myeloma. Journal of Clinical Oncology 2006; 24(6) 937-944.

[41] Lonial S et al. A Phase I/II Trial Combining High-dose melphalan and Autologous Transplant with Bortezomib for Multiple Myeloma: A Dose- and Schedule-Finding Study. Clinical Cancer Research 2010; 16(20) 5079-5086.

[42] Roussel M et al. Bortezomib and high-dose melphalan as conditioning regimen before autologous stem cell tranplantation in patients with de novo multiple myeloma: a phase 2 study of the Intergroupe Francophone du Myelome (IFM). Blood 2010; 115(1) 32-37.

[43] Sharma M et al. A Randomized Phase 2 Trial of a Preparative Regimen of Bortezomib, High-Dose Melphalan, Arsenic Trioxide, and Ascorbic Acid. Cancer 2012; 118(9) 2507-2515.

[44] Sahebi F et al. Sequential bortezomib, dexamethasone, and thalidomide maintenance therapy after single autologous peripheral stem cell transplantation in patients with multiple myeloma. Blood and Marrow Transplantation 2012; 18(3) 486-492.

[45] Garderet L et al. Superiority of the triple combination of bortezomib-thalidomide-dexamethasone over the dual combination of thalidomide-dexamethasone in patients with multiple myeloma progressing or relapsing after autologous transplantation: the MMVAR/IFM 2005-04 Randomized Phase III Trial from the Chronic Leukemia Working Party of the European Group for Blood and Marrow Transplantation. Journal of Clinical Oncology 2012; 30(20) 2475-2482.

[46] Mellqvist UH et al. Improved progression-free survival with bortezomib consolidation after high dose melphalan: results of a randomized phase III trial [abstract] . Haematologica 2011; 96(s1) S31 [abstract O-11].

[47] Nooka A et al. The Improved Efficacy of Bortezomib Containing Induction Regimens (BCIR) Versus Non-Bortezomib Containing Induction Regimens (NBCIR) in Transplant-Eligible Patients with Multiple Myeloma (MM): Meta-Analysis of Phase III Randomized Controlled Trials (RCTs) . 53rd ASH Annual Meeting and Exposition 2011; abstract 3994.

[48] Durie BGM et al. International uniform response criteria for multiple myeloma. Leukemia 2006; 20(9) 1467-1473.

[49] Rajkumar SV et al. Consensus recommendations for the uniform reporting of clinical trials: report of the International Myeloma Workshop Consensus Panel 1. Blood 2011; 117(18) 4691-4695.

[50] Paiva B et al. Multiparameter flow cytometric remission is the most relevant prognostic factor for multiple myeloma patients who undergo autologous stem cell transplantation. Blood 2008; 112(10) 4017-4023.

[51] Barlogie B,Tricot G. Complete response in myeloma: a Trojan horse? Blood 2006; 108(7) 2134.

[52] Walker R et al. Magnetic resonance imaging in multiple myeloma: diagnostic and clinical implications. Journal of Clinical Oncology 2007; 25(9) 1121-1128.

[53] Moreau P et al. Proteasome inhibitors in multiple myeloma: 10 years later. Blood 2012; 120(5) 947-959.

[54] Dick LR,Fleming PE. Building on bortezomob: second-generation proteasome inhibitors as anti-cancer therapy. Drug Discovery Today 2010; 15(5-6) 243-249.

[55] Sonneveld P et al. Carfilzomib combined with thalidomide and dexamethasone (CARTHADEX) as induction treatment prior to high-dose melphalan (HDM) in newly diagnosed patients with multiple myeloma (MM): a trial of the European Myeloma Network EMN . Blood (ASH Annual Meeting abstracts) 2011; 118(21) 290 [abstract 633].

[56] Jakubowiak AJ et al. Final results of a frontline phase 1/2 study of carfilzomib, lenalidomide, and low-dose dexamethasone (CRd) in multiple myeloma (MM). Blood (ASH Annual Meeting abstracts) 2011; 118(12) 288-289 [abstract 631].

Chapter 10

Autologous Stem Cell Transplantation for Acute Myeloid Leukemia

Pier Paolo Piccaluga, Stefania Paolini,
Giovanna Meloni, Giuseppe Visani and
Felicetto Ferrara

Additional information is available at the end of the chapter

http://dx.doi.org/10.5772/54280

1. Introduction

In the majority of patients, the therapy of acute myeloid leukemia (AML) has a curative intent and includes two phases, i.e. induction and consolidation. The former aims at complete remission (CR) achievement, the latter at the eradication of residual leukemic cells, which are undetectable at morphologic examination of bone marrow after induction therapy in patients in CR. Current induction regimens, conventionally based on the combination of daunorubicin and cytarabine result in CR rates of 60 – 70% of AML patients younger than 65 years; in order to improve both CR rate and quality, different studies tested alternative anthracyclines [1]-[5], higher schedules of Ara-C[6]-[10], the addition of a third cytotoxic drug [11]-[16] and, more recently, the combination with new agents. Overall, results have been disappointing even though the addition of gemtuzumab ozogamycin (GO), an antiCD33 monoclonal antibody conjugated with the cytotoxic agents chalicheamycin, has been reported to confer a significant advantage in selected patients with AML [17]-[21]. Notwithstanding, in absence of intensive post-induction therapy virtually all patients will ultimately relapse, therefore consolidation therapy is strictly needed. At present, after CR achievement all patients receive a consolidation chemotherapy based on intermediate or high dose ARA-C and then in young adult patients three options can be considered, i.e. allogeneic stem cell transplantation (allo-SCT), autologous SCT (ASCT) or repetitive intensive consolidation chemotherapy cycles (ICC) with high or intermediate dose ARA-C [22]-[37], depending on age, disease risk and donor availability. In particular, it is widely accepted that ICC and ASCT would be limited to patients with favorable risk, such as AML with t(8;21), AML with inv(16) or t(16;16) and AML with normal karyotype with NPM1 mutation in absence of mutations of FLT3/ITD gene [38], [39]. In the remaining patient population, allo-SCT must be considered when age and performance status result in an acceptable risk/benefit ratio. In this regard, it should be considered that in the last years morbidity and mortality from allo-SCT have been considerably reduced; in addition, the introduction into daily practice of reduced intensity conditioning (RIC) has allowed to offer the procedure to selected old and/or previously not eligible patient population.

Currently, even in patients with favourable prognostic factors at diagnosis, the role of ASCT remains unclear although most studies that have compared ASCT with ICC demonstrated a significantly lower rate of relapse following ASCT (5,6). Results in terms of survival were, however, less encouraging because of transplant-related deaths and the low rate of second CR in patients who relapsed after ASCT, therefore in the last year ASCT has become less popular, mainly in USA. Notwithstanding, different considerations should be made: first, both the occurrence of toxicity and mortality related to ASCT have greatly decreased since use of peripheral-blood stem cells was introduced, even in older patients. Second, reduction of relapse rate would represent a main therapeutic objective in the therapy of AML, just as it is any malignant disorder. Finally, consolidation therapy based on repeated courses of high-dose or intermediate-dose cytarabine is probably more toxic and costly than ASCT and is poorly feasible in patients aged over 55-60 years. In elderly patients, particularly, the dose intensification by either ASCT or ICC has failed so far to induce a significant benefit [40]-[42]. Therefore, novel more rational targeted agents are particularly warranted in this setting. On the other hand, two important conditions are necessary in order to perform ASCT: CR achievement and collection of an adequate number of $CD34^+$ cells ($> 2 \times 10^6$/Kg). As the latter aspect is concerned, it should be mentioned that a previous history of myeloid disorder (especially myelodysplastic syndrome), advanced age and the use of certain drugs during the induction and consolidation phases (e.g. fludarabine [43]) can significantly impair the possibility to collect an adequate number of cells.

Overall, data from the literature are controversial, but it has been definitively demonstrated that ASCT provides better results in patients with favorable risk diseases and low amount of minimal residual disease after induction/consolidation therapy. In the last years, a few complete meta-analyses and extensive reviews tried to draw some conclusions but were not able to indicate definite guidelines [44]-[46].

In this chapter, the authors review the current knowledge on the use of SCT in post-consolidation therapy of AML, based on their own experience and the most recent literature data, by mainly focusing on randomized clinical trials (RCT).

2. Randomized clinical trials comparing autologous stem cell transplantation and chemotherapy or no further therapy

In 1995 Zittoun et al. for the EORTC-GIMEMA groups reported on 941 AML patients treated with one or two cycles of standard Daunorubicine/Cytarabine schedule (3/7). Patients obtaining CR were submitted to one consolidation cycle including high-dose cytar-

http://dx.doi.org/10.5772/54280

abine (HD-AC) and Amsacrine. Subsequently, patients with HLA identical donor were allo-transplanted, whereas patients without HLA identical donor were randomized to receive ASCT or a second consolidation (ICC) with daunorubicine and HD-AC. The CR rate after induction therapy was 66%. The relapse rate were 40% in the two arms (ASCT) and 57% (ICC), respectively; DFS was longer for patients submitted to ASCT compared to patients submitted to ICC (48% vs 30%; p=0.05). However the OS was not significantly superior in the ASCT group, due to the greater ability of ASCT to rescue relapsed patients in the ICC arm [24].

In 1997 Harousseau and Colleagues reported data on 517 eligible patients (15-50 years of age) affected by previously untreated AML. Patients received 3 - 4 courses of conventional induction treatment (Ara-C: 200 mg/sqm/day for 7 consecutive days with either idarubicin administered intravenously on days 1 - 5 at a daily dose of 8 mg/sqm or rubidazone administered intravenously on days 1 - 4 at a daily dose of 200 mg/sqm). Patients aged 40 year or younger, in CR after induction therapy, were assigned to SCT if an HLA identical donor was available. All other patients received a first course of HD-Ara-C (3 gr/sqm) administered every 12 hours along 4 days (ICC) and then were randomized to receive either a second course of ICC or an ASCT. Eighty-eight patients out of 517 received an SCT, while 164 out of 517 were eligible for randomization (75 received ASCT, 71 received ICC). No differences in terms of OS and DFS were observed between the two arms: the 4 years DFS was 44 +/- 5.5% in ASCT group and 40.5 +/- 5.5% in ICC group (p value 0.41); the 4 years OS was 50 +/- 6% in ASCT group and 54.5 +/- 6% in ICC group (p value 0.72). The retrospective analysis of DFS and OS based on the cytogenetic risk could not detect any differences between the ASCT group and the ICC group [28].

In 1998 Cassileth et al. reported on 740 AML patients treated with standard 3/7 – 3/5 induction – consolidation chemotherapy cycles. Patients without an HLA identical donor were randomized between ASCT and HD-AC. The overall CR rate was 70%; the 4-years-DFS was 35% in both groups; the 4 years OS was 43% in ASCT group and 52% in ICC group respectively (p= 0.05) [25].

The first report on the MRC AML 10 trial was published in 1998 [29]. Patients were firstly randomly assigned to different induction chemotherapy regimens (DAT vs. ADE); all patients achieving CR after two induction courses received a third consolidation chemotherapy course (MACE). Patients who lacked an HLA-matched sibling donor were randomized to receive one more chemotherapy course (MidAC) followed by either ASCT or no further therapy; patients with an HLA-matched sibling donor were assigned to receive an SCT. Basis on the intention to treat analysis the number of relapses was significantly lower in the ASCT group than in the group assigned no further treatment (37% vs. 58%; p= 0.0007), resulting in superior DFS at 7 years (53% vs. 40%; p=0.04). No difference in terms of OS was observed. Of note, however, in this trial only 38% of patients available for randomization were randomized [29].

Tsimberidou et al. then reported data on 120 patients with de novo AML in 2003. All patients were treated with standard 3/7 regimen (2 courses) and if in CR underwent a first HD-AC course. All patients aged less than 50 years and with an HLA compatible donor received an SCT; patient aged more than 50 years or without an HLA-matched sibling donor were randomly assigned to receive a second HD-Ara course or an ASCT. With a median follow-up of 43 months the 3-year failure free survival rates was 42% for patients receiving ASCT and 33% for patients receiving conventional chemotherapy [33].

Subsequently, Breems et al in 2005 reported data on 646 patients enrolled in the HOVO/SAKK AML4 trial. After two cycle of induction therapy combining cytarabine with daunorubicine (first course) and amsacrine (second course), CR patients (75%) were addressed to a consolidation therapy with mitoxantrone and VP16. Eighty-one patients received SCT. Patients non eligible for SCT were randomized between ASCT (66 patients) and no further therapy (46 patients). After a median follow up of 154 months, there were no statistically significant differences concerning DFS, OS and relapse rate within the two randomization arms. There was a trend towards a better OS of the non-autografted patients. This was associated with a higher, though non significant, incidence of death in CR within the auto-transplanted group with respect to the no treatment group. The 5 years OS after relapse for patients previously auto-grafted was significantly shorter with respect to patients who received no further treatment [34].

A large European intergroup trial [47] later evaluated HD-AC induction and escalation of post-remission therapy in a 2-stage RCT. Patients under the age of 60 years were randomized to 1 of 2 induction courses (double HD-AC vs. standard cytarabine/HD-AC). Patients in remission received a third cycle of chemotherapy followed by a second randomization to ASCT or maintenance chemotherapy. Fifty-one percent assigned to maintenance received the assigned therapy, while only 24% received the assigned ASCT. Three-year remission duration was 50% versus 44%, 3-year relapse-free survival was 48% versus 43% for maintenance and ASCT, respectively, and there was no significant difference between the 2 arms when stratified according to cytogenetic risk profile [47].

An update of the AML10 study was then reported in 2006 [35]. Briefly, The overall survival of patients allocated to autologous transplantation was better than for those in the no-further-therapy arm (53% vs. 45%) at 10 years, with 165 patients at risk at that time point. Of note, although this difference was not statistically significant on a log-rank analysis (P=.09), the Kaplan-Meier plots clearly diverged after the first 3 years, the difference becoming significant. This was related to a highly significant reduction in relapse risk in the autograft arm (40% vs. 58%; P=.0005), with consequent improved DFS in the ASCT arm (50% vs. 39%; P=.03), a data which was partially obscured by a higher risk of death in remission (16% vs. 6%; P=.02). Overall, the study suggested a survival benefit with ASCT in patients in the good- and standard-risk groups but not in the poor-risk group. Conversely, it was unclear if any specific age group benefited [35].

Based on these studies, a couple of systematic meta-analyses and reviews, tried to delineate some possible indications. However, many data were conflicting a definitive recommendations appeared difficult. Particularly, Nathan and Colleagues performed a comprehensive meta-analysis on consolidation therapy for AML. In particular, they analyzed 6 studies including 1044 patients randomly assigned t receive ASCT vs. ICC (5 studies), or ASCT vs. no further treatment (1 study). Patients receiving ASCT had a better disease free but not different overall survival. Thus, they did not recommend ASCT as routine options for AML patients in first CR [45]. Thereafter, Visani and Colleagues, based on evidence based medicine (EBM) criteria, considered 6 RCT evaluating the role of ASCT and concluded that due to the heterogeneity of AML biology (i.e. molecular genetics), further studies specifically dedicated to the different entities were probably necessary to build robust recommendation according to EBM rules [46].

More recently, the HOVON Group reported the results of a prospective, randomized phase 3 trial evaluating ASCT vs. ICC in newly diagnosed AML patients in first CR (CR1) [48]. Patients with AML (16-60 years) in CR1 after 2 cycles of intensive chemotherapy and not eligible for allogeneic SCT were randomized between ICC (including etoposide and mitoxantrone) or ASCT (Bu/Cy). More than 90% of randomized patients received their assigned treatment (ICC, n = 259; ASCT, n = 258),. The 2 groups were comparable with regard to prognostic factors. The ASCT group showed a markedly reduced relapse rate (58% vs. 70%, P = 0.02) and better relapse-free survival at 5 years (38% vs. 29%,P =0.065) with non-relapse mortality of 4% vs. 1% in the chemotherapy arm (P =0.02). OS was similar (44% vs. 41% at 5 years, P =0.86), possibly because of more opportunities for salvage with second-line chemotherapy and SCT in patients relapsing on the chemotherapy arm. [48].

Finally, Pfirman et al reported the results of the AML96 trial [49], aiming to differentiate groups of patients according to the treatments that would provide them optimum benefit. Five hundred eighty six AML patients (aged below 60 years) - excluding those with t(8;21) – in CR1 after double induction treatment were consolidated with SCT or ASCT, or ICC containing HD-AC, in a priority-based and risk-adapted manner. The association between potentially prognostic variables and OS was assessed and a post-remission treatment (PRT) score was developed in 452 patients with a complete dataset. This score was then validated in additional 407 patients from the AML2003 trial. Age, percentage of CD34-positive blasts, FLT3-ITD mutant-to-wild-type ratio, cytogenetic risk, and de-novo or secondary AML were identified as independent prognostic factors, and included in the PRT score. Accordingly, patients were separated into three groups: favorable (N=190; 3-year survival 68%), intermediate (N=198; 49%), and unfavorable (n=64; 20%). These results were confirmed in the AML2003 trial dataset: 3-year survival for the favorable group (n=265) was 69%, for the intermediate group (n=114) it was 61%, and for the unfavorable group (n=28) it was 46%. Therefore, the 3 groups presented with significantly different survival probabilities (p=0.015). Additionally, the Authors found that in the favorable group, patients who received SCT (n=60) had higher survival probabilities (82%) than did those given chemotherapy (n=56, 55%; p=0.0012) or ASCT (n=74, 66%; p=0.044). In the intermediate PRT score group, patients receiving ASCT (n=69) had the best survival probabilities (62%) compared with those given chemotherapy (n=72, 41%; p=0.0006) or SCT (n=57, 44%; p=0.0045).

Overall, the study thus supported the use of autologous HSCT in patients aged 60 years or younger with an intermediate PRT score.

Results of the above mentioned studies on ASCT are summarized in Table 1.

Author	Population – Study design	Outcome		*Pvalues*
		Auto-SCT	**Chemotherapy/no further therapy**	
Zittoun et al	990 patients (< 59 y) previously untreated AML. (941 evaluable)	4 yrs DFS: 48 ± 5%	4 yrs DFS: 30 ± 5%	0.05
	Study design:	4 yrs OS : 56 ± 5%	4 yrs OS : 56 ± 5%	NS
	- Induction: cytarabine + doxorubicine			
	If PR: 2nd course of induction therapy Consolidation: HDAC+amsacrine			
	If CR, age<45 yrs and HLA compatible donor: allo-SCT (N= 144)			
	- If > 45 yrs and/or no HLA compatible donor: randomization (auto-SCT, N= 95 vs. 2nd course of intensive therapy, N=104)			
Harousseau et al	517 previously untreated AML patients (15-50 yrs)	4 yrs DFS: 44 ± 5.5%	4 y DFS: 40.5 ± 5.5%	NS
	Study design:	4 yrs OS: 50 ± 6%	4 y OS: 54.5 ± 6%	
	- Induction: cytarabine and idarubicine or rubidazone. If no CR: 2nd cycle	*Low risk group*		
	- Consolidation: HD-AC + Idarubicine or Rubidazone	4 yrs DFS: 50 ± 9%	4 yrs DFS : 56 ± 11%	NS
	If CR, age <40 yrs and HLA compatible donor: allo-SCT (N=88)	4 yrs OS: 59 ± 9%	4 yrs OS: 71 ± 8%	NS
	- If > 40 yrs and/or no HLA compatible donor: randomization (auto-SCT, N= 75 vs. ICC, N=71)			
		Intermediate risk group		
		- 4 yrs DFS: 38.5 ± 9%	- 4 yrs DFS: 31 ± 8.5%	NS
		- 4 yrs OS: 42.5 ± 9%	- 4 yrs OS: 55 ± 9%	NS
		High risk group		
		- 4 yrs DFS: 38 ± 10%	- 4 yrs DFS: 28.5 ± 10%	NS
		- 4 yrs OS: 46.5 ± 11%	- 4 yrs OS: 40 ± 11.5%	NS
Cassileth et al	772 previously untreated AML patients	4 yrs DFS: 35±9 %	4 yrs DFS: 35±9 %	NS
	(16-55 yrs)			
	Study design:	4 yrs OS: 43±9 %	4 yrs OS: 52±9 %	P=0.05

Author	Population – Study design	Outcome		*Pvalues*
		Auto-SCT	Chemotherapy/no further therapy	
	Induction: 2 cycles of idarubicine and cytarabine Consolidation: idarubicine and cytarabine - If CR and HLA compatible donor: allo-SCT (N=113) - If not HLA compatible donor: randomization auto-SCT (N =116) vs. HD-Cytarabine (N = 117)			
Burnett et al	1509 previously untreated AML patients aged less than < 56 yrs	10 yrs DFS: 50%	10 yrs DFS: 39%	0.03
	Study design:			
	- 2 Induction: Daunorubicine, Cytarabine, Thioguanine vs Daunorubicine, Cytarabine, VP-16	10 yrs OS: 53%	10 yrs OS : 45%	0.009
	- 1st Consolidation: Amsacrine, Cytarabine, VP-16			
	- Pts with HLA identical donor: 2nd consolidation (Mitoxantrone, Cytarabine) and allo-SCT	Relapse rate at 10 yrs: 40%	Relapse rate at 10 yrs: 58%	0.0005
	- Pts lacking HLA identical donor: 2nd consolidation (Mitozantrone, Cytarabine) and randomization to auto-SCT (N =190) vs. no further therapy (N =191)			
Tsimberidou et al	120 previously untreated AML patients (<60 yrs)	3 yrs OS: 58%	3 yrs OS: 46%	NS
	Study design:	3 yrs FFS: 42%	3 yrs FFS: 33%	NS
	- 2 Induction: Idarubicine, Cytarabine (3+7) - Consolidation: HD-AC - If < 50 y and HLA compatible donor : allo-SCT (N = 21) - If > 50 y and/or no HLA compatible donor: randomization (auto-SCT, N = 19 vs. 2nd HD-AC, N= 15)			
Breems et al	646 previously untreated AML patients (< 60 years)	5 yrs DFS: about 35%	5 yrs DFS: about 37%	NS
	Study design:	5 yrs OS : about 45%	5 yrs OS : about 55%	NS
	- Induction 1: Daunorubicine, Cytarabine (3+7)			

Author	Population – Study design	Outcome		*Pvalues*
		Auto-SCT	Chemotherapy/no further therapy	
	- Induction 2: Amsacrine, Cytarabine	7 pts died in CR within 9 months	1 pts died in CR within 9 months	NS
	- Consolidation: Mitoxantrone, VP16			
	- If eligible and compatible donor : allo-SCT (N = 81)	5 yrs OS after relapse: about 5%	5 yrs OS after relapse: about 25%	0.003
	- If non eligible: randomization (auto-SCT, N =66 vs. no therapy, N = 64)			
Buchner et al.	840 AML/high-risk MDS patients (age ≤ 60 years)	3 yrs DFS: 48%	3 yrs DFS: 46%	0.65
	Study design:			
	1st Randomization at induction: TAM-HAM vs. HAM-HAM	3 yrs OS : 43%	3 yrs OS : 41%	0.52
	TAD: thioguanine, cytarabine, and daunorubicin			
	HAM: cytarabine and mitoxantrone			
	Consolidation: TAD			
	2nd Randomization (auto-SCT, N=429 vs. manteinance, N = 411)			
	If eligible and compatible donor : allo-SCT (N= 128)			
Vellenga 2011	2,017 AML patients (age ≤ 60 years)			
	Induction 1: cytarabine and idarubicinInduction 2: cytarabine and amsacrine	5 yrs DFS: 38%	5 yrs DFS: 29%	0.065
	Consolidation: etoposide and mitoxantrone	5 yrs OS: 44%	5 yrs OS : 41%	0.86
	Randomization to ASCT (N=258) vs. Chemotherapy (N=259)	Relapse rate: 58%	Relapse rate: 70%	0.02
Pfirman 2012	1,151 AML patients (age ≤ 60 years)	*Favorable PRT:*		
	Assigment to ASCT (N=191) vs. Chemotherapy (N=223)	3 yrs OS: 66%	3 yrs OS : 55%	
	Assigment to SCT (N=172)	*Intermediate PRT:*		
		3 yrs OS: 62%	3 yrs OS : 41%	0.0006
		Adverse PRT:		
		3 yrs OS: 7%	3 yrs OS : 19%	

Table 1. Summary of the most relevant randomized clinical trials evaluating the role of ASCT in AML

3. Discussion and perspectives

Current intensive induction chemotherapy for patients with AML produces CR rates higher than 60-65 %; however, less than 30% of patients still survive for more 5 years free of disease. In this context, the aim of post-remission treatment is to eradicate clonogenic leukemic cells, which persists after induction and are ultimately able to induce disease relapse. Nonetheless, the optimal form of treatment is still under debate. As discussed, three main strategies are used to prevent relapse in patients with AML in first CR, including intensive chemotherapy based on intermediate-dose or high-dose cytarabine, and allogeneic and autologous hemopoietic stem cell transplantation. The choice among these approaches for an individual patient relies on two main factors, namely the expected risk of relapse as determined by biological features of leukemic cells and expected morbidity and mortality associated with a specific option, according to age and comorbidities [50].

Intensive chemotherapy (ICC) proved to be useful for improving AML patients outcome [17], [19]-[21], [51]-[55].

On the other hand, allogeneic SCT was demonstrated to be the most effective strategy to reduce the relapse risk [24], [25], [28], [29]. However, it is associated with a high-risk of treatment-related morbidity and mortality (TRM), and it is conventionally offered to younger patients with a HLA-matched sibling or unrelated donor. Of note, in the last years several evidences emerged that allogeneic SCT should not be offered as first option to patients with relatively favorable biological characteristics. The latter include a few genetic abnormalities – t(8;21)(q23;q22), inv(16)(p13q22), and t(15;17)(q22;q21) – as well as the presence of somatic mutations of *NMP1* and/or *CEBPA* genes in absence of other abnormalities. Therefore, for these patients, with the exception of M3 patients that can benefit from specific targeted agents, once achieved CR, the most suitable therapeutic options remain intensive chemotherapy and ASCT.

ASCT is an alternative approach to deliver an effective anti-leukemic myeloablative therapy to AML patients in CR, when a donor is not available. It has been demonstrated that ASCT is feasible and effective in AML, provided that an adequate induction/consolidation treatment has previously determined an effective in vivo purging. In fact, the results obtained with ASCT can be significantly affected by other relevant factors, including intensity of induction and consolidation chemotherapy as well as conditioning regimens, which strongly influencing the MRD burden before the procedure is performed [50]. Bearing this in mind, it is not surprising that the several RCT trying to define the role of ASCT as post-remission therapy in AML ended up with discrepant result. In particular, the nine largest studies, though considering 2,894 patients assigned to either ASCT or chemotherapy/no further therapy (among more than 8,000 enrolled ones) did not reach definitive conclusions (Table 1). In fact, although a reduced relapse risk was often recorded, only one study provided evidences of survival advantages for patients receiving ASCT, considering the whole population [35], while one assessed a significant advantage only in patients with an intermediate prognostic score [49]. Indeed, in most instances, the reduced leukemia recurrence was balanced by an increase TRM. In this regard, however, it should be mentioned that in the last years the mortality of ASCT has definitely declined, possibly challenging some of the results published so far. Moreover, reduction of the relapse rate is a pivotal objective in the treatment of AML, as the only way toward the cure. In addition, the continuous and very fast improvement in our knowledge of the biology of the disease on one hand clearly established that AML is not a unique disease, providing the basis for future more rationale therapies based on the specific molecular features, while on the other hand made more difficult to be interpreted results from most clinical trials, that were initiated when a comprehensive molecular characterization was not available. Accordingly, a modern view of the problem should consider these new elements and rather than debating whether ASCT is superior to SCT or ICC in AML, it would be more useful to identify those patients who would more benefit from the procedure.

Of note, one study (actually the most recently published) tried to identify the optimal post-remission strategy according to both clinical and biological features of the single case, recognizing three different groups based on an original post-remission treatment (PRT) score. Indeed, ASCT turned out to be the treatment of choice for the intermediate class, the outcome being quite favourable (Table 1). Therefore, although the proposed scoring systems will be probably modified/updated in the future, following, for example, the knowledge derived fro the most recent massive parallel sequencing studies [56] and the introduction of novel anti-leukemic compounds, an interesting scenario has probably (re)opened for ASCT. Finally, future research should focus on designing better ways to do autografts rather than conducting more trials comparing chemotherapy with the same autograft procedures currently in use, including the adoption of immunotherapy, the selection of patients based on the absence of a minimal residual disease [57] and/or of new biologic molecularly targeted compounds in the post-ASCT phase.

In conclusion, although evidence based indication cannot be offered for ASCT in AML, it is reasonable to consider it as a valid therapeutic option for AML patients at low-intermediate risk in CR1. Indeed, a main goal should be having optimal frontline genetic characterization, as well as MRD evaluation on the harvested cells. For high risk patients, unfortunately, SCT can be an option, if they achieve a good quality CR; otherwise, experimental procedures are mandatory.

Acknowledgements

Supported by BolognAIL, AIRC (IG4987; 5xMille 10007; IG10519), RFO (Prof. Piccaluga), Fondazione Cassa di Risparmio in Bologna, Fondazione della Banca del Monte e Ravenna, Progetto Strategico di Ateneo 2006 (Prof. Piccaluga).

http://dx.doi.org/10.5772/54280

Author details

Pier Paolo Piccaluga[1*], Stefania Paolini[1], Giovanna Meloni[2], Giuseppe Visani[3] and Felicetto Ferrara[4]

*Address all correspondence to: pierpaolo.piccaluga@unibo.it

1 Hematopathology and Hematology Sections, Department of Hematology and Oncology "L. and A. Seràgnoli", S. Orsola-Malpighi Hospital, University of Bologna, Italy

2 Hematology, Department of Cellular Biotechnologies and Hematology, Sapienza University, Rome, Italy

3 Haematology and Haematopoietic Stem Cell Transplant Center - San Salvatore Hospital – Pesaro, Italy

4 Division of Haematology and Stem Cell Transplantation Unit, Cardarelli General Hospital, Naples, Italy

References

[1] Yates J, Glidewell O, Wiernik P, et al. Cytosine arabinoside with daunorubicin or adriamycin for therapy of acute myelocytic leukemia: a CALGB study. Blood 1982;60(2):454-62.

[2] Arlin Z, Case DC, Jr., Moore J, et al. Randomized multicenter trial of cytosine arabinoside with mitoxantrone or daunorubicin in previously untreated adult patients with acute nonlymphocytic leukemia (ANLL). Lederle Cooperative Group. Leukemia 1990;4(3):177-83.

[3] Vogler WR, Velez-Garcia E, Weiner RS, et al. A phase III trial comparing idarubicin and daunorubicin in combination with cytarabine in acute myelogenous leukemia: a Southeastern Cancer Study Group Study. J Clin Oncol 1992;10(7):1103-11.

[4] Lowenberg B, Suciu S, Archimbaud E, et al. Mitoxantrone versus daunorubicin in induction-consolidation chemotherapy--the value of low-dose cytarabine for maintenance of remission, and an assessment of prognostic factors in acute myeloid leukemia in the elderly: final report. European Organization for the Research and Treatment of Cancer and the Dutch-Belgian Hemato-Oncology Cooperative Hovon Group. J Clin Oncol 1998;16(3):872-81.

[5] Visani G, Bernasconi P, Boni M, et al. The prognostic value of cytogenetics is reinforced by the kind of induction/consolidation therapy in influencing the outcome of acute myeloid leukemia--analysis of 848 patients. Leukemia 2001;15(6):903-9.

[6] Dillman RO, Davis RB, Green MR, et al. A comparative study of two different doses of cytarabine for acute myeloid leukemia: a phase III trial of Cancer and Leukemia Group B. Blood 1991;78(10):2520-6.

[7] Bishop JF, Matthews JP, Young GA, et al. A randomized study of high-dose cytarabine in induction in acute myeloid leukemia. Blood 1996;87(5):1710-7.

[8] Buchner T, Hiddemann W, Wormann B, et al. Double induction strategy for acute myeloid leukemia: the effect of high-dose cytarabine with mitoxantrone instead of standard-dose cytarabine with daunorubicin and 6-thioguanine: a randomized trial by the German AML Cooperative Group. Blood 1999;93(12):4116-24.

[9] Cahn JY, Labopin M, Sierra J, et al. No impact of high-dose cytarabine on the outcome of patients transplanted for acute myeloblastic leukaemia in first remission. Acute Leukaemia Working Party of the European Group for Blood and Marrow Transplantation (EBMT). British journal of haematology 2000;110(2):308-14.

[10] Bradstock KF, Matthews JP, Lowenthal RM, et al. A randomized trial of high-versus conventional-dose cytarabine in consolidation chemotherapy for adult de novo acute myeloid leukemia in first remission after induction therapy containing high-dose cytarabine. Blood 2005;105(2):481-8.

[11] Bishop JF, Lowenthal RM, Joshua D, et al. Etoposide in acute nonlymphocytic leukemia. Australian Leukemia Study Group. Blood 1990;75(1):27-32.

[12] Berman E, Arlin ZA, Gaynor J, et al. Comparative trial of cytarabine and thioguanine in combination with amsacrine or daunorubicin in patients with untreated acute nonlymphocytic leukemia: results of the L-16M protocol. Leukemia 1989;3(2):115-21.

[13] Rowe JM, Tallman MS. Intensifying induction therapy in acute myeloid leukemia: has a new standard of care emerged? Blood 1997;90(6):2121-6.

[14] Russo D, Pricolo G, Michieli M, et al. Fludarabine, arabinosyl cytosine and idarubicin (FLAI) for remission induction in poor-risk acute myeloid leukemia. Leukemia & lymphoma 2001;40(3-4):335-43.

[15] Robak T. Purine nucleoside analogues in the treatment of myleoid leukemias. Leukemia & lymphoma 2003;44(3):391-409.

[16] Holowiecki J, Grosicki S, Robak T, et al. Addition of cladribine to daunorubicin and cytarabine increases complete remission rate after a single course of induction treatment in acute myeloid leukemia. Multicenter, phase III study. Leukemia 2004;18(5): 989-97.

[17] Piccaluga PP, Martinelli G, Rondoni M, et al. Low dose gemtuzumab ozogamicin for relapsed acute myeloid leukaemia in elderly. Haematologica 2003;88(12):ECR37.

[18] Piccaluga PP, Martinelli G, Rondoni M, et al. Gemtuzumab ozogamicin for relapsed and refractory acute myeloid leukemia and myeloid sarcomas. Leukemia & lymphoma 2004;45(9):1791-5.

[19] Piccaluga PP, Martinelli G, Rondoni M, et al. First experience with gemtuzumab ozogamicin plus cytarabine as continuous infusion for elderly acute myeloid leukaemia patients. Leukemia research 2004;28(9):987-90.

[20] Castaigne S, Pautas C, Terre C, et al. Effect of gemtuzumab ozogamicin on survival of adult patients with de-novo acute myeloid leukaemia (ALFA-0701): a randomised, open-label, phase 3 study. Lancet 2012;379(9825):1508-16.

[21] Burnett AK, Russell NH, Hills RK, et al. Addition of Gemtuzumab Ozogamicin to Induction Chemotherapy Improves Survival in Older Patients With Acute Myeloid Leukemia. J Clin Oncol 2012.

[22] Reiffers J, Gaspard MH, Maraninchi D, et al. Comparison of allogeneic or autologous bone marrow transplantation and chemotherapy in patients with acute myeloid leukaemia in first remission: a prospective controlled trial. British journal of haematology 1989;72(1):57-63.

[23] Ferrant A, Doyen C, Delannoy A, et al. Allogeneic or autologous bone marrow transplantation for acute non-lymphocytic leukemia in first remission. Bone marrow transplantation 1991;7(4):303-9.

[24] Zittoun RA, Mandelli F, Willemze R, et al. Autologous or allogeneic bone marrow transplantation compared with intensive chemotherapy in acute myelogenous leukemia. European Organization for Research and Treatment of Cancer (EORTC) and the Gruppo Italiano Malattie Ematologiche Maligne dell'Adulto (GIMEMA) Leukemia Cooperative Groups. The New England journal of medicine 1995;332(4):217-23.

[25] Cassileth PA, Harrington DP, Appelbaum FR, et al. Chemotherapy compared with autologous or allogeneic bone marrow transplantation in the management of acute myeloid leukemia in first remission. The New England journal of medicine 1998;339(23):1649-56.

[26] Gale RP, Buchner T, Zhang MJ, et al. HLA-identical sibling bone marrow transplants vs chemotherapy for acute myelogenous leukemia in first remission. Leukemia 1996;10(11):1687-91.

[27] Isidori A, Bonifazi F, Visani G, Gherlinzoni F, Baccarani M, Lemoli RM. Autologous stem cell transplantation for acute myeloid leukemia patients in first complete remission: a 10-year follow-up study of 118 patients. Haematologica 2005;90(1):139-41.

[28] Harousseau JL, Cahn JY, Pignon B, et al. Comparison of autologous bone marrow transplantation and intensive chemotherapy as postremission therapy in adult acute myeloid leukemia. The Groupe Ouest Est Leucemies Aigues Myeloblastiques (GOELAM). Blood 1997;90(8):2978-86.

[29] Burnett AK, Goldstone AH, Stevens RM, et al. Randomised comparison of addition of autologous bone-marrow transplantation to intensive chemotherapy for acute myeloid leukaemia in first remission: results of MRC AML 10 trial. UK Medical Re-

search Council Adult and Children's Leukaemia Working Parties. Lancet 1998;351(9104):700-8.

[30] Burnett AK, Wheatley K, Goldstone AH, et al. The value of allogeneic bone marrow transplant in patients with acute myeloid leukaemia at differing risk of relapse: results of the UK MRC AML 10 trial. British journal of haematology 2002;118(2): 385-400.

[31] Gratwohl A, Baldomero H, Passweg J, et al. Hematopoietic stem cell transplantation for hematological malignancies in Europe. Leukemia 2003;17(5):941-59.

[32] Suciu S, Mandelli F, de Witte T, et al. Allogeneic compared with autologous stem cell transplantation in the treatment of patients younger than 46 years with acute myeloid leukemia (AML) in first complete remission (CR1): an intention-to-treat analysis of the EORTC/GIMEMAAML-10 trial. Blood 2003;102(4):1232-40.

[33] Tsimberidou AM, Stavroyianni N, Viniou N, et al. Comparison of allogeneic stem cell transplantation, high-dose cytarabine, and autologous peripheral stem cell transplantation as postremission treatment in patients with de novo acute myelogenous leukemia. Cancer 2003;97(7):1721-31.

[34] Breems DA, Boogaerts MA, Dekker AW, et al. Autologous bone marrow transplantation as consolidation therapy in the treatment of adult patients under 60 years with acute myeloid leukaemia in first complete remission: a prospective randomized Dutch-Belgian Haemato-Oncology Co-operative Group (HOVON) and Swiss Group for Clinical Cancer Research (SAKK) trial. British journal of haematology 2005;128(1): 59-65.

[35] Burnett AK, Wheatley K, Goldstone AH, Stevens R, Hann I, Hills RK. Long-term results of the MRC AML10 trial. Clin Adv Hematol Oncol 2006;4(6):445-51.

[36] Koreth J, Schlenk R, Kopecky KJ, et al. Allogeneic stem cell transplantation for acute myeloid leukemia in first complete remission: systematic review and meta-analysis of prospective clinical trials. Jama 2009;301(22):2349-61.

[37] Lioure B, Bene MC, Pigneux A, et al. Early matched sibling hematopoietic cell transplantation for adult AML in first remission using an age-adapted strategy: long-term results of a prospective GOELAMS study. Blood 2012;119(12):2943-8.

[38] Rosenblat TL, Jurcic JG. Induction and postremission strategies in acute myeloid leukemia: state of the art and future directions. Hematology/oncology clinics of North America 2011;25(6):1189-213.

[39] Rollig C, Bornhauser M, Thiede C, et al. Long-term prognosis of acute myeloid leukemia according to the new genetic risk classification of the European LeukemiaNet recommendations: evaluation of the proposed reporting system. J Clin Oncol 2011;29(20):2758-65.

[40] Thomas X, Suciu S, Rio B, et al. Autologous stem cell transplantation after complete remission and first consolidation in acute myeloid leukemia patients aged 61-70

years: results of the prospective EORTC-GIMEMA AML-13 study. Haematologica 2007;92(3):389-96.

[41] Gardin C, Turlure P, Fagot T, et al. Postremission treatment of elderly patients with acute myeloid leukemia in first complete remission after intensive induction chemotherapy: results of the multicenter randomized Acute Leukemia French Association (ALFA) 9803 trial. Blood 2007;109(12):5129-35.

[42] Hengeveld M, Suciu S, Karrasch M, et al. Intensive consolidation therapy compared with standard consolidation and maintenance therapy for adults with acute myeloid leukaemia aged between 46 and 60 years: final results of the randomized phase III study (AML 8B) of the European Organization for Research and Treatment of Cancer (EORTC) and the Gruppo Italiano Malattie Ematologiche Maligne dell'Adulto (GIMEMA) Leukemia Cooperative Groups. Annals of hematology 2012;91(6):825-35.

[43] Visani G, Lemoli RM, Tosi P, et al. Fludarabine-containing regimens severely impair peripheral blood stem cells mobilization and collection in acute myeloid leukaemia patients. British journal of haematology 1999;105(3):775-9.

[44] Reiffers J, Stoppa AM, Attal M, et al. Allogeneic vs autologous stem cell transplantation vs chemotherapy in patients with acute myeloid leukemia in first remission: the BGMT 87 study. Leukemia 1996;10(12):1874-82.

[45] Nathan PC, Sung L, Crump M, Beyene J. Consolidation therapy with autologous bone marrow transplantation in adults with acute myeloid leukemia: a meta-analysis. Journal of the National Cancer Institute 2004;96(1):38-45.

[46] Visani G, Olivieri A, Malagola M, et al. Consolidation therapy for adult acute myeloid leukemia: a systematic analysis according to evidence based medicine. Leukemia & lymphoma 2006;47(6):1091-102.

[47] Buchner T, Berdel WE, Schoch C, et al. Double induction containing either two courses or one course of high-dose cytarabine plus mitoxantrone and postremission therapy by either autologous stem-cell transplantation or by prolonged maintenance for acute myeloid leukemia. J Clin Oncol 2006;24(16):2480-9.

[48] Vellenga E, van Putten W, Ossenkoppele GJ, et al. Autologous peripheral blood stem cell transplantation for acute myeloid leukemia. Blood 2011;118(23):6037-42.

[49] Pfirrmann M, Ehninger G, Thiede C, et al. Prediction of post-remission survival in acute myeloid leukaemia: a post-hoc analysis of the AML96 trial. The lancet oncology 2012;13(2):207-14.

[50] Ferrara F. Renaissance of autologous stem cell transplantation for AML? The lancet oncology 2012;13(2):121-3.

[51] Rees JK, Gray RG, Wheatley K. Dose intensification in acute myeloid leukaemia: greater effectiveness at lower cost. Principal report of the Medical Research Council's AML9 study. MRC Leukaemia in Adults Working Party. British journal of haematology 1996;94(1):89-98.

[52] Hann IM, Stevens RF, Goldstone AH, et al. Randomized comparison of DAT versus ADE as induction chemotherapy in children and younger adults with acute myeloid leukemia. Results of the Medical Research Council's 10th AML trial (MRC AML10). Adult and Childhood Leukaemia Working Parties of the Medical Research Council. Blood 1997;89(7):2311-8.

[53] Mandelli F, Petti MC, Lo Coco F. Therapy of acute myeloid leukemia: towards a patient-oriented, risk-adapted approach. Haematologica 1998;83(11):1015-23.

[54] Stein EM, Tallman MS. Remission induction in acute myeloid leukemia. International journal of hematology 2012;96(2):164-70.

[55] Candoni A, Martinelli G, Toffoletti E, et al. Gemtuzumab-ozogamicin in combination with fludarabine, cytarabine, idarubicin (FLAI-GO) as induction therapy in CD33-positive AML patients younger than 65 years. Leukemia research 2008;32(12):1800-8.

[56] Patel JP, Gonen M, Figueroa ME, et al. Prognostic relevance of integrated genetic profiling in acute myeloid leukemia. The New England journal of medicine 2012;366(12):1079-89.

[57] Inaba H, Coustan-Smith E, Cao X, et al. Comparative Analysis of Different Approaches to Measure Treatment Response in Acute Myeloid Leukemia. J Clin Oncol 2012.

Chapter 11

Tumorablative Allogeneic Hematopoietic Stem Cell Transplantation in the Treatment of High-Risk and Refractory Leukemia — New Concepts and Clinical Practice

Wan-ming Da and Yong Da

Additional information is available at the end of the chapter

http://dx.doi.org/10.5772/53031

1. Introduction

The substance of bone marrow transplantation is the organ transplantation. Accurately, it is the grafting of hematopoietic and immunologic system. Comparing to the transplantation of solid organ, in the hematopoietic stem cell transplantation (HSCT), the ill organ, id est. hematopoietic and immunology system, is ablated by high-dose chemotherapy and total body irradiation(TBI) (conditioning regime). Thus, the normal hematopoietic stem cells could be engrafted and normal function of hematopoietic and immune system could be reconstituted. The standard myeloablative conditioning regimen would be reasonable or enough for the non-malignances of marrow, which needed by replacing therapy, such as marrow failure. However, for treatment of hematopoietic malignances, it maybe not cure the malignance diseases to ablate the normal hematopoietic, immune system and reconstitute the normal function of allogeneic hematopoietic and immune system of patients. Because the leukemic stem cells (LSC) are not only existence in the bone marrow, it might be occurrence in any site of body. For instance, the traditional myeloablative conditioning regimen to treat leukemia could have striking killing effects of leukemic cells, and residual leukemic cells further eradicated by effect of the graft versus leukemia (GVL), but the malignant cells are not always removed at all in the all patients, therefore, relapse post transplantation could be occurred in the some patients. In fact, the traditional allogeneic myeloablative HSCT could cure or improve outcome of acute leukemic patients with standard risk, however, the disease relapse after transplant for acute leukemia with high risk and refractory is 40% to 80% [1-4]. Moreover, the leukemic cells in the majority of relapsed cases originate from inceptive leukemic cells at initial diagnosis [5-7], which strongly indicated that the standard myeloablative conditioning regimen could remove the normal lymphohematopoietic system of the recipients and make grafts successfully ecize and proliferate, but could not always kill the residual leukemic stem cells in vivo, particularly the those in extramedullary sites. Those residual leukemic stem cells are the crime for the disease recurrence. We pioneered the tumorablative allogeneic hematopoietic stem cell transplantation (TAHSCT) for treatment of those patients with high-risk, refractory, even advanced-stage acute leukemia. The TAHSCT involve all parts in procedure of transplantation, the principal contents include two elements that are using the individual tumorablative conditioning regimen and enhancing the immunotherapy post-transplantation.

2. Indication of TAHSCT

The indication of TAHSCT is the patients with high-risk, relapsed, refractory, even advanced leukemia. On the one hand, the recurrence of disease post-transplantation in these patients is very high by standard myeloablative transplantation. In the recent years, with the development of immunosuppressant, antibiotic agents and effective supportive therapy, it makes significant improvement to reduce the morbidity and mortality of non-relapse, such as GVHD, infections and multiple organ failure, post allogeneic HSCT, how to prevention and treatment of relapse after allogeneic HSCT in these acute leukemia is the key point to increase the long-term survival. In a recent retrospective cohort from the Center for International Blood and Marrow Transplant Research, the 3-year overall survival rate only was 16% in patients who underwent allo-HSCT in relapse or primary induction failure of acute lymphoblastic leukemia (ALL) [4], for acute non-lymphoblastic leukemia (ANLL) with High-risk, refractory and relapsed, it could be up to 20%-40%. On the other hand, we are faced with more and more of those patients in the clinical transplantation. It is necessary to improve and optimize traditional procedure of HSCT.

3. Rationale of TAHSCT

The leukemia is the malignant clone disease derived from hematopoietic cell. The leukemic stem cell is quite different from the normal hematopoietic stem cell in the biocharacteristics [8]. Comparing to the latter, the former has strong growth vigor and tolerance in some degree to chemotherapy or radiotherapy. Furthermore, the leukemic stem cell is not only in marrow, but also infiltrates to any sites or organs besides hematopoietic system, including some sites in which the anti-leukemia drugs could but achieved to the treating concentration, such as central nerve system, skin and lung and so on. On account of the insight in bionomics of leukemic stem cell, and the results in clinical transplantation, it is demonstrated that standard myeloablative HSCT could not enough to root out of leukemic or leukemic stem cells, particularly the in extramedullary sites. Therefore, the myeloablative HSCT is not equal to TAHSCT, the residual leukemic or leukemic stem cell is the convict for relapse [9].

Although GVL effect after transplantation produces a marked effect, it is always later after transplant. Eventually, the residual leukemic stem cell could be proliferation and disease relapse occurrences [10].

The purpose of tumorablative tailored conditioning regimen is not only to suppress or destroy the immune and hematopoietic system to make space for engraftment, but also to ablate leukemic stem cell, especially the leukemic stem cells in the "asylum" of extramedullary sites, and to induce or enhance the GVL effect as far as possible [11].

Compared with myeloablative transplantation, besides removal of normal hematopoietic tissue, TAHSCT focuses more on killing residual tumor cells, especially elimination of extramedullary residual tumor cells. In the selection of drugs, it puts more emphasis on the killing intensity of drugs on leukemic cells, the maintenance effective concentration and enough time of killing effect, and reduction of post-transplant leukemia relapse to minimum [11]. Compared with non-myeloablative transplantation or reduced toxicity transplantation, the latter still retains hematopoietic stem cells of recipient, in some extent, for autologous hematopoietic reconstruction, but also residue a certain amount of leukemic stem cells which might cause relapse, therefore the reconstructed mixed hematopoietic chimerism often requires donor lymphocyte infusion (DLI) to ensure the possibility of a successful engraftment, and the prevention and treatment of relapse are also dependent on DLI and subsequent immunotherapy or targeted therapy[11]. Their comparison is showed in Figure 1.

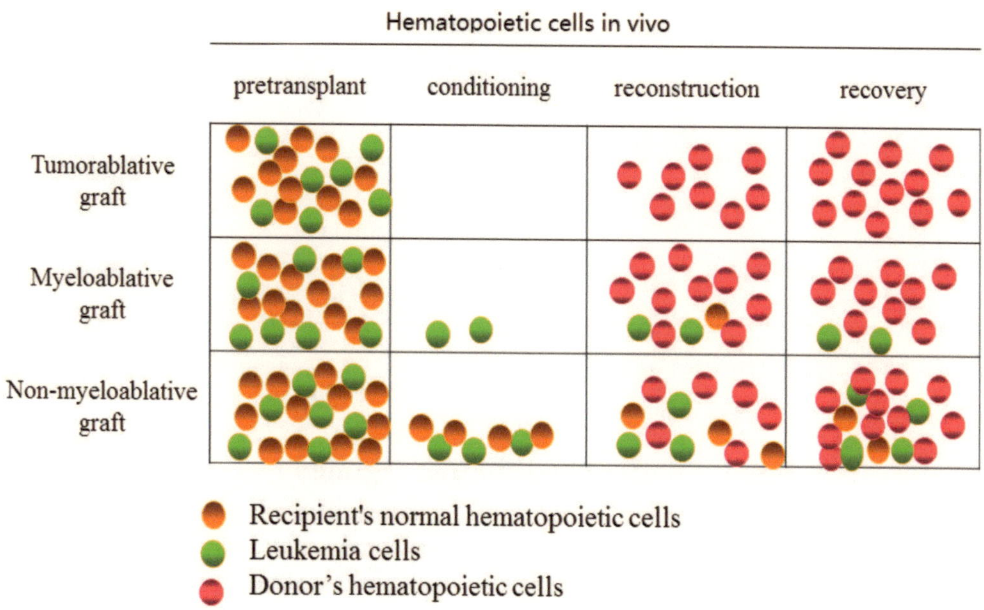

Figure 1. Comparison of tumorablative to myeloablative and non-myeloablative transplantation

4. Strategy for TAHSCT

Based on the regularity and characteristics of disease relapse after transplantation, we proposed a preventive pathway for leukemic recurrence post-transplantation in the early of 2007 years [12]. They are general prophylaxis, early intervention and clinical therapy. The general prophylaxis means to avoid the selection of high risks (shown in table 1) during the grafting procedure, the key points in early intervention are to institute a reasonable individual tumorablative conditioning regimen. The clinical therapy is to treat the leukemia in the early or frank relapse, including immunotherapy post transplantation. In the clinical practice for 5 years, the relapse rate of 85 and 83 cases with high risk, refractory or relapse received TAHSCT in 2008 and 2009 year was 2.3% and 5%, respectively [11]. It was strikingly advance; however, the challenge is still presence. Obviously, among the risks associated to relapse, the conditioning regimen and immunotherapy are more important.

Elements	High risk for relapse	Low risk for relapse
Recipient		
tumor cell burden	high	low
extramedullary disease	yes	no
unfavorable chromosome	Yes	no
unfavorable molecule	yes	no
sensitive to chemotherapy	insensitive	sensitive
performance status	worse	good
GVHD post grafting	no	yes
Grafts		
peripheral stem cell vs marrow cell	low	high
number of grating cells	low	high
T cell depleted	yes	no
Grafting technique		
conditioning regimen	non myeloablative	myeloablative
GVHD Prophylaxis	strong	fairly
immunosuppressive agent		reduce or stop early
interfere by immunotherapy		yes

Table 1. The risks associated to relapse in the transplantation

5. Approach to tumorablative conditioning regimen

Theoretically, the tumorablative conditioning regimen should contain drugs or TBI to ablate normal hematopoietic and immunologic tissue, also drugs or agents to get rid of leukemic or leukemic stem cell, particularly, those in the extramedullary sites. The ideal drugs should be high effective and targeted on the leukemic or leukemic stem cell, however, these special target agents have not been successfully used in clinical, and it should be exploited in the future. According to the clinical experiences for successful treatment of refractory or relapsed leukemia, and combining with standard myeloablative regimen, we selected some regimen with high effective and less toxicity, and establish a tumorablative conditioning regimens (shown in table 2).

Regimen	content	indication
HDAra-C+Bu/Cy	Ara-C 2.5g/m² IV, -11d- -9d Bu 1mg/kg.6hrs, -8d- -6d MCCNU 250mg/ m² (ANLL) or Vm26 300mg/ m² (ALL), -5d CY 50mg/kg.d IV, -3- -4d rest, -2 - -1d, HSCT d 0	High risk in CR
G-CSF primed HDAra-C+Bu/Cy	G-CSF 5μg/kg.d sc, -12- -9d Ara-C 3 g/ m² IV, -11d- -9d Bu 1mg/kg.6hrs, -8d- -6d MCCNU 250mg/ m² (ANLL) or Vm26 300mg/ m² (ALL), -5d CY 50mg/kg.d IV, -3- -4d Rest, -2 - -1d, HSCT d 0	Remission or early relapse in high risk patients with bone marrow hypoplasia or leukopenia
FLAG/RIT	G-CSF 5μg/kg.d sc, -14- -9d Ara-C 2 g/ m².d CI, -13- -9d FDL 30mg/ m².d IV, -13 --9d BSF 0.8mg/kg.6hr IV, -8 - -6d CY 50 mg/m².d IV, -5 - -4d MCCNU 250mg/m².d, -3d Rest, -2 - -1d, HSCT d 0	Progressive or advanced patients with ANLL
TBI/FLAG/CY	TBI 1.5 - 2 Gy, Bid, -13- -11d Vm26 300 mg/m ². IV, -10d G-CSF 5μg/kg.d sc, -10- -5d Ara-C 2 g/ m².d CI, -9- -5d FDL 30mg/ m².d, IV, -9 --5d CY 30 mg/kg.d IV, -4 - -3d Rest, -2 - -1d, HSCT d 0	Progressive or advanced patients with ALL

* All regimens can be used in transplantation of HLA matched unrelated and halo-identical HSCT, but ATG must be added. ATG, antithymocyte globulin; Ara-C, cytarabine; Bu, busulfan; BSF, busulfex; Vm26, teniposide; CY, cyclophosphamide; FDL, fludarabine; MCCNU, semustine; RIT, reduced intensive transplantation; TBI, total body irradiation.

Table 2. Some tumorablative conditioning regimen*

FLAG/reduced intensive transplantation (FLAG/RIT regimen) A large number of clinical practice confirmed that intravenous infusion of high-dose cytosine (Ara-C) was an effective rescue measure for the treatment of refractory or relapsed leukemia, about 40% refractory acute myeloid leukemia (AML) could achieve remission. Pharmacokinetic study on high-dose Ara-C intravenous infusion revealed that intravenous infusion of Ara-C (1.8-32.0) g/m^2 for 2 hours, every 12 hours, the plasma concentrations could reach (8-24) μg/ml, cerebrospinal fluid concentrations was about (10-15)% of plasma concentration. This high concentration of this drug in blood and cerebrospinal fluid was thought to be the pharmacological basis of significantly increased efficacy [13-15]. Ara-C combined with anthracycline (uniquinone) or acridines drugs could further improve the CR rate to 50 % [16, 17].

FLAG protocol consisting of fludarabine combined with Ara-C plus recombinant human granulocyte-stimulating factor (G-CSF) is currently a potent and well-tolerated treatment for refractory and relapsed AML. Fludarabine is a nucleotide analogue, acts as a ribonucleic acid inhibitor by phosphorylation to active triphosphate form F-ara-ATP. As a substrate for DNA synthesis in leukemic cells, F-ara-ATP has anti-leukemia activity by inhibition of DNA polymerase and ribose reductase, especially has a strong effect on quiescent cells. In vitro and in vivo studies proved that addition of fludarabine before Ara-C administration might increase the intracellular concentration of Ara-CTP, enhance the cytotoxicity and clinical efficacy of Ara-C, so that the CR rate of refractory and relapsed AML reaching 50% - 75%, CR period reaching 9 months and above [18]. Schmid et al [19] used combination chemotherapy with fludarabine and Ara-C for 4 days followed by reduced-toxicity allogeneic hematopoietic cell transplantation and post-transplant donor lymphocyte infusion in 103 refractory acute myeloid leukemia patients, followed up for a median period of 25 months. It was found that 1, 2 and 4-year overall survival rates were 54%, 40% and 32%, respectively. Therefore, the FLAG/RIT regimen is mainly used in treatment of ANLL with progressive or advanced patients with ANLL.

TBI/FLAG/CY regimen It consists of total body irradiation, FLAG and reduced cyclophosphamide, and always utilized to treat ALL in progressive or advanced phase. Because, TBI was more effectiveness in allo-HSCT for ALL.

G-CST priming regimen It is usually to treat the ANLL at remission or early relapse in high risk with bone marrow hypoplasia or leucopenia. Granulocyte-stimulating factor can induce the proliferation of AML cells and increase the proportion of S phase cells in vitro or in vivo, thereby enhancing cell sensitivity to chemotherapeutic drugs. Reasonable and sequential application of G-CSF and chemotherapeutics is another effective option for the treatment of refractory AML, such as the above-mentioned FLAG protocol and CAG protocol composed of low-dose Ara-C (LD-Ara-C), aclacinomycin and G-CSF. In fact a large number of experimental and clinical studies confirmed that pre-transplant application of G-CSF not only promoted the differentiation of T cells to TH2 and enhanced the function of regulatory T cells, but also amplified immature antigen-presenting cells and plasmacytoid dendritic cells, which was beneficial for maintenance of post-transplant T cells function and reduction the incidence of GVHD. Morris ES, et al. also confirmed that through modification of pegylation and combination with Flt-3L, G-CS might lead to activation and amplification of donor invariant NKT (iNKT) cells, a marked increase of post-transplant cell mediated $CD8^+$ T cytotoxicity, and enhancement of GVL effect [20]. Takahashi et al. had proved that application of G-CS together with conditioning regimen could reduce the post-transplantation relapse in refractory myeloid leukemia [21]. Ooi et al. used G-CSF + Ara-C or + total body irradiation and fludarabine as a conditioning regimen, and performed unrelated cord blood transplantation in adult AML patients, the results showed that 2-year disease-free survival was 76%[22]. Rational application of G-CSF in tumorablative conditioning regimen not only augmented anti-leukemia effect, but also separated GVHD and GVL effect to a certain degree, improved the safety of transplantation and reduced the relapse [23].

Regimen containing high Ara-C For the transplantation of ANLL with high risk in the complete remission, we used the regimen containing high Ara-C as tumorablative conditioning. As early as 2004, Lu DP et al. reported the application of GIAC protocol (Ara-C, busulfan, cyclophosphamide, MCCNU and G-CSF activated bone marrow and peripheral blood) including high-dose Ara-C, MCCNU and G-CSF for mobility of peripheral blood stem cells in donor-recipients HLA-unmatched or haploidentical hematopoietic stem cell transplantation. Post-transplant observation confirmed this protocol resulted in a higher disease-free survival rate (70%) and lower relapse rate (13%), further suggesting the necessity of intensified measures with direct anti-leukemic cell effects in the conditioning regimen [24].

Based on the above theory and the specific situation of individual patient, we have modified GIAC protocol and designed the HDAra-C+Bu/Cy. Preliminary clinical attempts have yielded encouraging results (Table 2).

In clinical practice, about one-third of AML and more than half of ALL patients relapsed firstly manifested as extramedullary relapse, such as leukemic sarcoma or infiltration into the central nervous system. So that, drugs with good liposolubility and ability to penetrate blood-brain barrier, such as Carmustine (BCNU), methyl cyclohexyl nitrosourea (MCCNU), teniposide (VM26) as well as high-dose Ara-C or MTX, should be chosen as a part of tumorablative regimen.

Our tumorablative conditioning regimen possess following features: The first, it could enhance the intensity of anti-leukemia chemotherapy. All regimens included continuous infusion of medium dose Ara-C for 72 hours, meanwhile drugs with good liposolubility were added such as MCCNU (acute myeloid leukemia) and teniposide (acute lymphocyte leukemia). The duration of the regimen extended to 11-14 days, which not only enhanced the anti-leukemia effects on leukemic (stem) cells in hematopoietic tissue, but also ensured a longer maintaining period of effective drug concentration in extramedullary tissue including central nervous system, and further depletion of leukemic (stem) cells in all tissues. Secondly, granulocyte-stimulating factor was added in some regimes. It not only recruited quiescent leukemic (stem) cells into proliferation cycle, increased the sensitivity to the killing effects of drugs, but also reduced or alleviates the incidence of post-transplant GVHD through regulation of immune cells, or might induce GVL effects. Third, the reduction of the dosage of alkylating agent decreased or alleviated the toxic and side effects, under the circumstances of depletion of normal hematopoietic tissue and effective immunosuppressant. Fourth, individualization was emphasized. In clinical, application of these regimens should focus on individualization, in view of the differences of cytogenetics and gene alterations in the pathogenesis of leukemic cell, clinical manifestations and prognosis, or pre-transplant disease, performance status and drug tolerance of patients. In addition, the regimens should also be adjusted in accordance with the donor source, for example, the transplantation of unrelated or haploididential donor, anti-lymphocyte globulin (ATG) should be included in the corresponding conditioning regimen [25]. We met a case of AML-M5 with primary resistance to chemotherapy, the blasts remain more than 50% in marrow after induction by daunomycin plus Ara-C (DA), idarubicin plus Ara-C (IA), mitoxantrone plus Ara-C and etoposide (MAE), CAG and FLAG regimen, but, after AE (amsacrine + Vm26) regimen, near CR was achieved. Then he received haplo-identical transplantation using TBI/FLAG/CY regimen in February, 2012, in which amsacrine + Vm26 instead of FLAG, because his leukemic cell is sensitive to amsacrine and Vm26. After successful engraftment, he is still alive in continual CR up to now.

6. Detection of minimal residual disease and immunotherapy post TAHSCT

Detection of minimal residual disease (MRD) and immunotherapy post transplantation are very impotent principle in the TAHSCT [25]. Although it is almost specific method to detect the marrow morphology, clone culture, immunophenotype, and abnormal gene or protein of leukemic cell, clinically, the flow cytometry (FCM) and polymerase chain reaction (PCR) are the more convenience, fast and sensitive. It should be routinely done post transplantation. In some patients, relapse proceeded in extramedullary sites, even sarcoma, especially, CNS, subperiosteum, skin, serous cavity, lung and intestinal tract, so image analysis also is necessity, such as, CT, MIR, PET or PET-CT. We had used a PET-CT to detect proceed relapse in extramedullary sites in an advanced case with ANLL after underwent unrelated HSCT, and successful pinpoint treated by the cyberknife.

With regard to immunotherapy, firstly, immunosuppressive agents should be decreased or even stopped as quickly as possible, when GVHD was strictly controlled. Then, if necessary, some immune modulators should be given, such as interferon, IL2 and thymopeptides. For the two latter, which should not use in T cell malignances. Finally, it is the cell therapy [25, 26].

Donor lymphocyte or G-CSF mobilized peripheral blood stem cell infusion (DLI/DSI) for treatment of leukemia relapse after allo-HSCT was introduced in early 1990s, being extremely effective in chronic myeloid leukemia. The DLI for AML relapse post-transplant has been questioned in general. Recently, Schmid C, et al retrospectively analyzed the data of 399 patients with AML in first hematological relapse after HSCT whose treatment did or did not include DLI. After correction for imbalances and established risk factors, the two groups were compared with respect to overall survival. Further, a detailed analysis of risk factors for survival among DLI recipients was performed. The results confirm a role for an allogeneic GVL effect in AML [27]. Various modifications of DLI have been investigated. These included the systematical use of mobilized donor PBSC concentrates instead of lymphocytes, or the systemic application of cytokine induced killer cells (CIK) for additional immunostimulation to increase GVL efficacy. In addition, infusion of allogeneic natural killer (NK) cells is also a promising innovative immunotherapy, being alloreactive NK cells reported to produce a strong GVL effect after haploidentical HSCT in patients with advanced AML, without causing GVHD [28]. In vitro studies have suggested the possibility to create specific antileukemic cytotoxicity by stimulation of donor lymphocytes using AML-derived dendritic cells. Porter and colleagues reported encouraging results from a phase I trial using conventional DLI, followed by an additional infusion of ex vivo activated donor T cells. Therefore, we used DLC, DSI or CIK as maintenance therapy after HSCT for patients in remission or in a minimal residual disease situation in our program to exploit the GVL efficacy, and got a ducky result. Recently, we treated 18 cases in relapse after allogeneic HSCT, including 11 of HLA matched sibling, 5 of haploidentical and 2 of matched unrelated donor by donor's dendritic cell-primed CIK (DC-CIK). After the median number of 3.6 X 10^9 DC-CIK infused, molecule complete remission was obtained in 12 cases (68%), and 11 of 12 cases are survival with a median follow-up of 12 (range 6-41) months, except 1 died of treatment related complication. It confirms that donor derived DC-CIK infusion is efficacious and safety in this setting [29]. However, DLI or CIK infusion was often associated with a considerable risk of GVHD, and clinically, we should be careful to assay and prevent from GVHD.

Along with screening and identification of new immunogenic tumor protein or peptides, anti-tumor specific functional T cells could be produced in vitro, the anti-leukemia specific immunotherapy would have more definite position in treatment of relapse post transplantation. Further experimental and clinical research are required to overcome the obviously high burden of leukemia blasts to escape from an allogeneic immunereaction in relapsing patients after allogeneic HSCT for refractory acute leukemia.

7. Clinical practice of TAHSCT

In fact, the first TAHSCT with HLA identical donor we preformed was in 2007 for a 54 years old female with resistant relapse, She was diagnosed as AML in July 2004 and obtained CR after 3 courses of chemotherapy by daunomycin plus Ara-C (DA), idarubicin plus Ara-C (IA) and mitoxantrone plus Ara-C (MA), and then received 13 courses of intensive consolidation chemotherapy including high dose of Ara-C, and autologous CIK infusion for 3 times. Her leukemia relapsed in the end of Dec 2006, and could not response to the several courses of reinduction chemotherapy. Before she received tumorablative allo-HSCT, there were 27% of leukemic blasts in marrow. The FLAG/RIT regimen was conditioned for HLA-identical sibling HSCT on February 22nd 2007. Her neutrophil and platelet were successfully engraftment on +18 days, chimerism analyses shown that full donor chimerism achieved by +30 days. Assay of MRD periodically by FCM monitoring after TAHSCT was zero. Grade I aGVHD of intestinal tract and liver was happened on +51 days, and thereafter invasive fungal infections in sinusitis, lung, liver on right-sidedness (pathological culture supported mucor infection) were happened, The Aspergillus was detected in sputum culture. All the complications above were controlled and cured after symptomatic treatment. She is still alive in continue complete remission up to now (more than 78 months).

The first TAHSCT with haploidentical donor we did was in 2008. The case with 22 years old male was diagnosed as AML with t (8; 21), AML1/ETO positive in August 2005. He had received several courses of intensive consolidation, and maintenance chemotherapy including high dose Ara-C. Three years later, he had leukemia relapse, and not obtained CR again after reinduction chemotherapy. Before transplantation, there were 75% of leukemic blasts in marrow. The patient received the FLAG/RIC/ATG conditioning regimen for HLA haploidentical TAHSCT from his sibling. On January 1st 2008, after TAHSCT, engraftment was durable with full donor chimerism, and detection of non MDR by FCM, chromosome, and realtime PCR for AML1/ETO fused gene monitoring. Limited cGVHD was controlled by CSA and prednisone in fewer months. The patient is still alive in disease-free survival (DFS) until now.

Between August 2006 and march 2007, a total of 57 patients with high risk/refractory leukemia were received tumorablative individualized conditioning regiments, included HDAra-C +Bu/Cy, G-CSF primed HDAra-C+Bu/Cy, and FLAG/RIT. Among 57, 20 patients of acute lymphoblast leukemia (ALL), 23 patients of acute myelogenous leukemia (AML), and 12 patients of chronic myeloid leukemia (CML) in accelerate or blast crises phase, and 2 patients of myelodysplastic syndrome–refractory anemia with excess of blasts (MDS-RAEB). 28 patients received haplo-identical transplantation, 17 patients HLA-identical unrelated donor transplantation and 12 patients HLA-identical sibling transplantation. The results showed that 56 patients, but one recovered with autologous hematopoiesis, attained durable engraftment. The median time to an absolute neutrophil count $>0.5x10^9/L$ was 16 (range: 12-21) days. The median time to a platelet count $>20x10^9/L$ was 18 (range: 12–32) days..With a median follow-up of 17.5 (2-34)months, the probabilities of OS and DFS were (74.7±6.1) % and (62.4±6.7) %, respectively. The incidence rate of aGVHD in grades II-IV and III-IV were (19.3±5.2)% and (12.3±4.3)% respectively. Extensive chronic GVHD was observed in 36 (64.3%) patients. Cytomegaloviremia (CMV) was observed in 39 (68.42%) patients. Hemorrhagic cystitis was observed in 13 (22.8%) patients. Fungous and bacterial infection occurred in 16 (28.07%) and 38 (66.67%) patients, respectively. The relapse in all patients occurred in 14 (24.6%). Among them, relapse rate in high risk and advanced group (blast cells were more than 20% in bone marrow) were 28.1% and 15.6%, respectively. 11 of 14 patients relapsed in marrow, 3 of 14 relapsed in extramedullary sites, 15 patients died (6 from hematological relapse, 5 from infection of bacterial and fungous, 4 from chronic GVHD) after 100 days. The toxicity in this TAHSCT could be tolerance, and overcame [30,31].

Recently, we reported forty-nine patients, from first affiliated hospital, Chinese PLA General Hospital, of hematological malignancy with high risk or refractory, including 24 AML, 14 ALL, 9 non-Hodgkin lymphoma (NHL), and 2 CML in blast crisis. All patients received haploidentical TAHSCT, in which umbilical cord mesenchymal stem cells were added. All patients achieved engraftment and complete remission after TAHSCT. Regimen-related toxicities were tolerable. Only five patients (10.2%) experienced relapse at a median time of 192 days after transplantation. The probability of 2 year leukemia free survival (LFS) in

AML and CML patients was 83.3%, which was significant higher than that in the ALL and NHL patients (40.0%, P<0.05) [32]. Our another 45 patients, form Beijing Dao-pei hospital, with refractory recurrent AML treated by TAHSCT [33]. The median blasts in marrow were 36% (20% to 92%) before transplantation, including 6 of HLA identical sibling, 9 of unrelated and 30 of haploindentical transplantation. All but 2 patients attained durable engraftment. The incidence of grade II to IV aGVHD and cGVHD were 34% and 59.1%, respectively. With median follow-up 30 (0.5 to 57) months, the relapse rate was 29.2%. Twenty nine (60.2%) patients remained CR since transplantation. Three years DFS and Overall survival (OS) were 60.2% and 62.6%, respectively. These data confirmed that the individualized tumorablative allogeneic hematopoietic stem cell transplantation is a promising and safety choice for treatment of high risk, refractory or relapse leukemia, even with high leukemia burden.

In the recent, these TAHSCT have being carried out in many hospitals in China. A total of 250 patients from 5 hospitals enrolled, the all patients with high-risk, resistant or relapse, even advanced hematological malignances, including leukemia and lymphoma [32,33-34]. The primary clinical observation revealed that the results are the similar to that above (data not published). Obviously, its efficacy must be confirmed by randomized, prospective clinical trials on a large population.

Actually, many investigators have being devoted to prevent from and treat recurrence post transplantation in refractory leukemia, including Schmid C, et al, who used a sequential treatment with chemotherapy and reduced-intensity conditioning for allogenic stem cell transplantation [19], and Takahashi S. et al, which used GCS-F combined regimen for allogeneic bone marrow transplantation shown above [21]. Recently, Eom KS, et al. reported that FLANG salvage chemotherapy as a safe bridge to transplantation for patients with relapsed or refractory acute myeloid leukemia is an effective regimen [35]. Arita K, et al. described that a sequential chemotherapy and myeloablative allogeneic hematopoietic stem cell transplantation for refractory acute lymphoblastic leukemia [36], and so on. All of them have obtained encouraging results. Comparing to them, our TAHSCT strategy emphasizes more entirety, individual and tumorablative efficacy in the sitting of tolerated toxicity.

8. Conclusion

In summary, the TAHSCT strategy is a primary entirety approach for treatment of hematopoietic malignances with high-risk, refractory or resistant relapse, based on the successful experiences either standard HSCT or chemotherapy for these patients. It is true, there were still some relapse post TAHSCT strategy, however, and it has reduced the relapse rate to about 20% in these patients, so it also highlights the need to improve. Theoretically, the residual leukemic stem cell is the chief offender in relapse post transplantation. We can utilize the differences in the biocharacteristics between normal and leukemic stem cells, which result of regulating disorder in proliferation, differentiation, apoptosis and ecize, signaling pathways, and so on [8,37,38], to exploit the targeted drugs with specific killing effects on LSC and niches for LSC, apoptosis-promoting and differentiation-inducing effects, such as tyrosine kinase inhibitor, FLT3 inhibitor [39], hypomethylating agent, and so on, together with the specific functional T cell adoptive immunotherapy. It should provide a broad prospect for the prevention and radical cure of relapse after TAHSCT in patients with refractory and relapsed leukemia.

Acknowledgements

I thank all members of hematology department, first affiliated hospital, Chinese PLA general hospital, and Beijing Dao-pei Hospital for their dedicated works in these studies. Thank Dr Haiyan Zhu for preparation of this manuscript.

Author details

Wan-ming Da[1] and Yong Da[2]

*Address all correspondence to: dwanming@medmail.com.cn, dayong2727@sohu.com

1 Department of Hematology, Unit of Bone Marrow Transplantation, and Department of Geriatric Hematology, Chinese PLA General Hospital & Chinese Medical College; The Department of Hematology, Unit of Bone Marrow Transplantation, First affiliated Hospital, Chinese PLA General Hospital & Chinese Medical College;, Beijing, China

2 The Affiliated Hospital of The Military Medical Science Academy of Chinese PLA, Beijing, China

References

[1] Appelbaum FR. Hematopoietic cell transplantation for adult acute myeloid leukemia. In: Appelbaum FR, Forman SJ, Negrin RS, Blume KG.(ed.). Thoma's hematopoietic cell transplantation: stem cell transplantation. 4th edition. Wiley–Blackwell, A John Wiley & Sons Ltd, Publication. p761-774.

[2] Lu Dao-pei. The major developments in hematopoietic stem cell transplantation—before 2003. J Peking University: Health Sciences 2003; 35(2) 113-114.

[3] Da Wan-ming, Wu Tong, Lu Dao-pei. Unceasingly improving outcomes of allogeneic hematopoietic cell transplantation. Chin J Hematol 2006; 7(8) 505-506.

[4] Duval M, Klein JP, He W,, et al. Hematopoietic stem cell transplantation for acute leukemia in relapse or primary induction failure. J Clin Oncol 2010; 28(23) 3730-3738.

http://dx.doi.org/10.5772/53031

[5] Da Wan-ming, Wu Tong, Lu Dao-pei. Prevention and treatment of leukemia relapse after allogeneic hematopoietic stem cell transplantation for acute leukemias . Chin J Hematol 2007; 28(12) 793-794.

[6] Flynn CM, Kaufman DS. Donor cell leukemia: insight into cancer stem cells and the stem cell niche. Blood 2007; 109(7) 2688-2692.

[7] Majhail NS. Old and new cancers after hematopoietic cell transplantation. Hematol Am Soc Hematol Educ Program 2008;142-149.

[8] Dick JE, Lapidor T. Biology of normal and acute myeloid leukemic stem cells. Int J Hematol 2008; 82(5) 389-396.

[9] Da WM. Strategies for prevention and management of leukemic relapse after allogeneic hematopoietic cell transplantation. Chinese Blood and Marrow Transplantation Forum and Annual Scientific Meeting of Hong Kong Society of Haematology. 2010; 37.

[10] Ringden O, Pavletic SZ, Anasetti C, et al. The graft-versus-leukemia effect using matched unrelated donors is not superior to HLA-identical siblings for hematopoietic stem cell transplantation. Blood 2009;113(13) 3110-3118.

[11] Da WM. Tumorablative allogeneic hematopoietc stem cell transplantation. In Da WM, Chen XQ (ed.) Hematology 2010. Beijing: Chinese Medical Multimedia Press. Beijing. 2010. p230-239.

[12] Da WM. Prevention and treatment for relapsed leukemia post allogeneic hematopoietic cell transplantation. Beijing: Chinese Medical Multimedia Press. 2008. p22.

[13] Breithaupt H, Pralle H, Eckhardt T, et al. Clinical results and pharmacokinetics of high-dose cytosine arabinoside(HD ARA-C) . Cancer. 1982, 50(7) 1248-1257.

[14] Capizzi RL, Poole M, Cooper MR, et al. Treatment of poor risk acute leukemia with sequential high-dose ARA-C and asparaginase. Blood 1984; 63(3) 694-700.

[15] Xie Xiao-tian, Li Ben-shang, Li L, et al. Determination of plasma and cerebrospinal fluid drug concentrations during high-dose cytosine arabinoside treatment[J]. Shanghai Medical Journal 2005; 28(3) 220-223.

[16] Kern W, Aul C, Maschmyer G, et al. Superiority of high-dose over intermediate-dose cytosine arabinoside in the treatment of patients with high-risk acute myeloid leukemia: results of an age -justed prospective randomized comparison. Leukemia 1998; 12(7) 1049-1055.

[17] Pastore D, Specchia G, Carluccio P, et al. FLAG-IDA in the treatment of refractory/ relapsed acute myeloid leukemia: single-center experience. Ann Hematol 2003; 82(4) 231-235.

[18] Clavio M, Carrara P, Miglino M, et al. High efficacy of fludarabine-containing therapy (flag-flang) in poor risk acute myeloid leukemia. Haematologica 1996; 81(6): 513-520.

[19] Schmid C, Schleuning M, Schwerdtfeger R, et al. Long-term survival in refractory acute myeloid leukemia after sequential treatment with chemotherapy and reduced - intensity conditioning for allogeneic stem cell transplantation. Blood 2006; 108(3) 1092-1099.

[20] Morris ES, MacDonald KP, Rowe V, et al. NKT cell- dependent leukemia eradication following stem cell mobilization with potent G-CSF analogs. J Clin Invest 2005; 115(11) 3093-3103.

[21] Takahashi S, Okamoto SI, Shirafuji N, et al. Recombinant human glycosylated granulocyte colony-stimulating factor (rhG-CSF) - combined regimen for allogeneic bone marrow transplantation in refractory acute myeloid leukemia. Bone Marrow Transplant 1994;13(3) 239-246.

[22] Ooi J, Iseki T, Takahashi S, et al. Unrelated cord blood transplantation for adult patients with de novo acute myeloid leukemia . Blood 2004; 103(2) 489-491.

[23] Morris ES, MacDonald KA, Hill GR. Stem cell mobilization with G-CSF analogs: a rational approach to separate GVHD and GVL? Blood 2006; 107(9) 3430-3435.

[24] Lu DP, Dong L, Wu T, et al. Conditioning including antithymocyte globulin followed by unmanipulated HLA-mismatched/haploidentical blood and marrow transplantation can achieve comparable outcomes with HLA-identical sibling transplantation. Blood 2006;107(8) 3065-3073.

[25] Da WM. Allogeneic tumorablative hematopoietc stem cell transplantation. BME & Clin Med 2012;16: 94-98.

[26] Mackall C, Fry T, Gress1 R, et al. Background to hematopoietic cell transplantation, including post transplant immune recovery. Bone Marrow Transplant 2009;44(8) 457-462

[27] Schmid C, Labopin M, Nagler A, et al. Donor Lymphocyte infusion in the treatment of first hematological relapse after allogeneic stem-cell transplantation in adults with acute myeloid leukemia: A retrospective risk factors analysis and comparison with other strategies by the EBMT acute leukemia working party. J Clin Oncol 2007; 25(31) 4938-4945.

[28] Nguyen S, Béziat V, Norol F, etal. Infusion of allogeneic natural killer cells in a patient with acute myeloid leukemia in relapse after haploidentical hematopoietic stem cell transplantation. Transpl Cellular Eng 2011;51(8) 1796-1778.

[29] Wang JB, Wu T, Yang JF et al. Management of early leukemia relapse after allogeneic hematopoietic stem cell transplantation by donor's dendritic cell-primed cytokine-induced killer cells. Blood 2008; 112: 307 (2008 ASH Annual Meeting Abstracts 829).

[30] Da WM, Wang JB, Zhang JP, et al. Tumorablative tailored allogeneic hematopoietic stem cell transplantation in treatment of high risk and reftractory leukemia. Chin J Experiment Hematol 2009;17:219-220.

[31] Wang JB, Da WM, Zhang JP, et al. Successful treatment of high risk/refractory leukemia by tumor-ablative individualized conditioning allogeneic hematopoietic stem cell transplantation. Chin J Hematol 2010; 31 (8) 505-509.

[32] Wu YM, Wang ZH, Cao YB, et al. Cotransplantation of haploidentical hematopoietic and umbilical cord mesenchymal stem cells with a modified myeloablative regimen for refractory/relapsed hematologic malignancy. Ann Hematol 2012 (in press).

[33] Wang JB, Wu T, Da WM, et al. Salvaged allgeneic hematopoietic stem cell transplantation for refractory recurrent acute myeloid leukemia. Chin J Hematol 2012; 33(6) 467-470.

[34] Wang ZH, Wu XX, Liu ZY, et al. Curative efficacy analysis of allo-HSCT using conditioning regimen of FLAG combined with modified BuCy for refractory and relapsed patients with hematologic malignancies. Chin J Experimental Hematol 2011; 19(1) 158-162.

[35] Eom KS, Min WS, Kim HJ,et al. FLANG salvage chemotherapy is an effective regimen that offers a safe bridge to transplantation for patients with relapsed or refractory acute myeloid leukemia. Med Oncol. 2011; 28 Suppl 1: S462–S470.

[36] Arita K, Kondo T, Sugita J, et al.Sequential chemotherapy and myeloablative allogeneic hematopoietic stem cell transplantation for refractory acute lymphoblastic leukemia. Int J Hematol 2011; 94(3) 291–295.

[37] Miller JS, Warren EH, van den Brink MRM, et al. NCI first international workshop on the biology, prevention, and treatment of relapse after allogeneic hematopoietic stem cell transplantation: Report from the committee on the biology underlying recurrence of malignant disease following allogeneic HSCT: graft-versus-tumor/leukemia reaction. Biol Blood Marrow Transplant 2010;16(5) 565-586.

[38] Abboud CN. Another nail in the AML coffin. Blood 2009;113(24) 6045-6046.

[39] Sorà F, Chiusolo P, Metafuni E, et al. Sorafenib for refractory FMS-like tyrosine kinase receptor-3 (FLT3/ITD+) acute myeloid leukemia after allogenic stem cell transplantation. Leukemia Research 2011; 35 , 422–423

Chapter 12

Stem Cell Transplantation in Chronic Lymphocytic Leukemia

Anna Gazzola, Alessandro Broccoli,
Vittorio Stefoni and Pier Paolo Piccaluga

Additional information is available at the end of the chapter

http://dx.doi.org/10.5772/53085

1. Introduction

Chronic lymphocytic leukemia (CLL) is the most common leukemia diagnosed in Western world, with an incidence of 3–6/10^5 per year, that increase to 12.8/10^5 at the age of 65 [1-3]. This disease is characterized by an extremely heterogeneous behavior, with the clinical course varying from patients who never require therapy to patients with a rapid disease progression and early resistance to treatment. In fact, about 20% of the patients show an aggressive course and die within few years from diagnosis [3, 4].

Molecular markers, such as cytogenetic alteration [5], immunoglobulin heavy chain (*IGH*) and *TP53* genes mutational status [6, 7], zeta associated protein 70 (ZAP70) expression [8] and CD38 expression [9], help to predict outcomes in CLL. However, their presence in the absence of clinical symptomatology is not a sufficient criterium to initiate therapy. Further, even in the absence of these factors, resistance to purine-analogue treatment can occur. This suggests that additional molecular mechanisms, which confer drug refractoriness in poor-risk CLL, do exist. In this regard, based on next generation sequencing studies, it was recently shown that additional genetic events can predict CLL patients outcome, including *NOTCH1*, *SF3B1*, and *BIRC3* mutations [10-15].

A major issue in CLL is the identification of the optimal timing and type of first-line treatment. In the current recommendation of International Workshop on CLL (IWCLL) updated guidelines for the diagnosis and treatment of CLL the therapy is reserved for advanced, symptomatic or progressive disease [16]. Numerous studies showed that, either as first-line therapy or in relapsed/refractory settings, treatment with new agents, such as alemuzumab, ofatumumab, lenalidomide, and flavoperidole [17-22] or combination of conventional chemotherapy to target immunotherapy lead to a better response in these patients [23-29]. These approaches significantly reduce the tumor load in refractory patients, even if the ultimate cure of disease has not yet been achieved. Therefore, CLL remains incurable outside the setting of allogeneic stem cell transplant (allo-SCT). In fact, up to date this is the only therapeutical approach that can potentially offer a curable solution to this disease [9]. The indications for SCT in CLL were established by European Bone Marrow Transplant (EBMT) [30]. Specifically, based on the evidence of efficacy and potential toxicity of SCT in CLL, these procedure is designated for high-risk CLL patents. These include: a) patients with *TP53* abnormalities, who fail to achieve complete remission (CR) or who progress within 12 months after purine analogues, b) those who relapse within 24 months after having achieved a response with purine-analogue-based combination therapy, c) those who relapsed after prior autologous SCT and d) patients who are fludarabine refractory [31, 32]. It should be noted that none of these categories requires assessment of biologic risk factors except cytogenetic detection of *TP53* deletions. Ongoing prospective clinical studies will determine the impact of biomarkers such as *IGH* mutational status and other cytogenetic abnormalities in identification of patients at sufficiently high risk for allogeneic SCT use at first CR. Several groups agree that allogeneic transplant early in the disease course is the best strategy for patients with high-risk or poor response to initial therapy. For those with durable first remissions, the timing of transplant is more controversial. The debate in *"when to proceed to a more aggressive treatment approach?"* in CLL is in part driven by the presence of new therapeutic strategies available for these patients. However, it is unknown how these therapies will change the indications for or the outcome following transplant in CLL. Nevertheless, these promising results have already started to impact the transplant recommendations in CLL patients in a similar manner to chronic myeloid leukemia (CML) patients in imatinib era.

In this chapter the Authors, based on their own experience as well as on the most updated literature, discuss the usage of autologous and allogeneic SCT in the clinical setting of CLL, also in the light of the novel biological prognostic indicators.

2. Autologous hematopoietic stem cell transplantation

Autologous stem cell transplantation (ASCT) has been extensively investigated as a treatment option for CLL patients during the last years.

Evidences from clinical and minimal residual disease (MRD) studies have suggested that ASCT has curative potential in only few patients. Nevertheless, ASCT might be capable of prolonged disease control even in CLL with poor-risk features.

Autologous transplantation consists in the collection of stem cells from the patient's marrow or peripheral blood before high-dose irradiation or chemotherapy and their subsequent reinfusion to guarantee a new blood production. The main problems with this procedure are the risk of re-infusion of leukemic cells that could potentially contaminate the stem cell population and the difficulty in mobilizing progenitor cell in patients who have received multiple previous treatments [33, 34]; particularly if purine analogs, have been administrated [35]. In addition, the outcome of ASCT is strongly correlated with the status of the disease: patients transplanted in CR have a much better outcome than those transplanted with active disease [36]. Therefore, optimal disease control prior to transplantation is mandatory [33, 34, 37].

Other factors that negatively influence the transplantation outcome and correlate with early relapse are: the interval between the diagnosis and the transplant, the number of prior lines of therapy, the presence of adverse cytogenetic abnormalities and of unmutated *IGH* genes [36, 38]. In addition, the detection of MRD by either polymerase chain reaction (PCR) or flow cytometry after transplantation anticipates clinical relapse [39, 40].

As mentioned above, different studies have investigated the role of ASCT in patients with CLL and the results were controversial. A retrospective matched-pair analysis suggested a survival advantage for ASCT in 66 patients who had undergone a uniform high dose therapy and transplantation over conventional therapy in 291 patients. With an overall median follow-up time of 70 and 86 months, survival was significantly longer for the patients who had undergone ASCT compared with conventionally treated patients [41]. However, in 2011, several prospective studies have failed to confirm the survival advantage of ASCT in advanced CLL patients [42, 43]. Brion et al. [43] published the results of a prospective multicenter randomized trial on the benefit of ASCT using a cyclophosphamide/TBI preparative regimen in advanced clinical-stage untreated CLL compared to conventional treatment. The conventional treatment was represented by 6 cicles of miniCHOP; for the ASCT cohort the scheduled therapy consisted of 3 miniCHOP cicles followed by immediate ASCT for patients with a very good partial remission (VGPR) or CR. This study highlights the absence of differences in median overall survival (OS) between the two groups thus denying the superiority of ASCT over conventional therapy.

The necessity of additional randomized studies to better clarify the role of ASCT in the management of patients with CLL was further emphasized by a comparative study conducted by the EBMT group in which 621 autografted patients were compared to 630 non-autografted patients. Patients autografted within 18 months of diagnosis had a better outcome that those treated with chemotherapy, but this was offset by an inferior outcome of patients autografted after 18 months [44]. In addition it was found a promising benefit by the T-cell mediated cytotoxicity via autologous transplantation in the high-risk CLL population.

Interestingly, Porter et al. [45] reported on the management of a chemo-refractory, CLL patient with del(17p) treated with autologous T-cells genetically modified to express anti-CD19; although the long-term disease control and late toxicities are not yet known, the patient was in remission [45].

Most of the studies published have relatively short follow up and therefore only focus on treatment related mortality (TRM) early after transplant, but the late consequences, particularly the development of secondary myelodysplasia and acute myeloid leukaemia (MDS/ AML), deserve some concern [37]. In fact, among 65 patients treated with fludarabine followed by ASCT, 8 developed MDS/ AML [37, 46]. Of note, in most studies, despite a high initial CR rate, relapse is common, suggesting that autologous transplant is unlikely to be curative in CLL [37]. However, based on the present literature, although ASCT cannot be considered as a standard treatment it should be considered in the context of clinical trials or as an innovative therapy to prolong survival in selected patients (i.e. those with chemosensitivity, absence of unfavorable factors, and transplanted early in the course of the disease).

3. Allogeneic hematopoietic stem cell transplantation

In recent years, allogeneic hematopoietic stem cell transplant (allo-SCT) was visibly emerged as the favorite treatment option for patients with high-risk CLL. In fact, in contrast with ASCT, allo-SCT can induce durable responses even in patients refractory to therapy [47-49]. Studies on the outcomes post ASCT failed to show a plateau effect on survival curves and resulted in a remarkably high incidence of secondary myelodysplastic syndromes (9% to 12%) [50]. On the contrary, in most series where allo-SCT has been carried out, a plateau is observed, with 40–60% of the patients remaining alive and free of disease 5–6 years after transplantation [39, 44, 46, 48, 49, 51-55]. Therefore, allo-SCT become, in the last two decades, the first treatment approach with curative potential in CLL.

The crucial anti-leukemic principle of allo-SCT in CLL appears to be the graft-versus-leukemia effect (GVL). The resultant GVL effect derived from alloreactive donor T cells is the key mechanism responsible for lowering relapse rates after allo-SCT. There is evidence that the GVL effect plays an essential role in controlling the disease and reverts poor prognostic biological variables such as unmutated *IGH* genes [56, 57]. In addition, one of the most important advantage of allo-SCT includes infusion of tumor-free hematopoietic progenitor and effector cells from healthy donors. Of note, it is important to exclude the presence in donor peripheral blood of a monoclonal population immunophenotypically identical to that of patients with CLL; in fact it was demonstrated that CLL clones were found in around 12% of the first-degree relatives of patients with CLL and in up to 3% of the general population [58]. Nevertheless, the use of allo-SCT is limited due to the advanced age of most patients with CLL and the high mortality associated with the procedure (in the range 24–47%), main causes for death being graft-vs host disease (GVHD) and infections.

At present, ongoing prospective clinical studies will determine the impact of biomarkers including *IGH* mutational status and other cytogenetic abnormalities in identification of patients with sufficiently high risk to deserve use of allo-SCT in first CR.

4. Myeloablative allogeneic stem cell transplantation

In myeloablative allo-SCT, patients are given extremely high doses of chemotherapy, with or without radiation, to wipe out, or "ablate," the marrow. Then they are given an infusion of donor stem cells to revive blood cell production and immunity.

Several theoretical advantages of myeloablative allo-SCT over ASCT are: a) none tumor contamination of the stem cell b) GVL effect to eliminate chemotherapy-resistant leukaemia cells by immune mechanisms c) better survival curves. In fact, studies from MD Anderson Cancer Center demonstrate improved outcome after allogeneic compared to ASCT [59] suggesting that myeloablative allo-SCT can induce durable remission even in patients with refractory disease. However, the major limitation of using myeloablative allo-SCT is the increased risk of transplant-associated morbidity and mortality, mostly from organ failure due to direct toxicity of the preparative regimen and/or development of GVHD [48, 60, 61].

Registry data from the International Bone Marrow Transplant Research (IBMTR) group and the EBMT group reported a transplant-related mortality (TRM) of 46% with mortality from GVHD of 20% [60]. These published data showed that approximately two-thirds of allo-transplanted CLL patients will succumb either to TRM or to recurrent disease, and approximately one-third will be cured of their disease [54, 60].

Active chronic GVHD is principal determinant of long-term morbidity and significantly reduced long-term health status in patients allografted for various hematological malignancies [62]. Indeed, transplant-related long-term morbidity after allo-SCT for CLL can be significant but is mainly restricted to those patients who have ongoing active chronic GVHD. However, in the majority of affected patients clinical symptoms of chronic GVHD resolved over time, allowing discontinuation of systemic therapeutic immunosuppression after a median of 25 months [63]. Further, a high graft rejection rates remain a relevant complication in myeloablative allo-SCT; a possible explanations could be the significant marrow infiltration in CLL patients at the time of transplantation, inversely correlated with outcome [64], and the role played by host dendritic cells, which are seriously defective in CLL patients [65]. Another problem is represented by the high infection rates, that correlated with preexisting immunosuppression. Infections are the cause of about 50% of all CLL-related deaths [62, 66] primarily in fludarabine and/or alemtuzumab-refractory patients [16, 67]. Moreover in recent reports the risk of infections has been clearly correlated with presence of GVHD [57, 63, 65, 68] and refractory disease [67, 69]. In addition, it is important to note that patients with chemosensitive disease have significantly better outcomes than patients with refractory disease, suggesting that an earlier application of allo-SCT may further improve transplantation outcomes [60, 70, 71].

In conclusion, allo-SCT is a therapy with curative potential in CLL and, in contrast to conventional treatment, with an high potential of providing long-term disease control even in patients with a very unfavorable biological and clinical risk profile. However, in addition to the disease risk, it is necessary to consider patient-related risk factors, such as age and comorbidity, when allo-SCT is performed [63].

5. Reduced-intensity conditioning stem cell transplantation (nonmyeloablative allo-SCT)

Although myeloablative allo-SCT in CLL can result in durable remissions, rates of TRM are after unacceptably and greatly reduced its application, even in the most refractory and high-risk individuals.

Reduced-intensity conditioning (RIC) regimens were introduced as a way to take advantage of GVL effect, reducing TRM and making transplant more approachable also in older or younger patients with comorbidities [72, 73]. These reduced regimens, are associated with improved TRM; in fact, in 2003, the EBMT reported outcomes of 77 CLL patients who received an allo-SCT [74]. The authors described an encouraging TRM rate of 18%, an impressive overall response rate of 91%, as well as a 69% complete response rate and a 22% partial response rate, associated with reduction in the ablative intensity of the preparative regimen. This lower TRM (18%), when compared with that linked to a myeloablative conditioning (46%), turned out to be extremely promising [74].

On the contrary, there were no significant differences in terms of OS or progression free survival (PFS) between these two groups [74]. In fact, although nonmyeloablative transplants may carry a stronger safety profile, the rate of relapse was higher than that associated with traditional myeloablative treatment [74]. Interestingly, instead, Sorror et al. have recently published data indicating that non-myeloablative transplants can provide a lower risk of relapse [63]. They reported encouraging long-term outcomes in 82 CLL patients who received RIC allo-SCT. In this study, at a median follow-up of 5 years, TRM, PFS, and OS were 23%, 39%, and 50%, respectively, suggesting a curative potential for RIC allo-SCT in patients with relapsed CLL, with a more favorable toxicity profile particularly in older patients who would not have been eligible to receive myeloablative conditioning regimens [63].

In contrast to ASCT where the efficacy relies exclusively on the cytotoxicity administered with the high-dose regimen, and in agreement to myeloablative allo-SCT, nonmyeloablative allo-SCT adds the immune-mediated anti-host activities conferred with the graft as a second fundamental principle of antileukemic efficacy: the GVL effect.

There is no doubt that the main therapeutic principle of allo-SCT in CLL is GVL activity and this evidence derives from some remarkable observations such as: 1) decreasing relapse incidence over time even in RIC allo-SCT, in contrast to ASCT or other intensive therapies [56, 60, 63, 70, 71, 75, 76], 2) durable clinical and molecular responses due to antitumor activity [77], 3) reduced relapse rates in patients with chronic GVHD [78], 4) increased relapse rates associated with T cell–depleted grafts [79, 80], 5) high efficacy of donor lymphocyte infusions (DLIs) in the post-transplant relapse [65, 80].

This finging supports alloreactivity as the principal mechanism responsible for GVL.

On the other hand, the most important cause of RIC allo-SCT failure in CLL patients is the disease relapse. Early relapses are correlated with chemorefractory disease at the time of transplantation, the most of time due to the unsuccessfulness of RIC regimens in controlling the disease before the GVL effect. The late relapse, instead, derives from different mechanism including: CLL clonal evolution, development of tolerance [80], presence of tumor cells in "GVCLL sanctuary sites"[63] and an insufficient GVL effect to produce a complete disease eradication. Interestingly, an high percentage of these late relapses occurred in lymph nodes without bone marrow or peripheral blood involvement, or even in patients with MRD negative status [40, 53, 55, 81, 82].

http://dx.doi.org/10.5772/53085

Quantitative MRD monitoring by RQ-PCR or flowcytometry is an essential tool to establish the clinical benefit of allo-SCT in CLL; in fact, the absence of detectable MRD, one year after allo-SCT, was strongly associated with a reduced risk of clinical relapse. In addition, there are evidences of a powerful correlation between MRD status and GVL activity, while its direct involvement for guiding GVL-inducing immunomodulation needs further evaluation [83]. Therefore quantitative MRD monitoring seems to be mandatory to assure safe and effective immunotherapy in the context of allo-SCT [83].

The best approach to post transplant immunotherapy in CLL includes monoclonal antibody (MoAbs). Some of them, although a still short follow-up, show very promising results and the use of MoAbs in the conditioning or just after transplant, could improve the results of allo-SCT. Initially, RIC allo-SCT was associated with the use of only fludarabine and cyclophosphamide. The CLL3X trial from the German CLL Study Group evaluated the long-term outcome of RIC allo-SCT in patients with poor-risk CLL who received allogeneic transplant following fludarabine and cyclophosphamide-based conditioning. The 4-year non relapse mortality (NRM), event-free survival (EFS), and OS were 23%, 42%, and 65%, respectively. To improve relapse-free survival following transplant and to modulate the impact of GVHD, MoAbs have been incorporated into transplant regimens [84]. Alemtuzumab, Rituximab are the most used MoAbs with recognized clinical activity in CLL. Alemtuzumab is a humanized anti-CD52 IgG1 MoAb with an activity in reducing the incidence of GVHD but, also, associated with an high risk of death from opportunistic infections [85]. Rituximab (anti-CD20 MoAb), instead, used in tandem with RIC preparative regimens, can induce response and help in disease control, decreasing the incidence of acute GVHD and modulating the GVL effect. [59]. However, there is no clear consensus concerning the optimal conditioning regimen to be used prior to allo-HCT. Using RIC regimens may reduce toxic deaths, but the success of non-myeloablative allo-SCT is highly dependent on the chemosensitivity of the disease.

6. Conclusion and future directions

Despite much progress in its treatment, CLL continues to be an incurable disease with standard treatments. SCT cell transplantation has changed the management of CLL patients with refractory disease or younger patients with aggressive disease. In particular, ASCT has partially failed in the treatment of advanced CLL: it prolongs survival in selected patients, but unfortunately do not cure the disease. In addition, secondary MDS/AML is one of major complication in autografted patients.

Allo-SCT, conversely, may be an acceptable option: myeloablative allo-SCT is an opportunity for younger patients with bulky, refractory, or aggressive disease; RIC allo-SCT, instead, is an emerging curative possibility for older patients with high-risk disease.

Although allo-SCT appears to result in high response rates and eradication of PCR detectable MRD, the follow up of most clinical trials is too short to assess whether allo-SCT can cure CLL.

Future approaches in menagment of CLL must take in consideration the balance between increased morbidity and mortality of SCT in CLL with the potentiality of new therapy in the setting of the improvements in outcome.

In the absence of any other treatment modalities currently capable of improving outcome in CLL, SCT should be considered the main option for patients with high-risk, refractory to standard therapy or with relapsed after prior ASCT.

Author details

Anna Gazzola, Alessandro Broccoli, Vittorio Stefoni and Pier Paolo Piccaluga

*Address all correspondence to: pierpaolo.piccaluga@unibo.it

Hematopathology and Hematology Sections, Department of Hematology and Oncology "L. and A. Seràgnoli", S. Orsola-Malpighi Hospital, University of Bologna, Italy

References

[1] O'Brien S, del Giglio A, Keating M. Advances in the biology and treatment of B-cell chronic lymphocytic leukemia. Blood 1995;85(2):307-18.

[2] Rozman C, Montserrat E. Chronic lymphocytic leukemia. The New England journal of medicine 1995;333(16):1052-7.

[3] Muller-Hermelink, Montserrat, Catovsky, Campo, Harris, Stein. Chronic lymphocytic leukaemia/small lymphocytic lymphoma. In: Swerdlow SH, Campo, E., Harris, N.L., Jaffe, E.S., Pileri, S.A., Stein, H., Thiele, J., Vardiman, J.W, ed. WHO Classification of Tumours of Haematopoietic and Lymphoid Tissues,. Lion: IARC; 2008:180-2.

[4] Dighiero G. Unsolved issues in CLL biology and management. Leukemia 2003;17(12):2385-91.

[5] Dohner H, Stilgenbauer S, Benner A, et al. Genomic aberrations and survival in chronic lymphocytic leukemia. The New England journal of medicine 2000;343(26): 1910-6.

[6] Damle RN, Wasil T, Fais F, et al. Ig V gene mutation status and CD38 expression as novel prognostic indicators in chronic lymphocytic leukemia. Blood 1999;94(6): 1840-7.

[7] Zenz T, Eichhorst B, Busch R, et al. TP53 mutation and survival in chronic lymphocytic leukemia. J Clin Oncol 2010;28(29):4473-9.

[8] Rassenti LZ, Huynh L, Toy TL, et al. ZAP-70 compared with immunoglobulin heavy-chain gene mutation status as a predictor of disease progression in chronic lymphocytic leukemia. The New England journal of medicine 2004;351(9):893-901.

[9] Rassenti LZ, Jain S, Keating MJ, et al. Relative value of ZAP-70, CD38, and immunoglobulin mutation status in predicting aggressive disease in chronic lymphocytic leukemia. Blood 2008;112(5):1923-30.

[10] Puente XS, Pinyol M, Quesada V, et al. Whole-genome sequencing identifies recurrent mutations in chronic lymphocytic leukaemia. Nature 2011;475(7354):101-5.

[11] Fabbri G, Rasi S, Rossi D, et al. Analysis of the chronic lymphocytic leukemia coding genome: role of NOTCH1 mutational activation. The Journal of experimental medicine 2011;208(7):1389-401.

[12] Quesada V, Conde L, Villamor N, et al. Exome sequencing identifies recurrent mutations of the splicing factor SF3B1 gene in chronic lymphocytic leukemia. Nature genetics 2011;44(1):47-52.

[13] Rossi D, Bruscaggin A, Spina V, et al. Mutations of the SF3B1 splicing factor in chronic lymphocytic leukemia: association with progression and fludarabine-refractoriness. Blood 2011;118(26):6904-8.

[14] Rossi D, Rasi S, Fabbri G, et al. Mutations of NOTCH1 are an independent predictor of survival in chronic lymphocytic leukemia. Blood 2012;119(2):521-9.

[15] Rossi D, Fangazio M, Rasi S, et al. Disruption of BIRC3 associates with fludarabine chemorefractoriness in TP53 wild-type chronic lymphocytic leukemia. Blood 2012;119(12):2854-62.

[16] Hallek M, Cheson BD, Catovsky D, et al. Guidelines for the diagnosis and treatment of chronic lymphocytic leukemia: a report from the International Workshop on Chronic Lymphocytic Leukemia updating the National Cancer Institute-Working Group 1996 guidelines. Blood 2008;111(12):5446-56.

[17] Lozanski G, Heerema NA, Flinn IW, et al. Alemtuzumab is an effective therapy for chronic lymphocytic leukemia with p53 mutations and deletions. Blood 2004;103(9): 3278-81.

[18] Moreton P, Kennedy B, Lucas G, et al. Eradication of minimal residual disease in B-cell chronic lymphocytic leukemia after alemtuzumab therapy is associated with prolonged survival. J Clin Oncol 2005;23(13):2971-9.

[19] Stilgenbauer S, Zenz T, Winkler D, et al. Subcutaneous alemtuzumab in fludarabine-refractory chronic lymphocytic leukemia: clinical results and prognostic marker analyses from the CLL2H study of the German Chronic Lymphocytic Leukemia Study Group. J Clin Oncol 2009;27(24):3994-4001.

[20] Wierda WG, Kipps TJ, Mayer J, et al. Ofatumumab as single-agent CD20 immunotherapy in fludarabine-refractory chronic lymphocytic leukemia. J Clin Oncol 2010;28(10):1749-55.

[21] Ferrajoli A, Lee BN, Schlette EJ, et al. Lenalidomide induces complete and partial remissions in patients with relapsed and refractory chronic lymphocytic leukemia. Blood 2008;111(11):5291-7.

[22] Phelps MA, Lin TS, Johnson AJ, et al. Clinical response and pharmacokinetics from a phase 1 study of an active dosing schedule of flavopiridol in relapsed chronic lymphocytic leukemia. Blood 2009;113(12):2637-45.

[23] Tam CS, O'Brien S, Wierda W, et al. Long-term results of the fludarabine, cyclophosphamide, and rituximab regimen as initial therapy of chronic lymphocytic leukemia. Blood 2008;112(4):975-80.

[24] Kay NE, Geyer SM, Call TG, et al. Combination chemoimmunotherapy with pentostatin, cyclophosphamide, and rituximab shows significant clinical activity with low accompanying toxicity in previously untreated B chronic lymphocytic leukemia. Blood 2007;109(2):405-11.

[25] Elter T, Borchmann P, Schulz H, et al. Fludarabine in combination with alemtuzumab is effective and feasible in patients with relapsed or refractory B-cell chronic lymphocytic leukemia: results of a phase II trial. J Clin Oncol 2005;23(28):7024-31.

[26] Wierda W, O'Brien S, Wen S, et al. Chemoimmunotherapy with fludarabine, cyclophosphamide, and rituximab for relapsed and refractory chronic lymphocytic leukemia. J Clin Oncol 2005;23(18):4070-8.

[27] Lamanna N, Kalaycio M, Maslak P, et al. Pentostatin, cyclophosphamide, and rituximab is an active, well-tolerated regimen for patients with previously treated chronic lymphocytic leukemia. J Clin Oncol 2006;24(10):1575-81.

[28] Flinn IW, Neuberg DS, Grever MR, et al. Phase III trial of fludarabine plus cyclophosphamide compared with fludarabine for patients with previously untreated chronic lymphocytic leukemia: US Intergroup Trial E2997. J Clin Oncol 2007;25(7):793-8.

[29] Catovsky D, Richards S, Matutes E, et al. Assessment of fludarabine plus cyclophosphamide for patients with chronic lymphocytic leukaemia (the LRF CLL4 Trial): a randomised controlled trial. Lancet 2007;370(9583):230-9.

[30] Dreger P, Corradini P, Kimby E, et al. Indications for allogeneic stem cell transplantation in chronic lymphocytic leukemia: the EBMT transplant consensus. Leukemia 2007;21(1):12-7.

[31] Dreger P, Brand R, Michallet M. Autologous stem cell transplantation for chronic lymphocytic leukemia. Seminars in hematology 2007;44(4):246-51.

[32] Gribben JG. Stem-cell transplantation in chronic lymphocytic leukaemia. Best practice & research 2007;20(3):513-27.

[33] Paneesha S, Milligan DW. Stem cell transplantation for chronic lymphocytic leukaemia. British journal of haematology 2005;128(2):145-52.

[34] Montserrat E. Role of auto- and allotransplantation in B-cell chronic lymphocytic leukemia. Hematology/oncology clinics of North America 2004;18(4):915-26, x.

[35] Tournilhac O, Cazin B, Lepretre S, et al. Impact of frontline fludarabine and cyclophosphamide combined treatment on peripheral blood stem cell mobilization in B-cell chronic lymphocytic leukemia. Blood 2004;103(1):363-5.

[36] Dreger P, van Biezen A, Brand R, et al. Prognostic factors for survival after autologous stem cell transplantation for chronic lymphocytic leukemia: the EBMT experience [abstract].. Blood 2002;96((Suppl. 1)):482a.

[37] Milligan DW, Fernandes S, Dasgupta R, et al. Results of the MRC pilot study show autografting for younger patients with chronic lymphocytic leukemia is safe and achieves a high percentage of molecular responses. Blood 2005;105(1):397-404.

[38] Ritgen M, Lange A, Stilgenbauer S, et al. Unmutated immunoglobulin variable heavy-chain gene status remains an adverse prognostic factor after autologous stem cell transplantation for chronic lymphocytic leukemia. Blood 2003;101(5):2049-53.

[39] Esteve J, Villamor N, Colomer D, et al. Stem cell transplantation for chronic lymphocytic leukemia: different outcome after autologous and allogeneic transplantation and correlation with minimal residual disease status. Leukemia 2001;15(3):445-51.

[40] Moreno C, Villamor N, Colomer D, et al. Clinical significance of minimal residual disease, as assessed by different techniques, after stem cell transplantation for chronic lymphocytic leukemia. Blood 2006;107(11):4563-9.

[41] Dreger P, Stilgenbauer S, Benner A, et al. The prognostic impact of autologous stem cell transplantation in patients with chronic lymphocytic leukemia: a risk-matched analysis based on the VH gene mutational status. Blood 2004;103(7):2850-8.

[42] Sutton L, Chevret S, Tournilhac O, et al. Autologous stem cell transplantation as a first-line treatment strategy for chronic lymphocytic leukemia: a multicenter, randomized, controlled trial from the SFGM-TC and GFLLC. Blood 2011;117(23): 6109-19.

[43] Brion A, Mahe B, Kolb B, et al. Autologous transplantation in CLL patients with B and C Binet stages: final results of the prospective randomized GOELAMS LLC 98 trial. Bone marrow transplantation 2011;47(4):542-8.

[44] Michallet M, Chevret S, R B. Comparative study between autologous or allogenic transplantations and conventional chemotherapy in chronic lymphocytic leukemia (CLL): A European Blood and Marrow Transplantation (EBMT) and French CLL Cooperative Group Study [abstract].. Bone marrow transplantation 2003.

[45] Porter DL, Levine BL, Kalos M, Bagg A, June CH. Chimeric antigen receptor-modified T cells in chronic lymphoid leukemia. The New England journal of medicine 2011;365(8):725-33.

[46] Gribben JG, Zahrieh D, Stephans K, et al. Autologous and allogeneic stem cell transplantations for poor-risk chronic lymphocytic leukemia. Blood 2005;106(13):4389-96.

[47] Montserrat E, Moreno C, Esteve J, Urbano-Ispizua A, Gine E, Bosch F. How I treat refractory CLL. Blood 2006;107(4):1276-83.

[48] Doney KC, Chauncey T, Appelbaum FR. Allogeneic related donor hematopoietic stem cell transplantation for treatment of chronic lymphocytic leukemia. Bone marrow transplantation 2002;29(10):817-23.

[49] Esteve J, Bosch F, C M. Prolonged response after allogeneic stem-cell transplantation (alloSCT) for high-risk chronic lymphoblastic leukemia (CLL): long term outcome of a series from a single institution [abstract].. Haematologica 2005;90((suppl 2)):41.

[50] Kharfan-Dabaja MA, Kumar A, Behera M, Djulbegovic B. Systematic review of high dose chemotherapy and autologous haematopoietic stem cell transplantation for chronic lymphocytic leukaemia: what is the published evidence? British journal of haematology 2007;139(2):234-42.

[51] Michallet M, Michallet AS, QH L. Conventional HLA-Identical Sibling Bone Marrow Transplantation is Able To Cure Chronic Lymphocytic Leukemia. A Study from the EBMTand IBMT Registries [abstract]. Blood 2003:102.

[52] Horowitz M, Montserrat E, K S. Haematopoietic stem cell transplantation for chronic lymphocytic leukemia [abstract]. Blood 2000:96.

[53] Khouri IF, Keating MJ, Saliba RM, Champlin RE. Long-term follow-up of patients with CLL treated with allogeneic hematopoietic transplantation. Cytotherapy 2002;4(3):217-21.

[54] Pavletic ZS, Arrowsmith ER, Bierman PJ, et al. Outcome of allogeneic stem cell transplantation for B cell chronic lymphocytic leukemia. Bone marrow transplantation 2000;25(7):717-22.

[55] Toze CL, Galal A, Barnett MJ, et al. Myeloablative allografting for chronic lymphocytic leukemia: evidence for a potent graft-versus-leukemia effect associated with graft-versus-host disease. Bone marrow transplantation 2005;36(9):825-30.

[56] Moreno C, Villamor N, Colomer D, et al. Allogeneic stem-cell transplantation may overcome the adverse prognosis of unmutated VH gene in patients with chronic lymphocytic leukemia. J Clin Oncol 2005;23(15):3433-8.

[57] Caballero D, Garcia-Marco JA, Martino R, et al. Allogeneic transplant with reduced intensity conditioning regimens may overcome the poor prognosis of B-cell chronic lymphocytic leukemia with unmutated immunoglobulin variable heavy-chain gene

and chromosomal abnormalities (11q- and 17p-). Clin Cancer Res 2005;11(21): 7757-63.

[58] Rawstron A, Hillmen P, Houlston R. Clonal lymphocytes in persons without known chronic lymphocytic leukemia (CLL): implications of recent findings in family members of CLL patients. Seminars in hematology 2004;41(3):192-200.

[59] Khouri IF, Przepiorka D, van Besien K, et al. Allogeneic blood or marrow transplantation for chronic lymphocytic leukaemia: timing of transplantation and potential effect of fludarabine on acute graft-versus-host disease. British journal of haematology 1997;97(2):466-73.

[60] Michallet M, Archimbaud E, Bandini G, et al. HLA-identical sibling bone marrow transplantation in younger patients with chronic lymphocytic leukemia. European Group for Blood and Marrow Transplantation and the International Bone Marrow Transplant Registry. Annals of internal medicine 1996;124(3):311-5.

[61] Khouri I, Champlin R. Allogenic bone marrow transplantation in chronic lymphocytic leukemia. Annals of internal medicine 1996;125(9):780.

[62] Fraser CJ, Bhatia S, Ness K, et al. Impact of chronic graft-versus-host disease on the health status of hematopoietic cell transplantation survivors: a report from the Bone Marrow Transplant Survivor Study. Blood 2006;108(8):2867-73.

[63] Sorror ML, Storer BE, Sandmaier BM, et al. Five-year follow-up of patients with advanced chronic lymphocytic leukemia treated with allogeneic hematopoietic cell transplantation after nonmyeloablative conditioning. J Clin Oncol 2008;26(30): 4912-20.

[64] Sorror ML, Maris MB, Sandmaier BM, et al. Hematopoietic cell transplantation after nonmyeloablative conditioning for advanced chronic lymphocytic leukemia. J Clin Oncol 2005;23(16):3819-29.

[65] Delgado J, Pillai S, Benjamin R, et al. The effect of in vivo T cell depletion with alemtuzumab on reduced-intensity allogeneic hematopoietic cell transplantation for chronic lymphocytic leukemia. Biol Blood Marrow Transplant 2008;14(11):1288-97.

[66] Oscier D, Fegan C, Hillmen P, et al. Guidelines on the diagnosis and management of chronic lymphocytic leukaemia. British journal of haematology 2004;125(3):294-317.

[67] Keating MJ, O'Brien S, Kontoyiannis D, et al. Results of first salvage therapy for patients refractory to a fludarabine regimen in chronic lymphocytic leukemia. Leukemia & lymphoma 2002;43(9):1755-62.

[68] Khouri IF, Saliba RM, Admirand J, et al. Graft-versus-leukaemia effect after non-myeloablative haematopoietic transplantation can overcome the unfavourable expression of ZAP-70 in refractory chronic lymphocytic leukaemia. British journal of haematology 2007;137(4):355-63.

[69] Perkins JG, Flynn JM, Howard RS, Byrd JC. Frequency and type of serious infections in fludarabine-refractory B-cell chronic lymphocytic leukemia and small lymphocytic

lymphoma: implications for clinical trials in this patient population. Cancer 2002;94(7):2033-9.

[70] Pavletic SZ, Khouri IF, Haagenson M, et al. Unrelated donor marrow transplantation for B-cell chronic lymphocytic leukemia after using myeloablative conditioning: results from the Center for International Blood and Marrow Transplant research. J Clin Oncol 2005;23(24):5788-94.

[71] Dreger P, Stilgenbauer S, Buttcher S, et al. Prognostic factors for outcome of nonmyeloablative allogeneic stem cell transplantation (NST) in poor-risk chronic lymphocytic leukemia (CLL): final results from a prospective multicenter trial (GCLLSG CLL3X study) [abstract]. Blood 2008:112: 212.

[72] Kolb HJ, Mittermuller J, Clemm C, et al. Donor leukocyte transfusions for treatment of recurrent chronic myelogenous leukemia in marrow transplant patients. Blood 1990;76(12):2462-5.

[73] Slavin S, Naparstek E, Nagler A, Ackerstein A, Kapelushnik J, Or R. Allogeneic cell therapy for relapsed leukemia after bone marrow transplantation with donor peripheral blood lymphocytes. Experimental hematology 1995;23(14):1553-62.

[74] Dreger P, Brand R, Hansz J, et al. Treatment-related mortality and graft-versus-leukemia activity after allogeneic stem cell transplantation for chronic lymphocytic leukemia using intensity-reduced conditioning. Leukemia 2003;17(5):841-8.

[75] Schetelig J, Thiede C, Bornhauser M, et al. Evidence of a graft-versus-leukemia effect in chronic lymphocytic leukemia after reduced-intensity conditioning and allogeneic stem-cell transplantation: the Cooperative German Transplant Study Group. J Clin Oncol 2003;21(14):2747-53.

[76] Dreger P, Brand R, Milligan D, et al. Reduced-intensity conditioning lowers treatment-related mortality of allogeneic stem cell transplantation for chronic lymphocytic leukemia: a population-matched analysis. Leukemia 2005;19(6):1029-33.

[77] Ritgen M, Bottcher S, Stilgenbauer S, et al. Quantitative MRD monitoring identifies distinct GVL response patterns after allogeneic stem cell transplantation for chronic lymphocytic leukemia: results from the GCLLSG CLL3X trial. Leukemia 2008;22(7): 1377-86.

[78] Farina L, Carniti C, Dodero A, et al. Qualitative and quantitative polymerase chain reaction monitoring of minimal residual disease in relapsed chronic lymphocytic leukemia: early assessment can predict long-term outcome after reduced intensity allogeneic transplantation. Haematologica 2009;94(5):654-62.

[79] Schetelig J, van Biezen A, Brand R, et al. Allogeneic hematopoietic stem-cell transplantation for chronic lymphocytic leukemia with 17p deletion: a retrospective European Group for Blood and Marrow Transplantation analysis. J Clin Oncol 2008;26(31):5094-100.

[80] Hoogendoorn M, Jedema I, Barge RM, et al. Characterization of graft-versus-leukemia responses in patients treated for advanced chronic lymphocytic leukemia with donor lymphocyte infusions after in vitro T-cell depleted allogeneic stem cell transplantation following reduced-intensity conditioning. Leukemia 2007;21(12):2569-74.

[81] Malhotra P, Hogan WJ, Litzow MR, et al. Long-term outcome of allogeneic stem cell transplantation in chronic lymphocytic leukemia: analysis after a minimum follow-up of 5 years. Leukemia & lymphoma 2008;49(9):1724-30.

[82] Tam CS, O'Brien S, Lerner S, et al. The natural history of fludarabine-refractory chronic lymphocytic leukemia patients who fail alemtuzumab or have bulky lymphadenopathy. Leukemia & lymphoma 2007;48(10):1931-9.

[83] Bottcher S, Ritgen M, Dreger P. Allogeneic stem cell transplantation for chronic lymphocytic leukemia: lessons to be learned from minimal residual disease studies. Blood reviews 2011;25(2):91-6.

[84] Hallek M, Fischer K, Fingerle-Rowson G, et al. Addition of rituximab to fludarabine and cyclophosphamide in patients with chronic lymphocytic leukaemia: a randomised, open-label, phase 3 trial. Lancet 2010;376(9747):1164-74.

[85] Delgado J, Thomson K, Russell N, et al. Results of alemtuzumab-based reduced-intensity allogeneic transplantation for chronic lymphocytic leukemia: a British Society of Blood and Marrow Transplantation Study. Blood 2006;107(4):1724-30.

Chapter 13

Current Status of Hematopoietic Stem Cell Transplantation in Patients with Refractory or Relapse Hodgkin Lymphoma

Leylagül Kaynar, Mustafa Çetin, Ali Ünal and
Taner Demirer

Additional information is available at the end of the chapter

http://dx.doi.org/10.5772/55106

1. Introduction

1.1. Current status of hematopoietic stem cell transplantation in patients with refractory or relapse hodgkin lymphoma

Although the high response rates, approximately 10% of patients with early-stage Hodgkin Lymphoma (HL) and 20% with advanced disease will be refractory to initial treatment or relapse after a first complete response [1-3]. The strategy for management of relapsed or refractory disease is to deliver salvage chemotherapy, followed by high-dose chemotherapy and autologous stem-cell transplantation (AutoSCT) in responding patients [4,5].

2. Autologous stem cell transplantation for Hodgkin lymphoma

2.1. Autologous stem cell transplantation at first-line therapy

The use of autoSCT for HL in first remission was wondered. There are only a few prospective randomized clinical trials focusing in this issue. Although historically controlled studies are promising, prospective controlled studies showed different results [6].

The HD01 trial included 163 patients achieving complete remission (CR) or partial remission (PR) with advanced HL after four cycles of ABVD (ABVD; doxorubicin, bleomycin, vinblastine, and dacarbazine) or other doxorubicin-containing regimens who had an unfavorable risk profile (at least two factors: high lactate dehydrogenase level, large mediastinal mass, more than one extranodal site of disease, low hematocrit, or inguinal involvement). The patients were randomly divided into two groups; AutoSCT and four additional cycles of conventional chemotherapy. There was no significant difference regarding complete remission rates, and the 5-year failure-free survival, overall survival and relapse-free survival rates between the groups [7]. Similarly, the recently published 10-year follow-up results could not demonstrate an advantage of the high dose therapy in terms of failure-free survival, overall survival and relapse-free survival rates between the groups [8]. The HD01 trial suggested that patients responding to an anthracycline- based regimen do not benefit from autoSCT at first line therapy.

In the GOELAMS Group's (Groupe Ouest Est d'Etude des Leucémies et Autres Maladies du Sang) randomized phase 2 study, H97-HR trial, the authors tested 2 intensive chemotherapy regimens in 158 patients with stage IIB - IV HL accompanied by high-risk factors. High-risk were defined by the presence of >5 involved lymphoid areas, and/or a mediastinal mass ratio > 0.45, and/or >2 extra lymph node sites affected by the disease. This study examined an early intensive chemotherapy and ABVD for 4 cycles followed by delayed myeloablative intensification. In one of the arms, patients received 3 courses of combined vindesine, doxorubicin, carmustine, etoposide, and methylprednisolone (VABEM) followed by low-dose lymph node irradiation. In the other arm, patients received 4 cycles of ABVD followed by myeloablative regimen containing carmustine, etoposide, cytarabine, and melphalan and underwent autoSCT. After the completion of treatment, the CR rates for both of arms were similar. The 5-year freedom from treatment failure and 5-year overall survival rates also were similar between the arms. Consequently, the authors recommended that conventional chemotherapy should remain the reference treatment in advanced and high risk HL [9].

In considering all these results, autoSCT for HL does not take place as a part of first line therapy even for high-risk patients.

2.2. Autologous stem cell transplantation for relapsed/refractory HL

Because conventional salvage chemotherapy and/or radiotherapy have poor results in first relapsed or progressive HL, autoSCT was evaluated as a curative approach for patients with relapsed or progressive disease. There were two prospective randomized clinical trials in the last twenty years. Firstly the British National Lymphoma Investigation performed a randomized smaller prospective study in 40 patients. The aim of the study was comparison of high-dose chemotherapy (BEAM = carmustine, etoposide, cytarabine, and melphalan) plus autoSCT (n=20) with the same drugs at lower doses (mini-BEAM) (n=20) in patients with primary refractory disease, early relapse, or prior failure of conventional therapy. All patients have been followed up for at least one year. Although there was no difference in overall survival (OS), both event-free survival and progression-free survival showed statistical significant differences in favour of BEAM plus autoSCT ($p = 0.025$ and $p = 0.005$, respectively) [10]. This study suggested that High-dose chemotherapy with autoSCT could provide better disease-free survival but not overall survival.

The second randomized multicenter trial (HD-R1) was performed by investigators of the German Hodgkin's disease Study Group and the Lymphoma Working Party (LWP) of the

EBMT to determine the benefit of HDCT in relapsed HL. Patients were randomly assigned to either four cycles of conventional chemotherapy (Dexa-BEAM: dexamethasone and carmustine, etoposide, cytarabine, and melphalan) or two cycles of Dexa-BEAM followed by autoSCT (n=73, n=88). After two cycles Dexa-BEAM chemotherapy, only 117 patients with chemosensitive disease proceeded to further treatment. Median follow-up was 39 months (IQR 3–78). Freedom from treatment failure at 3 years was significantly better for autoSCT patients (55%) than for those on Dexa-BEAM (34%; difference –21%, 95% CI –39 87 to –2 13; p=0 019), although OS was not different [11].

Data from randomized trials established autoSCT as standard therapy in relapsing HL patients responding to salvage therapy [11,12,13]. Moskowitz et al reported retrospective analysis of 75 consecutive patients with primary refractory HD, who were treated with high dose chemoradiotherapy and autoSCT. Median follow-up was 10 years. Only chemosensitivity to second-line chemotherapy predicted for a better survival, thus responding patients had an event-free survival (EFS), progression-free survival (PFS) and OS of 60%, 62% and 66%, respectively, versus 19%, 23% and 17% for patients who had a poor response to second-line chemotherapy ($P < 0.001$). Patients with disease refractory to first-line therapy but chemosensitive to standard-dose second-line therapy might have better outcome after an autoSCT [14]. Primary refractory patients or for patients in chemorefractory relapse, autoSCT has only a small likelihood to induce long-term remission. For these patients, autoSCT can be clinical option [13]. Patients with progressive disease after autoSCT have a poor outcome, and either allogeneic stem cell transplantation or other investigative approaches are necessary [14,15].

Nodular lymphocyte predominant HL (LPNHL) has to be accepted a complete different entity. There is almost no information in the literature about the impact of autoSCT in those groups of patients. Nevertheless, autoSCT can be considered a therapeutic option for LPNHL patients in advanced stages and relapsing after standard treatment [13].

2.3. High dose conditioning regimens for autologous stem cell transplantation

There is no randomized clinical trial comparing different high dose conditioning regimes for autoSCT. In retrospective analyses, the superiority of a specific regimen has not been demonstrated. Total body irradiation (TBI)-based regimens as high dose conditioning regimens were compared to chemotherapy combinations. There was no difference in efficacy and toxicity between the regimens [16,17,18]. German Hodgkin Study Group evaluated the impact of sequential high-dose therapy to increase the intensity of conditioning before autoSCT. Additional high-dose therapy did not improve the prognosis of patients with relapsed HL compared with the standard BEAM regimen and autoSCT, and was associated with increased toxic effects [19]. There was no benefit of increasing the intensity of conditioning regimen. Most of randomized trials of autoSCT in HL used BEAM regimen and this regimen is regarded, by most transplant centers, as the standard high dose-conditioning regimen [20].

2.4. Prognostic factors associated with outcome in relapse and refractor Hodgkin lymphoma before autologous stem cell transplantation

Several trials have identified prognostic factors in patients with RR-HL who have undergone subsequent salvage chemotherapy and autoSCT. These prognostic factors are summarized in table-I. Extra nodal disease, B symptoms at relapse and short remission duration after initial therapy have consistently been demonstrated to be predictors of poor outcome. Chemotherapy resistance prior to autoSCT is generally associated with poor outcome [21].

The depth of response to salvage chemotherapy before autoSCT is important. Detectable disease with functional imaging has predictive value for an unfavorable outcome [21, 22]. Jabbour et al suggested that positive functional pretransplant imaging (either gallium or Fluorine-18-Fluorodeoxyglucose Positron Emission Tomography; FDG-PET scan) can be predictive of poor outcome. The 3-year OS rates for patients with negative functional imaging were 87% but this was 58% in patients with positive functional imaging [22]. Moskowitz et al demonstrated that patients with negative functional imaging (either gallium or FDG-PET scan) had 75% EFS compared to 31% for patients with positive functional imaging [21]. Recently, Moskowitz et al reported a prospective phase 2 study. They used a risk-adapted approach to improve PFS after high-dose radio chemotherapy and autoSCT. First salvage chemotherapies were 2 cycles of ICE (ICE; ifosfamide, carboplatin, etoposide) in a standard or augmented dose, followed by restaging FDG-PET scan. Patients with a negative scan received a transplant. If the FDG-PET scan was still positive, patients received again chemotherapy (GVD; gemcitabine, vinorelbine, and liposomal doxorubicin). Patients without evidence of disease progression proceeded to high-dose chemotherapy (HDT)/ autoSCT. Patients transplanted with negative FDG-PET had an EFS of > 80%, versus 28.6% for patients with a positive scan (P <. 001). In that study, FDG-PET–negative status is a major factor in the determination of outcome. The finding that the outcome for patients receiving GVD and having a FDG-PET–negative result is indistinguishable from patients with ICE-based therapy induced FDG-PET–negative response argues that quality of response is an important determinant of outcome. The authors suggested that the goal of salvage chemotherapy in patients with HL should be a negative FDG-PET scan before HDT/AutoSCT [23].

3. Allogeneic stem cell transplantation

Although allogeneic stem cell transplantation (alloSCT) historically associated with significantly greater treatment-related mortality (TRM) than autologous stem cell transplantation (autoSCT), Because of the potential for graft versus lymphoma (GVHL) effects and the assurance of a tumor-free graft, alloSCT is carrying curative potential especially in patients with Hodgkin lymphoma (HL) who are younger than other lymphoma patients. Nevertheless, comparative studies between myeloablative alloSCT and autoSCT in refractory and relapsed HL patients provided evidence of GVHL effect. The widespread use of alloSCT in HL patients is still a matter of controversy because of the TRM and was quite limited until the advent of reduced intensity conditioning (RIC). Nowadays, increasing numbers of studies have been

Prognostic factors	Effects of prognostic factories	Reference
B symptoms at relapse, Extranodal disease at relapse, Initial remission duration of < 1 year	3-year PFS: No risk factor: 100%, 1 risk factor: 81%, 2 risk factors: 40% 3 risk factors: 0%	24
>2 prior chemotherapy regimens,	Patients who had received >2 chemotherapy regimens had a poorer DFS	17
Performance status; ECOG: 1-3 >1 chemotherapy regimens failed, The presence of mediastinal disease at AutoSCT	These factors significantly associated with FFS, Patients with >2 failed chemotherapy regimens have an estimated 4 year FFS of 10%.	25
Systemic symptoms at relapse, Disseminated pulmonary or bone marrow disease at relapse, More than minimal disease at AutoSCT	4-year FFP: No risk factor: 85% >1 risk factors: 41%	18
End-of-treatment to relapse interval < 12 months Presence of extranodal disease at relapse	4-year survival: No risk factor: 93%, 1 risk factor: 59%, 2 risk factors: 43%	26
Chemotherapy resistance prior to AutoSCT, Advanced disease stage at diagnosis (stage ≥III)	These factors associated with a higher risk of disease progression after transplantation	27
B symptoms, Extranodal disease, Complete remission duration of less than 1 year	EFS: 0-1 risk factor: 83% 2 risk factors: 27% 3 risk factors: 10%	28
Advanced stage at diagnosis, Radiotherapy before AutoSCT, A short first CR, Detectable disease at AutoSCT	5-year TTF: No risk factor: 71%±4% 3 or more risk factors: 18%±5%	29
Pre-autoSCT positive functional imaging (FDG-PET or gallium scan)	In this trial, the only factor that predicted an unfavorable outcome in the transplanted patients was a pre-HDT/AutoSCT–positive functional imaging	30
Pre-autoSCT, FDG-PET status	Persistent FDG-PET positivity after salvage therapy have a poor outcome with HDT/ AutoSCT	31

EFS: Event-free survival, FFP: Freedom from progression, FFS: Failure-free survival, PFS: Progression-free survival, TTF: Time to treatment failure, FDG-PET: Fluorine-18-Fluorodeoxyglucose Positron Emission Tomography, HDT: High-dose chemotherapy

Table 1. Prognostic factors in relapse and refractory HL

investigating the role of especially RIC-alloSCT in patients with refractory or relapsed HL, most of whom have previously received autoSCT. [32]

3.1. Myeloablative conditioning regimens

Myeloablative Conditioning (MAC) alloSCT derives its benefit from both the high-dose conditioning regimen and the GVHL effect. To determine whether this approach would benefit patients with advanced HL, proof of concept was reported by Appelbaum et al. in Seattle [33]. Eight patients with disseminated HL resistant to MOPP chemotherapy were treated with high-dose chemoradiotherapy and marrow transplantation from an HLA-match sibling donor (MSD). Two patients remain alive in unmaintained complete remission (CR) at 38 and 39 months after transplant. In the other six patients, reasons for failure included relapse of lymphoma (two patients), or death due to complications of the transplant procedure. They suggest that some patients with MOPP-resistant HL can obtain prolonged CR following intensive chemoradiotherapy followed by alloSCT.

In Johns Hopkins Oncology Center, the first prospective study to compare autoSCT with allogeneic marrow transplantation (alloBMT) was done by Jines RJ. et al. [34]. Patients with HL who have failed two or more chemotherapy regimens or who have relapsed after an initial chemotherapy-induced remission of less than 12 months are seldom cured with conventional salvage therapies. They studied the effect of high-dose cytoreductive therapy followed by alloBMT in 50 patients with relapsed HL. Twenty-one patients with HLA-matched donors had alloBMT, one patient received marrow from an identical twin, and 28 patients received autologous grafts. The authors demonstrated that autoSCT and alloSCT yield similar EFS. However, they did see a difference in relapse rates between recipients of alloSCT (17%) and autoSCT recipients with chemosensitive disease (34%), which indicated the possibility of a GVHL effect [34]. In another study Anderson et al. reported from Seattle, between 1970 and 1991, 127 patients (median age, 29 years) with relapsed or refractory HL received high-dose chemotherapy with or without irradiation, followed by autoSCT (n=68), alloSCT (n=53), or syngeneic transplantation (n=6) [35]. The 5-year actuarial probabilities of OS, EFS, relapse, and nonrelapse mortality (NRM) for the entire group were 21%, 18%, 65%, and 49%, respectively. HLA-match allogeneic marrow recipients had a statistically lower relapse rate compared with recipients of autologous marrow, but OS, EFS, and NRM rates were not significantly different. They described that the use of HLA-MSD marrow results in a lower relapse rate and, thus, for some individuals, may be preferable to the use of autologous marrow. The European Group for Blood and Marrow Transplantation (EBMT) directed a retrospective study of 45 patients (median age, 29 years) with refractory or relapsed HL, who received matched sibling allogeneic versus autoSCT [36]. The 4-year actuarial probabilities of OS, PFS, relapse, and NRM were 25%, 15%, 61%, and 48% and 37%, 24%, 61%, and 27% after alloBMT and autoBMT, respectively. The 4-year actuarial probability of survival was 30% after alloBMT and 64% after autoBMT (P =.007). This difference is mainly due to a higher TRM rate after alloBMT (65% v 12%, P =.005). The authors suggested that patients with relapsed or resistant HL derive no benefit from alloSCT over autoSCT. However, lower relapse rate (13%) among patients with acute GVHD, supporting a substantial GVHL effect with alloSCT. The second EBMT registry reported by Peniket et al. in 2003, One-hundred-sixty-seven poor-prognosis HL patients who underwent myeloablative transplantation (77% with stage III or IV at diagnosis and 42% with chemoresistant disease) were assessed. [37] These patients received allogeneic transplants as their first transplant procedure. Actuarial OS at 4 years from transplantation was 24.7% years. These outcomes are relatively poor because of the high TRM associated with these procedures in patients with HL (51.7% actuarial procedure-related mortality at 4 years). The authors concluded that the high TRM was probably a reflection of the large percentage of patients with resistant disease. Gajewski et al. reviewed IBMTR data on 100 consecutive patients with HL who received HLA-match sibling BMT between 1982 and 1992. All patients had advanced disease [38]. Eighty-nine of 100 patients were not in remission at the time of transplant. Fifty had pretransplant Karnofsky scores less than 90% and 27 had active infection in the week before transplant. The 3-year probability of relapse and the probability of OS were 65% and 15, respectively. They concluded that the role of HLA-identical sibling BMTs have a limited therapeutic effect in advanced HL. Akpek et al. evaluated the long-term outcome after allogeneic and autologous blood or marrow transplantation in patients with relapsed or refractory HL [39]. They analyzed the outcome of 157 consecutive patients with relapsed or refractory HL, who underwent BMT between 1985 and 1998. There was a trend for probability of relapse in sensitive patients to be less after alloBMT at 34% (range, 8% to 59%) versus 51% (range, 36% to 67%) for the auto patients (HR = 0.51, P =.17). There seems to be a clinical GVHL effect associated with alloBMT. Allogeneic BMT for HL also seems to have a lower risk of secondary AML/MDS than autoBMT. Thus, alloBMT warrants continued study in HL.

Actually, comparisons of MAC versus autologous transplantation are problematic, because MAC regimens are favored in patients with extensive prior therapy or comorbidities. In general, higher NRM has been associated with myeloablative regimens and a greater risk of relapse associated with autologous approaches, leading to similar long-term outcomes for patients treated with each approach. The poor outcomes for allogeneic stem cell transplantation in HL reported by the IBMTR and the EBMT may result from the selection of patients with unfavorable risk for these studies. At that point, however, high rate of the TRM with MAC in Hodgkin lymphoma is considered is rarely pursued [40]. Because of the most of this studies have been reported at least couple decade ago, the effect of myeloablative transplantation in HL may reevaluate with the developed supportive treatment modalities and new drugs.

3.2. Reduced-intensity conditioning

Last decade has seen a radical change in the approach to alloSCT. Previously, MAC regimens were thought to be necessary for preventing graft rejection, making marrow space, and providing antitumor activity. Interest in exploring transplantation with RIC in relapsed or refractory HL arise from evidence of GVHL effect in these studies comparing allogeneic with autologous transplantation. In an attempt to decrease TRM, authors increasingly have been using alloSCT with RIC to patients with HL. Reduced-intensity conditioning could be sufficient to restore allogeneic engraftment, and allow graft versus host reactions could eliminate host hematopoiesis and provide antitumor effects. This has allowed treatment to be considered in older patients, or patients with co-morbidities, including those with HL for whom prior autoSCT failed.

The EBMT analyzed the first retrospective analysis comparing RIC-alloSCT (n=89) with MAC-alloSCT (n=79) in patients with relapsed or refractory HL [41]. In the RIC group NRM was significantly decreased, OS was better and there was a trend for better PFS. Results demonstrated nearly twice the relapse incidence in the RIC group (57% vs. 30%), but the 5-year OS was significantly higher in the RIC group (28% vs. 22%, P=0.003). They also indicate that the existence of a GVHL effect correlated to the development of GVHD and additional efforts to reduce the high relaps rate seen in both groups of patients. The centers reported the outcomes of 143 patients undergoing unrelated donor reduced-intensity and nonmyeloablative (RIC/NST) SCT for relapsed and refractory HL between 1999 and 2004 reported to the CIBMTR [42]. They analyzed Patients were heavily pretreated, including autoSCT in 89%. With a median follow-up of 25 months, the probability of TRM at day 100 and 2 years was 15% and 33%, respectively. The probabilities of PFS and OS were 30% and 56% at 1 year and 20% and 37% at 2 years. The presence of extranodal disease and the Karnofsky Performance Scale (KPS) <90 were significant risk factors for TRM, PFS, and OS, whereas chemosensitivity at transplantation was not. Dose intensity of the conditioning regimen (RIC versus NST) did not impact outcomes.

Peggs et al. undertook RIC-alloSCT in 49 patients with multiply relapsed HL, 44 (90%) of whom had progression of disease after previous autoSCT, number of previous treatment courses was five [range 3-8], and time from diagnosis 4.8 years [range 0.6-4.8]) [43]. Thirty-one patients had HLA matched donors who were related and 18 had donors who were unrelated. All patients engrafted. Eight of 49 (16%) had grade II-IV acute GVHD and seven (14%) had chronic GVHD before DLIs. Sixteen (33%) patients had DLI from 3 months after transplantation for residual disease or progression. Six (38%) of the 16 developed grade II-IV acute GVHD and five developed chronic GVHD. Nine (56%) showed disease responses after infusion (eight complete, one partial). Non-relapse-related mortality was 16.3% at 730 days (7.2% for patients who had related donors vs. 34.1% for those with unrelated donors, p=0.0206). Projected 4 year OS and PFS was 55.7% and 39.0%, respectively (62.0% and 41.5% for related donors). In this prospectively study, authors showed that the potential for durable responses in patients who have previously had substantial treatment for HL. In another prospective study, Alvarez et al. described the results of RIC-alloSCT in patients with advanced HL. Forty patients with relapsed or refractory HL were homogeneously treated with an RIC protocol (fludarabine 150 mg/m(2) intravenously plus melphalan 140 mg/m(2) intravenously) and cyclosporin A and methotrexate as GVHD prophylaxis [44]. Twenty patients (50%) were allografted in resistant relapse, and 38 patients received hematopoietic cells from an HLA-match sibling. Five patients (12%) died from early TRM (before day +100 after allo-RIC). One-year TRM was 25%. Acute GVHD developed in 18 patients (45%). Chronic GVHD developed in 17 (45%) of the 31 evaluable patients. The response rate 3 months after the allo-RIC was 67% (21 [52%] CRs and 6 [15%] partial remissions). Eleven patients received DLIs for disease relapse. The response rate after DLI was 54% (3 complete remissions and 3 partial remissions). Overall survival and PFS were 48% +/- 10% and 32% +/- 10% at 2 years, respectively. They suggest that RIC-alloSCT

http://dx.doi.org/10.5772/55106

is feasible in heavily pretreated HL patients and has an acceptable early TRM. Results are better in patients allografted in sensitive disease. Both responses observed after the development of GVHD and DLI may suggest a GVHL effect. Allogeneic RIC has to be considered an effective therapeutic approach for patients who have had treatment failure with previous autoSCT. Anderlini et al. reported that a total of 40 patients with relapsed/refractory HL underwent RIC-alloSCT [45]. Disease status at alloSCT was refractory relapse (n=14) or sensitive relapse (n=26). The conditioning regimens were fludarabine-cyclophosphamide+/-antithymocyte globulin (n=14), a less intensive regimen, and fludarabine-melphalan (FM) (n=26), a more intensive one. The two groups had similar prognostic factors. Day 100 and cumulative TRMs (18-month) were 5 and 22%. Twenty-four patients (60%) are alive (14 in CR or CR-unconfirmed) with a median follow-up of 13 months (4-78). In all, 16 patients expired (TRM n=8, disease progression n=8). FM patients had better OS (73 vs. 39% at 18 months; P=0.03), and a trend towards better PFS (37 vs 21% at 18 months; P=0.2). Reduced intensity alloSCT is feasible in relapsed/refractory HL patients with a low TRM. This group updated the results comparing outcomes of 58 patients with HL underwent RIC-alloSCT from a MRD (n=25) or a MUD (n=33) [46]. Forty-eight (83%) had undergone prior autoSCT. Disease status at transplant was refractory relapse (n=28) or sensitive relapse (n=30). Cumulative day 100 and 2-year TRM rates were 7% and 15%, respectively (day 100 transplant-related mortality MRD vs. MUD 8% vs. 6%, p=ns; 2-year MRD vs. MUD 13% vs. 16%, p=ns). Projected 2-year overall and progression-free survival rates are 64% (49-76%) and 32% (20-45%), with 2-year disease progression/relapse at 55% (43-70%). There was no statistically significant difference in OS, PFS, and disease progression/relapse between MRD and MUD transplants. They also suggested that FM as a preparative regimen for RIC-alloSCT in progression-free survival HL is associated with a significant reduction in TRM, with comparable results in MRD and MUD allograft.

In a recent study, Sarina et al., using RIC as a salvage option, to evaluate the role of alloSCT in patients HL relapsing after autoSCT [47]. In this retrospective study based on the commitment of attending physicians to perform a salvage alloSCT; thus, only HL patients having human leukocyte antigen-typing immediately after the failed autoSCT were included. Of 185 patients, 122 found an identical sibling (55%), a matched unrelated (32%) or a haploidentical sibling (13%) donor; 63 patients did not find any donor. Clinical features of both groups did not differ. Two-year PFS and OS were better in the donor group (39.3% vs 14.2%, and 66% vs 42%, respectively, P <.001) with a median follow-up of 48 months. In multivariable analysis, having a donor was significant for better PFS and OS (P <.001). Patients allografted in complete remission showed a better PFS and OS. They concluded that, that: (1) HL patients relapsing after an autoSCT have a survival advantage if they undergo RIC alloSCT; (2) CR achievement before RIC alloSCT is very important and influences patients' clinical outcome; and (3) NRM after RIC alloSCT is rather low; therefore, this procedure can be considered a feasible option in the clinical setting. Stephen et al. [48] investigated the role of RIC-alloSCT in the management of patients with HL. To further define its role they conducted a retrospective analysis of 285 patients with HL who underwent a RIC-alloSCT in order to identify prognostic factors that predict outcome. Eighty percent of patients had undergone a prior autoSCT and 25% had refractory disease at transplant. Non-relapse mortality was associated with chemorefractory disease, poor performance status, age >45 and transplantation before 2002. For patients with no risk factors the 3-year non-relapse mortality rate was 12.5% compared to 46.2% for patients with 2 or more risk factors. The use of an unrelated donor had no adverse effect on the non-relapse mortality. Acute GVHD grades II-IV developed in 30% and chronic GVHD in 42%. The development of chronic GVHD was associated with a lower relapse rate. The disease progression rate at one and five years was 41% and 58.7% respectively and was associated with chemorefractory disease and extent of prior therapy. Donor lymphocyte infusions were administered to 64 patients for active disease of whom 32% showed a clinical response. Eight out of 18 patients receiving DLIs alone had clinical responses. Progression-free and OS were both associated with performance status and disease status at transplant. Patients with neither risk factor had a 3- year PFS and OS of 42% and 56% respectively compared to 8% and 25% for patients with one or more risk factors. Relapse within six months of a prior autologous transplant was associated with a higher relapse rate and a lower progression-free. They recommend that for patients deemed to be at high risk of failing an autologous transplant a RIC-alloSCT may represent a more effective therapy and prospective comparative studies in this setting should be considered.

On the basis of current data, some authors recommend RIC-allo for HL only in the context of prospective clinical trials because allo-SCT trials continue to report disappointing relapse rates. However, if clinicians feel strongly about proceeding with this strategy, patients with refractory disease should be excluded and opportunities to exploit the GVHL effect should be used [20].

3.3. The role of functional imaging in response assessment before RIC allo-SCT

Dodero et al. investigated the prognostic value of PET scanning in 80 patients who had chemosensitive disease (34 patients with HG-NHL and 46 patients with HL before undergoing to RIC and followed by alloSCT [49]. Positron emission tomography was used to assess before they underwent alloSCT: 42 patients had negative PET studies, and 38 patients had positive PET studies. Patients underwent allograft from MSD (n = 41) or alternative donors (n = 39). At the time of the last follow-up, 48 patients were alive (60%), and 32 had died. The 3-year cumulative incidence of nonrecurrence mortality and disease recurrence was 17% and 40%, respectively. The cumulative incidence of disease recurrence was significantly lower in the PET-negative patients (25% vs. 56%; P =.007), but there was no significant difference between the patients with or without chronic GVHD (P =.400). The patients who had negative PET studies before undergoing alloSCT also had significantly better outcomes in terms of 3-year OS (76% vs. 33%; P =.001) and 3-year PFS (73% vs. 31%; P =.001). On multivariate analysis, OS was influenced by PET status (hazard ratio [HR], 3.35), performance status (HR, 5.15), and type of donor (HR, 6.26 for haploidentical vs. MSD; HR, 1.94 for MUD vs. MSD)l. The current results indicated that PET scanning appears to be an accurate tool for assessing prognosis in patients who are eligible for RIC allografting.

In summary, autoSCT and alloSCT may be curative therapy options for patients with relapse/ refractory HL. AutoSCT is a standard therapy in relapsing HL patients especially responding to salvage therapy. However, there is no place of autoSCT even in high-risk patients with first remission. If patients with refractory disease after first-line treatment respond second line

http://dx.doi.org/10.5772/55106

standard-dose therapy, these patients can benefit from an autoSCT. BEAM regimen is used as the standard high dose-conditioning regimen. Detectable disease with functional imaging before autoSCT can be an indicator of poor prognosis. Patients with relaps or progressive disease after autoSCT need either alloSCT or other investigative approaches. The review of the literature of alloSCT in HL patients has showed that in the last decade an increasing interest has been raised on this topic. The results that are now available allow the following considerations: (i) Although myeloablative allogeneic SCT has lower relaps rate still has been associated high non relaps mortality; (ii) The effect of myeloablative transplantation in HL may reevaluate with the developed supportive treatment modalities and future studies should be aimed at integrating intensified therapies and/or new drugs into the treatment plan to pursue the best response before allografting; (iii) RIC alloSCT is a feasible option in relapsed/refractory HL patients even if they were heavily pretreated; (iv) 20–30% of the allografted patients can be long-term survivors after RIC alloSCT; (v) Complete response is the most important predictor of a favorable outcome; (vi) PET scanning appears to be an accurate tool for assessing prognosis in patients who are eligible for RIC allografting; thus (vii) Since the relapse risk after RIC-alloSCT is still high maintenance treatment or immunological methods should be explored in prospective clinical trials.

Author details

Leylagül Kaynar[1*], Mustafa Çetin[2], Ali Ünal[3] and Taner Demirer[4]

*Address all correspondence to: drleylagul@gmail.com

1 Erciyes University, Faculty of Medicine, Deparment of Hematology, Erciyes Transplant Center, Kayseri, Turkey

2 Erciyes University, Faculty of Medicine, Deparment of Hematology, Erciyes Transplant Center, Kayseri, Turkey

3 Erciyes University, Faculty of Medicine, Deparment of Hematology, Erciyes Transplant Center, Kayseri, Turkey

4 Ankara University, Medical School, Deparment of Hematology, Ankara, Turkey

References

[1] Engert, A, Plütschow, A, Eich, H. T, Lohri, A, Dörken, B, Borchmann, P, et al. Reduced treatment intensity in patients with early-stage Hodgkin's lymphoma. N Engl J Med. (2010)., 363, 640-652.

[2] Diehl, V, Franklin, J, Pfreundschuh, M, Lathan, B, Paulus, U, Hasenclever, D, et al. Standard and increased-dose BEACOPP chemotherapy compared with COPP-ABVD for advanced Hodgkin's disease. N Engl J Med. (2003). , 348, 2386-2395.

[3] Eich, H. T, Diehl, V, Görgen, H, Pabst, T, Markova, J, Debus, J, et al. Intensified chemotherapy and dose-reduced involved-field radiotherapy in patients with early unfavorable Hodgkin's lymphoma: final analysis of the German Hodgkin Study Group HD11 trial. J Clin Oncol. (2010). , 28, 4199-4206.

[4] Linch, D. C, Winfield, D, Goldstone, A. H, Moir, D, Hancock, B, Mcmillan, A, et al. Dose intensification with autologous bone-marrow transplantation in relapsed and resistant Hodgkin's disease: results of a BNLI randomised trial. Lancet (1993). , 341, 1051-1054.

[5] Schmitz, N, Pfistner, B, Sextro, M, Sieber, M, Carella, A. M, Haenel, M, et al. Aggressive conventional chemotherapy compared with high-dose chemotherapy with autologous haemopoietic stem-cell transplantation for relapsed chemosensitive Hodgkin's disease: a randomized trial. Lancet (2002). , 359, 2065-2071.

[6] Nademanee, A, Molina, A, Fung, H, Stein, A, Parker, P, Planas, I, et al. High-dose chemo/radiotherapy and autologous bone marrow or stem cell transplantation for poor-risk advanced-stage Hodgkin's disease during first partial or complete remission. Biol Blood Marrow Transplant. (1999). , 5(5), 292-298.

[7] Federico, M, Bellei, M, Brice, P, Brugiatelli, M, Nagler, A, Gisselbrecht, C, et al. High-dose therapy and autologous stem-cell transplantation versus conventional therapy for patients with advanced Hodgkin's lymphoma responding to front-line therapy. J Clin Oncol. (2003). , 21(12), 2320-2325.

[8] Carella, A. M, Bellei, M, Brice, P, Gisselbrecht, C, Visani, G, Colombat, P, et al. High-dose therapy and autologous stem cell transplantation versus conventional therapy for patients with advanced Hodgkin's lymphoma responding to front-line therapy: long-term results. Haematologica. (2009). , 94(1), 146-148.

[9] Arakelyan, N, Berthou, C, Desablens, B, De Guibert, S, Delwail, V, Moles, M. P, et al. Early versus late intensification for patients with high-risk Hodgkin lymphoma-3 cycles of intensive chemotherapy plus low-dose lymph node radiation therapy versus 4 cycles of combined doxorubicin, bleomycin, vinblastine, and dacarbazine plus myeloablative chemotherapy with autologous stem cell transplantation: five-year results of a randomized trial on behalf of the GOELAMS Group Cancer. (2008). , 113, 3323-3330.

[10] Linch, D. C, Winfield, D, Goldstone, A. H, Moir, D, Hancock, B, Mcmillan, A, et al. Dose intensification with autologous bone-marrow transplantation in relapsed and resistant Hodgkin's disease: Results of a BNLI randomized trial. Lancet (1993). , 341, 1051-1054.

[11] Schmitz, N, Pfistner, B, Sextro, M, Sieber, M, Carella, A. M, Haenel, M, et al. Aggressive conventional chemotherapy compared with high-dose chemotherapy with au-

tologous haemopoietic stem-cell transplantation for relapsed chemosensitive Hodgkin's disease: a randomised trial. Lancet. (2002). Jun 15;, 359(9323), 2065-2071.

[12] Linch, D. C, Winfield, D, Goldstone, A. H, Moir, D, Hancock, B, et al. Dose intensification with autologous bone- marrow transplantation in relapsed and resistant Hodgkin's disease: Results of a BNLI randomized trial. Lancet (1993). , 341, 1051-1054.

[13] Ljungman, P, Bregni, M, Brune, M, Cornelissen, J, De Witte, T, Dini, G, et al. Allogeneic and autologous transplantation for haematological diseases, solid tumours and immune disorders: current practice in Europe 2009. Bone Marrow Transplanst. (2010). , 45(2), 219-234.

[14] Moskowitz, C. H, Kewalramani, T, Nimer, S. D, Gonzalez, M, Zelenetz, A. D, & Yahalom, J. Effectiveness of high dose chemoradiotherapy and autologous stem cell transplantation for patients with biopsy-proven primary refractory Hodgkin's disease. Br J Haematol (2004). , 124, 645-652.

[15] Kewalramani, T, Nimer, S. D, Zelenetz, A. D, Malhotra, S, Qin, J, Yahalom, J, et al. Progressive disease following autologous transplantation in patients with chemosensitive relapsed or primary refractory Hodgkin's disease or aggressive non- Hodgkin's lymphoma. Bone Marrow Transplant (2003). , 32, 673-679.

[16] Gutierrez-delgado, F, Holmberg, L, Hooper, H, Petersdorf, S, Press, O, Maziarz, R, et al. Autologous stem cell transplantation for Hodgkin's disease: busulfan, melphalan and thiotepa compared to a radiation-based regimen. Bone Marrow Transplant. (2003). , 32(3), 279-285.

[17] Nademanee, A, Donnell, O, Snyder, M. R, Schmidt, D. S, Parker, G. M, & Stein, P. M. AS, et al. High-dose chemotherapy with or without total body irradiation followed by autologous bone marrow and/or peripheral blood stem cell transplantation for patients with relapsed and refractory Hodkin's disease: results in 85 patients with analysis of prognostic factors. Blood (1995). , 85, 1381-1390.

[18] Horning, S. J, Chao, N. J, Negrin, R. S, Hoppe, R. T, Long, G. D, Hu, W. W, et al. High-dose therapy and autologous hematopoietic progenitor cell transplantation for recurrent or refractory Hodgkin's disease: analysis of the Stanford University results and prognostic indices. Blood (1997). , 89, 801-813.

[19] Josting, A, Müller, H, Borchmann, P, Baars, J. W, Metzner, B, Döhner, H, Aurer, I, et al. Dose intensity of chemotherapy in patients with relapsed Hodgkin's lymphoma. J Clin Oncol. (2010). Dec 1;, 28(34), 5074-5080.

[20] Kuruvilla, J, Keating, A, & Crump, M. How I treat relapsed and refractory Hodgkin lymphoma. Blood. (2011). Apr 21;, 117(16), 4208-4217.

[21] Moskowitz, C. H, Yahalom, J, Zelenetz, A. D, Zhang, Z, Filippa, D, Teruya-feldstein, J, et al. High-dose chemo-radiotherapy for relapsed or refractory Hodgkin lympho-

ma and the signifi- cance of pre-transplant functional imaging. Br J Haematol. (2010). , 148(6), 890-897.

[22] Jabbour, E, Hosing, C, Ayers, G, Nunez, R, Anderlini, P, Pro, B, et al. Pretransplant positive positron emission tomography/gallium scans predict poor outcome in patients with recurrent/refractory Hodgkin lymphoma. Cancer. (2007). , 109(12), 2481-2489.

[23] Moskowitz, C. H, Matasar, M. J, Zelenetz, A. D, Nimer, S. D, Gerecitano, J, Hamlin, P, et al. Normalization of pre-ASCT, FDG-PET imaging with second-line, non-cross-resistant, chemotherapy programs improves event-free survival in patients with Hodgkin lymphoma. Blood. (2012). , 119(7), 1665-1670.

[24] Reece, D. E, Connors, J. M, Spinelli, J. J, Barnett, M. J, Fairey, R. N, Klingemann, H. G, et al. Intensive therapy with cyclophosphamide, carmustine, etoposide +/- cisplatin, and autologous bone marrow transplantation for Hodgkin's disease in first relapse after combination chemotherapy. Blood (1995). , 83, 1193-1119.

[25] Bierman, P. J, Bagin, R. G, Jagannath, S, Vose, J. M, Spitzer, G, Kessinger, A, et al. High dose chemotherapy followed by autologous hematopoietic rescue in Hodgkin's disease: long term follow-up in 128 patients. Ann Oncol (1993). , 4, 767-773.

[26] Brice, P, Bouabdallah, R, Moreau, P, Divine, M, André, M, Aoudjane, M, et al. Prognostic factors for survival after HDT and ASCT for patients with relapsed HD: analysis of 280 patients from the French Registry. Bone Marrow Transplant (1997). , 20, 21-26.

[27] Popat, U, Hosing, C, Saliba, R. M, Anderlini, P, Van Besien, K, Przepiorka, D, et al. Prognostic factors for disease progression after high-dose chemotherapy and autologous hematopoietic stem cell transplantation for recurrent or refractory Hodgkin's lymphoma. Bone Marrow Transplant. (2004). , 33(10), 1015-23.

[28] Moskowitz, C. H, Nimer, S. D, Zelenetz, A. D, Trippett, T, Hedrick, E. E, Filippa, D. A, et al. A 2-step comprehensive high-dose chemotherapy second-line program for relapsed and refractory Hodgkin disease: analysis by intent to treat and development of a prognostic model. Blood (2001). , 97, 616-623.

[29] Sureda, A, Constans, M, Iriondo, A, Arranz, R, Caballero, M. D, Vidal, M. J, et al. Prognostic factors affecting long-term outcome after stem cell transplantation in Hodgkin's lymphoma autografted after a first relapse. Ann Oncol (2005). , 16, 625-633.

[30] Moskowitz, A. J, Yahalom, J, & Kewalramani, T. Pretransplantation functional imaging predicts outcome following autologous stem cell transplantation for relapsed and refractory Hodgkin lymphoma. Blood (2010). , 116(23), 4934-4937.

[31] Moskowitz, C. H, Matasar, M. J, Zelenetz, A. D, Nimer, S. D, Gerecitano, J, Hamlin, P, et al. Normalization of pre-ASCT, FDG-PET imaging with second-line, non-cross-

resistant, chemotherapy programs improves event-free survival in patients with Hodgkin lymphoma. Blood. (2012). , 119(7), 1665-70.

[32] Salit, R. B, Bishop, M. R, & Pavletic, S. Z. Allogeneic Hematopoietic Stem Cell Transplantation: Does It Have a Place in Treating Hodgkin Lymphoma? Curr Hematol Malig Rep (2010). , 5, 229-238.

[33] Appelbaum, F. R, Sullivan, K. M, Thomas, E. D, Buckner, C. D, Clift, R. A, et al. Allogeneic marrow transplantation in the treatment of MOPP-resistant Hodgkin's disease. J Clin Oncol (1985). , 3, 1490-1494.

[34] Jones, R. J, Piantadosi, S, Mann, R. B, Ambinder, R. F, Seifter, E. J, Vriesendorp, H. M, et al. High-dose cytotoxic therapy and bone marrow transplantation for relapsed Hodgkin's disease. J Clin Oncol (1990). , 8, 527-537.

[35] Anderson, J. E, Litzow, M. R, Appelbaum, F. R, Schoch, G, Fisher, L. D, Buckner, C. D, et al. Allogeneic, syngeneic, and autologous marrow transplantation for Hodgkin's disease: the year Seattle experience. J Clin Oncol (1993). , 21.

[36] Milpied, N, Fielding, A. K, Pearce, R. M, Ernst, P, & Goldstone, A. H. Allogeneic bone marrow transplant is not better than autologous transplant for patients with relapsed Hodgkin's disease. European Group for Blood and Bone Marrow Transplantation. J Clin Oncol (1996). , 14, 1291-1296.

[37] Peniket, A. J. Ruiz de Elvira MC, Taghipour G, Cordonnier C, Gluckman E, et al. An EBMT registry matched study of allogeneic stem cell transplants for lymphoma: allogeneic transplantation is associated with a lower relapse rate but a higher procedure-related mortality rate than autologous transplantation. Bone Marrow Transplant (2003). , 31, 667-678.

[38] Gajewski, J. L, Phillips, G. L, Sobocinski, K. A, Armitage, J. O, Gale, R. P, Champlin, R. E, et al. Bone marrow transplants from HLA-identical siblings in advanced Hodgkin's disease. J Clin Oncol (1996). , 14, 572-578.

[39] Akpek, G, Ambinder, R. F, Piantadosi, S, Abrams, R. A, Brodsky, R. A, Vogelsang, G. B, et al. Long-term results of blood and marrow transplantation for Hodgkin's lymphoma. J Clin Oncol (2001). , 19, 4314-4321.

[40] Corradini, P, Sarina, B, & Farina, L. Allogeneic transplantation for Hodgkin's lymphoma. Br J Haematol. (2011). , 152(3), 261-272.

[41] Sureda, A, Robinson, S, Canals, C, Carella, A. M, Boogaerts, M. A, Caballero, D, et al. Reduced-intensity conditioning compared with conventional allogeneic stem-cell transplantation in relapsed or refractory Hodgkin's lymphoma: an analysis from the Lymphoma Working Party of the European Group for Blood and Marrow Transplantation. J Clin Oncol (2008). , 26, 455-462.

[42] Devetten, M. P, Hari, P. N, Carreras, J, Logan, B. R, Van Besien, K, et al. Unrelated donor reduced-intensity allogeneic hematopoietic stem cell transplantation for re-

lapsed and refractory Hodgkin lymphoma. Biol Blood Marrow Transplant (2009). , 15, 109-117.

[43] Peggs, K. S, Hunter, A, Chopra, R, Parker, A, Mahendra, P, Milligan, D, et al. Clinical evidence of a graft-versus-Hodgkin's-lymphoma effect after reduced intensity allogeneic transplantation. Lancet (2005). , 365, 1934-1941.

[44] Alvarez, I, Sureda, A, Caballero, M. D, Urbano-ispizua, A, Ribera, J. M, et al. Non-myeloablative stem cell transplantation is an effective therapy for refractory or relapsed Hodgkin lymphoma: results of a Spanish prospective cooperative protocol. Biol Blood Marrow Transplant (2006). , 12, 172-183.

[45] Anderlini, P, Saliba, R, Acholonu, S, Okoroji, G. J, Donato, M, et al. Reduced-intensity allogeneic stem cell transplantation in relapsed and refractory Hodgkin's disease: low transplant-related mortality and impact of intensity of conditioning regimen. Bone Marrow Transplant (2005). , 35, 943-951.

[46] Anderlini, P, Saliba, R, Acholonu, S, Giralt, S. A, Andersson, B, et al. Fludarabine-melphalan as a preparative regimen for reduced-intensity conditioning allogeneic stem cell transplantation in relapsed and refractory Hodgkin's lymphoma: the updated M.D. Anderson Cancer Center experience. Haematologica (2008). , 93, 257-264.

[47] Sarina, B, Castagna, L, Farina, L, Patriarca, F, Benedetti, F, Carella, A. M, et al. Allogeneic transplantation improves the overall and progression-free survival of Hodgkin lymphoma patients relapsing after autologous transplantation: a retrospective study based on the time of HLAtyping and donor availability. Blood. (2010). , 115(18), 3671-3677.

[48] Robinson, S. P, Sureda, A, Canals, C, Russell, N, Caballero, D, Bacigalupo, A, et al. Reduced intensity conditioning allogeneic stem cell transplantation for Hodgkin's lymphoma: identification of prognostic factors predicting outcome. Haematologica (2009). , 94, 230-238.

[49] Dodero, A, Crocchiolo, R, Patriarca, F, Miceli, R, Castagna, L, Ciceri, F, et al. Pretransplantation [18 F]Fluorodeoxyglucose Positron Emission Tomography Scan Predicts Outcome in Patients With Recurrent Hodgkin Lymphoma or Aggressive Non-Hodgkin Lymphoma Undergoing Reduced-Intensity Conditioning Followed by Allogeneic Stem Cell Transplantation. Cancer (2010). , 116, 5001-5011.

Chapter 14

Iron Overload and Hematopoetic Stem Cell Transplantation

Zeynep Arzu Yegin, Gülsan Türköz Sucak and
Taner Demirer

Additional information is available at the end of the chapter

http://dx.doi.org/10.5772/53819

1. Introduction

Hematopoietic stem cell transplantation (HSCT) is an established treatment modality with a curative potential in a variety of hematological disorders. Although remarkable advances in transplant immunology and supportive care allowed widespread use of HSCT, transplant related morbidity and mortality remain as a problem [1-7]. Early complications including sinusoidal obstruction syndrome (SOS), hemorrhagic cystitis, engraftment syndrome, idiopathic pneumonia syndrome (IPS), infections and graft versus host disease (GVHD) are the major causes of morbidity and non relapse mortality (NRM). High doses of radiotherapy and chemotherapy of the conditioning regimen have adverse effects on all organs and tissues of the recipient, which also triggers several early and late effects of variable intensity [1, 3, 5-8]. Iron overload (IO) is a relatively common condition in patients with hematological malignacies and HSCT recipients. Free iron which accompanies IO might contribute to the already existing prooxidant state in HSCT recipients by inducing the formation of reactive oxygen species (ROS). Tissue peroxidation and organ damage, as a consequence, contribute to the development of some early transplant complications [2, 4, 5, 9]. Increasing number of transplants performed each year and improved transplant techniques result in a rise in the number of long term survivors. The primary goal of HSCT is to cure the primary disease. However long term transplant related morbidity might be very challenging and might significantly impair the quality of life. Late effects might be the consequence of the direct toxicity of chemoradiotherapy and/or the immunologic complications mainly consisting of GVHD. Besides the secondary late effects including osteoporosis and dental caries, very late effects, namely cardiovascular toxicity considered as tertiary late effect may also occur. Among this wide spectrum of complications, IO has a substantial role as a contributor to liver toxicity, infections and SOS and as a predictor of transplant outcome. Hematopoietic SCT recipients have been demonstrated to have a high degree of liver iron content (LIC) almost in the range of hereditary hemochromatosis (HH) and IO was shown to cause liver fibrosis, heart failure, hypogonadism, diabetes and endocrinopathy in HSCT recipients in the long run [4, 6, 7, 10].

Iron is an essential element which plays a key role in several biochemical reactions including oxygen transport and electron transfer. It mediates the conversion of hydrogen peroxyde (H2O2) to highly toxic free radicals leading to tissue damage by oxidation of proteins, peroxidation of membrane lipids and modification of nucleic acids [4]. Under normal circumstances, an appreciable concentration of free iron does not exist outside physiological sinks. Any released ferrous iron (Fe+2) is immediately chelated in cells by compounds such as citrate or adenosine diphosphate. Thus, labile iron could not participate in the Haber–Weiss reaction, which catalyses the formation of ROS. Free iron may directly initiate lipid peroxidation which destroys membrane structure resulting in increased oxidative stress and cellular damage. Excess iron accumulation causes chronic free radical induced tissue damage in multiple organs and leads to progressive organ dysfunction, which results in significant morbidity and mortality. In this respect, IO should be prevented in order to preclude the adverse impact of free iron on natural homeostasis [9, 11].

This chapter will focus on iron balance and the course of excess iron in HSCT recipients. The adverse impact of IO on transplant outcome and the preventive strategies will also be discussed.

2. Body

2.1. Iron homeostasis

Iron is vital for all living organisms and takes part in several metabolic processes, including DNA synthesis, oxygen and electron transport. Although iron is a critical element in cell growth and multiplication, it is potentially toxic in excess amounts by generating ROS [5, 11-13]. Reactive oxygen species have a potential to damage DNA and proteins by lipid peroxidation. Labile iron participates in free radical formation via Fenton reaction which was first recognized in 1894. Namely, trace amounts of iron as Fe^{+2} could catalyze the oxidation of tartrate by H_2O_2. Consequently, superoxide anion (O_2^-) or H_2O_2 is converted to toxic free radicals such as hydroxyl radical (OH^-). This process is mediated by the Fenton reaction catalyzed by iron, where O_2^- reduces ferric iron (Fe^{+3}) to produce oxygen and Fe^{+2}. This reduced iron becomes reoxidized by H_2O_2 to produce OH^- [5, 11].

a. $Fe^{+2} + H_2O_2 = Fe^{+3} + OH^- + {}^{\cdot}OH$

b. $O_2^{\cdot -} + H_2O_2 \rightarrow Fe \rightarrow OH^{\cdot} + OH^- + O_2$

Figure 1. a. Fenton reaction; b. Iron catalyzed Haber–Weiss reaction or the superoxide driven Fenton reaction [5].

http://dx.doi.org/10.5772/53819

There are no physiological mechanisms in humans to excrete excess iron and iron homeostasis is primarily regulated at the level of absorbtion [4, 9, 11, 14-16]. The majority of iron absorbtion occurs via enterocytes in the proximal small intestine. The conversion of dietary inorganic non–heme iron to Fe^{+2} is facilitated by the brush border ferri reductases. Iron is transported across the cellular membrane by the divalent metal transporter 1 (DMT1) which transfers Fe^{+2} across the apical membrane and into the cell through a proton coupled process [9, 15, 16]. Ferroportin is an iron efflux pump that mediates the export of Fe^{+3} from the enterocyte. Prior to transport, Fe^{+2} is converted to Fe^{+3} by either hephaestin or ceruloplasmin both of which have ferroxidase activity. Subsequently, iron is uploaded to transferrin which is the primary iron transporter in the circulation. Ferric iron bound to transferrin is soluble and non reactive. The majority of iron (60–70%) is incorporated into hemoglobin while the rest is stored in hepatocytes, myoglobin and reticuloendothelial macrophages [9]. Hepcidin, the main regulator of iron absorbtion, inhibits intestinal absorbtion and release of storage iron in iron-overloaded states, whereas its expression is markedly decreased in iron deficiency states. Hepcidin interacts directly with ferroportin, causing its internalization, degradation and blocking iron release from cells to plasma. Hepcidin acts as an acute phase reactant which is responsible for the anemia of inflammation. Its production is upregulated by body iron excess and inflammation whereas downregulated by anemia and hypoxia [9, 14, 16].

Cell survival depends on the balance between the destructive and beneficial effects of iron [9, 12]. Natural iron homeostasis comprises regulation mechanisms to control iron excess. The primary protective pathway is the sequestration of iron in ferritin or transferrin. Ferritin is the chief storage molecule while transferrin is functionary for the transport of iron. Ferritin captures and buffers the intracellular iron pool, thus it makes iron available for critical cellular processes while protecting lipids, DNA and proteins from potentially toxic effects of iron. Iron stored in ferritin is not capable of catalyzing radical reactions and is considered as safe. It is well known that serum ferritin concentration closely parallels body iron reserves. However, as free iron is the main form of iron which can precipitate in oxidative stress, any measure of unbound iron will result in deleterious effects. The balance of free iron to bound iron changes and free iron becomes available to catalyze free radical reactions in iron overloaded states [5, 9]. Large amounts of excess iron in the circulation are likely to exceed the serum iron binding capacity (SIBC) and non transferrin bound iron (NTBI) will emerge eventually. Non transferrin bound iron bypasses the normal regulatory mechanism of receptor mediated iron uptake and is able to stimulate the peroxidation of membrane lipids and the formation of ROS. The intracellular counterpart of NTBI is considered as labile iron pool (LIP) which is bound mainly to low molecular weight compounds. Labile iron pool is catalytically active and capable of initiating free radical reactions. The expansion of the LIP and simultaneously increased NTBI may trigger cell toxicity. Generation of LIP leads to unregulated iron uptake and subsequent intracellular storage either within ferritin molecules or as hemosiderin. The adverse effects of IO can arise from the elevation of NTBI and LIP in plasma and might as well cause organ damage mediated by the accumulation of tissue iron in target organs. The equilibrium between the LIP and iron locked in the ferritin shell is critical to maintain the normal function of cellular iron enzymes. Imbalance in this equilibrium results in the uncontrolled loading of organs, such as the liver, heart and endocrine glands, with free iron which generates free radicals and causes cell damage [12, 17]. Eventually, NTBI and LIP may be more relevant iron markers than serum ferritin and transferrin as a predictor of IO induced tissue damage. Alterations in ferritin levels are seen commonly in clinical practice often reflecting perturbations in iron homeostasis or metabolism. Serum ferritin differs markedly from tissue ferritin in molecular weight, iron and carbonhydrate content, subunit size and amino acid sequence. The extracellular form of ferritin, termed as serum ferritin, is used as a clinical marker of iron status. Tissue ferritin is the more efficient storage form of iron than is serum ferritin and the function of serum ferritin has to be clarified in these circumstances [9, 12]. Serum ferritin is usually correlated with NTBI, whereas inflammation, acute and chronic liver diseases and malignancies may also cause elevated serum ferritin levels regardless of the iron stores [12].

2.2. Iron overload and stem cell transplantation

Iron overload is a significant problem in autologous (auto) and allogeneic (allo) HSCT recipients and may adversely affect transplant outcome [4, 18]. The diagnosis of IO has been reported in up to 88% of long term survivors of HSCT on the basis of serum ferritin levels [19]. Iron overloaded state may last for a long time after transplantation. In a cross sectional study by Majhail et al, in which LIC on MRI was used for diagnosis, the prevalence of IO was reported to be 32% in allo-HSCT recipients who had survived 1 year or more following HSCT [20]. In another study by the same group, serum ferritin levels were found to be above 1000 ng/ml in 34% of allo-HSCT and 13% of auto-HSCT recipients. Thus, IO may be less prevalent among recipients of auto-HSCT compared to allo-HSCT as expected [21].

The main causes of IO in HSCT are prolonged dyserythropoiesis, increased intestinal iron absorbtion due to anemia and chemotherapy associated mucositis which leads to increased iron absorbtion, transfusion burden and release of iron from injured tissues [8, 22].

Iron overload is particularly common in HSCT recipients with hemoglobinopathies and hematological malignancies which require frequent transfusions and is associated with ineffective erythropoesis such as acute leukemia and myelodysplastic syndrome (MDS). Transfusion load is considered to be the principal cause of IO in this group, as each unit of packed red blood cells (PRBC) contains approximately 200–250 mg iron. Since there is no physiological mechanism for excreting excess iron, iron accumulation is inevitable after 10–20 transfusions [22-24]. Ineffective erythropoiesis might be a contributing factor leading to excessive iron absorbtion particularly in MDS and thalassemia which is mediated by erythroid regulators of iron metabolism which suppress hepcidin and result in increased iron absorbtion. Elevated growth differentiation factor 15 (GDF–15) levels are considered to be the initiating event in this context. Ineffective erythropoiesis either as a feature of the underlying disease or a consequence of intensive treatment leads to inhibition of hepcidin possibly due to overexpression of GDF–15 and thus increases iron absorbtion and toxicity. Hematopoietic SCT recipients are at risk of IO due to prior transfusion load, increased iron absorbtion related to elevated GDF–15 levels and peri–tansplant transfusions [22, 24, 25].

Bone marrow (BM) and tumor cell destruction which occurs as a consequence of high dose therapy and release of iron from damaged cells as well as underutilization of iron due to the inhibition of erythropoiesis as a result of cytotoxic therapy are important factors in the etiology of IO. Erythropoiesis, which is the main route of iron utilization, is temporarily halted by the conditioning regimen [8, 22, 23, 26]. Conditioning treatment with chemo/radiotherapy during HSCT causes toxicity and immunosuppression leading to organ damage and infectious complications mainly in the first 3 months of the procedure [27]. Free iron, which acts as a free radical catalyser, might increase the toxicity of the conditioning regimen during HSCT. Serum iron parameters were demonstrated to be elevated 2–3 days during conditioning chemotherapy prior to stem cell infusion in a report by Gordon et al [13]. Non transferrin bound iron appears shortly after conditioning regimen and remains detectable in most patients throughout the peri–transplant period. Transferrin saturation (TS) increases during the conditioning regimen, often reaching to levels above 80% with the consequent emergence of NTBI [28]. The ability of ferritin to sequestrate iron and binding of iron to transferrin is exhausted in HSCT recipients receiving conditioning regimen, thus leading to excess NTBI formation. The extent of BM suppression caused by the conditioning regimen is correlated with the elevation of NTBI [27]. A substantial decrease in plasma anti-oxidant defense has also been demonstrated in HSCT recipients, and NTBI levels were found to be inversely correlated with plasma antioxidant capacity in a report by Yegin et al [29]. A derangementof the prooxidative/antioxidative balance was demonstrated as antioxidants only partially recover to baseline values until day 14 after HSCT [30, 31].

Hepatic toxicity due to chemotherapy and radiation might lead to hepatocellular damage with subsequent further release of hepatic iron stores. Liver damage may also disturb transferrin synthesis [28, 30]. A decrease in transferrin due to hepatic toxicity, stored iron leaking from injured liver to blood and a suppression of erythropoietic activity during treatment may causes elevated TS levels. Thus, increasing TS succeeds and contributes to the appearance of potentially toxic NTBI in the circulation. Iron in its NTBI form is a potent catalyst in Fenton's reaction which produces ROS capable of causing cellular damage through various mechanisms. Tissue damage such as mucositis and liver injury is common after HSCT and may be partly mediated by NTBI during cytotoxic chemoradiotherapy [28, 29, 32]. It is indicated that increased NTBI levels may contribute to organ toxicity and infectious complications in the early post–transplant period [29].

Complication	Incidence	Mechanism of Injury
Infection	Variable	Immune dysregulation, mediated in part by IO, iron-rich microbial environment
Chronic liver disease	Common	Multifactorial, including IO
SOS	Common (up to 54%)	Conditioning regimen, prior irradiation, possibly IO
IPS	Uncommon (2-8%)	Pro-inflammatory events and increased ROS (mediated by IO)

Table 1. Complications of IO in patients undergoing HSCT [24]

Complication	Comments
Early (<1 year)	
Infections	Mucormycosis, invasive aspergillosis, listeria monocytogenes and other infections
Acute GVHD	No clear evidence available, elevated ferritin might increase risk
SOS	Iron overload might increase risk
NRM	Elevated ferritin associated with increased risk in allo and auto-HSCT recipients
Late ("/>1 year)	
Infections	Mucormycosis, invasive aspergillosis and other infections
Chronic GVHD	No clear evidence available, decreased risk reported with elevated ferritin
Liver Function Abnormalities	Iron overload increases risk
Cardiac Late Effects	Iron overload might increase risk
NRM	No clear evidence available

Table 2. The Role of IO in Early and Late Complications of HSCT [4]

Iron toxicity may play an important role in the pathogenesis of transplant related complications [Table 1, 2]. In a series of 25 patients who underwent HSCT, very high levels of ferritin (>3000 ng/ml) and TS (>100%) dramatically increased transplant related mortality (TRM) and decreased overall survival (OS) which was particularly attributed to infections [32]. As iron is an essential element for all pathological microorganisms, excess amounts of free iron might increase microbial growth and the probability of severe infections [33]. The coexistence of excess plasma iron with the damage to the mucosal barrier may also predispose to infectious events with bacterial translocation. Hypoferraemia is a normal response to infection and appears to be a part of a natural resistance mechanism whereas hyperferremia can predispose to bacterial and fungal infections. In this context, elevated TS and ferritin levels are proven risk factors for the development of systemic fungal infections in patients with hematological malignancies [1, 33, 34]. Furthermore, an increase in late fungal infections, especially mucormycosis, has been reported in iron loaded patients after HSCT [35]. Elevated pre–transplant ferritin levels seem to effect prognosis adversely in myeloablative HSCT primarily due to increased NRM. On the other hand, elevated iron stores apart from providing a milieu for infection and organ toxicity, may also be in relevance to tumor growth. Thus elevated ferritin levels might be in association with relapse and relapse mortality [36]. Mahindra et al reported that elevated pre–transplant serum ferritin level was an independent adverse risk factor for OS in patients undergoing non myeloablative HSCT. Inferior survival in patients with elevated ferritin was related to both higher rates of treatment related mortality and relapse mortality [37]. On the other hand it should also be noted that ferritin is an acute phase reactant and a marker of inflammation besides its role as a surrogate marker of iron status. Thus, elevated ferritin levels might as well indicate a group of patients with more agressive primary disease biology and a subgroup of patients who are already more likely to experience disease relapse. Thus the association of elevated ferritin levels with relapse might be unrelated to IO.

The adverse impact of IO on transplant outcome has been demonstrated most convincingly in patients with thalassemia where class III patients with extensive liver damage had higher TRM [38]. Besides increased TRM, other complications attributed to IO includes fungal infections, hepatic dysfunction and hepatic SOS/Veno occlusive disease (VOD) [4, 27, 38, 39]. In fact, thalassemia is a benign disorder and ferritin is directly a marker of excess iron and elevated levels could not be attributed to the biology of an underlying malignant pathology. As a result of the above mentioned data, pre–transplant serum ferritin was included in a prognostic scoring system for acute leukemia and MDS patients undergoing allo–HSCT [40]. The late morbidity of IO is primarily due to the involvement of heart and liver. Although iron related liver function test (LFT) abnormalities have been reported, there are no studies that describe the role of IO in late onset cardiomyopathy and hepatic fibrosis/cirrhosis in patients transplanted for diseases other than thalassemia. Post–transplant iron depletion therapy has been shown to reverse hepatic fibrosis and cardiomyopathy in children with thalassemia who have undergone allo–HSCT [4].

2.3. Iron overload and transplant complications

2.3.1. Liver complications

Liver disease is a frequent cause of morbidity and mortality following allo–HSCT and affects 90% of recipients and up to 5–10% of toxic deaths are liver related. Liver injury in the early post–transplant period may be secondary to drug toxicity, SOS, acute GVHD, opportunistic infections, total parenteral nutrition, tumor invasion and cholestatic disorders [3, 41]. Long term liver disease is also a common complication of HSCT, as 57, 5% of survivors developed chronic liver disease (CLD) at 2 years after transplantation in a retrospective series of 106 patients reported by Tomas et al. In this retrospective study, the combination of chronic hepatitis C and IO was presented as the most frequent cause of CLD [41]. On the other hand, chronic GVHD also contributes to liver toxicity. The timing and pattern of LFT abnormalities, history of pre or post transplantation hepatitis, presence of GVHD at other sites and transfusion burden might be helpful in determining the etiology of liver disease. Accurate diagnosis of the etiology of liver dysfunction is generally problematic even though the patterns of biochemical, clinical and histological abnormalities can aid diagnosis. Liver biopsy in patients following HSCT is not without risks, particularly due to thrombocytopenia during the early post–transplant period. The most common indication for liver biopsy is to assess the possibility of GVHD in allo–HSCT in the late post–transplant period with persistently abnormal LFTs and no evidence of GVHD on other sites. In this clinical setting, the sensitivity and specifity of serum ferritin as a marker of IO is not well defined due to its concomittant role as an acute phase reactant [3, 5, 8, 24, 41-43]. Liver biopsy may be performed when atypical clinical features are present or multiple disease processes are likely to occur simultaneously or when there is poor response to therapy that has been instituted [44]. The management of liver dysfunction under these conditions may be complicated as overlapping features often complicate the diagnosis and establishing the correct diagnosis is crucial to institute disease specific therapy. Autopsies performed in 10 patients who died early after HSCT showed iron accumulation in a range equivalent to that of patients suffering from HH [26]. A cumulative cirrhosis incidence of 3, 8% by 20 years after HSCT has been reported previously [8]. This rate seems to be an underestimation as the majority of long term survivors have not been subjected to liver biopsy. In a retrospective study by Sucak et al, severe IO was demonstrated in 75% of 24 liver biopsies which were performed with the presumptive diagnosis of hepatic GVHD in 20 patients with persistent elevation of liver enzymes in the post–transplant setting. The initial clinical diagnosis of GVHD was refuted in 43, 5% of the patients. Median number of post–transplant transfusions, TS and ferritin levels were found to be significantly higher in patients who had histologically proven hepatic IO. A significant correlation between serum ferritin levels and histological grade of iron in the hepatocytes was also demonstrated [10]. In another study by Iqbal et al, the diagnosis obtained at laparoscopic liver biopsies altered targeted therapy in 31% of patients. Iron overload was found in 81, 25% of a total of 32 biopsies [45]. A diagnosis of IO after HSCT was demonstrated based on histological evidence of siderosis found in 52, 4% of liver biopsies performed at 15–110 days post-transplant in another study. Liver biopsies were performed for diagnostic purposes in patients with chronic liver dysfunction. An improvement in LFT was observed in 21 of the 23 patients (91%) with IO who underwent phlebotomy [41]. Namely, IO seems to be underestimated as a cause of liver dysfunction in HSCT setting and liver biopsy which allows disease specific therapy could be life saving.

Hepatic IO may also worsen the natural course of chronic viral hepatitis and the response to antiviral therapy. Fujita et al demonstrated that liver iron deposition was more common in chronic hepatitis C compared to hepatitis B and was associated with liver disease progression. Increased hepatic iron stores in chronic hepatitis C were related to resistance to Interferon/Ribavirin treatment [46]. Thalassemic patients with liver fibrosis and hepatomegaly who undergo HSCT, have a markedly reduced OS and event free survival compared to patients without evidence of liver disease. The liver disease in these patients is due to a combination of severe IO and chronic viral hepatitis both of which improve with effective iron chelation therapy [19, 26, 47]. Iron is also deposited in other tissues such as myocardium or BM. Slow and spontaneous decrease in iron stores has been reported in thalassemic children in the years following HSCT. This natural iron depletion could normalize iron stores in individuals with mild siderosis. However, in patients with moderate to severe IO this slow depletion could not prevent the development of liver dysfunction. For this reason, iron depletion protocols have been developed for patients with severe IO [19, 23, 26, 47].

2.3.2. Sinusoidal obstruction syndrome (veno occlusive disease)

Sinosoidal obstruction syndrome is a treatment related toxicity associated with auto and allo–HSCT which is seen in 6–54 % of the recipients. The severity of SOS ranges from a mild reversible to a progressive course with a mortality rate close to 100% [5, 24].

The role of pre–transplant hyperferritinemia in the development of SOS was first demonstrated by Morado et al in a cohort of 180 auto–HSCT recipients. In this prospective study, SOS was defined in 12, 2% of patients based on McDonald criteria. Patients with pre–transplant ferritin levels above 300 mg/dl were shown to have a higher risk of developing SOS [48]. In a recent report by Maradei et al, a pre–transplant serum ferritin level above 1000 ng/dl was identified as an independent risk factor for the development of SOS [39]. A retrospective study of 250 HSCT recipients by Sucak et al, in which SOS incidence was reported to be 29, 7%, demonstrated significantly higher pre–transplant serum ferritin levels in patients with SOS [49]. In another study reported by Sucak et al, pre–transplant ferritin levels were found to be higher in HSCT recipients who developed SOS in the post–transplant setting [50]. Serum ferritin may be increased in conditions other than IO in this particular group of patients, including chronic inflammation and infection. Nevertheless, values higher than 1000 ng/ml were rarely reported in these inflammatory conditions [1, 25, 29, 39, 48-51].

Iron induced hepatotoxicity is multifactorial which involves oxidative stress and modulation of gene expression of Kuppfer cells. Cellular injury is induced by iron generated ROS and peroxidation of lipid membranes [39]. Risk factors associated with the development of SOS are defined as preexisting liver dysfunction, previous abdominal irradiation, high dose total body irradiation, high dose preperative regimens, advanced disease and HLA mismatch or unrelated HSCT. The typical hepatocellular lesion of SOS mainly occurs in zone 3 of hepatic acines including a characteristic endothelial lesion which is shown to be associated with hypercoagulability. The oxidant effect of iron on endothelial and and hepatocyte membranes mediated by ROS contributes to the development of these typical lesions of SOS [48, 50]. The risk of SOS is higher in carriers of at least one allele of the hemochromatosis gene, HFE, which predisposes to iron deposition in the liver [24].

2.3.3. *Infections*

Patients with HH and other diseases with IO are considered to be more susceptible to infections, as iron adversely affects the phagocytic, chemotactic and bactericidal capacity of granulocytes and monocytes and inhibits the activity of natural killer cells and macrophages [35, 52]. A number of studies have demonstrated the adverse impact of IO on the development infections in HSCT recipients. Tachibana et al observed an association between IO and blood stream infections (BSI) in 114 patients who underwent allo–HSCT. They found that pre–transplant serum ferritin levels significantly predicted BSI within the 100–day period after allo–HSCT [1]. A direct correlation between hepatic IO and BSI was demonstrated in a retrospective cohort of 154 allo – HSCT recipients, as patients with hepatic IO tended to experience more frequent and prolonged episodes of lethal BSI [53]. Altes et al reported a ferritin level above 1500 μg/l was associated with the occurence of bacteremia and febrile days in first 3 months after auto–HSCT [27]. A prospective study investigated the risk factors for 140 early infection episodes which occured in 367 multiple myeloma (MM) patients undergoing auto–HSCT. Bone marrow iron stores were identified as significant risk factors for early severe infections [54]. Pre–transplant serum ferritin levels were demonstrated to be associated with fungal infections after allo–HSCT in several studies [33-35, 49, 55, 56]. Tunçcan et al identified the predictive role of pre–transplant serum ferritin level in the development of hepatosplenic candidiasis among 255 HSCT recipients. Hepatosplenic candidiasis was diagnosed in 6 (2, 3%) patients. Pre–transplant serum ferritin levels were significantly higher in patients with hepatosplenic candidiasis [55]. Özyilmaz et al studied the relationship between serum ferritin level and pulmonary fungal infections in 148 allo – HSCT recipients. In this study, the sensitivity and specifity of ferritin > 1000 ng/ml for the prediction of fungal pulmonary infections were found to be 67% and 70%, respectively [56].

2.3.4. Idiopathic Pneumonia Syndrome (IPS)

Idiopathic pneumonia syndrome comprises a group of disorders that result in interstitial pneumonitis and/or widespread alveolar injury with an incidence of 2–8 % and a mortality of up to 70% in the HSCT setting. There is increasing evidence implicating ROS and pro–inflammatory events as major contributing factors to IPS [5, 24]. The mechanism of iron induced IPS probably involves endothelial injury by catalytically active iron released from heme groups, which can trigger a cascade of events leading to acute lung injury and pulmonary fibrosis [24]. Currently, there are no studies regarding the direct association of IO and IPS, except the oxidative milieu, which is partly a consequence of IO.

2.3.5. Graft-versus-host disease (GVHD)

The role of IO in the pathogenesis of GVHD has been evaluated in a number of studies. There are conflicting results regarding the relationship between IO and GVHD in HSCT recipients. In a prospective cohort of 190 allo – HSCT recipients reported by Pullarkat et al, the effect of elevated pre–transplant ferritin on acute GVHD was assessed. Grade 2 or above acute GVHD was diagnosed in 48% of patients. Acute GVHD was more frequent in patients with high ferritin levels (≥1000 ng/ml). This was attributed to the increased ROS mediated injury on exposure to the conditioning regimen in iron overloaded patients, as antigen exposition following tissue injury was indicated to be the initiating event in the pathogenesis of GVHD [38]. Similarly in a report by Platzbecker et al, which was performed in 172 patients with MDS, transfusion burden reflected by ferritin levels, was found to be correlated with a higher probability of acute GVHD [57]. On the other hand, Mahindra et al investigated 222 patients who underwent myeloablative allo–HSCT and demonstrated that pre–transplant ferritin level >1910 μg/l was associated with decreased incidence of chronic GVHD [58]. Furthermore, in a study of 264 patients who underwent allo–HSCT for various hematological malignancies, no significant difference in the cumulative incidence of acute and chronic GVHD was demonstrated in high (≥599 ng/ml) and low (<599 ng/ml) ferritin groups [59]. Alessandrino et al reported that transfusion dependency was an independent risk factor for the development of acute GVHD, but not for chronic GVHD [60]. On the other hand, IO might as well mimic GVHD resulting in unnecessary continuation or intensification of immunosuppressive therapy for GVHD [18]. Apart from hepatocellular, cardiac and other organ dysfunction, IO may worsen the natural course of liver GVHD, similar to the status with chronic hepatitis and its response to therapy [3, 18, 23, 51, 57]. It is speculated that intestinal iron absorbtion is increased as a result of epithelial injury related to chemotherapy or GVHD. Suggesting that IO might be the consequence rather than being the cause of intestinal GVHD [23]. The liver and the intestinal mucosa, which express essential iron regulatory genes including hepatic antimicrobial protein (HAMP), the gene that encodes hepcidin and ferroportin 1, are targets of conditioning related toxicity as well as GVHD, initiated by donor derived T lymphocytes. The ensuing release of cytokines including IL-6, might directly affect the expression of hepcidin as IL-6 is a potent inducer of hepcidin via STAT3 [61]. Graft versus host disease also involves the interaction of Fas ligand expressed on activated donor T lymphocytes with host tissue including enterocytes and hepatocytes. T lymphocyte induced tissue damage disrupts iron homeostasis leading to uncontrolled iron accumulation which may aggravate tissue damage related to the development of GVHD and infections [15]. The pattern of the relationship between IO and GVHD remains to be confirmed in future studies.

2.4. Prognostic role of iron overload in stem cell transplantation

Several recent reports demonstrated that IO is an adverse prognostic factor for patients undergoing allo–HSCT [1, 17, 22, 36, 59, 62-66]. In a retrospective cohort of 114 AML and MDS patients, the OS rate at 5 years was found to be significantly better in patients with ferritin levels < 1000 ng/ml [1]. Tanaka et al evaluated the outcome of 47 patients with acute leukemia or MDS who underwent reduced intensity HSCT. High ferritin level which was defined as >1000 ng/ml was associated with worse 2 year OS on multivariate analysis [62]. Another study by the same group demonstrated the adverse impact of elevated ferritin levels on 5 year OS in a cohort of 143 patients with acute lymphoblastic leukemia (ALL) and acute myeloblastic leukemia (AML) who received allo–HSCT with myeloablative and non myeloablative conditioning regimens [63]. Transfusion dependency, predicted by serum ferritin levels, was found to be independently associated with reduced OS and increased NRM in a retrospective cohort of 357 MDS patients undergoing allo–HSCT [60]. The transplant iron score which included serum ferritin level above 1000 ng/ml was tested in 78 patients who received allo or auto–HSCT. The independent impact of IO on transplant survival was indicated with the most pronounced predictive power of the iron score restricted to allo–HSCT recipients. A high iron score (≥2) was associated with 50% absolute decrease in OS at 1 year [67]. Lim et al reported the adverse impact of elevated serum ferritin on OS in 99 MDS patients who underwent reduced intensity HSCT [64]. Altes et al demonstrated that serum ferritin levels ≥3000 μg/l and TS ≥100% were associated with a decreased OS and increased TRM, which was attributed to a high infectious mortality [32]. On the other hand Pullarkat et al analyzed 190 patients and demonstrated that elevated pre–transplant ferritin levels were associated with increased risk of death and day 100 mortality, mainly due to acute GVHD and infections [38]. Mahindra et al demonstrated a pre–transplant serum ferritin > 685 ng/ml was associated with lower OS and relapse free survival in 315 patients with Hodgkin and non Hodgkin lymphoma who received auto–HSCT, whereas same ferritin level exhibited a higher incidence of relapse and relapse mortality. They identified the baseline ferritin level was best correlated with poor survival. They concluded that elevated iron stores may also increase tumor growth, as tumor cells require more iron for DNA synthesis due to rapid proliferation [36]. Same group confirmed their results in a study of 222 allo–HSCT recipients with a serum ferritin level >1910 µg/l associated with lower OS, lower relapse free survival and higher NRM rates [58]. Furthermore they demonstrated inferior survival rates related to higher rates of TRM and relapse mortality in patients with elevated ferritin levels who received non myeloablative conditioning [37]. In a large retrospective study by Armand et al, an elevated pre–transplant serum ferritin level was significantly associated with lower OS and disease free survival. This association was particularly restricted to patients with acute leukemia and MDS which was particularly attributed to transfusion load. They suggested a possible role of iron chelation therapy in the pre and post – transplant setting, as they showed an absolute difference of 37% in 5–year OS for patients with MDS between the highest and lowest ferritin quartiles [66]. Sucak et al demonstrated an adverse impact of a pre–transplant serum ferritin level >500 ng/ml on OS and TRM in 250 patients who received auto and allo–HSCT, underscoring the prognostic effect of IO in auto transplants [49]. The same group confirmed their results with a more toxic form of iron, NTBI, in a retrospective cohort of 149 patients. In concordance with the previous report, a significant impact of NTBI on day 30 and day 100 survival was shown in auto–transplanted patients for the first time in iron and transplant connection [29]. Notwithstanding, in a prospective study by Armand et al, pre–transplant IO predicted by LIC which is considered to be the gold standard indicator of IO, was not found to be associated with increased mortality, relapse, SOS or GVHD [68]. Therefore, they assumed that the adverse prognostic impact of pre–transplant hyperferritinemia may be related to factors independent of IO. Taken together, it is speculated that ferritin may be prognostic not because it reflects iron stores but because it is an acute phase reactant [68, 69].

2.5. Diagnosis of iron overload

2.5.1. Liver biopsy

Liver remains to be the most accessible parenchymal organ that can be used to estimate tissue iron load after HSCT. Iron overload is not uncommonly seen in various other primary liver diseases such as alcoholic liver disease, chronic viral hepatitis, non alcoholic steatohepatitis, liver cirrhosis and HH. Histological evaluation of liver specimens is essential in the management of these disorders. The reported incidence of significant liver fibrosis in HSCT recipients varies from 5% to 80% and LIC has been demonstrated to have a particular role in the progression of fibrosis [26, 41, 70]. Though ferritin continues to be the mainstay for the initial clinical evaluation of IO, liver biopsy is still the gold standard for quantifying iron. Measurement of hepatic iron stores provides the most reliable estimate of body iron burden. Liver iron content exceeding 80 mcmol/g of liver dry weight was found to be consistent with IO with a hepatic index greater than 1, 9 mmol/kg/year. However, the need for a relatively large volume of tissue as well as its invasive nature has made this procedure less appealing to most clinicians and patients [4, 9, 53]. Although liver biopsy is an invasive procedure and can not be safely administered in patients with very low platelet counts, a liver biopsy can be advantageous in some HSCT recipients as it can also exclude alternative causes of hepatic dysfunction, such as infections and GVHD. In high risk patients, liver biopsy using a transjuguler approach may be a feasible alternative to percutaneous biopsy [4, 17].

2.5.2. Non-invasive procedures

Superconducting quantum interference device (SQUID) assesses total body iron by using biomagnetic susceptometry. Ferritin and hemosiderin are the only paramagnetic materials in the human body, thus the magnitude of these parameters is directly related to the amount of iron in a certain volume of tissue. The device utilizes the magnetic property of iron in ferritin and hemosiderin to estimate hepatic iron stores. Furthermore, it is considered to be the non invasive reference standard for estimation of LIC as it has an excellent correlation with liver biopsy. However, widespread clinical use is limited by its cost, complexity and very limited availability [4, 9, 17].

Liver iron content measurement has limited predictive value for extrahepatic iron deposition. The liver is the dominant iron reservoir for the body, accounting for more than 80% of the total body iron and has high capacity mechanisms for clearing both transferrin and NTBI species from the circulation. The heart and endocrine tissues have tightly regulated transferrin uptake and develop IO only when there is circulating NTBI. High liver iron (15-20 mg/g dry weight) damages liver parenchyma and increases circulating NTBI levels dramatically. As no liver iron can be considered safe from a cardiac and endocrinological perspective, extrahepatic monitoring by magnetic resonance imaging (MRI) is essential [71]. Magnetic resonance imaging becomes increasingly important in the evaluation of iron status as it is non invasive, more rapidly and widely available. Designating liver iron by older MRI techniques and equipment showed variable correlation with the biopsy estimates of LIC. More recent MRI techniques T2* and R2* MRI are reproducible methods for non invasive estimation of LIC with reported sensitivity and specifity of 89% and 80%, respectively [4, 17, 72-74]. It has the additional benefit of identifying relatively early IO within organs prior to the onset of dysfunction. Magnetic resonance imaging can be used to co-measure iron deposition within the heart, liver and pituitary gland as it does not appear that a single organ gives the full picture of total body IO. In fact, patients can accumulate cardiac iron, despite apparently normal hepatic iron levels and thus be at risk for arrhythmia or congestive heart failure. The discordance of values in two tissues can be resolved with the use of MRI to detect cardiac iron. Cardiovascular MRI could potentially be used not only to determine myocardial iron content but also cardiac function and therefore could be used to investigate the effects of iron mediated organ damage. Non invasive measurement of LIC has also been achieved using an MRI technique based on the proton transverse relaxation rates within the liver. The technique can be implemented on, most clinical 1, 5–T MRI measurements, making it readily available to the clinical community. This technique resulted in a high specifity and sensitivity over a greater range of LIC than any other MRI–based method of LIC assessment [9].

2.5.3. Ferritin

High prevalence of IO in long term survivors of HSCT emphasizes the need for routine screening for IO in this population. Ferritin is a cellular iron storage protein that buffers iron in a soluble and non toxic form. Under normal conditions ferritin levels in the serum are low but steadily increase in conditions of IO. Therefore, assessment of serum ferritin levels serves as a simple and widely used surrogate marker for IO. Serum ferritin levels are however subject to natural fluctuation and can also be greatly affected by a range of inflammatory conditions that are particularly relevant in HSCT recipients. Although being a useful test for initial screening of IO in HSCT recipients, serum ferritin is not a reliable indicator of total body iron burden particularly in patients who have ongoing acute infections or inflammatory diseases [2, 4, 17, 20, 22, 23, 38, 75, 76]. Serial serum ferritin measurements can compensate the potential fluctuations and help to establish a general picture of IO over time. Nevertheless, at 1 year after–transplantation when inflammatory stress has largely subsided, most patients have a serum ferritin of <1000 ng/ml and no clinical evidence of IO; serum ferritin in these patients decline slowly with time [23]. Unlike tissue ferritin a substantial proportion of serum ferritin is glycosylated which suggests that plasma ferritin is actively secreted from reticuloendothelial system or parencymal cells. Serum ferritin in contrast to tissue ferritin was claimed to have a low iron content even in iron loaded patients in some earlier studies. It is therefore claimed that serum ferritin does not provide a major source of hepatic iron either in normal individuals or in patients with IO diseases [4, 20, 22, 23, 75]. On the contrary a direct correlation between serum ferritin levels and transfusion burden has been observed with a level of 1000 ng/ml after a median of 21 PRBC transfusions. Thus repeated measurement of serum ferritin levels seems to be a valid method to monitor secondary IO in patients with transfusion dependent anemias and MDS [17]. Majhail et al studied the prevalence of IO in 56 allo–HSCT recipients and demonstrated the poor predictive value of ferritin for estimating LIC. The overall prevalence of IO was 32%. Clinically significant IO (LIC>7 mg/g) was uncommon in patients with serum ferritin levels less than 1000 ng/ml. However, the LIC on MRI was moderately correlated with serum ferritin. As a result, they indicated ferritin to be a good screening test but a poor predictor of tissue IO and recommended estimation of LIC before initiating chelation therapy. They considered that this lack of association between ferritin and LIC might be related to the variability in ferritin levels because of ineffective erythropoiesis or underlying inflammation or infection [20]. Whereas in a study by Bazuave et al, serum ferritin, transferrin, TS, iron, soluble transferrin receptor (sTfR) and C reactive protein levels in 230 HSCT recipients were measured. All iron parameters were found to be significantly associated with survival. A combination of ferritin and TS was shown to have the highest prognostic power. They concluded that the predictive power of ferritin was derived from its association with IO rather than inflammation. Inferior survival in patients with IO was related to both TRM and relapse. As sTfR and TS were found to have superior prognostic value when compared to ferritin, they suggested to combine serum ferritin with TS for prediction of IO [2].

Recent evidence suggests that the determination of iron status before HSCT has important prognostic implications. There is a gap between the time that patients are identified for HSCT and the time that actual transplant takes place. During this period, most patients stay transfusion dependent. After patients are exposed to conditioning regimen and stem cell infusion, serum ferritin levels are prone to a false elevation due to its role as an acute phase reactant. Thus, accurate evaluation and diagnosis of iron toxicity after HSCT remains as a challenge [53, 67] [Table 3].

Diagnostic Test	Advantages	Disadvantages
Liver Biopsy	Reference method, can assess degree of hepatic fibrosis, can evaluate other causes of hepatic dysfunction (GVHD)	Invasive procedure, not feasible in patients with thrombocytopenia or coagulopathy
SQUID	Good correlation with liver biopsy, noninvasive	Very limited availability
MRI	Good correlation with liver biopsy (T2 or R2 MRI), noninvasive, widely available	Variety of MRI techniqueshave not been validated with liver biopsy, contraindications (metal implants, claustrophobia)
Serum ferritin and TS	Noninvasive, widely available	Sensitive but not specific for IO, poor correlation with liver biopsy

Table 3. Diagnostic Tests for Assessment of Body Iron Stores in HSCT Recipients [4]

2.5.4. Non Transferrin Bound Iron (NTBI)

Non transferrin bound iron is toxic to living systems because it can act as a catalyst in the formation of ROS which in turn stimulate lipid peroxidation in membranes. In iron-overloaded states when SIBC becomes fully saturated, NTBI complexes appear in the serum. In a study by Harrison et al, serum ferritin was raised in 21 of 28 patients following treatment for hematological malignancy, whereas only 16% of them had LFT abnormalities. However, NTBI was detected in 4 of 6 patients with an unexplained elevated LFTs. Therefore, they considered that NTBI might be a more specific indicator of IO than the serum ferritin concentrations [77]. Assessment of NTBI is a potentially useful approach that allows the estimation of toxic iron levels. However, the methods for determining this free fraction of body iron and its precise prognostic significance require fine tuning [17].

2.6. Treatment of iron overload

The current paradigm of managing post–transplant IO is based on extensive experience in children with transfusion dependent anemias [4]. Post–transplant iron depletion therapy has been shown to reverse hepatic fibrosis and cardiomyopathy in patients with thalassemia [4, 78]. However, there is no published data indicating the benefit of iron removal therapy on long term morbidity and mortality in HSCT recipients, especially for diseases other than thalassemia [4].

Decisions regarding the management of IO should be individualized and based on a review of several factors including the need for ongoing PRBC transfusion therapy, time since transplantation, ability to tolerate iron depleting therapy and urgency to reduce body iron stores [Table 4]. For instance, coexisting anemia can preclude the use of phlebotomy whereas renal impairment might increase the risk of toxicity from iron chelating drugs. Also depletion of iron stores would be more imperative in patients with IO related liver test abnormalities or cardiac dysfunction compared to those without end organ toxicites [4].

Modality	Advantages	Disadvantages
Phlebotomy	Extensive experience with proven efficacy, no significant side effects	Not feasible in patients with anemia or poor venous access
Deferoxamine	Extensive experience with proven efficacy	Inconvenient administration route and schedule, side effects (ototoxicity, growth retardation)
Deferiprone	Oral iron chelator	Unproven efficacy, side effects (neutropenia, hepatic fibrosis)
Deferasirox	Oral iron chelator, efficacy similar to deferoxamine	Long term toxicity profile not established, side effects (nephrotoxicity)

Table 4. Treatment Options for Iron Overload after HSCT [4]

Iron overload may be a cause of persistent hepatic dysfunction after HSCT. Patients with LIC>15 mg/g dry weight should be treated aggresively with both phlebotomy and chelation; when LIC is 7–15 mg/g dry weight, phlebotomy is indicated; when LIC is under 7 mg/g dry weight treatment is indicated only if there is evidence of liver disease. Mobilization of iron from heavily overloaded patients improves cardiac function, normalizes serum alanine transaminase levels and results in improved liver histology [24, 79].

In patients with extreme IO, effective pre–transplant chelation therapy is suggested to improve post–transplant survival, as IO is clearly related to treatment related morbidity and mortality after HSCT [4, 24, 67, 79]. In the pre–transplant period vigorous iron chelation may be important but prospective studies are required to prove a survival benefit after HSCT. In the post–transplant period phlebotomy sometimes combined with erythropoiesis stimulating agents (ESA) may be successfully applied in thalassemia. For those patients who can not be phlebotomized iron chelation can be considered. Prospective studies of the impact of iron chelation therapy before and after HSCT on post–transplant morbidity and mortality are mandatory [4, 24].

The American Society for Blood and Marrow Transplantation (ASBMT) 2012 guidelines recommend annual serum ferritin measurement in patients who received PRBC transfusions pre or post–transplantation. Subsequent monitoring with serum ferritin should be considered among patients with elevated levels, especially in the presence of abnormal LFTs, PRBC transfusions or HCV infection. Additional diagnosting testing including liver biopsy, MRI or SQUID may be indicated if therapy is intended for presumptive IO. Current prescribing guidelines recommend continuation of iron reduction till ferritin levels are below 500 ng/ml [3, 9, 51, 60, 72].

2.6.1. Phlebotomy

Phlebotomy is a feasible option for the treatment of IO following HSCT. Many studies have documented its efficacy in early and late post–transplant setting. It has been shown that subclinical left ventricular diastolic dysfunction and impaired left ventricular contractility in patients with thalassemia may be reversed by phlebotomy initiated after HSCT [51]. Iron

http://dx.doi.org/10.5772/53819

overload should be treated by means of phlebotomy and/or chelation therapy especially when IO coexists with chronic viral hepatitis. Phlebotomy has the advantage over chelation of better compliance, fewer side effects and lower costs. The use of ESA may facilitate the success of this strategy in patients with low hemoglobin levels [4, 19, 22, 26, 70].

After normalization of transaminases and serum ferritin with aggressive phlebotomy, maintenance phlebotomy is required every 3-6 months to prevent iron reaccumulation and keep serum ferritin in a low normal range. The gradual rise in ferritin after successful iron depletion suggests that there is a signal for increased iron absorbtion and the signal persists well beyond the peri–transplant period. It may be that post–transplant immunosuppressants reduce the level of cytokines that normally stimulate hepcidin production and allow increased absorbtion of dietary iron. In addition hepatic GVHD may result in disordered hepcidin regulation, as it likely does in chronic viral hepatitis and might explain increased risk of IO and the need for maintenance phlebotomy after successful iron depletion [23].

2.6.2. *Iron chelation*

Treatment with phlebotomy is not possible in patients who are transfusion dependent. Chelation may be preferred for iron depletion [9]. There are limited data on the pharmacological chelation of iron during the post–transplant period including the safety, optimal dose, time for initiation of treatment and duration of therapy [51, 80, 81].

Deferoxamine, the first available iron chelator, has a proven efficacy and safety with decades of experience and has also been studied in HSCT recipients. Recommended treatment schedule is at least 5 nights per week subcutaneous delivered via a pump for 8-12 hours [4, 9]. It is effective in lowering serum ferritin levels and LIC and prevents endocrinological complications. Long term treatment is also associated with a reduction in cardiac complications and improved survival. Redness and induration at the infusion site are the most common side effects. Audiological, ophthalmological, growth and bone toxicities may be minimized by avoiding overchelation. Deferoxamine treatment in the HSCT setting is complicated by the short half life and the ability to release iron to bacteria and fungi. Deferoxamine supports the growth of zygomycetes because it acts as xenosidephore delivering iron to iron uptaking molecules of the species [22, 51, 81]. The greatest challenge with DFO is patient adherence with therapy because the need for parenteral administration is cumbersome, uncomfortable, inconvenient and time consuming [51]. Cardiac morbidity and mortality continue to occur in patients treated with DFO, likely related to difficulties with adherence [4, 9, 22, 51, 81].

Deferiprone is an oral iron chelator which was first identified in 1980s and subsequently approved for clinical use in Canada and Europe especially when DFO is contraindicated. Deferiprone is not commercially available in all countries and has not been investigated in HSCT recipients. It has a short half life of only 1, 5 hours and thus requires 3 times daily dosing. Unfortunately, it does not control liver iron as effective as DFO even after years of continued treatment. In contrast, a recent study in patients with thalassemia showed better myocardial function in those receiving Deferiprone. Retrospective studies have also demonstrated reduced cardiac morbidity and mortality and lower myocardial iron deposition among patients treated with Deferiprone compared with DFO and Deferasirox (DFX). A reduction or stabilization of serum ferritin levels and LIC in most patients with transfusional IO was demonstrated. The high risk of agranulocytosis necessitates weekly blood monitoring. Thus, toxicity profile of the drug may be inappropriate for transplant recipients [4, 9, 81].

A novel oral iron chelator, DFX was approved by the US Food and Drug Administration in 2005 and represents a significant advancement in the treatment of IO. It is a tridentate oral iron chelator which is lipid soluble but highly protein bound. It has a plasma half life about 12 hours and thus is ideal for once daily dosing. It binds iron in a 2/1 ratio. It is excreted by the hepatobiliary system and the chelated iron is excreted via the feces. The effective dose is between 20-40 mg/kg. It is generally well tolerated by patients although some dose modifications may be necessary for diarrhea. Phase III trials demonstrated that DFX at 20-30 mg/kg/day led to the maintenance or reduction of iron burden as measured by LIC in chronically transfused patients. Reductions in LIC and serum ferritin are similar to those found in the subcutaneous use of DFO. Commonly reported side effects include skin rash, nausea, vomiting and diarrhea and elevations in serum creatinine levels, which may be important in patients treated with calcineurin inhibitors. Gastrointestinal disturbances often improve with continued administration of the drug. Elevations in serum creatinine occur in approximately 1/3 of subjects. Side effects associated with DFX therapy may overlap or exacerbate early complications such as calcineurin induced renal injury seen after allo–HSCT, which Mkes it complicated to use early after HSCT. The availability of an oral iron chelator has simplified the treatment of IO, but more experience with its use in HSCT recipients is needed [4, 9, 22, 80, 81].

3. Conclusion

The role of IO in HSCT recipients and guidelines for screening strategies warrants further studies. The value of routine screening for IO, the method of determining it, whether it should be with serum ferritin, by determining LIC with non invasive MRI or biopsy and identifying a subgroup of patients who might benefit from phlebotomy and/or iron chelating agents requires future prospective studies. The possibility of IO should be considered in patients who are candidates for HSCT. Red blood cell transfusion should be limited whenever possible and chelation and/or phlebotomy should be considered in the course of documented IO. pre–transplant preventive measures should also be adopted to avoid IO and improve survival in these patients.

Author details

Zeynep Arzu Yegin[1], Gülsan Türköz Sucak[1] and Taner Demirer[1]

*Address all correspondence to: Taner.Demirer@medicine.ankara.edu.tr

1 Gazi University Faculty of Medicine, Department of Hematology, Ankara, Turkey

Ankara University Faculty of Medicine, Department of Hematology, Ankara, Turkey

References

[1] Tachibana T, Tanaka M, Takasaki H, Numata A, Ito S, Watanabe R, Hyo R, Ohshima R, Hagihara M, Sakai R, Fujisawa S, Tomita N, Fujita H, Maruta A, Ishigatsubo Y, Kanamori H. Pretransplant serum ferritin is associated with bloodstream infections within 100 days of allogeneic stem cell transplantation for myeloid malignancies. Int J Hematol 2011;93(3):368-374.

[2] Bazuaye GN, Buser A, Gerull S, Tichelli A, Stern M. Prognostic impact of iron parameters in patients undergoing allo-SCT. Bone Marrow Transplant 2012; 47(1):60-64.

[3] Majhail NS, Rizzo JD, Lee SJ, Aljurf M, Atsuta Y, Bonfim C, Burns LJ, Chaudhri N, Davies S, Okamoto S, Seber A, Socie G, Szer J, Van Lint MT, Wingard JR, Tichelli A; Center for International Blood and Marrow Transplant Research (CIBMTR); American Society for Blood and Marrow Transplantation (ASBMT); European Group for Blood and Marrow Transplantation (EBMT); Asia-Pacific Blood and Marrow Transplantation Group (APBMT); Bone Marrow Transplant Society of Australia and New Zealand (BMTSANZ); East Mediterranean Blood and Marrow Transplantation Group (EMBMT); Sociedade Brasileira de Transplante de Medula Ossea (SBTMO). Recommended screening and preventive practices for long-term survivors after hematopoietic cell transplantation. Biol Blood Marrow Transplant 2012;18(3):348-371.

[4] Majhail NS, Lazarus HM, Burns LJ. Iron overload in hematopoietic cell transplantation. Bone Marrow Transplant 2008;41(12):997-1003.

[5] Evens AM, Mehta J, Gordon LI. Rust and corrosion in hematopoietic stem cell transplantation: the problem of iron and oxidative stress. Bone Marrow Transplant 2004;34(7):561-571.

[6] Carreras E. Early complications after HSCT. In: Apperley J, Carreras E, Gluckman E, Masszi T. Haematopoietic Stem Cell Transplantation. Chapter 11; 2012 Revised Edition. p177-194.

[7] Tichelli A, Socie G on behalf of the Late Effects Working Party of the EBMT. Late effects in patients treated with HSCT. In: Apperley J, Carreras E, Gluckman E, Masszi T. Haematopoietic Stem Cell Transplantation. Chapter 15; 2012 Revised Edition. p249-269.

[8] Strasser SI, Sullivan KM, Myerson D, Spurgeon CL, Storer B, Schoch HG, Murakami CS, McDonald GB. Cirrhosis of the liver in long-term marrow transplant survivors. Blood 1999;93(10):3259-3266.

[9] Knovich MA, Storey JA, Coffman LG, Torti SV, Torti FM. Ferritin for the clinician. Blood Rev 2009;23(3):95-104.

[10] Sucak GT, Yegin ZA, Ozkurt ZN, Aki SZ, Karakan T, Akyol G. The role of liver biopsy in the workup of liver dysfunction late after SCT: is the role of iron overload underestimated? Bone Marrow Transplant. 2008;42(7):461-467.

[11] Emerit J, Beaumont C, Trivin F. Iron metabolism, free radicals, and oxidative injury. Biomed Pharmacother 2001;55(6):333-339.

[12] Lee DH, Jacobs DR Jr. Serum markers of stored body iron are not appropriate markers of health effects of iron: a focus on serum ferritin. Med Hypotheses 2004;62(3): 442-445.

[13] Gordon LI, Brown SG, Tallman MS, Rademaker AW, Weitzman SA, Lazarus HM, Kelley CH, Mangan C, Rubin H, Fox RM, et al. Sequential changes in serum iron and ferritin in patients undergoing high-dose chemotherapy and radiation with autologous bone marrow transplantation: possible implications for treatment related toxicity. Free Radic Biol Med 1995;18(3):383-389.

[14] Deugnier Y, Brissot P, Loréal O. Iron and the liver: update 2008. J Hepatol 2008;48 Suppl 1:S113-123.

[15] Deeg HJ, Spaulding E, Shulman HM. Iron overload, hematopoietic cell transplantation, and graft-versus-host disease. Leuk Lymphoma 2009;50(10):1566-1572.

[16] Ma AD, Gordeuk VR. Iron metabolism, iron overload and the porphyrias. In: American Society of Hematology Self Assessment Program. Chapter 4; 2010. p93-108.

[17] Malcovati L. Impact of transfusion dependency and secondary iron overload on the survival of patients with myelodysplastic syndromes. Leuk Res 2007;31 Suppl 3:S2-6.

[18] Kamble RT, Selby GB, Mims M, Kharfan-Dabaja MA, Ozer H, George JN. Iron overload manifesting as apparent exacerbation of hepatic graft-versus-host disease after allogeneic hematopoietic stem cell transplantation. Biol Blood Marrow Transplant 2006;12(5):506-510.

[19] Socié G, Salooja N, Cohen A, Rovelli A, Carreras E, Locasciulli A, Korthof E, Weis J, Levy V, Tichelli A; Late Effects Working Party of the European Study Group for Blood and Marrow Transplantation. Nonmalignant late effects after allogeneic stem cell transplantation. Blood 2003;101(9):3373-3385.

[20] Majhail NS, DeFor T, Lazarus HM, Burns LJ. High prevalence of iron overload in adult allogeneic hematopoietic cell transplant survivors. Biol Blood Marrow Transplant 2008;14(7):790-794.

[21] Majhail NS, DeFor TE, Lazarus HM, Burns LJ. Iron-overload after autologous hematopoietic cell transplantation. Leuk Res 2009;33(4):578-579.

[22] Pullarkat V. Iron overload in patients undergoing hematopoietic stem cell transplantation. Adv Hematol 2010;2010. pii: 345756.

[23] Kamble R, Mims M. Iron-overload in long-term survivors of hematopoietic transplantation. Bone Marrow Transplant 2006;37(8):805-806.

[24] De Witte T. The role of iron in patients after bone marrow transplantation. Blood Rev 2008;22 Suppl 2:S22-28.

[25] Koreth J, Antin JH. Iron overload in hematologic malignancies and outcome of allogeneic hematopoietic stem cell transplantation. Haematologica 2010;95(3):364-366.

[26] Strasser SI, Kowdley KV, Sale GE, McDonald GB. Iron overload in bone marrow transplant recipients. Bone Marrow Transplant 1998;22(2):167-173.

[27] Altes A, Remacha AF, Sarda P, Baiget M, Sureda A, Martino R, Briones J, Brunet S, Canals C, Sierra J. Early clinical impact of iron overload in stem cell transplantation. A prospective study. Ann Hematol 2007;86(6):443-447.

[28] Sahlstedt L, Ebeling F, von Bonsdorff L, Parkkinen J, Ruutu T. Non-transferrin-bound iron during allogeneic stem cell transplantation. Br J Haematol 2001;113(3): 836-838.

[29] Yegin ZA, Paşaoğlu H, Aki SZ, Özkurt ZN, Demirtaş C, Yağci M, Acar K, Sucak GT. Pro-oxidative/antioxidative imbalance: a key indicator of adverse outcome in hematopoietic stem cell transplantation. Int J Lab Hematol 2011;33(4):414-423.

[30] Dürken M, Nielsen P, Knobel S, Finckh B, Herrnring C, Dresow B, Kohlschütter B, Stockschläder M, Krüger WH, Kohlschütter A, Zander AR. Nontransferrin-bound iron in serum of patients receiving bone marrow transplants. Free Radic Biol Med 1997;22(7):1159-1163.

[31] Dürken M, Herrnring C, Finckh B, Nagel S, Nielsen P, Fischer R, Berger HM, Moison RM, Pichlmeier U, Kohlschütter B, Zander AR, Kohlschütter A. Impaired plasma antioxidative defense and increased nontransferrin-bound iron during high-dose chemotherapy and radiochemotherapy preceding bone marrow transplantation. Free Radic Biol Med 2000;28(6):887-894.

[32] Altès A, Remacha AF, Sureda A, Martino R, Briones J, Canals C, Brunet S, Sierra J, Gimferrer E. Iron overload might increase transplant-related mortality in haematopoietic stem cell transplantation. Bone Marrow Transplant 2002;29(12):987-989.

[33] Altes A, Remacha AF, Sarda P, Sancho FJ, Sureda A, Martino R, Briones J, Brunet S, Canals C, Sierra J. Frequent severe liver iron overload after stem cell transplantation and its possible association with invasive aspergillosis. Bone Marrow Transplant 2004;34(6):505-509.

[34] Kontoyiannis DP, Chamilos G, Lewis RE, Giralt S, Cortes J, Raad II, Manning JT, Han X. Increased bone marrow iron stores is an independent risk factor for invasive aspergillosis in patients with high-risk hematologic malignancies and recipients of allogeneic hematopoietic stem cell transplantation. Cancer 2007;110(6):1303-1306.

[35] Maertens J, Demuynck H, Verbeken EK, Zachée P, Verhoef GE, Vandenberghe P, Boogaerts MA. Mucormycosis in allogeneic bone marrow transplant recipients: report of five cases and review of the role of iron overload in the pathogenesis. Bone Marrow Transplant 1999;24(3):307-312.

[36] Mahindra A, Bolwell B, Sobecks R, Rybicki L, Pohlman B, Dean R, Andresen S, Sweetenham J, Kalaycio M, Copelan E. Elevated ferritin is associated with relapse af-

ter autologous hematopoietic stem cell transplantation for lymphoma. Biol Blood Marrow Transplant 2008;14(11):1239-1244.

[37] Mahindra A, Sobecks R, Rybicki L, Pohlman B, Dean R, Andresen S, Kalaycio M, Sweetenham J, Bolwell B, Copelan E. Elevated pretransplant serum ferritin is associated with inferior survival following nonmyeloablative allogeneic transplantation. Bone Marrow Transplant 2009;44(11):767-768.

[38] Pullarkat V, Blanchard S, Tegtmeier B, Dagis A, Patane K, Ito J, Forman SJ. Iron overload adversely affects outcome of allogeneic hematopoietic cell transplantation. Bone Marrow Transplant 2008;42(12):799-805.

[39] Maradei SC, Maiolino A, de Azevedo AM, Colares M, Bouzas LF, Nucci M. Serum ferritin as risk factor for sinusoidal obstruction syndrome of the liver in patients undergoing hematopoietic stem cell transplantation. Blood 2009;114(6):1270-1275.

[40] Armand P, Kim HT, Cutler CS, Ho VT, Koreth J, Ritz J, Alyea EP, Antin JH, Soiffer RJ. A prognostic score for patients with acute leukemia or myelodysplastic syndromes undergoing allogeneic stem cell transplantation. Biol Blood marrow Transplant 2008;14(1);28-35.

[41] Tomás JF, Pinilla I, García-Buey ML, García A, Figuera A, Gómez-García de Soria VGG, Moreno R, Fernández-Rañada JM. Long-term liver dysfunction after allogeneic bone marrow transplantation: clinical features and course in 61 patients. Bone Marrow Transplant 2000;26(6):649-655.

[42] Ho GT, Parker A, MacKenzie JF, Morris AJ, Stanley AJ. Abnormal liver function tests following bone marrow transplantation: aetiology and role of liver biopsy. Eur J Gastroenterol Hepatol 2004;16(2):157-162.

[43] Choi SW, Islam S, Greenson JK, Levine J, Hutchinson R, Yanik G, Teitelbaum DH, Ferrara JL, Cooke KR. The use of laparoscopic liver biopsies in pediatric patients with hepatic dysfunction following allogeneic hematopoietic stem cell transplantation. Bone Marrow Transplant 2005;36(10):891-896.

[44] Strasser SI. When should liver biopsy be performed after hematopoietic stem cell transplantation? J Gastroenterol Hepatol 2008;23(2):167-169.

[45] Iqbal M, Creger RJ, Fox RM, Cooper BW, Jacobs G, Stellato TA, Lazarus HM. Laparoscopic liver biopsy to evaluate hepatic dysfunction in patients with hematologic malignancies: a useful tool to effect changes in management. Bone Marrow Transplant 1996;17(4):655-662.

[46] Fujita N, Sugimoto R, Urawa N, Araki J, Mifuji R, Yamamoto M, Horiike S, Tanaka H, Iwasa M, Kobayashi Y, Adachi Y, Kaito M. Hepatic iron accumulation is associated with disease progression and resistance to interferon/ribavirin combination therapy in chronic hepatitis C. J Gastroenterol Hepatol 2007;22(11):1886-1893.

[47] Gaziev J, Sodani P, Polchi P, Andreani M, Lucarelli G. Bone marrow transplantation in adults with thalassemia: Treatment and long-term follow-up. Ann N Y Acad Sci 2005;1054:196-205.

[48] Morado M, Ojeda E, Garcia-Bustos J, Aguado MJ, Arrieta R, Quevedo E, Navas A, Hernandez-Navarro F. Serum Ferritin as Risk Factor for Veno-occlusive Disease of the Liver. Prospective Cohort Study. Hematology 2000;4(6):505-512.

[49] Sucak GT, Yegin ZA, Ozkurt ZN, Aki SZ, Yağci M. Iron overload: predictor of adverse outcome in hematopoietic stem cell transplantation. Transplant Proc 2010;42(5): 1841-1848.

[50] Sucak GT, Aki ZS, Yagcí M, Yegin ZA, Ozkurt ZN, Haznedar R. Treatment of sinusoidal obstruction syndrome with defibrotide: a single-center experience. Transplant Proc 2007;39(5):1558-1563.

[51] Unal S, Kuskonmaz B, Hazirolan T, Eldem G, Aytac S, Cetin M, Uckan D, Gumruk F. Deferasirox use after hematopoietic stem cell transplantation in pediatric patients with beta-thalassemia major: preliminary results. Pediatr Hematol Oncol 2010;27(6): 482-489.

[52] Marx JJ. Iron and infection: competition between host and microbes for a precious element. Best Pract Res Clin Haematol 2002;15(2):411-426.

[53] Ali S, Pimentel JD, Munoz J, Shah V, McKinnon R, Divine G, Janakiraman N. Iron overload in allogeneic hematopoietic stem cell transplant recipients. Arch Pathol Lab Med 2012;136(5):532-538.

[54] Miceli MH, Dong L, Grazziutti ML, Fassas A, Thertulien R, Van Rhee F, Barlogie B, Anaissie EJ. Iron overload is a major risk factor for severe infection after autologous stem cell transplantation: a study of 367 myeloma patients. Bone Marrow Transplant 2006;37(9):857-864.

[55] Tunçcan OG, Yegin ZA, Ozkurt ZN, Erbaş G, Aki SZ, Senol E, Yağci M, Sucak G. High ferritin levels are associated with hepatosplenic candidiasis in hematopoietic stem cell transplant candidates. Int J Infect Dis 2010;14 Suppl 3:e104-107.

[56] Özyilmaz E, Aydogdu M, Sucak G, Aki SZ, Ozkurt ZN, Yegin ZA, Kokturk N. Risk factors for fungal pulmonary infections in hematopoietic stem cell transplantation recipients: the role of iron overload. Bone Marrow Transplant 2010;45(10):1528-1533.

[57] Platzbecker U, Bornhäuser M, Germing U, Stumpf J, Scott BL, Kröger N, Schwerdtfeger R, Böhm A, Kobbe G, Theuser C, Rabitsch W, Valent P, Sorror ML, Ehninger G, Deeg HJ. Red blood cell transfusion dependence and outcome after allogeneic peripheral blood stem cell transplantation in patients with de novo myelodysplastic syndrome (MDS). Biol Blood Marrow Transplant 2008;14(11):1217-1225.

[58] Mahindra A, Bolwell B, Sobecks R, Rybicki L, Pohlman B, Dean R, Andresen S, Sweetenham J, Kalaycio M, Copelan E. Elevated pretransplant ferritin is associated

with a lower incidence of chronic graft-versus-host disease and inferior survival after myeloablative allogeneic haematopoietic stem cell transplantation. Br J Haematol 2009;146(3):310-316.

[59] Kataoka K, Nannya Y, Hangaishi A, Imai Y, Chiba S, Takahashi T, Kurokawa M. Influence of pretransplantation serum ferritin on nonrelapse mortality after myeloablative and nonmyeloablative allogeneic hematopoietic stem cell transplantation. Biol Blood Marrow Transplant 2009;15(2):195-204.

[60] Alessandrino EP, Della Porta MG, Bacigalupo A, Malcovati L, Angelucci E, Van Lint MT, Falda M, Onida F, Bernardi M, Guidi S, Lucarelli B, Rambaldi A, Cerretti R, Marenco P, Pioltelli P, Pascutto C, Oneto R, Pirolini L, Fanin R, Bosi A. Prognostic impact of pre-transplantation transfusion history and secondary iron overload in patients with myelodysplastic syndrome undergoing allogeneic stem cell transplantation: a GITMO study. Haematologica 2010;95(3):476-484.

[61] Andrews NC. Forging a field: the golden age of iron biology. Blood 2008;112:219-230.

[62] Tanaka M, Tachibana T, Numata A, Takasaki H, Matsumoto K, Maruta A, Ishigatsubo Y, Kanamori H. A prognostic score with pretransplant serum ferritin and disease status predicts outcome following reduced-intensity SCT. Bone Marrow Transplant 2012;47(4):596-597.

[63] Tachibana T, Tanaka M, Takasaki H, Numata A, Maruta A, Ishigatsubo Y, Kanamori H. Pre-SCT serum ferritin is a prognostic factor in adult AML, but not ALL. Bone Marrow Transplant 2011;46(9):1268-1269.

[64] Lim ZY, Fiaccadori V, Gandhi S, Hayden J, Kenyon M, Ireland R, Marsh J, Ho AY, Mufti GJ, Pagliuca A. Impact of pre-transplant serum ferritin on outcomes of patients with myelodysplastic syndromes or secondary acute myeloid leukaemia receiving reduced intensity conditioning allogeneic haematopoietic stem cell transplantation. Leuk Res 2010;34(6):723-727.

[65] Armand P, Kim HT, Cutler CS, Ho VT, Koreth J, Ritz J, Alyea EP, Antin JH, Soiffer RJ. A prognostic score for patients with acute leukemia or myelodysplastic syndromes undergoing allogeneic stem cell transplantation. Biol Blood Marrow Transplant 2008;14(1):28-35.

[66] Armand P, Kim HT, Cutler CS, Ho VT, Koreth J, Alyea EP, Soiffer RJ, Antin JH. Prognostic impact of elevated pretransplantation serum ferritin in patients undergoing myeloablative stem cell transplantation. Blood 2007;109(10):4586-4588.

[67] Storey JA, Connor RF, Lewis ZT, Hurd D, Pomper G, Keung YK, Grover M, Lovato J, Torti SV, Torti FM, Molnár I. The transplant iron score as a predictor of stem cell transplant survival. J Hematol Oncol 2009; 2:44.

[68] Armand P, Sainvil MM, Kim HT, Rhodes J, Cutler C, Ho VT, Koreth J, Alyea EP, Neufeld EJ, Kwong RY, Soiffer RJ, Antin JH. Does iron overload really matter in stem cell transplantation? Am J Hematol 2012;87(6):569-572.

[69] Carreras E. Risk assessment in haematopoietic stem cell transplantation: the liver as a risk factor. Best Pract Res Clin Haematol 2007;20(2):231-246.

[70] Liatsos C, Mehta AB, Potter M, Burroughs AK. The hepatologist in the haematologists' camp. Br J Haematol 2001;113(3):567-578.

[71] Wood JC. Impact of iron assessment by MRI. Hematology Am Soc Hematol Educ Program 2011;2011:443-450.

[72] Gandon Y, Olivié D, Guyader D, Aubé C, Oberti F, Sebille V, Deugnier Y. Non-invasive assessment of hepatic iron stores by MRI. Lancet 2004;363(9406):357-362.

[73] Voskaridou E, Douskou M, Terpos E, Papassotiriou I, Stamoulakatou A, Ourailidis A, Loutradi A, Loukopoulos D. Magnetic resonance imaging in the evaluation of iron overload in patients with beta thalassaemia and sickle cell disease. Br J Haematol 2004;126(5):736-742.

[74] St Pierre TG, Clark PR, Chua-anusorn W, Fleming AJ, Jeffrey GP, Olynyk JK, Pootrakul P, Robins E, Lindeman R. Noninvasive measurement and imaging of liver iron concentrations using proton magnetic resonance. Blood 2005;105(2):855-861.

[75] Nielsen P, Günther U, Dürken M, Fischer R, Düllmann J. Serum ferritin iron in iron overload and liver damage: correlation to body iron stores and diagnostic relevance. J Lab Clin Med 2000;135(5):413-418.

[76] Butt NM, Clark RE. Autografting as a risk factor for persisting iron overload in long-term survivors of acute myeloid leukaemia. Bone Marrow Transplant 2003;32(9): 909-913.

[77] Harrison P, Neilson JR, Marwah SS, Madden L, Bareford D, Milligan DW. Role of non-transferrin bound iron in iron overload and liver dysfunction in long term survivors of acute leukaemia and bone marrow transplantation. J Clin Pathol 1996;49(10): 853-856.

[78] Muretto P, Angelucci E, Lucarelli G. Reversibility of cirrhosis in patients cured of thalassemia by bone marrow transplantation. Annals of Internal Medicine 2002;136(9):667-672.

[79] McDonald GB. Review article: management of hepatic disease following haematopoietic cell transplant. Aliment Pharmacol Ther 2006;24(3):441-452.

[80] Yeşilipek MA, Karasu G, Kazik M, Uygun V, Ozturk Z. Posttransplant oral iron-chelating therapy in patients with beta-thalassemia major. Pediatr Hematol Oncol 2010;27(5):374-379.

[81] Kwiatkowski JL. Real-world use of iron chelators. Hematology Am Soc Hematol Educ Program 2011;2011:451-458.

Chapter 15

Sickle Cell Disease (SCD) and Stem Cell Therapy (SCT): Implications for Psychotherapy and Genetic Counselling in Africa

Oluwatoyin Olatundun Ilesanmi

Additional information is available at the end of the chapter

http://dx.doi.org/10.5772/53082

1. Introduction

Sickle Cell Disorder (SCD) is an inherited disease of red blood cells which has no widely available cure (Bernaudin, Socie, Kuentz, et al., 2007). While current medical therapies can make a significant difference in short-term effects (i.e. to relieve pain symptoms, prevent infections and manage complications such as eye damage, and strokes; and control complications), the progressive deterioration in organ function results in increased mortality and decreased quality of life among affect persons in Nigeria. Presently, blood and bone marrow stem cell transplant appear to be the only viable option for its eliminating. This option is hugely expensive and unaffordable for the vast majority of the affected Nigerian families since most of them could barely provide for the general routine medication therapies of the patient. Little attention is being given to the management of this disorder in Nigeria as compared to diseases such as malaria and polio myelitis. Institutional research attention and international funding support towards the search for ways to predict the severity of and for curative therapies of this disorder are also limited in Africa.

Globally, sickle cell disorders (SCD) affect millions of people of all races throughout the world. About 80% of affected children are born in developing countries and about 50 – 80% of children with SCD die each year in low – middle income countries. Nonetheless, its magnitude in Nigeria and Africa on the whole is alarming. Nigeria has the largest burden of SCD in Africa *(see table 1 for a presentation of the progress report)*. At least 40 million Nigerians are carriers (AS) versus 2 million Americans. Over 150,000 Nigerians are born each year with sickle cell anaemia (SS) versus 2,000 in America (Akinyanju, 2009). Numerous families in Nigeria have lost loved ones to this red blood cell disorder. About 80,000 people are living with SCD in USA versus estimated ±1,000,000 in Nigeria (Akinyanju, 2009).

1916 - 1945	Virtually nil > 4y
1946 – 1965	Virtually no adults. Known as paediatric disease
1966 – 1985	Many adolescents and young adults
1986 – 2010	Many adults as parents and in workforce

Table 1. Progress of Nigerians with SS

Molineaux et al (1979) noted that there is no other known inherited disorder present at such high frequency in a large population and of comparable severity as sickle cell anaemia in Africa. With rising standards of living and control of malaria, sickle cell anaemia may become an immense medical, social and economic problem all over Africa (see table 2 below).

Indigenous	**Imported**
• Sub Saharan countries north of Zambesi River	• Mediterranean Basin e.g. Greece,Sardinia
• Eastern Saudi Arabia	• USA & Canada
• Some States in India	• Brazil, Belize Columbia etc
	• Cuba, Jamaica, Haiti, Barbados, Trinidad etc
	• UK, France, Holland,
Distribution and Names of Indigenous Sickle Gene Haplotypes	
Region	Haplotype
• Western West Africa	• Senegal – 3
• *Cameroon (Ekona)	• Cameroon – 17
• Central West Africa	• Benin – 19
• Central & East Africa	• Bantu – 20
• East Saudi Arabia & India	• Arab/India – 31
Note: * Most Cameroonians have the Benin haplotype. The Ekona haplotype is a recent discovery among the small population of the Ekona ethnic group	

Table 2. Where SCD is Found

The symptoms of SCD are seen predominantly in one-third of all aboriginal inhabitants (or their descendants) of parts of tropical and sub-tropical regions where malaria is or was common and in people from parts of the Middle East, Central India, Spanish-speaking regions (South America, Cuba, Central America); Saudi Arabia; and countries bordering the Mediterranean Sea, especially Turkey, Greece, and Italy (Akinyanju, 2009). This is because in

http://dx.doi.org/10.5772/53082

areas where malaria is common, there is a survival value in carrying only a single sickle-cell gene (sickle cell trait) (Akinyanju, 2009).

In the US, SCD affects around 72,000 people, most of whose ancestors come from Africa (Benjamin & Payne, 2007). It occurs in about 1:500 African-American births and 1:1000-1400 Hispanic-American births. About 2 million Americans, or 1:12 African Americans carry the sickle cell trait. Its occurrence among the Hispanic-Americans is about 5%, their median survival based on 1991 national data was 42 years for males, 48 years for females (California Institute For Regenerative Medicine, 2009). By twenty years of age, about 15% of children with SCD suffer major strokes and by 40 years of age, almost half of the patients have had central nervous system damage leading to significant cognitive dysfunction. These patients suffer significant damage to lungs and kidneys as well as severe chronic pain that impacts on quality of life. In Brazil, SCD is considered the most common monogenic disease seen predominantly in the black population as well as among individuals from parts of the Middle East, Central India and countries bordering the Mediterranean Sea, especially Italy and Greece.

2. What is Sickle Cell Disease (SCD)?

Sickle cell disorders (SCD) is a group of inherited autosomal recessive disorder characterized by production of abnormal of haemoglobin (Hb), resulting in anemia, susceptibility to pneumococcal and other infections, pain, stroke, and multiple organ dysfunctions. Normal red blood cells are soft, smooth, round and flexible and last about 120 days. It flows easily through blood vessels, but the abnormal hemoglobin which causes the red blood cells to be hard and sticky looks like a C-shaped farm tool called a sickle under the microscope (see figure 1).

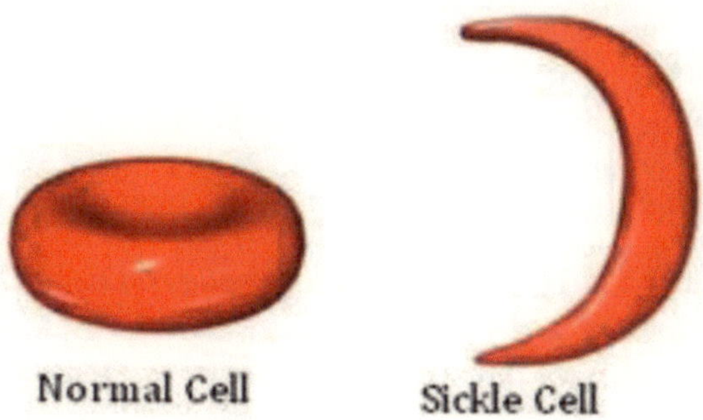

Figure 1. Shapes of Red Blood Cells

Sickle cell disorders, there are two main types of heamoglobin S- fetal Hb (F) in the unborn child (fetus) and adult Hb (A) after birth. It encompasses > 960 variants of sickling syndromes caused by abnormal sickle hemoglobin. Some are harmless; few are incompatible with life and some like Hb S can make life more challenging. The most common and most severe variant of SCD is hemoglobin SS (homozygous) disease. Other forms include Sickle cell/C disorder (Hb SC), Sickle cell/β thalassaemia (Hb Sβ^{+}thal or Hb Sβ^{0}thal), SD- Punjab, SO Arab, S Lepore and SE disease (NIH, 2010a, b).

Sickle cell trait (also known as being a carrier) occurs when a person has one gene for sickle hemoglobin and one gene for normal hemoglobin. Approximately one in ten African-Americans carries sickle cell trait. People who are carriers generally do not have any medical problems and lead normal lives. If you are a carrier you cannot develop sickle cell disease.

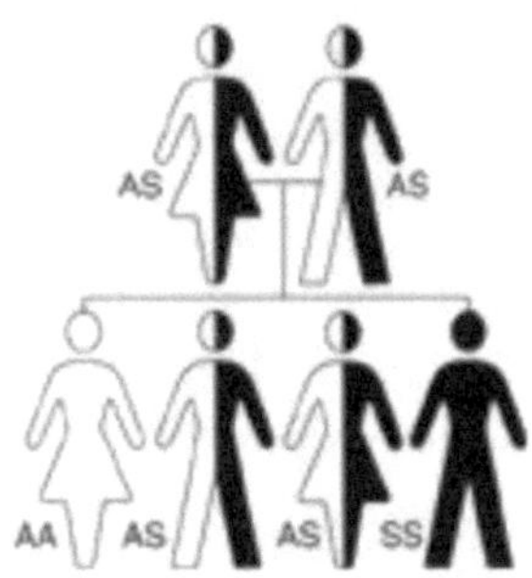

Figure 2. Types of Genotype Status

3. Symptom of Sickle Cell Disease: Related to painful and painless complication

Symptoms of sickle cell disease vary, ranging from mild to severe or life-threatening crises. It may occur without warning, and may go away and then come back many times. The signs and symptoms are linked to anaemia, pain and disease's complications. SCD anaemia related symptoms (lack of RBC) range from mild to very severe symptoms. The symptoms of anemia are fatigue (feeling tired or weak), shortness of breath, dizziness, headaches, coldness in the hands and feet, and paler than normal skin or mucous membranes (the tissue that lines your nose, mouth, and other organs and body cavities).

The pain related symptoms are debilitating pain episode or crisis which can affect bones, lungs, abdomen, and joints; as well as damage organs and increase the risk of stroke. The pain can be acute or chronic, but acute pain is more common. Acute pain is sudden and can range from mild to very severe. The pain usually lasts from hours to as long as a week or more. Chronic pain often lasts for weeks or months. Such pains can be limiting, unbearable and mentally draining. The painful crises are the leading cause of emergency room visits and hospital stays for people who have sickle cell anaemia.

Disease complication related of SCD crises are painful episodes (crises), acute chest syndrome, anemia (low hemoglobin), organ damage due to iron overload, infections, lung

http://dx.doi.org/10.5772/53082

problems, leg ulcers, bone damage, strokes, and premature death. It can cause hand-foot syndrome which is a blockage of the small blood vessels in the hands and feet in children (usually those younger than 4 years of age) leading to pain, swelling, and fever. SCD can also initiate splenic crisis in the abdomen, and pulmonary hypertension.

4. Etiology of SCD crises

The exact cause of episodic painful crisis is unknown. However, more than one factor is involved. First, the crises can occur whenever sickled red blood cells form clumps or abnormal curved shapes called sickles in the bloodstream. These clumps of cells stick to small blood vessels and block blood flow and oxygen to the limbs and organs. This can result in pain and damage to body organs, such as kidneys. It can trigger a stroke and other medical problems. For instance, the Hand-Foot Syndrome usually occurs whenever sickled RBCs block the small blood vessels in the hands and feet in children (usually those younger than 4 years of age). It can lead to pain, swelling, and fever. Swelling often occurs on the back of the hands and feet and moves into the fingers and toes. One or both hands and/or feet may be affected at the same time.

Gallstones usually develop in the gallbladder whenever there is too much bilirubin in the body. Gallstones may cause steady pain that lasts for 30 minutes or more in the upper right side of the belly, under the right shoulder, or between the shoulder blades. The pain may happen after eating fatty meals. People who have gallstones may have nausea (feeling sick to the stomach), vomiting, fever, sweating, chills, clay-coloured stools, or jaundice (a yellowish colour of the skin or whiteness of the eyes).

Ulcers on the Legs (sores) usually begin as small, raised, crusted sores on the lower third of the leg. Leg sores may occur more often in males than in females. These sores usually develop in people who are aged 10 years or older. The cause of sickle cell ulcers isn't clear. The number of ulcers can vary from one to many. Some heal quickly, but others persist for years or come back after healing.

Splenic Crisis can also occur whenever the spleen traps red blood cells that should be in the bloodstream. This causes the spleen to grow large and leads to anaemia. Acute Chest Syndrome may be caused by infection or sickle cells trapped in the lungs. People who have this condition often have chest pain, shortness of breath, and fever. They also often have low oxygen levels and abnormal chest x- ray results. Pulmonary Hypertension occurs as a result of damage to the small blood vessels in the lungs which make it hard for the heart to pump blood through the lungs. This causes blood pressure in the lungs to rise. Increased blood pressure in the lungs is called pulmonary hypertension (PH). Shortness of breath and fatigue are the main symptoms of PH. Priapism which is painful, unwanted erections may occur whenever sickle cells block blood flow out of an erect penis. Over time, priapism can damage the penis and lead to impotence.

SCD crises may also be caused by factors such as dehydration, infections, hypoxia, cold temperature, surgery and emotional stress. Dehydration often increases the risk of a sickle cell crisis. Drinking plenty of fluids can lower the risk of a painful crisis. Other factors include bacterial infections. Infants and young children with sickle cell disease are especially vulnerable to serious infections, such as those that cause meningitis (infection of the lining of the brain) and blood infection.

5. Treatment goals

The goals of treatment options in SCD are symptom control; prevention of infections and stroke; detection and management of disease complications such as vaso-occlusive crisis, chronic pain syndromes, chronic hemolytic anemia, pulmonary hypertension, and the various organ damage syndromes. Accepted treatment options include narcotic pain killers, drugs, chronic blood transfusions, hydroxyurea, and stem cell transplantation (SCT) for selected children and young adults. Narcotic pain killers are used to treat the severe pain. Drugs are used to stimulate production of additional blood cells. Transfusions are used to treat the anemia and to dilute the sickle cells with normal red blood cells.

6. Stem cell transplantation treatment for SCD

Stem cells are parent cells found in all tissues and organs of the body, such as the bone marrow, skin, muscles, brain, peripheral blood, umbilical cord blood, and, rarely, fetal liver. The early cells nurtured in the bone marrow or less frequently from umbilical cord blood that mature into red and white blood cells and platelets are called multi-potent stem cells or immature cells. Stem cells produce erythroid cells, granulocytes, lymphoid cells, megacaryocytes and monocytes by a number of differentiation steps. Stem cells maintain normal cell populations in a healthy bone marrow controlled by haemopoietic growth factors, and stem cells have the capacity for self-renewal. Haemopoietic growth factors include erythropoietin, interleukins, glucocorticoids, sex hormones and thyroid hormones.

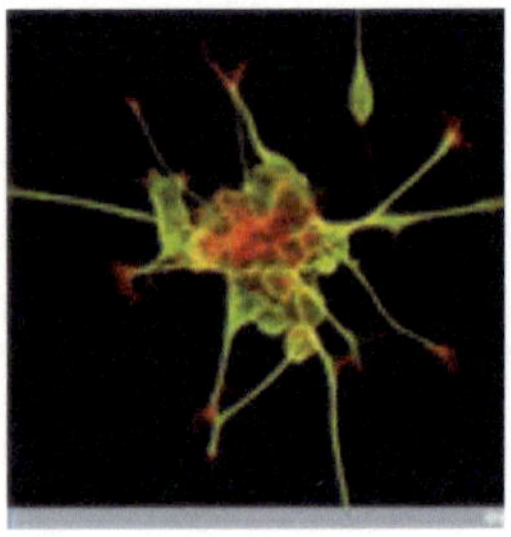

Figure 3. Stem Cell

Stem cell transplantation (SCT) refers to transplantation of the hematopoietic stem cells (HSCs) from a donor into an individual. Transplanted human bone marrow or stem cells are dynamic biological entities that interact intimately with—and are influenced by—the physiology of the recipient. It is a very risky procedure. Before they are transplanted, cultured human stem cells are maintained under conditions that promote either the self-renewing expansion of undifferentiated progenitors or the acquisition of differentiated properties indicative of the phenotype the cells will assume (The National Institutes of Health resource for stem cell research, 2010). After incompletely differentiated human stem cells are transplanted, additional fine-tuning occurs as a consequence of instructions received from the cells' physiologic microenvironments within the recipient.

The goal of SCT is elimination of the sickle erythrocyte and its cellular progenitors and replacement with donor hematopoietic pluripotent stem cells which give rise to erythrocytes that express no sickle hemoglobin (HbS). This will eventually reduce Hb S levels to those associated with the trait condition. It has the possibility of preventing serious complications from SCD which can cause extensive morbidity and early death.

The donor sources of hematopoietic stem cells transplantation (HSCT) include cells obtained from another person (sibling or unrelated donor), termed allogeneic transplant; an identical twin, termed syngeneic transplant; or the patient, termed autologous transplant (Samavedi, 2011). The autologous HSCT (using the individual's own stem cells) involves using peripheral blood stem cell transplantation (PBSCT) to treat disorders such as multiple myeloma, non-hodgkin lymphoma, hodgkin disease, acute myeloid leukemia, neuroblastoma, germ cell tumors, autoimmune disorders (systemic lupus erythematosus [SLE], systemic sclerosis) and amyloidosis (Samavedi, 2011). It has not been used in the treatment of SCD.

Myeloablative allogeneic hematopoietic stem-cell transplantation (HSCT) is the only potentially curative treatment option for selected individuals with sickle cell anemia or thalassemia major (Samavedi, 2011; Krishnamurti, 2008; Walters, 2004). According to Samavedi (2011) and Doubek, Folber, Koristek, et al. (2009), it can also be used in the treatment of conditions such as, leukemia, myeloproliferative disorders and myelodysplastic syndromes. Successful allogeneic SCT not only eliminates the sickle-cell-induced vaso-occlusive symptomatology, but also leads to reversal of some of the end organ damage that occurred prior to the procedure.

In allogeneic HSCT, it is preferable for donors to have a human leukocyte antigen (HLA) type that is identical to the recipient. Matching is performed on the basis of variability at three of more loci of the HLA gene (e.g., HLA-A, HLA-B, HLA-DRB1). Usage of a non-myeloablative conditioning regimen prior to allogeneic SCT for transplantation of pediatric patients with SCD have been largely unsuccessful due to high rates of graft rejection (Bernaudin, Vannier, et al., 1997). Thus, current opinion is that children with high-risk SCD and a suitably matched donor should be offered allogeneic SCT using a conventional myeloablative conditioning regimen. To date, nearly all transplants have utilized HLA-identical sibling donors, which have limited the number of eligible sickle cell patients.

With HLA variability and lack of appropriate donors, there are increases in transplant-related morbidity and mortality, including graft rejection and graft-versus-host disease (GVHD). All donor-to-patient stem cell transplants use material which contains donor T-cells. These donor T-cells react to the patient's body as foreign and causes GVHD which is a significant cause of illness and even death in stem cell transplants. It is generally avoided by using a donor as closely matched to the patient as possible. Usage of closely matched donors reduce will the risk of GVHD. In the normal population, a patient has about a 30% chance of having a matched sibling donor. However, SCD is a genetic disease, passed on from parents to children. A brother or sister who is a close match to the patient is very likely to also have SCD, making them inappropriate as a donor. The chance that an SCD patient has a matched sibling donor is less than 15%.

The optimal timing for marrow transplantation in the course of SCD remains uncertain because of the unpredictable nature and clinical heterogeneity of the disease (Walters, 2005). Selection criteria for optimal candidates continue to evolve; however, children and young adults, generally before the age of 21 years are considered the most appropriate candidates. Indications for HSCT have been empirically determined from prognostic factors derived from studies of the natural history of SCD. The most common indications for which patients with SCD have undergone HSCT are a history of stroke, recurrent acute chest syndrome, or frequent vaso-occlusive episodes (Novelli, Kato, Ragni, Zhang, Hildesheim, Nouraie, Barge, Meyer, Hassett, Gordeuk, Gladwin & Isenberg, 2012).Children and young adults who have severe complications (e.g. stroke, recurrent acute coronary syndrome [ACS], refractory pain) and have a human-leukocyte antigen (HLA)-matched donor are the best candidates for transplantation (Panepinto, Walters, Carreras, Marsh, Bredeson, Gayle, et al 2007). Very few adults are considered for transplantation due to existing comorbidities and toxicity of treatment (Walters, 2005).

7. Indications for stem cell transplantation

Sickle cell disorder (SCD) has no widely available cure. Its current medical therapies have only being relieving pain symptoms, preventing infections and managing complications such as eye damage, and strokes; and control complications. The progressive deterioration in organ function has being resulting in increased mortality and decreased quality of life. Some severe cases are resistant to existent therapies and can cut life even shorter.

Presently, Blood and Marrow Stem Cell Transplant appear to be the only viable option for eliminating SCD, especially in high risk patients. Patients with SCD are characterized as high-risk if they have central nervous system pathology (clinical or subclinical stroke, seizures), recurrent severe acute chest syndrome, chronic unremitting pain, or early evidence of end organ damage such as pulmonary hypertension. The appropriateness of SCT can be more firmly established in the presence of these high-risk features.

The bone marrow nurtures stem cells, which are early cells that mature into red and white blood cells and platelets. By destroying the sickle cell patient's diseased bone marrow and stem cells and transplanting healthy bone marrow from a genetically-matched donor, normal hemoglobin may be produced. Clinical studies using a few carefully selected patients have reported very successful results (Harvey Simon, 2009). Unfortunately, only about 7% of patients with sickle cell meet the criteria for transplantation, including those who:

- Are age 16 or younger (generally considered the better candidates, but patients in their 20s have had successful transplants)
- Have severe symptoms but no long-term organ or neurologic damage
- Have a genetically matched brother or sister who will donate their marrow

The clinical indicator for stem cell application for SCD is based on stem cells' biological properties of self-renewal and their capability to give rise to differentiated cell progenies that maintain tissue homeostasis in physiological and pathological conditions (Lindvalland and Kokaia, 2010; Orlacchio et al., 2010; Sendtner, 2009; Yu and Silva, 2008). Thus, neural stem cells in the subventricular zone of the lateral ventricles and the subgranular zone of the dentate gyrus in the hippocampus of the mammalian brain maintain the capability to generate new neural cells throughout the lifetime (Conti and Cattaneo, 2010; Ma et al., 2009; Galli et al., 2008).

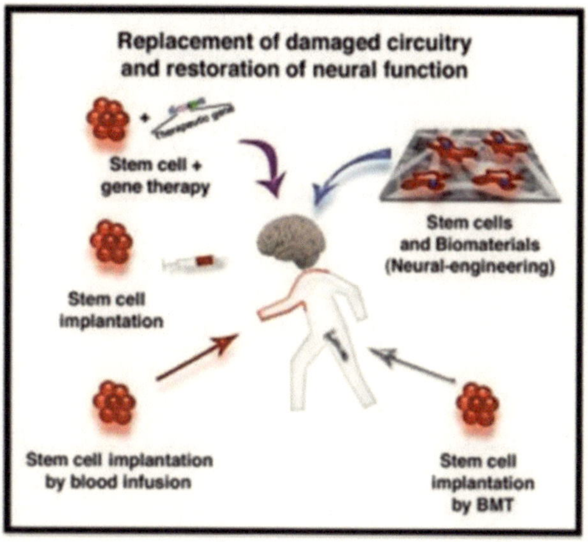

Figure 4. Stem cells replacement therapy for neurological diseases. Cartoon schematizes the different strategies for stem cell delivery in order to repair the degenerated tissue.

8. Risks and benefits of stem cell transplantation for sickle cell disease management in Nigeria

There are two major barriers to stem cell transplants to treat SCD. First is the risk of serious illness associated with donor-to-patient stem cell transplant; and 2) the lack of appropriate donors. For these reasons, only about 300 of these transplants have been performed to date. All donor-to-patient stem cell transplants use material which contains donor T-cells. These donor T-cells react to the patient's body as foreign, causing graft-versus-host disease (GVHD). GVHD is a significant cause of illness and even death in stem cell transplants, and is generally avoided by using a donor as closely matched to the patient as possible, and by appropriate care after the transplant to quickly address symptoms of GVHD when they arise.

Another problem with stem cell transplant for SCD is a lack of donors. Usually, a patient's brother or sister who is a close genetic match to the patient is the preferred donor. Using closely matched donors reduces the risk of GVHD: the closer the donor cells are to the patient's cells, the less severe the immune reaction is likely to be.

9. Emerging therapies and promising research in Nigeria

In Sub-Saharan Africa, human stem cells offer new opportunities and promise of a remarkable array of novel therapeutics for the management of Sickle Cell Disorders (SCD) (The National Institutes of Health, 2009). In Nigeria, the conventional pharmacotherapeutic treatment options for SCD have been extremely limited in part to the management of its painless and painful crises. That is the current medical therapies for SCD only have significant short-term effects on affected persons in Nigeria. Nonetheless, recently a couple of medical Scientists at the University of Benin Teaching Hospital (UBTH), Benin-City, led by Nosakhare Bazuaye recorded a major scientific breakthrough in having a successful stem cell transplant in October 2011 on a 7 year old patient with sickle cell anemia (who had suffered stroke) after an appropriate allogeneic 14-year sibling donor was identified. This feat was the first of its kind in Nigeria and third in Africa. It came on the heels of earlier ones carried out in Egypt and South Africa (Sun Editorial, 2011).

10. The silence of literature on stem cell therapy for SCD in Africa

Numerous reasons could be advance for the silence of literature in Africa on stem cell transplantation. Firstly, Africa is a continent consisting of many developing nations. In most of these nations, particularly the tropic region, falciparum mosquito is highly endemic. The focus of such nations like Nigeria and Kenya in research and governmental policies has been on malaria. Although SCDs are highly prevalent in these nations, the attention of the government and policy makers have not been fully gained for the management of SCDs. Even among the medical practitioners, sheer magnitude of SCDs induces apathy and or feeling of helplessness. The availability of limited resources for the management of sickle cell crises and complications in these nations could also be attributable to the silence of literature in Africa on stem cell therapy for the management of SCDs. Currently, in Nigeria, attention is being drawn towards genetic counselling for affected individuals and their relations as well as to those intending to marry each other.

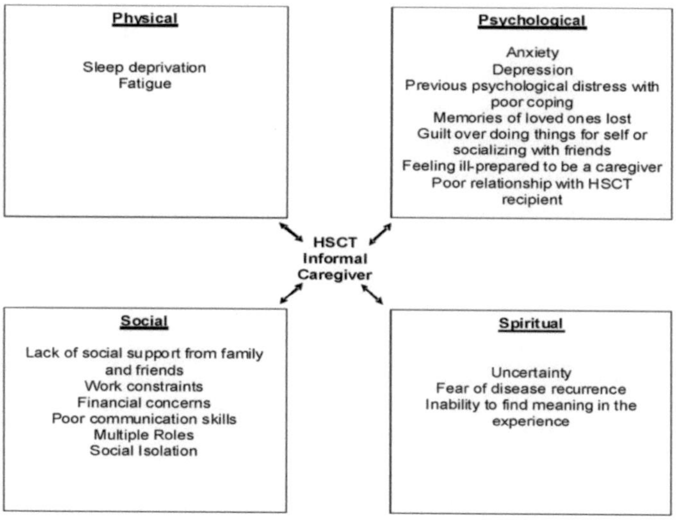

Scheme 1.

The political instability in most African nations has made it difficult for policy makers to present bills on the effective management of sickle cell disorders in Africa. The economic and emotional burden of this disorder is huge on the affected individuals and families in Africa. Many of the affected families are poor. Poverty is another reason for the relative silence of stem cell transplantation in Africa. This is coupled with the inequitable distribution of resources such as money, education, information and health care services in African nations. There is also the issue of lack of Respect and Support for research - molecular, clinical and operational- in Africa. The governments and corporate bodies in Africa do not fund research. There is usually low political will (conflict too) and no funding of research related to SCD management in Africa.

11. Implications for psychotherapy and genetic counselling

Although stem cell transplant has curative potentials for sickle cell anaemia or thalassemia, the physical side effects and psychological distress related to this treatment could be severe and even life threatening for the patients, the donor, and family members. First, for the patient, the transplant procedure is very risky and may be psychologically devastating and traumatizing.It can lead to serious physical and psychological side effects or even death. Approximately 5 percent of patients do not survive and it is used only in very severely affected children and young adults for whom there is a donor who is an appropriate genetic match. For instance, Greenfield (2007) in his eloquent essay and personal reflection as both a psychologist and transplant patient described the reality that the "powerful experience" of transplant caused him to re-experience psychological vulnerabilities despite years of psychoanalysis and therapy to address his past issues.

Secondly, bone marrow transplant carries its own dangers and limitations, especially for patients who do not receive a bone marrow transplant from a well-matched brother or sister donor. About 10% of those who have bone marrow transplants die from the treatment. In patients who do not receive a bone marrow donation from a matched sibling, the transplanted cells from a donor (called allogeneic grafts) may attack the patient's own tissues, a potentially fatal condition called graft-versus-host disease (GVHD). Drugs that destroy bone marrow and suppress immunity must be administered before the procedure so that the body's immune system does not attack the transplanted tissue. Still, this does not always prevent the problem.

Other very serious complications include bleeding, pneumonia, and severe infection. Those who live but are not cured face long-term problems caused by the drugs used in transplantation and by the disease itself. Even in those who are cured, long-term consequences may include a higher risk for cancer and infertility.

Psychologically, all the physical complications that patients face after transplantation may have significant impact on their daily and cognitive functioning. The patients may experience significant global psychological distress encompassing areas of existential concerns, obsessive-compulsiveness, loneliness, and ongoing health concerns such as memory loss (Rusiewitcz, et al, 2008). They may also experience post-traumatic depression as a result of chronic graft versus host disease (cGVHD), long-term issues of ongoing medical appointments, and side-effects of medications (Sherman, Cooke & Grant, 2005; Syrjala, Langer, Abrams, et al., 2004). They may also experience challenging cognitive changes and post-transplant sexual difficulties such as vaginal dryness for women and erectile dysfunction in men (Sherman, Cooke & Grant, 2005). The patient may be unprepared for post-transplant life. In line with the time trajectory of HSCT as indicated in the table below, the physical and distressing psychological effects of stem cell transplantations have serious implications for genetic counseling and psychotherapy.

Diagnosis	HSCT	Short-term follow-up	Long-term follow-up
Decision to transplant	In-hospital treatment	Frequent controls	Quality of life
Donor search	Side effects, Toxicity	Tx-related mortality	Return to "normal life"
	Engraftment	GvH disease	Relapse

Table 3. Time Trajectory of HSCT

For the psychotherapists, genetic counsellors, psychologist, (particularly clinical psychologists) and informal help-givers including the spiritual help-givers., the pathological basis of SCD and stem cell transplantation have generated such questions as: Are the psychotherapists and genetic counselors in Nigeria sufficiently equipped to adequately meet the psychological needs of the individuals living with SCD in Nigeria after stem cell transplantation? Are the professionals (such as psychologists and counsellors) seeing the need to explore the possibility of blending pharmacotherapy with culturally accepted psychotherapeutic interventions for pain coping and increase of steady state among individuals living with SCD in this part of the world? As a result, professionals should focus on the development of cross-culturally relevant psychotherapeutic measures that will address specific needs of patients, donors and their families prior to and after stem cell transplantation. Such cross-culturally sensitive psychotherapeutic programmes will take into consideration the psychological, cultural and spiritual aspects of individuals living with SCD in order to provide them with holistic care. In other words an eclectic but harmonious combination of behavioural techniques (therapeutic interventions) and cross-cultural therapeutic techniques could be more potent in achieving desirable therapy- outcomes.

There is also the urgent need for research that would assess the level of genetic counselling and psychotherapy being offered individuals living with SCD during their crisis state and steady state whether they are appropriate and whether they are being implemented properly. Appropriate psychological interventions can profoundly alter sets of beliefs, ways of thinking, affective states and patterns behaviour.

Added to this is the urgent need to train enough genetic counselors and psychotherapists with special focus on SCD and stem cell transplantation across sub-Saharan Africa where the disorder is prevalent, and most especially Nigeria which has the largest burden of SCD globally.

The hall mark of psychotherapy and genetic couselling in Nigeria for SCD and SCT shall be meeting the psychosocial health needs of the patients, the donor and family members. Because of the unique nature of the transplant experience, psychosocial assessment and interventions should be a high priority. The transplant procedure itself is complex and although the mortality has improved over the years since transplants began in the 1970's it continues to be a significant stressor. The recovery after transplant can come with prolonged physical and psychological set-backs, and extreme social strain on the patient's caregiver, friends and family members. In addition, the transplant experience can include multiple hospital readmissions for acute complications, slow recovery and long-term issues (Eldredge, Nail, Maziarz, Hansen, Ewing & Archbold, 2006).

12. Conclusion

Sickle Cell Disorder (SCD) is an inherited disease of red blood cells characterized by pain episodes, anemia (shortage of red blood cells), serious infections and damage to vital organs which vary greatly from one person to the next. This disorder has no widely available cure in Nigeria. Allogenic stem cell transplantation is the only treatment option with curative potentials, but is not readily affordable to most SCD affected families in sub-Saharan Africa.

Generally, hematopoietic cell transplantation (HCT) for sickle cell disease (SCD) has a strong track record of efficacy and there is growing appreciation that its benefits exceed its risks in selected individuals. The results of transplantation are best when performed in children with a sibling donor who is HLA-identical. Globally, Nigeria has the largest burden of this disorder. Government's commitment and strong political will are needed to support and fund all activities and research geared towards effective management of SCDs in Nigeria so that stem cell transplantation can become a clinical curative and affordable reality for patients with sickle cell disorders and their families.

While there appears to be a considerable benefit to those who survive with stable engraftment of donor cells, there are also significant health risks to those who undergo this treatment. Therefore, engagement of trained psychotherapists and genetic counselors with focus on SCD and HCT with the patients, donors and their families should be conducted to ensure informed consent for this procedure. Presently, HCT is reserved for patients who have experienced significant complications of sickle cell disease, such as stroke, recurrent episodes of acute chest syndrome or intractable vaso-occlusive pain. Consequently, Nigeria is in dire need of strong institutional support for and training of psychotherapists and genetic counselors for the psychosocial management of patients with SCD, allogenic stem cell donors and their families.

Author details

Oluwatoyin Olatundun Ilesanmi

Address all correspondence to: toytundun@yahoo.com, toytundun@aol.com

Obafemi Awolowo University, Ile-Ife, Osun State, Nigeria

References

[1] Adler, Nancy. E., Page, Ann. E. K., & editors, . Institute of Medicine (IOM) Cancer Care for the Whole Patient: Meeting Psychosocial Health Needs. Washington, DC: The National Academies Press; (2008).

[2] Andreani, M., Testi, M., Gaziev, J., et al. Quantitatively different red cell/nucleated cell chimerism in patients with long-term, persistent hematopoietic mixed chimerism after bone marrow transplantation for thalassemia major or sickle cell disease. Haematologica. (2011). , 96, 128-133.

[3] Akinyanju, Olu(2009). The National Burden of Sickle Cell Disorder And The Way Forward. Retrieved from http://www.sicklecellfoundation.com/Nat%20Burden%20SCD.pdf

[4] Bernaudin, F., Socie, G., Kuentz, M., et al. Long-term results of related myeloablative stem-cell transplantation to cure sickle cell disease. Blood. (2007). http://bloodjournal.org/cgi/content/full/110/7/2749, 110(7), 2749-2756.

[5] Bernaudin, F. S. G., Vannier, J. P., et al. Report of the French experience concerning 26 children transplanted for severe sickle cell disease. Bone Marrow Transplant. (1997). suppl , 2, 112-115.

[6] Bhatia, M., & Walters, M. C. Hematopoietic stem cell transplantation for thalassemia and sickle cell disease: past, present and future. Bone Marrow Transplant. (2008). Jan;, 42(2), 109-17.

[7] Long-term follow-up of adults with severe sickle cell disease after hematopoietic stem cell transplantation using reduced intensity conditioning. Blood. (ASH Annual Meeting Abstract). 2010;116:261.Bernaudin F, Socie G, Kuentz M, et al. Long-term results of related myeloablative stem-cell transplantation to cure sickle cell disease. Blood.;, 110, 2749-2756.

[8] California Institute For Regenerative Medicine(2009). State stem cell agency application DRDR1-01452: Stem Cell Gene Therapy for Sickle Cell Disease. retrieved from http://www.cirm.ca.gov/ReviewReports_DR1-01452., 1-01452.

[9] Conti, L., & Cattaneo, E. Neural stem cell systems: physiological players or in vitro entities? Nat Rev Neurosci (2010). , 11(3), 176-87.

[10] Doubek, M., Folber, F., Koristek, Z., et al. Autologous hematopoietic stem cell transplantation in adult acute lymphoblastic leukemia: still not out of fashion. Ann Hematol. Sep (2009). Medline]., 88(9), 881-7.

[11] Eldredge, D. H., Nail, L. M., Maziarz, R. T., Hansen, L. K., Ewing, D., & Archbold, P. G. Explaining family caregiver role strain following autologous blood and marrow transplantation. J Psychosoc Oncol. (2006). PubMed], 24(3), 53-74.

[12] Galli, R., Gritti, A., & Vescovi, A. L. Adult neural stem cells. Methods Mol Biol 438 2008. , 67-84.

[13] Greenfield RM.Reflections on leukemia and adult stem-cell transplantation: some personal psychological factors. Palliat Supportive Care. (2007). , 5, 415-418.

[14] Harvey Simon,(2009). Sickle cell disease- In-Depth Report. 2011 University of Maryland Medical Center (UMMC).. The New York Times. Health Guide. retrieved on Monday, August 1 3, 2012 from Patient Education (http://www.umm.edu/patiented/index.htm)

[15] Horan, J. T., Liesveld, J. L., Fenton, P., et al. Hematopoietic stem cell transplantation for multiply transfused patients with sickle cell disease and thalassemia after low-dose total body irradiation, fludarabine, and rabbit anti-thymocyte globulin. Bone Marrow Transplant. (2005). , 35, 171-177.

[16] Iannone, R., Ohene-Frempong, K., Fuchs, E. J., Casella, J. F., & Chen, A. R. Bone marrow transplantation for sickle cell anemia: progress and prospects. Pediatr Blood Cancer. (2005). May;, 44(5), 436-40.

[17] Iannone, R., Casella, J. F., Fuchs, E. J., et al. Results of minimally toxic nonmyeloablative transplantation in patients with sickle cell anemia and β-thalassemia. Biol Blood Marrow Transplant. (2003). , 9, 519-528.

[18] Krishnamurti, L., Karbanda, S., Biernacki, Zhang. W., Baker, K. S., Wagner, J. E., et al. Stable long-term donor engraftment following reduced-intensity hematopoietic cell transplantation for sickle cell disease. Biol Blood Marrow Transplant. (2008). Nov;, 14(11), 1270-8.

[19] Lindvall, O., & Kokaia, Z. Stem cells in human neurodegenerative disorders- time for clinical translation? J Clin Invest (2010). , 120(1), 29-40.

[20] Bonaguidi, D. K., Ming, G. L., & Song, H. Adult neural stem cells in the mammalian central nervous system. Cell Res (2009). , 19(6), 672-82.

[21] Molineaux et al ((1979).). Annals of Trop Medicine & Parasitology , 73, 301-310.

[22] Novelli, E. M., Kato, G. J., Ragni, M. V., Zhang, Y., Hildesheim, Nouraie. M., Barge, S., Meyer, M. P., Hassett, A. C., Gordeuk, V. R., Gladwin, M. T., & Isenberg, J. S. Plasma thrombospondin-1 is increased during acute sickle cell vaso-occlusive events and associated with acute chest syndrome, hydroxyurea therapy, and lower hemolytic rates. Am J Hematol. (2012). Mar; , 87(3), 326-30.

[23] National Institutes of Health,[a] Medical Encyclopedia. Thalassemia. Updated 2010 Jan 31. Accessed(2011). Jun 7. Available at URL address: http://www.nlm.nih.gov/medlineplus/ency/article/000587.htm

[24] National Institutes of Health [b], National Human Genome Research Institute.Learning about thalassemia. Updated 2010 Jun 28. Accessed (2011). Jun 7. Available at URL address: http://www.genome.gov/10001221

[25] Orlacchio, A., Bernardi, G., Orlacchio, A., & Martino, S. Stem cells: an overview of the current status of therapies for central and peripheral nervous system diseases. Curr Med Chem (2010). , 17(7), 595-608.

[26] Panepinto, J. A., Walters, M. C., Carreras, J., Marsh, J., Bredeson, C. N., Gayle, R. P., et al. Matched-related donor transplantation for sickle cell disease: report from the Center for International Blood and Transplant Research. Br J Haematol. (2007). Jun;Epub 2007 Apr 24., 137(5), 479-85.

[27] Quinn CT, Miller ST.Risk factors and prediction of outcome in children and adolescents who have sickle cell anemia. Hematol/Oncol Clin. (2004). Dec;18(6).

[28] Rusiewicz, A., Du, Hamel. K. N., Burkhalter, J., et al. Psychological distress in long-term survivors of hematopoietic stem cell transplantation. Psychooncology. (2008). PubMed], 17, 329-337.

[29] Samaved VenkataHematopoietic Stem Cell Transplantation. Edited by Emmanuel C Besa, Medscape Drug, Disease and Procedure. Reference © (2011). WebMD, LLC retrieved from http://emedicine.medscape.com/article/ -overviewon 7/29/2012 2:34 AM

[30] Sendtner, M., Stem, cells., tailor-made, diseased., & neurons, Nature. (2009). 457(7227), 269-70.

[31] Sherman, R. S., Cooke, E., & Grant, M. Dialogue among survivors of hematopoietic cell transplantation support-group themes. J Psychosoc Oncol. (2005). PubMed], 23, 1-24.

[32] Sun Editorial(2011). Transparency for Nigeria. retrieved on Friday, 14 October 2011 08:59 from http://www.transparencyng.com/index.php/news-categories/ -health/ 5494-ubth-nigeriasfirst-stem-cell-transplant.source: http://www.sunnewsonline.com/ webpages/opinion/editorial/2011/oct/14/editorial-14-10-2011-001.html

[33] Syrjala KL, Langer SL, Abrams JR, et al.2004Recovery and long-term function after hematopoietic cell transplantation for leukemia or lymphoma. JAMA.;PubMed], 291, 2335-2343.

[34] The National Institutes of Health resource for stem cell research(2010). Scientific Progress and Future Research Directions Arrow10. Assessing Human Stem Cell Safety. In Stem Cell Information [World Wide Web site]. Historical content: June 17, 2001. Bethesda, MD: National Institutes of Health, U.S. Department of Health and Human Services, 2010 [Retrieved on Monday, August 06, 2012] from http://stemcells.nih.gov/ info/2001report/chapter10.asp.

[35] The National Institutes of Health(2009). Stem Cell Basics. retrieved from http://stemcells.nih.gov/staticresources/info/basics/SCprimer2009.pdf

[36] Vermylen, C., Cornu, G., Ferster, A., et al. Haematopoietic stem cell transplantation for sickle cell anaemia: the first 50 patients transplanted in Belgium. Bone Marrow Transplant. (1998)., 22, 1-6.

[37] Impact of bone marrow transplantation for symptomatic sickle cell disease: an interim report. Blood. ., 95, 1918-1924.

[38] Walters MC.(2005). Stem cell therapy for sickle cell disease: transplantation and gene therapy. Hematology Am Soc Hematol Educ Program. 2005; , 66-73.

[39] Yu, D., & Silva, G. A. Stem cell sources and therapeutic approaches for central nervous system and neural retinal disorders. Neurosurg Focus 24(3-4):(2008)., E11 EOF.

Chapter 16

Alternative Donor Sources for Hematopoietic Stem Cell Transplantation

Francisco Barriga, Nicolás Rojas and
Angélica Wietstruck

Additional information is available at the end of the chapter

http://dx.doi.org/10.5772/53083

1. Introduction

Allogeneic hematopoietic stem cell transplantation (HSCT) has become a common procedure for the therapy of hematologic malignancies, immune disorders and many other blood related disorders. Over 18,000 procedures are performed yearly in the US and Europe. The donor of choice for allogenictransplantation is a fully HLA matched sibling, which is available for only 20 to 25% of patients. Alternate donor sources have been developed and in the past few years transplant using these sources have surpassed the ones from sibling donor. These alternate sources are: adult volunteer donors which have been organized in large national registries; umbilical cord blood that is stored in blood banks worldwide; and manipulated stem cells grafts from haploidentical relatives. There is a wide variation in the transplant procedures, complications and outcomes between these sources, as well as debate over which one is the best source for each given patient, with few prospective comparative trials reported or in progress to settle this issue. We review the development and present status of each alternate source along with reported comparisons of properties and outcomes.

2. Hematopoietic stem cell transplantation: Purpose and indications

HSCT is a procedure where the entire hematopoiesis and immune system are replaced by the donor's cells [1]. HSCT can be classified according to its purpose, HSC donor type and HSC origin

The purposes of HSCT are:

1. Rescue a cancer patient from the effects of high dose chemotherapy and total body radiation. The most common indications are leukemia and lymphomas, which account for more than two thirds of transplants.

2. Correct a congenital or acquired cell disorder of the hematopoietic system (i.e., severe aplastic anemia and immune deficiencies, some inborn errors of metabolism)

3. Control the proliferation of cancer cells through immune mediated mechanisms that from part of the graft versus host reaction

4. Reset the immunological system, which had proven useful in patients with severe autoimmune disorders

Donor types are autologous, where stem cells are obtained from the patient, and allogeneic where stem cells are obtained form a donor. Autologous cells are only used in the treatment of malignant disorders that do not involve the bone marrow and autoimmune diseases.

An ever growing list of malignant and nonmalignant disorders is treated with HSCT (Table 1). It has grown at a rapid pace in the past two decades. Annual procedures in the US and Europe have gone from a few hundred in the early 90´2 to over 18,000 in 2011 [2, 3](figure 1).

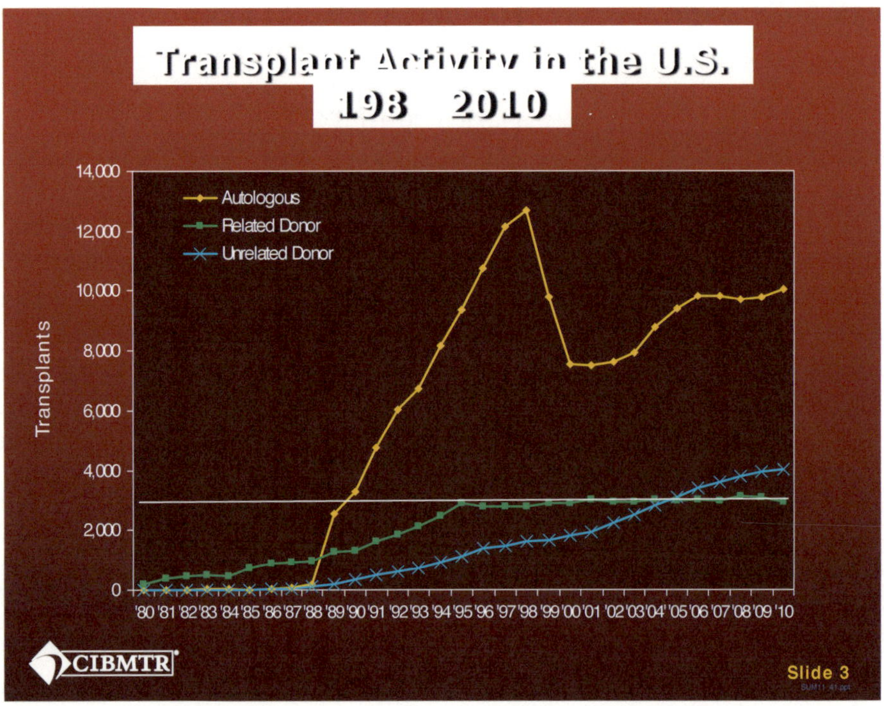

Figure 1. Transplant activity in the US, 1980 to 2010, by type of transplantation. Since 2005 unrelated donor transplants surpassed sibling donor procedures [2].

http://dx.doi.org/10.5772/53083

Leukemia and lymphomas with specific clinical or biological characteristics, including:

- Acute high risk myelogenousleukemia (AML):
    - Antecedent hematological disease (e.g., myelodysplasia (MDS))
    - Treatment-related leukemia
    - Induction failure
    - First complete remission with intermediate- or poor-risk cytogenetics or molecular markers
    - AML after relapse
    - Second complete remission and beyond
- Acute high risk lymphoblastic leukemiaincluding:
    - Poor-risk cytogenetics (e.g., Philadelphia chromosome (t(9;22)) or 11q23 rearrangements)
    - High White cell blood count ("/>30,000 - 50,000) at diagnosis in adults
    - t(11;22) in infants
    - Central Nervous system NS or testicular involvement
    - No complete remission within 4 weeks of initial treatment
    - Second complete remission and beyond
- Chronic myelogenousleukemia:
    - No hematologic or response post-tyrosine kinase inhibitor (TKI) initiation
    - Disease progression or intolerance to TKI
    - Accelerated phase or blast crisis (myeloid or lymphoid)
- Chronic lymphocytic leukemia
- Juvenile myelomonocyticleukemia
- Hodgkin lymphoma
- Non-Hodgkin lymphoma

Multiple myeloma and other plasma cell disorders

Severe aplastic anemia and other marrow failure states, including:

- Severe aplastic anemia
- Fanconianemia
- Paroxysmal nocturnal hemoglobinuria (PNH)
- Pure red cell aplasia
- Amegakaryocytosis / congenital thrombocytopenia

SCID and other inherited immune system disorders, including:

- Severe combined immunodeficiency (SCID, all sub-types)
- Wiskott-Aldrich syndrome

Hemoglobinopathies, including:

- Beta thalassemia major
- Sickle cell disease

Hurler's syndrome and other inherited metabolic disorders, including:

- Hurler's syndrome (MPS-IH)
- Adrenoleukodystrophy
- Metachromatic leukodystrophy

Myelodysplastic and myeloproliferative disorders, including:

- Refractory anemia (all types)
- Chronic myelomonocyticleukemia
- Agnogenic myeloid metaplasia (myelofibrosis)

Familial erythrophagocyticlymphohistiocytosis and other histiocytic disorders

Other malignancies

Table 1. Current indications of allogeneic stem cell transplantation

3. Sources of hematopoietic stem cells

3.1. Bone marrow

Marrow tissue obtained by repeated bone punctures and filtered to eliminate bone particles and fat was the original source of HSC. It contains 1 to 15 % of CD34+ cells, the marker by which HSC are identified. Bone marrow transplantation was performed successfully as a result of the studies done by Donnall Thomas and the group at the Fred Hutchinson Cancer Center during the 1960s [4]. Early studies demonstrated the effect of high radiation therapy doses and chemotherapy in the bone marrow as well as the capacity to regenerate the individual's hematopoietic function by reinfusion of stored bone marrow cells from himself or a donor. Bone marrow as a source continues to be widely used but it has not increased due to the inherent nature of the procedure that includes general anesthesia, results in considerable blood loss and is often followed with significant donor discomfort.

3.2. Mobilized peripheral stem cells

Donors treated with hematopoietic colony stimulating factors, mainly G-CSF, will mobilize large amounts of CD34+ cells to their peripheral blood. These cells can be recovered by leucopheresis, a procedure that circulates the blood of the patient/donor through a centrifuge, separates white blood cells and reinfuse the remaining blood back to the donor. This is the preferred source today for adult donors, which results in the harvest of large quantities of both CD34+ cells and other mononuclear cells, mainly T lymphocytes. Both hematopoietic and immune reconstitution are faster with PBSC than with bone marrow and less opportunistic infections have been reported in patients receiving them [5, 6]. In patients with leukemia, they have also been associated with higher incidence of chronic graft versus host disease and improvements in survival but direct comparisons in a single center have been few. In one of the few randomized trials comparing both stem cell sources, Storek et al reported a fourfold increase of post transplant circulating CD45RA (naïve T cell precursors) in recipients of PBSC, as well as a significant decrease in fungal an bacterial infections. In this report survival was improved in PBSC recipients. Although earlier reports found that the incidence of chronic graft versus host disease in patients receiving higher doses of CD34+ in a PBSC graft more recent studies in larger number of patients have shown an overall benefit of the CD34+ dose [7].

3.3. Umbilical cord blood

Blood obtained from the placenta at birth is rich in high quality HSCs and can reconstitute the hematopoietic function in a patient just like bone marrow or mobilized peripheral stem cells [8, 9]. These cells have to be cryopreserved right after collection and stored for latter use in liquid nitrogen. Cord blood banks have been established worldwide to provide this stem cell source (see below) Umbilical cord blood grafts contain fewer HSCs than other sources and because of this its use was initially limited in adult patients [10, 11]. Ways to circumvent this limitation have been developed using pooled cells from two cord blood units. This modality was first done by the group in University of Minnesota looking to expand a cord blood unit

http://dx.doi.org/10.5772/53083

while using a second one to increase the cell dose. Patients transplanted in this fashion had quicker hematopoietic cell recovery compared to those who received a single cord blood unit and transplant related mortality was greatly reduced [12]. An intriguing result was that only one unit of cord blood was identified in the peripheral blood of the patient, a phenomenon yet to be fully explained. These early results gave way to widespread use of two cord grafts in adult patients [13, 14]

A second alternative to increase the cell dose content has been expanding the cells before use. Many studies to accomplish this are on the way but it has not yet reached clinical use [15, 16]

3.4. Donor sources for allogeneic transplantation

Donors for allogeneic HSCT are matched in 3 to 6 loci of the human major histocompability complex (HLA, see below). Matching criteria are very strict due to the risk of acute and chronic graft versus host disease, the most common complication of HSCT, which can result in significant morbidity and mortality. Based on their origin and match grade donors can be divided into:

- Fully matched relative, almost always a sibling and rarely other family members. As HLA loci are inherited in a Mendelian fashion, the chances of a patient having a matched sibling are 25% with each sibling, which determines that only 20 to 25% of the patients have this type of donor. The chances improve in larger families.

- Partially matched relative: the donor shares at least one haplotype with the patient (haploidentical). HSC grafts need to be manipulated either with positive selection of CD34+ cells o negative selection of T lymphocytes.

- Matched or partially matched unrelated donor: presently there are more than 20 million unrelated donors listed in registries worldwide (see below) which are accessible for patients needing a transplant. These include adult volunteer donors and cord blood units stored in public access blood banks. The match grade accepted for a transplant depends on the criteria of the transplant center.

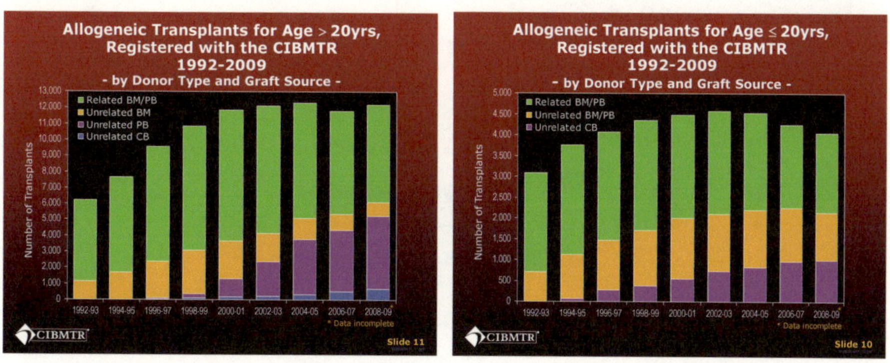

Figure 2. Stem cells sources and donor types in the US allogeneic transplantation, 1992 to 2009.

4. HLA typing in stem cell trasnplantation

The HLA system, also referred to the major histocompatibility complex, is a series of genes that expressed in the surface of immune and non immune cells represent a keystone in immune regulation and mediate graft acceptance or rejection in human allogeneic transplantation. First described in the 1950s as leukoagglutinin antibodies that appeared in the serum of pregnant women after blood transfusion, in 1967 the first nomenclature for HLA antigens was developed after initial efforts of systematization and standardization. Initially HLA antigens were described by serologic reaction with standard antibodies but as the genes encoding these antigens were sequenced, DNA techniques were adopted to increase the repertoire and further understand the polymorphic structure of the complex.

The antigens of the HLA system are encoded in genes located in the short arm of chromosome 6 (6p21.3). Their mission is to orchestrate the humoral and cellular immune responses, a basic issue in self and non-self molecular recognition. HLA antigens are localized on cell surface membranes and they form part of the antigen presenting complex with T cells receptors. The HLA/MHC region is inherited as a haplotype, which means that one person inherits 50% of the genetic information for MHC from the mother and the other half from the father, and shares a codominant expression. The most significant characteristic of this zone is its high polymorphism, which confers a huge variation between individuals.

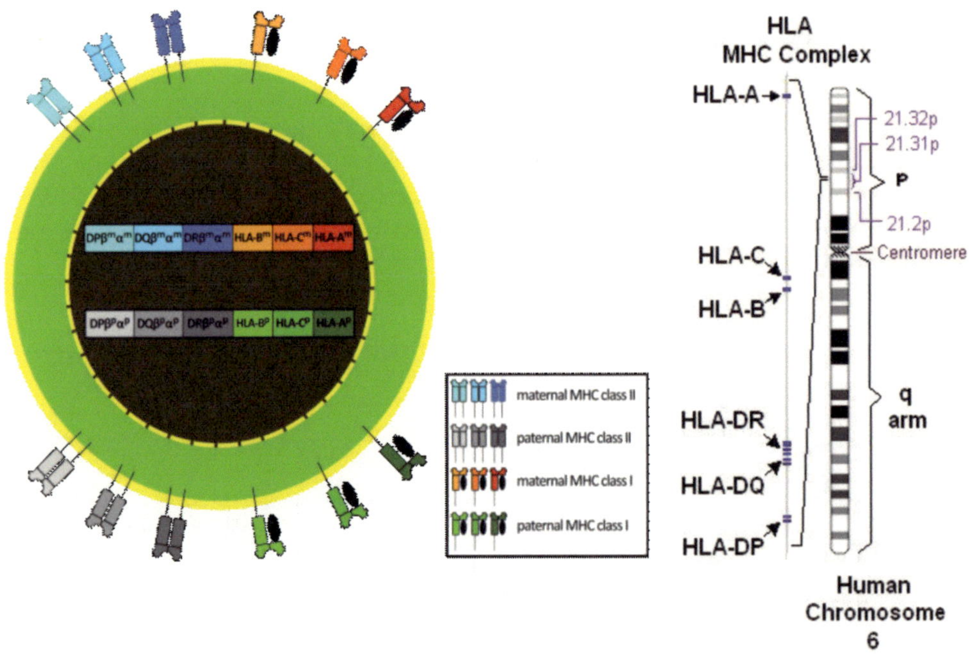

Figure 3. A Inherited MHC I and II complex antigens expressed on leucocyte membrane B. HLA encoding regions in Cr6.

There are two distinct classes of HLA molecules, named I and II; genes HLA-A, B and C encode homonym antigens (A, B and C) and conform class I molecules. They are expressed in all cells and mediate antigen recognition which triggers activation of cytotoxic lymphocytes. Class II antigens are HLA-DR, DQ and DP and its corresponding antigens. They are expressed in professional antigen presenting cells and together with the T cell receptor they form the complex that activates T helper cells.

It has been widely described that the one of the main prognostic factors in HSCT is HLA matching, which plays a significant role in engraftment, overall survival, transplant related infections, and leukemia control [17].

5. Development of donor registries and cord blood banks

Large registries of volunteer donors were the natural solution to the need of patients who lacked a matched sibling for transplantation. Because of the highly polymorphic nature of the HLA system, thousands of donors had to be recruited to find matches for a sizable population of patients. This required the development of large organizations which recruit donors, obtain all the necessary information along with blood samples for HLA typing and enter all this information in searchable registries that can identify and contact the donor in case their stem cells are required. Registries work with donor centers which perform all the necessary medical tests and, if the transplant goes through, harvest stem cells from the bone marrow or peripheral blood.

Most of this activity started around blood banks that had leucopheresis programs and volunteer donors for platelets products with HLA typing done. Most registries are national, government supported organizations that work with their transplant and donor centers. Once they became established it was also natural that international collaboration soon commenced and stem cell products traveled between countries and continents. The first successful unrelated donor transplant took place in 1973 in New York when a young boy with an inherited immunodeficiency received multiple marrow transplants from a donor identified as a match through a blood bank in Denmark. Driven by the need of a single patient with Wiskott Aldrich disease, a congenital immune deficiency that could only be cured with a transplant, the Anthony Nolan Registry was started in England in 1974. The first unrelated donor transplant for a patient with advanced leukemia was done in 1979 in Seattle and spurred the formation of the National Bone Marrow Donor Registry, which later became the National Marrow Donor Program (www.nmdp.org). NMDP has grown to recruit over 5,000,000 volunteer donors. Their vast experience in donor selection is summarized in periodical guidelines and recommendations [18]. The first transplants facilitated through these registries were done in the mid 80′s. Soon many more registries around the world would follow; increasing the donor pool from a few thousand in 1980 to over 20 million by 2012 (figure 4).

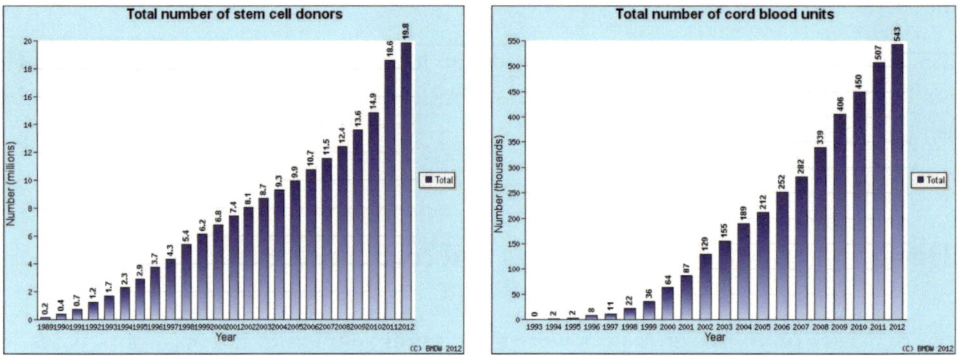

Figure 4. Stem cell donors and cord blood units listed in BMDW (www.bmdw.org)

The largest donor registries are located in the US and Europe, accounting for more than 60% of the donor pool. Based on the finding of large amounts of high quality HSC in newborn blood the first HSCT with umbilical cord blood was done in France in 1988 in a child with Fanconi Anemia who received the cord blood cells of his matched newborn sibling [19]. Since cord blood cells can be frozen and stored in liquid nitrogen for very long periods of time without losing their properties, cord blood Banks were established in the early 90s with blood units collected from the placenta at birth. HLA typing is done in these units and the cord blood bank acts as a donor registry, increasing furthermore the donor pool.

In 1988 the Europdonor Foundation was started in the Netherlands to facilitate access to donor registries around the world in a single site. Their network site, Bone Marrow Donors World-wide (www.bmdw.org) works as a registry of registries and allows for search among all available donors. Presently BMDW lists donors from 112 registries in 50 countries.

Their mission is listed as:

- To maximize the chance of finding a stem cell donor or cord blood unit by providing access to all stem cell donors and cord blood units available in the world.
- To minimize the effort required for stem cell donor or cord blood unit searches: only registries with potential stem cell donors or cord blood units need to be contacted.

Two consequences are derived from this significant increase in the donor pool:

- Unrelated donor transplant activity has increased at a parallel pace (figure 1). Annual procedures in the US and Europe have gone from a few hundred in the early 90´2 to over 18,000 in 2011 (CIBMTR, EBMT) and in both cases have surpassed the number of sibling donor patients transplants, which have remained constant. Despite this, and taking into account that needing a transplant and having a sibling donor are independent variables, there is room for much improvement in making this therapy available to all who need it.

- The larger donor pool coupled to improvement in HLA typing with identification of a growing number of alleles allow for much better matching between donor an patient and this likely accounts for the improvement in transplant results. A recent CIBMTR report showed that the difference in one year survival of patients transplanted for leukemia or myelodysplasia comparing those with a sibling donor to an unrelated one was reduced from 20% to less than 10% in the past two decades [2, figure 4]. We cannot rule out an effect of improved transplant center experience in this result, but other smaller reports have confirmed that donor source (sibling vs. unrelated) is less relevant to outcome.

In 1990 the World Marrow Donor Association was born to help coordinate international searches and transplant of hematopoietic stem cells, keeping track of all the products facilitated inside participating countries and those exported to other countries. According to their 2010 annual report [20], 15,256 patients were transplanted during that year with stem cells form unrelated donors. Of those, 7183 stem cells products (45,7%) were imported to the country where the patient received their transplant, that is, every day 20 stem cell products travel from one country to another. 12,237 products were obtained from adult donors and 3028 were cord blood grafts (19,4%) making cord blood the fastest growing stem cell source, even though it only represents 2,5% of the donor pool. The reasons favoring this are a shorter search time, immediate availability and less strict HLA match requirements. Also, more centers are become familiarized with this type of transplant procedure, accounting for its increased use.

A general overview at the global map displayed by CIBMTR, EBMT and WMDA immediately highlights the large difference of access and activity of stem cell transplantation among different regions of the world. In general the size of the national registries mirrors the transplant activity for each region. By far, Europe and North America have the larger registries (16,2 million donors) and account for the highest transplant activity (18,500 in 2010), followed by some countries in Asia. South America and Africa lag behind. The top 5 countries shipping marrow or peripheral stem cells products are Germany, USA, Japan, United Kingdom and China, accounting for 83% of shipments. The registries recruiting the largest amount of donors in that year were REDOME (Brazil), NMDP (USA), ZKRD (Germany), CMDP (China) and CRIR (USA), accounting for 79% of the donors recruited. The five largest suppliers of cord blood units were USA, Japan, Spain, France and Italy.

6. Haploidentical stem cell transplantation

Haploidentical stem cell transplantation consists in the use of a graft from a related donor, usually parents or siblings, with whom the patient shares at least 50% (up to 80%) of the MHC alleles. The graft itself can be collected by apheresis or bone marrow aspiration and it has to be manipulated to allow for engraftment and prevent graft versus host disease.

Two main advantages of transplantation from a full haplotype mismatched family member are evident:

1. Most, if not all, patients have an HLA-partially matched relative who is available to serve as a donor. In fact most patients will have more than one donor, allowing the possibility of switching to another relative if more than one graft is required [21, 22].

2. More frequent than not the best donor can be chosen between many candidates. The graft is immediately available once the best candidate is chosen, as is the case in sibling transplantation

Haploidenticaltransplantation has been limited by historically high rates of graft rejection, GVHD, TRM, and poor immune reconstitution, resulting in a high incidence of serious opportunistic infection. Both myeloablative and reduced intensity conditioning transplant strategies have been attempted looking for better outcomes, with diverse results. The first attempts of HLA-non identical stem cell transplantation were reported in 1985 by Beatty et al [23], who described the problems and adverse effects derived from unmanipulatedhaploidentical grafts using myeloablative conditioning regimens. This study reported non permissive toxicity and mortality with type II HLA mismatch as well as higher rates of GVHD with class I antigen mismatch. It also set the stage for graft manipulation, which has improved outcomes. Some of the strategies involved are:

1. Ex vivo T cell depletion, that resulted in improving acute and chronic GVHD, overall and event free survival [24].

2. Ex vivo positive selection of CD34+ cells resulting in a T cell reduced graft [25].

3. In vivo immune suppression with anthymocyte globulin and post transplant high dose cyclophosphamide [26]

4. Ex vivo induction of alloantigen specific anergy by coculturing host and donor BM mononuclear cells with either CTLA-4-IG or anti–B-7.1 and B7.2 antibodies [27]

Delayed immune reconstitution after haploidentical HSCT is the main contributor to morbidity and mortality of this technique. The reasons for this are T cell depletion of the graft, thymic dysfunction induced by pretransplant chemotherapies and conditioning regimens, and GVHD occurrence and its treatment [28].The other major challenge for haploidentical HSCT is the high relapse rate, and several strategies are been developed like the use of tumor specific T cells and the use of NK from the donor as shown below.

Intense pretransplant conditioning and graft manipulation to rid of T lymphocytes is associated with delayed hematological and immune recovery, resulting in an increased rate of infection. To circumvent this drawback, large doses of CD34+cells have been used to improve the speed of hematological recovery with success [29]. To hasten immune recovery and also make the procedure tolerable to older patients, less intense or reduced conditioning regimens have been tried [30] but the effect on improving immune recovery have been modest.

Perhaps the most disturbing side effect of T cell depletion to allow a haploidentical graft is the abolition of the graft versus tumor effect with the increased rate of post transplant recurrence of leukemia. This was observed in the first attempts with haploidentical grafts. Despite this a substantial graft versus tumor effect has been attributed to the infusion of natural killer (NK) cells, which are not depleted with T lymphocytes [31]. The best described element regarding NK cell activity is the inhibitory killer cell immunoglobulin-like receptor (KIR), which helps prevent NK cells from damaging host tissues, [32]. KIRs are expressed by NK cells from the donor and interact with host HLA class I epitopes (HLA-C) in the recipient. If the KIR-HLAC is mismatched, the inhibitory action of the receptor fails and the alloreactive NK cell is activated against the host cell. KIR mismatch between donor and recipient has been associated with improved survival after HSCT in AML, appearing to promote engraftment, reduce GVHD and decrease leukemic relapse [33, 34].

Further attempts to "engineer" the graft has been made to improve results. Handgretinger [24] developed a protocol based on animal models, using NK cell enriched CD3+ depleted stem cells, with either myeloablative or reduced intensity conditioning regimens, plus anti CD-20 for in vivo B cell depletion. Assessment of immune reconstitution by flow cytometry showed a faster recovery of CD4+, CD56+ and thymic precursors measured by TREC analysis. The protocol reported significant reduction in transplant related mortality as well as incidence of cytomegalovirus and adenoviral infections,

7. Donor search algorithms

It is widely recognized that the HLA matching level is the most important factor for transplant outcome [16, 35, 36]. Thus, fully matched siblings are the best source of HSC for transplantation, also due to their immediate availability, lower transplant related complications and mortality, and reduced costs in obtaining stem cells. Nevertheless a fully matched HLA graft also implies a reduced alloreactive effect of donor T cells against tumor cells in patients transplanted for malignant diseases and this can reflect on a higher rate of relapse, which has to be weighed against the reduced transplant related mortality.

Several aspects can be taken into consideration when choosing an unrelated donor among the different alternatives and they all come into play simultaneously. A very important one is center experience, which in itself accounts for most of the improvement in outcome [37]. Large transplant programs usually have preferences regarding the donor chosen based in their experience. The search process, stem cell procurement, and previous results weigh in their policy. Some programs only use one source of stem cells (i.e. adult donor or cord blood) and establish search and procurement protocols based on this choice. Programs with a preference for cord blood grafts will consider using less compatible cord blood units (4/6 match) or resource to double cord blood grafts for adult patients [11,12, 38, 39] before considering an adult donor with a single high resolution HLA mismatch. Other programs with no cord blood transplant experience will either resource to a partially matched donor or forfeit transplantation altogether. Perhaps the most center-dependent modality is haploidentical transplantation.

Few centers have the infrastructure and professional teams trained in T cell depletion or CD34+ enrichment and despite its obvious appeal and having been around for a long period of time the procedure has not reached wide acceptance. The total number of haploidentical transplants reported to EBMT in the past decade has remained almost unchanged [3].

Despite different preferences in donor selection some points are generally agreed upon in the transplant community, which rely in overall experience and careful review of multiple published reports [40].

1. The best alternate donor for unrelated transplantation in a patient who can wait for the search process to be completed is a fully matched adult with at least 8 high resolution (i.e. 4 digit or similar) matched alleles [34, 41]. Some centers will require a 10/10 match, usually including DQB1, for donor acceptance. Unfortunately, and despite the massive recruitment of donors worldwide, we are still far from securing a fully matched donor for every patient. A 2004 report by the National Marrow Donor Program in the US, with over 4 million recruited donors, projected that only white and hispanic patients would have an over 50% chance for a fully matched donor by 2007, with other ethnic groups faring much worse [42]. When more than one fully matched donor is available other secondary aspects can be taken into consideration: younger age, male sex, CMV serology referred to the patient, ABO compatibility, larger weight and rapacity. Despite this, only HLA matching and donor age affect patient survival [17].

2. If no such donor is available or the patient cannot wait, most centers will opt for fully matched or single mismatched cord blood unit (6/6 or 5/6; HLA-A and B in low resolution and DRB1 in high resolution), provided it reaches a total nucleated cell dose of at least 3,0 x 10^7 per kg of the patient [43]. This is readily available for most children up to 40 kg. [44]but can be difficult for large adults. In this situation most programs recur to a double cord blood unit graft, a modality that has gained wide acceptance [37,38]. If no highly matched cord blood units are available the options mentioned are either an adult donor with a single major locus mismatch or a single or double 2 mismatched cord blood unit (4/6). This situation is generally decided upon center experience and bias towards one or the other graft source.

3. It has been difficult to place haploidentical transplantation in donor selection algorithms since most of these procedures are done in few highly specialized centers that have the facilities and trained staff for it. Recently, new approaches to avoid graft rejection and GVHD by in vivo T cell depletion with potent immune suppression and chemotherapy have been tried with reported results that are similar to the use of double cord blood grafts [45]. In general transplant related mortality in haploidentical transplantation has been reportedly lower than using cord blood but this advantage has been offset with the higher risk of relapse, which makes this source less recommendable for patients with high risk disease. Longer follow up will be needed to address the question of which particular patient benefits from which particular donor.

	Adult donor	Cord blood	Haploidentical
Donor availability adults	Fully matched 50% One mismatch 70%	5/6 or 6/6 : 85% * 4/6: 100% *	80%
Donor availability pediatric	Fully matched 50% One mismatch 70%	5/6 or 6/6: 90% 4/6: 100%	100%
Average time from search to transplant	3 months (0,5-6)	21 days (7-60)	7 days
Target CD34+ dose /kg	> 2 x 106	> 0,1 x106	>10x106
Graft manipulation	Not required	Not required	Required
T cells in the graft	Replete (PBSC "/> BM)	Partially depleted	Depleted
T cell immune reconstitution	3 months	6-9 months	6-9 months
Acute Graft versus host	Higher depending on mismatch	Less depending on mismatch	Rare
Chronic graft versus host	Higher depending on match	Similar depending on match	Rare
Relapse risk	Similar to less	Similar to less	Higher
CMV reactivation **	Similar	Frequent	Frequent
Post transplant cell infusion	Possible	Not possible	Possible

*single or double unit graft

**depends on the donor and patient serology results

Table 2. Comparison between alternative donor sources

8. Transplant outcomes: Comparison among donor sources

Large registries have tracked the progress of HSCT results in the past decades. The information obtained from them allows comparing in an extensive number of patients the impact of disease type and stage, the donor source and donor type in transplant outcomes. Analysis of the data from CIBMTR has shown that transplant results in young patients with hematological malignancy in early stages of the disease comparing related unrelated donors have improved consistently in the past 20 years, reducing a 20% difference in one year survival to less than 10% (figure 5). This data strongly supports the use of a matched adult volunteer donor as the first choice when one is available, and this is something most centers will agree upon. The challenge and controversy comes from selecting between a cord blood graft, a mismatched unrelated adult donor or ahaploidentical donor [46]. Several studies have addressed this issue for different graft sources in patients with different diseases, both in adults and in children, based on registry data or comparing published reports using a single donor source. Very few clinical trials have attempted to compare graft sources and none have been randomized [44]. Center preference and the difficulty involved in search logistics will make very unlikely that a randomized trial will ever be accomplished.

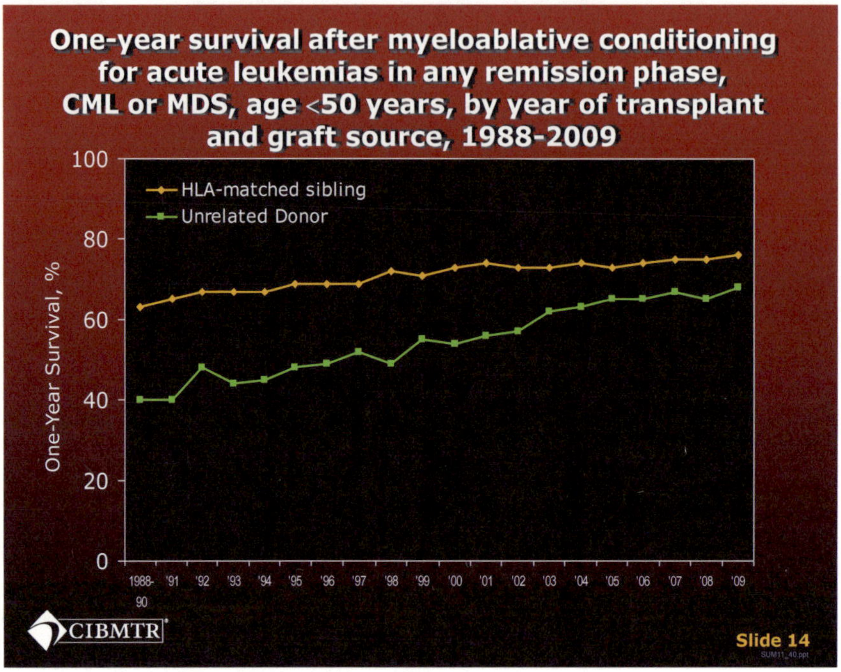

Figure 5. Improvement in one year survival of HSCT from related and unrelated donors in patients with hematological malignancies.

Several large studies have compared umbilical cord blood with mismatched unrelated donors in patients with hematologic malignancies:

Laughlin et al compared results of a single cord blood unit graft versus a 7/8 HLA matched unrelated donor in 233 patients from the databases of CIBMTR and the National Cord Blood Program in New York and found similar outcomes when measuring transplant related mortality, event free survival and overall survival. Survival was a sobering 26% to 20% in cord blood versus mismatched donor, and did not reach significance [47].

Eapen published in 2007 the results in a large group of children with acute leukemia transplanted with a single cord blood unit, a fully matched unrelated donor or a mismatched unrelated donor [48]. The measured outcome was leukemia free survival, also assessing the relative effect of cell dose and HLA matching in the outcome of cord blood transplants. The study included 785 patients younger than 16 years at transplantation with acute lymphoblastic or acute myeloid leukemia who received either a single-unit cord-blood or a bone-marrow graft from an HLA-matched or HLA-mismatched unrelated donor in the USA. The comparisons were made between six groups: HLA-matched cord blood, one-antigen mismatched high-cell-dose cord blood, one antigen mismatched low cell dose cord blood, two antigen mismatched cord blood (any dose), allele-mismatched bone marrow, and allele-matched bone marrow. Early transplant related mortality was significantly less in patients who received fullymatched marrow and cord blood, or a high dose one antigen mismatched cord. This advantage was offset by a higher incidence of relapse in the first group with similar leukemia free survival among all groups analyzed. These data support the use of HLA-matched and one- or two-antigen HLA-mismatched umbilical cord blood in children with acute leukemia who need transplantation. The higher risk of non-relapse mortality associated with unrelated bone marrow and cord blood transplantation raises anxiety among pediatric oncologists when considering these donor sources for their patients. Nevertheless, the ever growing number of donor registries and cord blood banks will improve the chances of finding the best suited donor.

Another study by Eapen et al in 2010 [49] reported the outcome on 1525 adult patients transplanted for acute leukemia with unrelated matched or mismatched donors comparing them to single cord blood unit recipients. Transplant related mortality, leukemia free survival and overall survivals were almost identical among cord blood recipients and mismatched unrelated donor recipients. Overall survival 43-44%, a significant improvement from previous studies.

Trying to address the question whether a more mismatched stem-cell source will give better disease control due to a potential increased graft versus leukemia effect, Zhang 50 et al compared leukemia free and overall survival among 348 children with leukemia registered with CIBMTR who were transplanted with unrelated donor bone marrow, unrelated cord blood and HLA-matched sibling bone marrow. 3-year leukemia free survival was comparable among all groups, despite higher risks of acute and chronic GVHD after unrelated donor transplantation and higher non relapse mortality after mismatched unrelated donor BM and cord blood transplantation. The pattern of treatment failure differed by donor type. Whereas nonrelapse mortality was higher after unrelated donor transplantation, they observed a higher, but not statistically significant, risk of relapse after HLA-matched sibling donor transplantation. A logical conclusion to this and other reports is that as transplant related mortality is curbed with better control on infections, a more mismatched graft may be better for high risk leukemia. Similar results were published previously by Minnesota group [51]comparing single center transplant outcomes by HSC source for children less than 18 years with ALL in second complete remission. In a more limited sample of patients, their results also suggest that transplant outcomes are remarkably similar in recipients of matched sibling, matched unrelated or umbilical cord donor grafts.

Very few studies have compared outcomes of unrelated donors with haploidentical transplantation. Most reports come from single center studies and they are difficult to interpret due to the different techniques employed for haploidentical donor selection and graft manipulation. A recent study compared the results of two large parallel clinical trials: one, using haploidentical donors with in vivo treatment of the recipient with post transplant high dose cyclophosphamide ; two, using a double 0 to 2 antigen mismatched cord blood graft [44]. One year survival in both groups was similar around 50%. Nevertheless large differences in outcome were noted: non relapse mortality was higher in the cord blood group (24% vs. 7%) but relapse was lower (31% vs. 45%).

Different considerations apply for patients with nonmalignant disease. The emphasis is put on engraftment, quick immune recovery and avoidance of graft versus host disease. In this regard most patients can wait and receive alternative therapy until a suitable donor is found and therefore the search process can be prolonged as much as needed. In some cases the transplant has to proceed more urgently to avoid organ damage, chronic blood transfusion or repeated infections. As most patients with nonmalignant diseases are children a well matched cord blood unit is usually available for almost every patient. The challenge of cord blood transplantation in children with nonmalignant diseases is that with the exception of severe immune deficiencies, the rate of graft failure is much higher than those with leukemia, making strict HLA matching more necessary [52]. A plausible explanation is that children with leukemia almost always receive chemotherapy to induce a remission before transplantation so their immune system is greatly impaired before they start the transplant conditioning regimen. Total body radiation is also extensively used in transplantation for malignancy resulting in complete lymphodepletion in this patient population. Moreover patients with nonmalignant disease usually receive either anti thymocyte globulin or as part of their conditioning regimen to prevent graft versus host procedure. An approach that would merit consideration is the delivery of chemotherapy which in itself prolongs or hampers immune reconstitution.

9. Donor search in Chile: Progress in a developing country

Several shortcomings apply to the development of transplant programs in developing countries. Lack of resources, shortage of trained staff and poor understanding of the benefits of transplantation by the medical community all play into this reality. In 2010 WMDA reported that out of the 4054 unrelated cord blood units that were shipped worldwide, 2706 were provided by Europe, Australia and North America, 1324 by all Asia, and only 24 by South America and none by Africa. Only 206 were transplanted in South America, a continent that harbors more than 300 million inhabitants.

In Chile our transplant program was started in 1989 with sibling donors. As we were able to successfully treat patients, the problem of those without a family match became compelling. Our initial efforts to conduct searches for unrelated donors in the international registries were hampered by the difficulty of implementing high quality HLA typing in our country, the relatively small size of the donor pool and the restrictive policy of most international registries in Europe and the US to work with transplant centers outside their network. This reality changed in 1996 when Cord Blood Banks were implemented and the first procedures using this source were done worldwide. Despite the initial small number of cord units started at that time, we were able to find one or two antigen mismatched cord blood units for most of our patients and through collaboration with the National Cord Blood Program in New York the first procedures were done in 1997. Discouraged by poor results and high transplant related mortality mainly caused by infection we decided to consider 0 to 1 antigen mismatched cord blood units only. Initially we could only find such a donor for 50% of our patients [53], but that percentage increased steadily during the next years. Cord blood transplantation gave us the initial experience we needed and in 2009 our program started to recur to unrelated adult donors facilitated first through NMDP and latter by registries and cord blood banks in the US,

Germany, Spain, France, Italy, Netherlands, Australia among others. In the last 4 years the proportion of unrelated donor transplants doubled the matched sibling procedures. A recent review of our data showed that out of 108 completed unrelated donor searches we were able to identify a fully matched adult donor in 18 patients, and a 0 to 1 antigen mismatched cord blood unit with > 3x 10^7 cells /kg in 73 patients (84% of the total). In only 9 patients we were not able to find a suitable donor, most of them adults. In summary, despite our mixed native American and Spanish ascent almost all our patients in Chile are able to identify an unrelated donor for stem cell transplantation.

10. Conclusion

Substantial biases in donor selection are the result of center preference and it is not forthcoming that controlled clinical trials will be conducted to demonstrate superiority of one source above the other. On the other hand much work is being carried to improve the donor pool in all three donor sources:

1. As registries continue to expand the chances for patients with uncommon HLA alleles to find a donor will improve steadily, especially for those from ethnic communities under-represented in the registries

2. Work in expanding cord blood cells and understanding and manipulating their homing properties will result in safer transplantation of larger amounts of cells and faster hematopoietic reconstitution.

3. Groups developing haploidentical transplantation have worked hard in graft manipulation testing strategies of adding back alloreactive lymphocytes to reduce the risk of relapse while maintaining a low incidence of graft versus host disease.

In this scenario transplant physicians will be confronted with multiple choices when they plan a procedure in a patient lacking a sibling donor, especially when they are able to find highly matched adult volunteers, cord blood units of high quality and ever better matched with the patient, and the infrastructure and experience to perform haploidentical transplantation. In an ideal world where all of them are available, the disease and stage, the age of the patient and the perceived or proven risk for a prolonged or partial immune reconstitution will come into play.

Author details

Francisco Barriga, Nicolás Rojas and Angélica Wietstruck

Hematology Oncology Section, Division of Pediatrics, Pontificia Unversidad Católica de Chile, Chile

References

[1] Copelan EA Hematopoietic stem-cell transplantation. N Engl J Med 2006; 354:1813-1826.

[2] Pasquini MC, Wang Z . Current use and outcome of hematopoietic stem cell transplantation: CIBMTR Summary Slides 2011. Available at: http://www.cibmtr.org

[3] Passweg JR, Baldomero H, Gratwohl The EBMT activity survey: 1990–2010. Bone Marrow Transplantation 2012; 47: 906–923.

[4] Thomas ED, Storb R, Clift RA, Fefer FA, Johnson FL, Neiman PE, Lerner KG, Glucksberg H, Buckner CD Bone-Marrow Transplantation. N Engl J Med 1975; 292:832-843.

[5] Storek J, Dawson MA, Storer B et al: Immune reconstitution after allogeneic marrow transplantation compared with blood stem cell transplantation Blood June 1, 2001 vol. 97 no. 11 3380-3389

[6] Ottinger HD, Beelen DW, Scheulen B, et al: Improved immune reconstitution after allotransplantation of peripheral blood stem cells instead of bone marrow. Blood. 1996; 88:2775-2779.

[7] Pulsipher MA, Chitphakdithai P, Logan BR: ,Donor, recipient, and transplant characteristics as risk factors after unrelated donor PBSC transplantation: beneficial effects of higher CD34+ cell dose Blood 2009: 114:2606-2616

[8] Kurtzberg J, Laughlin M, Graham ML, et al. Placental blood as a source of hematopoietic stem cells for transplantation into unrelated recipients. N Engl J Med 1996;335:157-166.

[9] Laughlin MJ, Barker J, Bambach B, et al. Hematopoietic engraftment and survival in adult recipients of umbilical-cord blood from unrelated donors. N Engl J Med 2001;344:1815-1822.

[10] Rocha V, Gluckman E. Improving outcomes of cord blood transplantation: HLA matching, cell dose and other graft- and transplantation-related factors. Br J Haematol 2009;147(2):262-274.

[11] Wagner JE, Barker JN, DeFor TE, et al. Transplantation of unrelated donor umbilical cord blood in 102 patients with malignant and nonmalignant diseases: influence of CD34 cell dose and HLA disparity on treatment-related mortality and survival. Blood 2002;100(5):1611-1618.

[12] Barker JN, Weisdorf DJ , DeFor TE ,. Blazar BR, Miller JS, Wagner JE Rapid and complete donor chimerism in adult recipients of unrelated donor umbilical cord blood transplantation after reduced-intensity conditioning Blood 2003 102:1915-1919

[13] Brunstein CG, Gutman JA, Weisdorf DJ, et al. Allogeneic hematopoietic cell transplantation for hematological malignancy: relative risks and benefits of double umbilical cord blood. Blood 2010;116(22):4693-4699.)

[14] Ballen KK, Spitzer TR, Yeap BY, et al. Double unrelated reduced-intensity umbilical cord blood transplantation in adults. Biol Blood Marrow Transplant 2007;13(1):82-89

[15] Hofmeister CC, Zhang J, Knight KL, Le P, Stiff PJ. Ex vivo expansion of umbilical cord blood stem cells for transplantation: growing knowledge from the hematopoietic niche. Bone Marrow Transplant 2007;39:11-23.

[16] Shpall EJ, Quinones R, Giller R, et al. Transplantation of ex vivo expanded cord blood. Biol Blood Marrow Transplant 2002;8:368-376.

[17] Kamani N, Spellman S, Hurley CK, et al. State of the art review: HLA matching and outcome of unrelated donor umbilical cord blood transplants. Biol Blood Marrow Transplant.2008;14(1):1-6.

[18] Bray RA, Hurley CK, Kamani NR, et al. National Marrow Donor Program HLA matching guidelines for unrelated adult donor hematopoietic cell transplants. Biol Blood Marrow Transplant 2008;14(S9):45-5

[19] Gluckman E, Broxmeyer HA, Auerbach AD, et al. Hematopoietic reconstitution in a patient with Fanconi's anemia by means of umbilical-cord blood from an HLA-identical sibling. N Engl J Med 1989;321(17):1174-1178.

[20] Foeken LM, Green A, Hurley CK, Marry E, Wiegand T, Oudshoorn M. Monitoring the international use of unrelated donors for transplantation: the WMDA annual reports. Bone Marrow Transplantation 2010: 45, 811–818.

[21] Reisner, Y.; Hagin, D.; Martelli, MF. Haploidentical hematopoietic transplantation: current status and future Perspectives. Blood 2011: 118; 6006-6017.

[22] Koh, L-P.; Rizzieri, DA.; Chao, NJ. (2007) Allogeneic Hematopoietic Stem Cell Transplant Using Mismatched/ Haploidentical Donors. Biology of Blood and Marrow Transplantation 2007: 13;1249-1267

[23] Beatty, PG., Clift, RA., Mickelson, EM., Nisperos, BB., Flournoy, N., Martin, PJ., Sanders, JE., Stewart, P., Buckner, CD., Storb, R., Thomas, ED., Hansen, A. Marrow Transplantation from Related Donors Other Than HLA-Identical Siblings. New England Journal of Medicine 1985:. 313:765-71.

[24] Aversa, F., Terenzi, A., Tabilio, A., Falzetti, F., Carotti, A., Ballanti, S., Felicini, R., Falcinelli, F., Velardi, A., Ruggeri, L., Aloisi, T., Saab, JP., Santucci, A., Perruccio, K., Martelli, MP., Mecucci, C., Reisner, Y., Martelli, MF:Full haplotype-mismatched hematopoietic stem-cell transplantation: a phase II study in patients with acute leukemia at high risk of relapse. Journal of Clinical Oncology 2005; 23;3447-3454.

[25] Aversa, F., Tabilio, A., Velardi, A., Cunningham,I., Terenzi, A., Falzetti, F., Ruggeri, L., Barbabietola, G., Aristei, C., Latini, P., Reisner, Y., Martelli, MF., Felicini, R., Falci-

nelli, F., Carotti, A., Perruccio, K., Ballanti, S., Santucci, A., Gambelunghe, C . Treatment of high-risk acute leukemia with T-cell-depleted stem cells from related donors with one fully mismatched HLA haplotype. New England Journal of Medicine. 1998; 339:1186-1193.

[26] Huang, X-J., Liu, D-H., Liu, K-Y: Treatment of acute leukemia with unmanipulated HLA-mismatched/ haploidentical blood and bone marrow transplantation. Biology of Blood Marrow Transplantion 2009;15:257-265.

[27] Davies, JK., Gribben, JG., Brennan, LL., Yuk, D., Nadler, LM., Guinan, EC. Outcome of alloanergized haploidentical bone marrow transplantation after ex vivo costimulatory blockade: results of 2 phase 1 studies. Blood 2008;112:2232-2241.

[28] Handgretinger, R., Chen, X., Pfeiffer, M., Schumm, M., Mueller, I., Feuchtinger, T., Hale, G., Lang, P. Cellular Immune Reconstitution after Haploidentical Transplantation in Children. Biology of Blood and Marrow Transplantation 2008; 14:59-65

[29] Lang, P., Bader, P., Schumm, M., Feuchtinger, T., Einsele, H., Fuhrer, M., Weinstock, C., Handgretinger, R., Kuci, S., Martin, D., Niethammer, D., Greil, J. Transplantation of a combination of CD133+ and CD34+ selected progenitor cells from alternative donors. British Journal of Haematology 2004; 124: 72–79

[30] Rizzieri, DA., Koh, LP., Long, GD., Gasparetto, C., Sullivan, KM., Horwitz, M., Chute, J., Smith, C., Gong, JZ., Lagoo, A., Niedzwiecki, D., Dowell, JM., Waters-Pick, B., Liu, C., Marshall, D., Vredenburgh, JJ., Gockerman, J., Decastro, C., Moore, J., Chao, NJ. Partially matched, nonmyeloablative allogeneic transplantation: clinical outcomes and immune reconstitution. Journal of Clinical Oncology 2007; 25:690-697.

[31] Ruggeri, L., Capanni, M., Urbani, E., Perruccio, K., Shlomchik, WD., Tosti, A., Posati, S., Rogaia, D., Frassoni, F., Aversa, F., Martelli, MF., Velardi, A. Effectiveness of donor natural killer cell alloreactivity inmismatched hematopoietic transplants. Science 2002; 295:2097–2100.

[32] Davies, SM., Ruggieri, L., DeFor, T., Wagner, JE., Weisdorf, DJ., Miller, JS., Velardi, A., Blazar, BR. Evaluation of KIR ligand incompatibility in mismatched unrelated donor hematopoietic transplants. Killer immunoglobulin-like receptor. Blood 2002; 100:3825–3827

[33] Moretta, A., Pende, D. Locatelli, F., Moretta, L.. Activating and inhibitory killer immunoglobulin-like receptors (KIR) in haploidentical haemopoietic stem cell transplantation to cure high-risk leukaemias. Clinical and Experimental Immunology 2009; 157: 325–331

[34] Velardi, A. Role of KIRs and KIR ligands in hematopoietic transplantation. Current Opinion in Immunology 20: 2008; 581–587

[35] Lee SJ, Klein J, Haagenson M, et al. High-resolution donor-recipient HLA matching contributes to the success of unrelated donor marrow transplantation. Blood 2007;110(13):4576-4583.

[36] Petersdorf EW, Hansen JA, Martin PJ. Major-histocompatibility-complex class I alleles and antigens in hematopoietic-cell transplantation. N Engl J Med. 2001; 345(25): 1794-1800.

[37] Bacigalupo A, Sormani MP, Lamparelli T, et al. Reducing transplant-related mortality after allogeneic hematopoietic stem cell transplantation. Haematologica, 2004; 89(10):1238-1247.

[38] Rocha V, Gluckman E. Clinical use of umbilical cord blood hematopoietic stem cells. Biol Blood Marrow Transplant. 2006; 12(1, Suppl. 1):34-41.

[39] Majhail NS, Brunstein CG, Wagner JE. Double umbilical cord blood transplantation. Curr Opin Immunol. 2006; 18(5):571-575

[40] Grewal SS, Barker JN, Davies SM, Wagner JE. Unrelated donor hematopoietic cell transplantation: marrow or umbilical cord blood? Blood. 2003; 101(11):4233-4244.

[41] Hurley CK, Wagner JE, Setterholm MI, Confer DL. Advances in HLA: Practical implications for selecting adult donors and cord blood units. Biol Blood Marrow Transplant. 2006; 12(1, Suppl. 1):28-33.

[42] Kollman C, Abella E, Baitty RL, et al. Assessment of optimal size and composition of the U.S. National Registry of hematopoietic stem cell donors. Transplantation 2004;78(1):89-95.

[43] Barker JN, Scaradavou A, Stevens CE. Combined total effect of total nucleated cell dose and HLA-match on transplant outcomes in 1061 cord blood recipients with hematologic malignancies. Blood 2010;115(9):1843-1849.

[44] Kurtzberg J,Prasad VK, Carter SL, et al. Results of the Cord Blood Transplantation Study (COBLT): clinical outcomes of unrelated donor umbilical cord blood transplantation in pediatric patients with hematologic malignancies. Blood 2008;112(10): 4318-4327

[45] Brunstein CG, Fuchs EJ, Carter SL, et al. Alternative donor transplantation: results of parallel phase II trials using HLA-mismatched related bone marrow or unrelated umbilical cord blood grafts. Blood 2011;118(2):282-288

[46] Ballen KK, Koreth j, Chen Y, Dey BR, Spitzer TR: Selection of optimal alternative graft source: mismatched unrelated donor, umbilical cord blood, or haploidentical transplant. Blood March 1, 2012 vol. 119 no. 9 1972-1980

[47] Laughlin MJ, Eapen M, Rubinstein P, et al. Outcomes after transplantation of cord blood or bone marrow from unrelated donors in adults with leukemia. N Engl J Med 2004;351(22):2265-2275

[48] Eapen M, Rubinstein P, Zhang MJ et al. Outcomes of transplantation of unrelated donor umbilical cord blood and bone marrow in children with acute leukaemia: a comparison study. Lancet. 2007;369:1947-19542

[49] Eapen M, Rocha V, Sanz G, Scaradavou A, Zhang MJ, Arcese W et al. Effect of graft source of unrelated donor hematopoietic stem-cell transplantation in adults with acute leukemia: a retrospective analysis. Lancet 2010; 11: 653–660.

[50] Zhang MJ, Davies S, Camitta B, et al. Comparison of Outcomes after HLA-Matched Sibling and Unrelated Donor Transplantation for Children with High-Risk Acute Lymphoblastic Leukemia. Biol Blood Marrow Transplant. 2012;18:1204-1210

[51] Smith A, Baker K, DeFor T, et al. Hematopoietic Cell Transplantation for Children with Acute Lymphoblastic Leukemia in Second Complete Remission: Similar Outcomes in Recipients of Unrelated Marrow and Umbilical Cord Blood versus Marrow from HLA Matched Sibling Donors. Biol Blood Marrow Transplant. 2009; 15:1086-10934

[52] Horan J, Wang T, Haagenson M, Spellman SR, Dehn J, Eapen M, Frangoul H, Gupta V, Hale GA, Hurley CK, Marino S, Oudshoorn M, Reddy V, Shaw P, Lee SJ, Woofrey A.Evaluation of HLA matching in unrelated hematopoietic stem cell transplantation for nonmalignant disorders. Blood. 2012 Jul 24

[53] Barriga F, Wietstruck A: Search for unrelated donor umbilical cord blood units for allogeneic stem cell transplantation: results in two time periods. Transplant Proc. 2007 Apr;39(3):629-30.